ANGER AND ANGST

Jason Kenney's Legacy and Alberta's Right

Trevor W. Harrison and Ricardo Acuña, eds.

Montréal • New York • London

To the people of Alberta

"Fear is the anticipation of the pain in the future.
Anger is the remembrance of pain in the past.
Hostility is wanting to get even."

—Deepak Chopra

Black Rose Books No. XX435

Library and Archives Canada Cataloguing in Publication

Title: Anger and angst : Jason Kenney's legacy and Alberta's right / Trevor W. Harrison and Ricardo Acuña, eds.
Other titles: Jason Kenney's legacy and Alberta's right
Names: Harrison, Trevor W., 1952- editor. | Acuña, Ricardo, editor.
Identifiers: Canadiana (print) 20230161219 | Canadiana (ebook) 20230161243 | ISBN 9781551648088 (hardcover) | ISBN 9781551648064 (softcover) | ISBN 9781551648101 (PDF)
Subjects: LCSH: Kenney, Jason, 1968- | LCSH: United Conservative Party. | LCSH: Right and left (Political science)—Alberta. | LCSH: Populism—Alberta. | LCSH: Alberta—Politics and government—1971- | CSH: Alberta—Politics and government—2015-
Classification: LCC FC3676.1.K46 A54 2023 | DDC 971.23/04—dc23

C.P. 35788 Succ. Léo Pariseau
Montréal, QC H2X 0A4
CANADA
www.blackrosebooks.com

ORDERING INFORMATION

USA/INTERNATIONAL
University of Chicago Press
Chicago Distribution Center
11030 South Langley Avenue
Chicago IL 60628
(800) 621-2736 (USA)
(773) 702-7000 (International)
orders@press.uchicago.edu

CANADA
University of Toronto Press
5201 Dufferin Street
Toronto, ON
M3H 5T8
1-800-565-9523
utpbooks@utpress.utoronto.ca

UK/EIRE
Central Books
50 Freshwater Road
Chadwell Heath, London
RM8 1RX
+44 20 8525 8800
contactus@centralbooks.com

CONTENTS

PUBLIC SECTOR

SOCIAL ALBERTA

ACKNOWLEDGEMENTS

THIS BOOK IS the product of three years of hard work and planning. Beginning in early 2020, several meetings were held among keen political observers to discuss Alberta's emerging political landscape under the United Conservative Party. Not surprisingly, these discussions took many twists and turns over the next two-and-one-half years.

We want to thank the authors for their cooperation and dedication in writing and re-writing their chapters arising out of those meetings. Several answered our pleas late in the process to join the project. A few others dropped out along the way, whose contributions we want nonetheless to also acknowledge. We want also to thank those individuals who attended our early meetings as special guests and who provided useful feedback to the ideas expressed.

Sarah Pratt deserves special thanks for her excellent work in copy-editing the manuscript. Jackie Flanagan, Gordon Laxer, and Graham Thompson read drafts of the book in its late stages and provided comments for which we are also indebted.

We also want to acknowledge the organizing and convening support provided by the Parkland Institute, where Trevor was Director when the project started and Ricardo continues as Executive Director to this day.

We want to thank our publisher, Dimitrios Roussopoulos, whose enduring support of independent, left progressive ideas remains a singular beacon on a diminishing landscape; and to Rodolfo Borello, Clara-Swan Kennedy, and the rest of the staff of Black Rose Books.

Finally, we want to thank on a personal note our partners, closest friends, parents, and children for their helpful feedback, unwavering support, and constant encouragement over the past few years.

Any errors or omissions are, of course, the responsibility of the editors.

T.W.H. and R.A.

NOTES ON CONTRIBUTORS

Ricardo Acuña is executive director of the Parkland Institute, a public policy research institute in the Faculty of Arts at the University of Alberta.

Laurie Adkin is a professor of political science at the University of Alberta.

Robert L. (Bob) Ascah has held posts with Alberta Federal and Intergovernmental Affairs and Alberta Treasury, and ATB Financial. From 2009-2013, he was Director of the Institute for Public Economics. His blog is Abpolecon.ca.

Bob Barnetson is a professor of labour relations at Athabasca University.

Dr. Yale D. Belanger (Ph.D.) is professor of Political Science at the University of Lethbridge (Alberta), and a Member, Royal Society of Canada, College of New Scholars, Artists, and Scientists.

Janet Brown is one of Alberta's most recognized pollsters and political analysts, and a regular media commentator on Alberta's political trends.

Dr. Susan Cake is an Assistant Professor in Human Resources and Labour Relations at Athabasca University.

Laticia Chapman is a PhD Candidate in Political Science at the University of Alberta. Laticia's doctoral research on libraries in rural Canada examines the meaning of the public sphere in small communities.

John Church, Ph.D., is a professor in the Department of Political Science at the University of Alberta.

David J. Climenhaga is a journalist, political commentator, trade union activist, and the author of the AlbertaPolitics.ca blog.

Brooks DeCillia spent 20 years reporting and producing news for CBC. These days, he's an assistant professor with Mount Royal University's School of Communication Studies.

Roger Epp is professor emeritus of Political Science at the University of Alberta.

Nick Falvo is a Calgary-based research consultant with a PhD in Public Policy.

Jason Foster is the director of Parkland Institute and an associate professor of Human Resources and Labour Relations at Athabasca University.

Lise Gotell is Professor of Women's and Gender Studies and the Landrex Distinguished Professor, University of Alberta.

Trevor W. Harrison is a professor of sociology at the University of Lethbridge and former director of the Parkland Institute.

Ben Henderson is a former member of Edmonton City Council (2007-21) and currently co-chair of End Poverty Edmonton.

David Newhouse is Onondaga from the Six Nations of the Grand River community near Brantford, Ontario, and Professor of Indigenous Studies and Director of the Chanie Wenjack School for Indigenous Studies, Trent University.

Marc Spooner is a professor in the Faculty of Education at the University of Regina. Find him on Twitter @drmarcspooner.

Gillian Steward is a Calgary-based journalist, whose regular column appears in the Toronto Star, and former managing editor at The Calgary Herald.

Bridget Stirling is a PhD candidate in educational policy studies at the University of Alberta and former Edmonton Public School Board trustee and vice-chair.

Kevin Taft was elected for three terms in the Alberta Legislature and served as Leader of the Opposition from 2004 to 2008.

CHAPTER 1

INTRODUCTION

Trevor Harrison and Ricardo Acuña

"The crisis consists precisely in the fact that the old is dying and the new cannot be born; in this interregnum a great variety of morbid symptoms appear."

— Antonio Gramsci (1971), *Prison Notebooks*, 1930

ELECTION NIGHT in Alberta April 16, 2019. Jason Kenney looked out on the gathering crowd of United Conservative Party supporters, and it was good. His supporters were loud, even ecstatic, if still angry, despite the party's overwhelming victory: 63 of 87 seats and nearly 55 per cent of the popular vote.

Kenney smiled broadly, basking in the belligerent energy. And why shouldn't he? More than anyone else, this victory was his. Over the two previous years, he had knit together Alberta's disparate conservative elements to defeat Rachel Notley's New Democrats. Four years earlier, Notley's "outsiders" had stolen the crown, a crown — as Kenney repeatedly implied — properly worn by conservatives. The NDP was, in the dismissive words used by its opponents, "an accidental government."

Fomenting a fortress mentality in Alberta has long served conservatives' political interests. But the degree of anger directed at the province's perceived enemies rose to new heights in the years after 2015, surging throughout the 2019 campaign, and finally consuming the party itself in the years after taking office.

Kenney stoked feelings of grievance, victimhood, and alienation, while offering himself up as a strong and decisive leader who would take on and win against the demonic forces attacking the province. Righteous vengeance would soon be visited upon Alberta's enemies, within and without: teachers, labour unions, academics, environmentalists, Quebec and British Columbia, the big city mayors — but especially, the federal Liberals led by Justin Trudeau, son of the long dead, but still despised, Pierre Trudeau.

The grievous accident of the NDP's victory had now been redressed. For the moment, at least, the universe seemed once more to spin on its natural axis. Yet, three years later, Jason Kenney resigned as premier, rejected not only by the wider public but also half of UCP members. The leadership race that followed laid bare for all the public to see the party's internal divisions. The election of Danielle Smith as party leader and premier in October 2022 papers over these divisions, while opening up even larger fault-lines in the province. The crisis grows.

This book examines critically the extraordinary years of the UCP's time in office, 2019-2023; a period arguably the most chaotic in Alberta's political history, challenged only perhaps by William Aberhart's Social Credit during the immediate years after 1935. Inevitably, Jason Kenney and COVID-19 cast a large shadow over the story and are recounted in the chapters that follow, whether as foreground or background. But they are not the entire story, and then only a surface one. We argue the UCP's problems of governance are a symptom of a long-term illness afflicting Alberta's body politique, one nurtured by a political elite. Kenney, Smith, and the UCP could not have arrived on Alberta's scene separate from a set of historical and material conditions.

Conventional (conservative) accounts view Alberta as struggling to escape the clutches of a colonialist Confederation bent on holding the province down. Irrespective of any historical truths these accounts might hold — Western Canada was indeed founded as a colonial extension of Central Canada — we argue instead that Alberta's larger struggle over the past several decades has been an internal one. Central to this struggle has been an entrenched elite, based largely in the oil and gas sector, that is increasingly fearful of losing its power; fearful, more broadly, of the province that Alberta is struggling to become. Since the 1980s, the political coalition underpinning this historic bloc has been rent by a series of economic crises and broad social changes. In an effort to maintain its power, this elite has constructed an elaborate mythology based on a culture of grievance and victimhood. Both Kenney's electoral victory in 2019 and Smith's leadership win in 2022 successfully employed appeals to these myths. But while useful as a short-term, cynical, political tactic, it is a major obstacle to dealing with Alberta's deep economic and social divisions, and to efforts to become a functioning democracy.

This book has its origins in the aftermath of the 2019 election. In February 2020, a small group of academics and other political observers gathered in Edmonton for a free-wheeling discussion of Alberta's unfolding political scene. Though no one could predict the twists and turns that followed, it was clear to everyone assembled that Alberta was entering a critical, if uncertain, period of political turmoil. Based on that initial meeting, several other (Zoom) gatherings were held over the next two years.

At this book's heart lies an account of how the UCP has governed; the ideas, personalities, and social forces that have driven its agenda. It follows in the vein of numerous others that have detailed Alberta's unique political culture (Macpherson,

1953; Richards and Pratt, 1979; Finkel, 1989; Lisac, 1995, 2005; Bratt, Brownsey, Sutherland, and Taras, 2019). It is a direct sequel to two earlier books, *The Trojan Horse: Alberta and the Future of Canada* (Laxer and Harrison, 1995) and *The Return of the Trojan Horse: Alberta and the New World (Dis)Order* (Harrison, 2005). The first of these books examined the early years of Ralph Klein's premiership in the context of the rise of the New Right. Against this backdrop, Alberta exuded a triumphalist spirit extolling small government, private markets, and possessive individualism. There was, as Margaret Thatcher had intoned years earlier, "no alternative." Liberal democracy, dressed now in the garb of neo-liberalism, had seemingly put an end to history (Fukuyama, 1992).

Almost immediately, however, globalist celebrations dimmed. The world's markets stumbled. Climate change moved from a theoretical to a real problem. Popular movements took to the streets. Then, on September 11, 2001, Islamic terrorists struck New York and Washington. History, as it turned out, had not ended; it was very much alive and kicking. Borders shutdown; conflict and war — especially in the Middle East — flared anew. *The Return of the Trojan Horse* in 2005 detailed the impact upon Alberta of the neoliberal global order's unravelling — an unravelling that picked up steam in 2008 with the onset of the Great Recession. The stage was set for the multiple economic, environmental, social, and political convulsions that have beset Alberta — and the world — ever since.

The Conservative Party's Long Descent into Dysfunction

When it was defeated in 2015, Alberta's Progressive Conservative Party had been in office just short of 44 years, surpassing its Social Credit predecessor which had governed for a then-record 36 years. While travelling under the same name, the PCs had changed a great deal over that time in response to internal party dynamics, external pressures, and profound social changes within Alberta itself.

Peter Lougheed's PCs, elected in 1971, were a collective of young, urban — and urbane — politicians. They heartily disliked Social Credit's overt religiosity, parochial thinking, and rentier attitude to economic development. The Lougheed government sought to modernize Alberta through state-driven investments. The OPEC crisis of 1973 provided the financial heft necessary to meet these aims.

The Iranian Revolution of 1979 at first provided a further boost to Alberta's Treasury. Lougheed had famously argued that Albertans were the owners of the resource, and deserved a fair share of the revenues, at least 35 per cent. The prospect of US$100 per barrel oil made PC politicians and oil producers drool with anticipation.

In a world dominated by oil scarcity and restricted options, higher prices are a good thing for oil producers, oilfield suppliers, and labourers. But they are a decidedly bad thing for consumers and manufacturers faced with rising inflation, and employees not tied to the industry's fortunes. In order to ward off a potentially massive destabilization of Canada's economy — boom in one region, bust in another — Pierre Trudeau's federal Liberal government introduced the National Energy Program which set a ceiling, but also a floor, on the price of energy, while encouraging Canadian ownership of the oil and gas industry, as well as resource exploration and conservation. But the Lougheed government, and those of other oil producing regions, complained the program trod on provincial jurisdiction.

Eventually, the two levels of government came to a revenue-sharing plan, only to see the agreement's financial underpinnings collapse overnight when the price of oil fell, courtesy of OPEC. While every oil producing region was negatively impacted, Alberta suffered particularly because of oil's increasingly disproportionate role in its economy. The federal Liberals were blamed for the collapse; Prime Minister Trudeau's effigy was hung on more than one occasion. But the collapse of oil was a world-wide phenomenon. The oil riggers heading south found no better opportunities in the Texas or Oklahoma fields.

As the economy declined, public anger also increasingly focused on the Alberta government. Amidst eroding support, and growing populist anger (see below), Lougheed stepped down in 1984, and was succeeded by Don Getty. Getty attempted to revive the economy through a combination of public sector austerity, direct investments in diversification, and loans to business. The oil and gas sector remained favoured, however, in the form of subsidies, foregone taxes, and reduced royalty rates. Where the Alberta government had previously operated with a modicum of control over the oil industry, the situation was now reversed. The economy's dependence upon the oil industry was replicated at the political level by the PC's growing dependence upon the same industry to remain in office.

The Getty government's efforts at diversification failed spectacularly, leaving Alberta taxpayers on the hook, and putting to an end — at least for a time — notions of direct government ownership. Getty resigned the fall of 1992, amidst another economic downturn.

On the political ropes, the Progressive Conservative party morphed into a new version of itself. The party anointed a new leader, Ralph Klein. Klein's populist appeal resuscitated the party's fortunes. His party rode to victory in 1993 on one of Alberta's now recurrent waves of anger, and a platform of harsh austerity. The Klein government adopted *en masse* the emergent neo-liberal orthodoxy of privatization, deregulation, low taxes, and small government. The mantra of small government and low taxes was music to the ears of Alberta's entrenched corporate elite, but also found support among conservative rural voters.

How long Albertans would have tolerated a ratchetting down of provincial services and the wholesale selling off of public assets, we will never know. Within two years, the price of oil and natural gas rebounded. Budget surpluses of $7-10 billion per year became an annual, and anticipated, event, like the coming of Christmas. Flush once more with money, the Klein government was able to assuage discontent while claiming a wholly unearned reputation for fiscal competence. A political catchphrase — "The Alberta Advantage" — was coined alleging the province's wealth was a product of low tax rates, and not the reverse. The day of reckoning for both the party and Alberta's finances was thus put off a little longer.

Klein's disinterested and increasingly desultory performance as premier came to a head in 2006. He resigned after delegates to the party convention gave him a less than overwhelming show of confidence. The search for a successor began. Alberta's corporate leaders viewed only two candidates as acceptable: Jim Dinning and Ted Morton[1], both former Alberta finance ministers, both Calgary-based. In a surprise vote, however, the party's rural constituents threw their support behind Ed Stelmach, an unassuming MLA from the farming community of Vegreville, north-east of Edmonton.

The 2006 PC leadership race happened at the height of one of Alberta's most significant oil booms. Inflation was running rampant, small and large businesses were having difficulty finding or affording staff, and the gap between the outrageous profits of the oil companies and provincial revenue from royalties was astounding, even for Albertans; so much so, that the issue of reviewing royalty rates became an important part of debates and discussions during the race. Stelmach promised that, should he become premier, he would take steps to ensure Albertans were receiving a fair share of oil wealth in the province. That promise boosted his fortunes, and helped him win the leadership.

Soon after taking office, Premier Stelmach launched a royalty review. That review recommended some significant changes to the structure and amount of provincial royalties. Faced with an aggressive campaign of protests, lobbying, and electoral threats, Stelmach stopped short of fully embracing the panel's recommendations. He did, however, make some moderate changes that made royalty rates more sensitive to oil prices than they had been in the past. Despite Stelmach's compromise, the oil industry was apoplectic and decided to send him a message. It found a messenger in the form of the Wildrose Alliance Party of Alberta, a right-wing fringe party founded in early January 2008 from remnants of several other fringe parties. Corporate donations to the Wildrose Alliance surged, especially during the 2008 provincial election.

But for the Great Recession that began in 2008, Stelmach's PCs might have weathered the discontent. At first, Alberta seemed to escape the crisis' worst impacts, but the recession caused a downturn in global oil consumption, resulting in a drop in Alberta's revenues. In the spring of 2009, Stelmach's government announced it would run a record deficit for the year. Despite the fact the new royalty regime only kicked in on January 1, 2009, at which point, due to low prices, no one in the province was paying higher royalties than they had the day before, Alberta's petroleum industry blamed the new royalty regime for the deficit. No government since, including the NDP, has seriously attempted to raise royalty rates on oil and gas.

There are two political certainties arising from any economic downturn in Alberta. First, many Albertans will turn their anger against a grab-bag of perceived enemies, mainly the federal government and Central Canada elites, resulting in the rise of populist parties and protest movements.[2] Second, Alberta's governing coalition will unravel, giving rise to internal protest movements or third parties upset with their own government (see Hiller, 1977). Pummeled politically from all sides, Stelmach stepped down in October 2011.

Once again, Alberta's corporate elite had its choice as successor: Gary Mar, a career politician who had held numerous posts in the Klein government and had strong business connections. But, instead, a newly-minted MLA, Alison Redford, cobbled together a coalition of urban liberals and progressives to win the leadership. The coalition held together long enough to defeat the Danielle Smith-led libertarian-populist Wildrose party in the April 2012 election that followed, but a fault-line had emerged in the broader conservative coalition. While the PCs won, taking 61 seats, the Wildrose party took 17 seats, mainly in rural Alberta. More worrying for the PCs, the party garnered only 44 per cent of the vote compared to Wildrose' 34 per cent, a fissure that grew larger over the next year. Redford already had little support within caucus, which had favoured Mar. The knives quickly came out after a series of political blunders saw her poll numbers fall to Stelmach-like levels.

Redford's resignation in March 2014 sent the PCs once more searching for a saviour. He was found this time in the person of Jim Prentice. Prentice was a prominent Conservative MP who had held several distinguished ministries in Stephen Harper's government. Prentice exuded something of the aura of Lougheed. He promised a new style of politics, one that would be honest and devoid of scandal. Above all, he promised to heal the split on the right between the PCs, Wildrose, and other fringe elements.

In December 2014, Prentice and Smith made a secret agreement whereby she and eight other Wildrose members crossed the floor to join the PCs. Far from being a brilliant political maneuver, many Albertans — Wildrose supporters in particular — viewed the unprecedented action as an example of cynical and corrupt politics. Both Prentice's and Smith's lustre dimmed, while the remainder of the Wildrose faction held firm under a new leader, Brian Jean, a former Conservative MP from Fort McMurray-

Athabasca.[3] The subsequent election of May 5, 2015, saw Rachel Notley's New Democrat's take 54 of 87 seats (nearly 41 per cent of the vote), compared to 21 seats (24 per cent) for Wildrose, which had rebuilt itself under a new leader, Brian Jean, and nine seats (28 per cent) for the PCs. Prentice immediately resigned as party leader.

The New Democrat Interregnum

To paraphrase Marc Antony (via Shakespeare), the alleged harm that governments do is later repeated ad nauseum by their opponents, while the good they do is later forgotten, erased, or adopted without attribution. Reviewing the NDP's time in office, Ricardo Acuña (2019, p. 28) termed the NDP the most "activist government in recent Alberta history," yet far from radical. Among its first acts, the new government banned corporate and union donations to political parties, replaced the 10 per cent flat rate tax with a progressive tax featuring four new rates[4], and approved interim spending in the key areas of health, education, and social services. Though facing an enormous drop in revenue, due to shrunken oil prices, it declined to slash public services, and later increased the minimum wage to $15 per hour — fairly standard Keynesian practices during a recession. The new government also revised the Alberta Labour Relations Code and the Employment Standards Code, created Alberta's first ministry for the status of women, increased funding for women's shelters, introduced a pilot project for $25-a-day daycare[5] (see Cake, Chapter 16), and introduced protections for gay-straight alliances in schools.

The NDP's time in office was not without controversy. Though there were no major scandals, and little discord — Notley ran a tight and focused ship — there were policy mistakes and errors in judgement. An effort to extend occupational health and safety provisions, and WCB coverage, to farm workers failed to adequately consult the farming community. Following a review headed by the former head of ATB Financial, the government did not raise oil and gas royalty rates, a decision that confused and angered many of the party's supporters. While the party's climate plan had some bold elements — the accelerated phasing-out of coal-fired plants to 2023, setting an absolute cap on oil sands emissions, and introducing a carbon levy — the energy sector's usual suspects condemned it as going too far, while some party supporters complained it did not go far enough. Finally, the NDP's policies on long-term care proved inadequate during the pandemic of 2020-22.

Dealt a difficult hand, Notley's NDP was a generally moderate and capable government. Arguably, it was similar to the early Lougheed PCs. (The latter was more leftwing on economic policy, for example, in enacting public ownership of key sectors, as in the case of Pacific Airline, but less supportive of labour.) Both Lougheed and Klein benefited from rapidly rising oil revenues early in their terms, but Notley's government received no such advantage. The price of Western Canada Select oil remained low throughout her term. Her party received no credit either for holding Alberta steady in the face of the resultant rise in unemployment and provincial debt. Once again, despite the fact that oil and gas markets are global, and that the oil sands in particular are a less attractive long-term financial investment, radio talk shows and Postmedia – which owns all of Alberta's major newspapers — blamed the province's difficulties on socialism, an epithet directed not only at the NDP but previous Conservative governments. The anger provoked by social media and politicians found its apogee in physical threats made to Notley and several of her cabinet ministers, especially women MLA's, requiring the RCMP to provide police protection. As during past economic downturns, conspiracy theories flourished.

In late 2018, a few hundred yellow-vested protestors, drawn primarily from the oil and gas sector, took to Alberta's streets (Harrison, 2019). The orchestrated protests gained traction just before Christmas when some of the same yellow vests joined a convoy of 1200 trucks, driven by oil patch workers and their supporters, that had assembled in the town of Nisku, just south of Edmonton. Blocking traffic as they went, the convoy drove slowly to the capital, making known as they did their demand that oil pipelines be built to "free" the resource from its land-locked status. The spectacle gave birth three days later to similar pro-pipeline protests in other Alberta towns and cities, including Calgary, Brooks, Edson, Grande Prairie, and Medicine Hat. Hopping the provincial border, a trucker protest also took place in Estevan, Saskatchewan that same day. The protests were a dry run for the Freedom Convoy occupations of Ottawa and several border crossings in early 2022 (see Harrison, Chapter 5).

While blocked pipelines were the immediate cause of fear and loathing, the narrative echoed a familiar search for enemies who — so the argument goes — are bent on keeping "the West" in perpetual servitude. Similarly, the NDP's introduction of a carbon tax, a measure recommended by most economists, including those of a conservative bent, further added to a chorus of theft orchestrated by external enemies. Despite its aggressive and unapologetic cheerleading for pipelines and for the oil industry — a stance much criticized by many of its own supporters — the NDP continued to be blamed for its handling of the economy and for simply being the NDP. For many conservative voters, the NDP was merely one of a host of usual enemies, the federal Liberals, and Quebec, to which some now added Canada's Supreme Court and the United Nations, attempting to take away Alberta's economic and political birth-right.

The protesters' arguments were rarely coherent, and often contradictory; anger, the common glue. The protesters received sympathetic support from some high-profile individuals, like Danielle Smith, Ted Morton, and University of Calgary economist,

Jack Mintz. But the rallies also frequently attracted members of far-right hate groups, such as the Proud Boys and the Soldiers of Odin. A whirlwind of anger and angst swept the province.

Re-Uniting the Right

To say that Alberta conservatives were stunned by the New Democrat win in 2015 is an understatement. Not only was Alberta viewed as a safely conservative place provincially, since the late 1980s it had also been the beachhead for the New Right's transformation of conservatism federally. Through first the Reform Party and later the Canadian Alliance Party, the New Right in 2003 had orchestrated a hostile takeover of the Progressive Conservatives. With the exception of a few urban ridings, the rebuilt federal Conservatives could count on winning — often without bothering to campaign in the province — nearly every Alberta seat. The 2015 provincial election shook this confidence. If the NDP could win provincially, might the federal party's hold on the province also be waning?

It was comforting for conservatives to believe the NDP's victory was the result of vote-splitting between Wild Rose and PC supporters, and not deeper social and political changes occurring within the province. The solution was thus to broker a marriage between the conservative factions, a solution for which there was already a blueprint: the merger/takeover of the federal Alliance and Progressive Conservative parties fourteen years earlier. A prominent federal politician, Jason Kenney, played a major role in that event.

As the party had when it welcomed a prominent federal Conservative, Jim Prentice, in 2014, Alberta's conservative brain-trust turned to Kenney in its hour of desperation. It had long been thought he harbored ambitions of succeeding Stephen Harper as federal Conservative leader, a belief that grew following the latter's resignation after the party's defeat in the fall 2015 election. Instead, Kenney announced in July 2016 that he would run for leadership of Alberta's Progressive Conservatives. He seemed an inspired choice. He had the kind of background that spelled "winner."

Like many prominent Alberta conservative politicians, Kenney was actually born elsewhere — Oakville, Ontario in 1968 — though raised in nearby Wilcox, Saskatchewan. He graduated from Athol Murray College of Notre Dame, a private Catholic High School. (He remains a devout Catholic.) Kenney later attended St. Michael's University School in Victoria, BC, and still later the University of San Francisco, a Jesuit-university where he was attracted to the ideas of prominent neo-conservative theorists. He became a noted anti-abortionist and opponent of Gay rights and free speech on campus, but left after one year without completing a degree. Kenney is a "movement conservative" (Lambert, 2022), meaning that he sees the role of government being to radically transform society; in the case of Alberta, following an ideologically-driven agenda of corporatism mixed with social conservatism (see Stewart, Chapter 4).

Still in his early twenties, Kenney got a job as executive assistant to Ralph Goodale, the Saskatchewan Liberal party's leader at the time. He soon left, however, to become the Alberta Taxpayer's Federation's first executive director, followed the next year (1989) by being named president and chief executive of the newly minted Canadian Taxpayer's Federation. In his role with the CTF, Kenney made a name for himself in a shouting match with then Alberta Premier Ralph Klein in 1993 when he confronted Klein over MLA's "gold-plated pensions." Facing an election, Klein relented; days later, the pension plan was eliminated. Kenney's political star was ascendant.

In 1997, at the age of 29, Kenney was elected to the House of Commons as a member of the Reform Party for the riding of Calgary Southeast. Three years later, he became chief advisor and speech-writer for Stockwell Day during Day's successful bid for the leadership of the Canadian Alliance party. Kenney remained a key Day supporter and alleged speech-writer during the Alliance party's much less successful federal election campaign that same year.

To a remarkable degree, Kenney's political trajectory and later problems mirror that of Day's rise and eventual fall, which paved the way for Stephen Harper taking the Alliance party's helm in the spring of 2002 (Harrison, 2002), and a year-and-a-half later becoming leader of the newly-formed Conservative Party of Canada. By this time, Kenney had established himself as a loyal foot-soldier for the party.

Like Harper, Kenney was a vocal supporter of the US-led invasion of Iraq in 2003 and remains an equally staunch supporter of Israel. Known for his dedication and hard work, it was expected that Kenney might be given a cabinet post when the Conservatives won office in 2006, but Harper instead assigned him the task of building party support within the ethnic communities (see Acuña, Chapter 3). That community was traditionally viewed as part of the Liberal support base due to its long-established support for immigration, but many conservatives viewed the religious background of many immigrants as fertile ground for recruitment. Kenney took on the task of political conversion with enthusiasm and some success.

Now, in 2016, Kenney — the prototype of a career politician — was called upon to take on another task: to re-unite Alberta's conservative factions and thus save the party — and the province — from the socialist threat. The task proceeded with military precision. Kenney declared he would run for leadership of the PC party. Given his existing profile, and the fact the demoralized PCs were devoid of any genuine challengers, he easily won the leadership the following March.

Following brief negotiations, Kenney and Brian Jean agreed upon a plan to unite the PC and Wildrose parties. In July 2017, members of both the Progressive Conservative and Wildrose parties voted overwhelmingly in support of a merger. The leadership race quickly followed.

Despite Kenney's credentials, it seemed at least possible that Jean might defeat him. A ThinkHQ poll released in May 2017 showed that Jean was heavily preferred over Kenney by Albertans and that Jean held a substantial lead even among likely UCP voters (Julie, 2017).

The leadership race was bitter. Kenney, like Dinning, Mar, and Prentice before him, was the establishment candidate; Jean was the candidate of the rural populists. Rumours of dirty tricks by the Kenney team abounded, rumours that continued to dog the party in the years that followed.[6] But, on October 28, 2017, Kenney secured an easy first ballot victory over Jean to take the UCP leadership. The final tally showed he took 61 per cent of the vote to Jean's almost 32 per cent, with third place finisher Doug Schweitzer taking just seven per cent. Kenney's path to the premiership was set.

Conservative voters in Alberta like a winner, and Kenney seemed best positioned to deliver victory. Besides his political background in federal politics, Kenney — as described by journalist Don Martin (2016) — was "brilliantly analytical," "fiercely articulate," "flawlessly bilingual," and "tirelessly energetic," while keeping "his social conservative beliefs under a kimono that's never to be lifted."

Even at that time, however, many UCP party supporters had concerns about Kenney. They viewed him as something of an outsider, even a carpetbagger, insufficiently disposed to grassroots democracy. It was widely believed that he was simply using Alberta as a launching pad for his federal ambitions. In a province, moreover, whose political culture embraces genuineness, Kenney seemed profoundly inauthentic. He was clearly uncomfortable wearing a cowboy hat or sitting in the cab of a Ford F-150. While he talked endlessly about "the people," he did not seem one of them.

For the moment, however, any such concerns were set aside. Over the next year and a half, Kenney travelled the province polishing his and the party's profile. In his speeches, Kenney gave no quarter. Alberta was beset with enemies, both within and without, but he would slay them. A vote for the UCP meant a return to prosperity. A majority of Albertans believed Kenney's promises.

The UCP's victory came as conservatism in Canada was reaching its high-water mark. Following the spring 2019 election, six of Canada's ten provinces were led by Conservative governments, including not only Alberta, but Saskatchewan, Manitoba, and Ontario. Only months earlier, in December 2018, the cover of *Maclean's* showed Rob Ford, Brian Palliser, Jason Kenney, and Scott Moe, bracketing federal leader

Andrew Scheer, under the title, "The Resistance." The sunny ways of Justin Trudeau's victory four years earlier had given way to growing clouds, giving conservatives hopes of having another powerful ally in confronting Ottawa.

But the UCP's relatively easy victory in 2019 papered-over some hard realities that Jason Kenney's self-satisfied smile on election night could not hide. Kenney returned to an Alberta of his imagination; indeed, one that conservative elites, comfortable in their idea of the province's unchanging nature, had not taken time to recognize or understand. Kenney returned to an Alberta that is largely urban, young, and socially liberal; an Alberta that, if fitfully, is trying to come to terms with the decline of oil and gas dependency; an Alberta in which issues of climate change and Indigenous rights increasingly hold centre stage; an Alberta where even rebounding oil prices are no guarantee of increased employment; and an Alberta that would soon face a pandemic that shook the province's economy and conservative notions of small government, even as it drove a spike down the centre of definitions of what "is" conservatism.

The United Conservative Party Takes the Reins

The UCP quickly appointed a cabinet. It consisted of 22 members (seven women and 15 men). A dearth of elected members from Edmonton meant a severe over-representation of Calgary MLAs in cabinet (all details in Canadian Press, 2019). While the next few years saw the cabinet's expansion, frequent recycling, and some departures, the key members of that first cabinet — those whom Kenney trusted — remained largely in place throughout. Those key members included Kenney's former leadership opponent, Schweitzer (Calgary-Elbow), a lawyer, appointed to the ministry of justice and solicitor general; Tyler Shandro (Calgary-Acadia), a lawyer, minister of health; Ric McIver (Calgary-Hays), a former Calgary councillor, minister of transportation; Adriana LaGrange (Red Deer-North), a Catholic School division trustee and rehab practitioner, minister of education; Travis Toews (Grande-Prairie-Wapiti), an accountant, minister of finance; Jason Nixon (Rimby-Rocky Mountain House-Sundre), non-profit sector, minister of environment and parks; Sonya Savage (Calgary North-West), oil and gas executive, minister of energy; Demetrios Nicolaides (Calgary-Bow), communication background, minister of advanced education; Jason Copping (Calgary Varsity), lawyer, minister of labour and immigration; Rebecca Schulz (Calgary-Shaw), minister of children's services; and Kaycee Madu (Edmonton-South West), lawyer, minister of municipal affairs. Of these key ministers, LaGrange, Toews, Nixon, Nicolaides, Schulz, and Savage held their posts throughout the UCP mandate. A few played musical ministries; Schweitzer, Shandro, and Copping, in particular. Others, for reasons of scandal (Devin Dreeshen, minister of agriculture and forestry, and Tracy Allard, minister of municipal affairs) or disloyalty to Kenney (Leela Aheer, minister of culture, multiculturalism, and status of women) were dropped.

The UCP's first hundred days in office saw a whirlwind of legislation designed to polish the government's activist agenda but also to erase, in policy if not memory, all traces of the NDP (Wright, 2019, pp. 24-28; Braid, 2019); for example, reversing changes to Alberta's labour laws (see Foster, Cake, and Barnetson, Chapter 10). Much of the legislation (repealing the carbon tax, declaring Alberta open for business, reducing red tape, restoring the election of senators) was meant to appeal to the UCP's base. Several other measures were directed at the party's supporters in the corporate sector; for example, a drop in corporate taxes, later speeded up, from 12 per cent to 8 per cent. The energy sector, in particular, was assisted by allowing municipalities to give property tax exemptions to energy companies and guaranteeing that no changes to the oil and gas royalty structure would occur for ten years. In a measure directed at its social conservative base, the government amended legislation regarding gay-straight alliances in schools and embarked on a curriculum review (see Stirling, Chapter 12). The new government also created a "war room" — later formally named the Canadian Energy Centre — to combat the bad press generated, in the eyes of government supporters, by anti-oil, anti-Alberta environmentalists (see Adkin, Chapter 7). In the same vein, the government also launched an investigation into foreign funding received by those anti-oil, anti-Alberta environmentalists for the purpose of "landlocking Alberta's oil" (see Taft, Chapter 8). Much of its legislation was bathed in symbolism, as was the government's rhetoric which, when not attacking the former NDP government, was focused on the Liberal government in Ottawa.

Kenney's own focus on the federal Liberals bordered on an obsession, leading to renewed speculation that he viewed the premiership as a stepping stone to becoming prime minister. His hatred of the Trudeau government precluded giving Ottawa any credit, even when it provided controversial support for the TransMountain pipeline and financial assistance for cleaning up orphan wells; or provided Albertans with more financial support per capita during the COVID-19 pandemic than given to people in any other province — $11,410 per person, compared to Ontario, the second highest, at $9,940 (von Scheel, 2021). Kenney's reluctance to engage with the Trudeau Liberals often harmed Albertans, as when his government refused for nine months to participate in a cost-share agreement with Ottawa to provide a one-time payment of $1200 to frontline workers or to participate in the federal government's national child care program (see Cake, Chapter 16). Beyond personal animus, ambition, or stubbornness, however, Kenney's relentless attack on the federal Liberals played well with the UCP's base, reinforcing the belief they are an isolated and persecuted victim of Confederation.

The political attacks were directed not only at the Trudeau Liberals, however. The UCP also kept focused on a host of perceived enemies at home. It is traditional political practice to attack and hobble one's real and imagined political enemies. The

UCP quickly set its sights on public institutions and public sector workers within them, as well as K-12 teachers (see Stirling, Chapter 12) and faculty within post-secondary institutions (Spooner, Chapter 14). Remarkably, even in the midst of the COVID-19 pandemic, the government continued its assault on the public health care system and its nurses and doctors whose support was essential to dealing with the crisis (Church, Chapter 13).

Kenney's own pugnacious character found its echo in several ministers. Kaycee Madu, Jason Nixon, Sonya Savage, and Tyler Shandro, in particular, tended to angrily ignore criticisms and view everything through a political lens. In short order, the government made enemies not only of the usual suspects, but also ranchers (over coal development in the Rockies), farmers, municipal politicians (over EMS restructuring and policing) (see Henderson, Chapter 17), and doctors (cancellation of a contract). As Mueller relates (Chapter 9), the government's actions also increasingly alienated young people who no longer saw a future in the province. The list of organizations angry and distrustful of the government grew steadily, as did the number of pointless issues whose chief appeal was that they were favoured by the UCP's narrow base. Scarce were the government's major files that did not rile some group. Former allies — like the Justice Centre for Constitutional Freedoms and conservative journalists Licia Corbella and Rick Bell — increasingly found Jason Kenney's leadership wanting (see Climenhaga, Chapter 6).

Alberta's economy remained a key part of the story. The election campaign's slogan was "jobs, economy, pipelines," accompanied by a frequent choral background of "build that pipe, build that pipe." But the price of oil did not immediately recover, nor did the jobs return. As in the past, the harsh realities were brought home of living in a single-resource economy, dependent upon external forces, and — for political reasons — reluctant to examine its finances (see Ascah, Chapter 11). As the hopes and dreams of election night began fading, old divisions reemerged. Like topsoil in a Prairie windstorm, the idea that replacing the NDP would return Alberta to prosperity quickly vanished. By the fall of 2019, Kenney's — and the party's — appeal was already in decline (see Brown, Chapter 2). Over the course of three years, the party stumbled repeatedly on those areas in which it showed particular interest, such as education, energy, and health, while ignoring other important areas, such as housing (see Falvo, Chapter 15) and race relations (Chaudry, Chapter 20), especially the rise in hate speech during the pandemic that followed. The issues of some specific groups — women[7] (see Gotell, Chapter 21) and Indigenous peoples (Belanger and Newhouse, Chapter 18) — disappeared almost entirely from the government's agenda.

In the midst of everything, COVID-19 came along. The two years from spring 2020 until spring 2022 saw infection and death rates soar throughout much of the world, its variations remaining still a threat. Though Canada as a whole did better than most countries in dealing with the pandemic waves washing its shores, Alberta's efforts proved weak, vacillating, and desultory.

From the pandemic's start, Kenney placed himself squarely at the centre of its handling. More than any Canadian premier or even Prime Minister Trudeau, Kenney chose to dominate the COVID-19 updates, though some noticed he tended to show up only when there was good news. Early on, Kenney often held a pointer to alert viewers to numbers on a graph. Other times, he expounded on the virus' causes, virality, and lethality. Soon, he acquired the epithet, "Professor Kenney." Viewers sometimes wondered whether health officials, as they stepped to the podium, were left anything to say except, "What he said."

The pandemic continued. The UCP's handling of the pandemic, now centred on Kenney, was increasingly condemned on all sides as slow, weak, uneven, and contradictory (see Brown, Chapter 2; and Harrison, Chapter 5). Many Albertans wanted tougher measures; many UCP backbenchers and supporters in rural communities, mistrustful of government and of science, wanted few, if any. The UCP, itself, manifested the split. Worse, several UCP MLA's, including members of cabinet, were caught flaunting health care restrictions, resulting in accusations of entitlement that had brought down earlier conservative regimes. Kenney, who exuded a sense of righteous authority and resolve when denouncing Alberta's illusory enemies, proved weak and indecisive when tasked with handling actual problems within his caucus and his own office.

As 2020 turned to 2021, Alberta's economy still languished, the pandemic still lingered, and Kenney's leadership leaked support. A palace revolt stirred. By spring 2021, the revolt was in full bloom.

The Insurrection Grows

In early April 2021, 17 UCP MLAs wrote a public letter denouncing the imposition of new lockdowns.[8] A month later, some members openly called for Kenney to resign. Two rural MLA's, Todd Loewen and Drew Barnes, were removed from caucus, but the action did not stem the revolt (Braid, 2021). Calls for Kenney to step down as leader grew throughout the summer and fall of 2021, with the result that the UCP ceased almost entirely to govern during the year that followed, consumed instead by internal conflicts and the premier's efforts to hold on to his leadership.

Amidst growing internal discontent, the UCP board decided in December to move up a leadership review to the spring of 2022. Party dissidents wanted a provincewide virtual vote open to all UCP members to be held by early March. It was agreed instead that an in-person leadership review would be held at a Special General Meeting in

Red Deer on April 9. On March 15, however, Brian Jean — former Wildrose leader, and co-founder of the UCP — won a byelection victory in Fort McMurray-Lac La Biche. Jean immediately called for Kenney to resign, saying that he was prepared to take on the leadership of a renewed party.

As April 9 drew closer, the vote to remove Kenney as leader gained steam. A delegate attendance in Red Deer of perhaps 20,000 people was predicted. Citing logistical problems, UCP President Cynthia Moore announced on March 23 that the in-person leadership vote was changed to a mail-in vote of all party members, the results of which would be announced on May 11.

Kenney was clearly rattled. In a secretly recorded speech to his party's caucus staff in March 2022, he said he would not let "the mainstream conservative party become an agent for extreme, hateful, intolerant, bigoted and crazy views," adding that, "The lunatics are trying to take over the asylum. And I'm not going to let them" (story and quotes in von Scheel, 2022). The leaked recording solidified opposition to Kenney among those who viewed his leadership as top-down and disrespectful of grassroots members.

On April 9, the original date for the leadership review, Kenney spoke in Red Deer to a small gathering of hand-picked supporters, recounting a litany of "promises made, promises kept" and, as he saw it, the UCP's successes under his leadership. He warned against division, what would happen if the NDP were ever returned to power, and — by implication — the need to support him in the leadership review (Kaiser, 2022).

In the weeks leading up to May 11, some members spoke openly of a culture of fear and intimidation that Kenney and his staff had created within the party. Still, or perhaps because of this, many observers thought Kenney would squeak out a victory; and, in any case, would not step down as leader. But when the final votes were announced on May 11, he received only 51.4 per cent support, and immediately said he was quitting as leader. Kenney later blamed his demise on "a small but highly motivated, well-organized and very angry group of people who believe that I and the government have been promoting a part of some globalist agenda, and vaccines are at the heart of that" (Braid, 2022). In the eyes of many, that explanation left out much. While there was surprise, there were few tears shed; in the backrooms of UCP detractors to his leadership, happiness reined.

The UCP's executive decided the new leader would be chosen by a vote of party members using a preferential voting system whereby the new leader would have to achieve the support of at least 51 per cent plus one of the members. The new leader would be announced on October 6. In the meantime, Kenney remained as party leader and premier. Polls showed a sudden bump in support for the UCP who — even absent a decided leader — would defeat the NDP.

The search began for a new saviour of Alberta's dis-united right.

The Race and Smith's Coronation

Seven candidates vied for the UCP leadership: five UCP MLA's (Leela Aheer, Brian Jean, Rajan Sawhney, Rebecca Shulz, and Travis Toews), one independent MLA, Todd Loewen, and one unelected individual, former Wildrose leader, Danielle Smith. Despite the large number of entries, however, most observers quickly viewed the race as likely to come down to a choice of Jean, Smith, or Toews.

Each of the three had notable strengths and weaknesses. Toews had the vast support of sitting MLAs, including — though not formerly announced — Kenney, and had the cachet of being the party's former finance minister, but was also viewed as dull and unspiring. Smith, after her political self-immolation as Wildrose leader in 2014, had rebuilt her profile as a talk-show host where she garnered a large number of loyal listeners, but also acquired a reputation for taking extremist positions. Jean, the party's sympathetic everyman, was lauded for having saved the Wildrose party from oblivion in 2015, but like Toews was considered uninspiring. None of the others was viewed as likely to win, though some hoped Aheer, Sawhney, and Shulz might move women's issues to the centre of debate and also encourage a more civil and collegial form of politics than exercised under Kenney.

Smith quickly seized the high-ground, setting the pace and direction for the others. She singled out Alberta's traditional enemies, the Liberals and Central Canada (the "Laurentian elites"), but added Alberta Health Services in an appeal to her anti-vaxxer supporters; and, going even farther afield, attacked such organizations as the United Nations, the World Economic Forum (WEF), and the World Health Organization (WHO), viewed by conspiracy theorists as part of an alleged globalist agenda to replace capitalism with socialism (the "Great Reset").

By the week of August 8, Trump-like hysteria was in full-flight. An online message sent by Smith's team read:

> [T]he WEF is an anti-democratic group of woke elites that advocate for dangerous socialist policies that cause high inflation, food shortages and a lack of affordable energy, which in turn, leads to mass poverty, especially in the developing world.
>
> There is no question what their agenda is — they want to shut down our energy and agriculture industries as fast as they can.[9]

Conspiracy mongering aside, Smith's most politically astute move came in the form of the Free Alberta Strategy. The strategy's centrepiece is the Sovereignty Act, co-authored with former Wildrose MLA, Rob Anderson.

The proposed Act vaguely wavers between greater autonomy and outright secession, and reminds of comedian Yvon Deschamps' oft quoted joke about Quebec:

That all the province wants is to be independent within a strong and united Canada. Many UCP supporters believe that Alberta's problems would magically melt away if Ottawa would just "butt out." Encouraged by some conservative politicians, some even suggest the province, like Quebec, constitutes a nation.

Smith's Sovereignty Act harkens back to Social Credit's efforts under Premier William Aberhart to enlarge Alberta's jurisdictional autonomy, efforts later ruled by the Supreme Court as unconstitutional (Conway, 2014, p. 108; Finkel, 1989). But like so many ideas espoused by Canada's current conservatives — right to work, charter schools, privatized health care[10] — the Act also echoes state's-rights arguments in the U.S.; historically, in particular, that individual states could ignore and refuse to enforce within their borders any act passed by Congress or the Federal government which it viewed as transgressing rights reserved to itself.[11]

Howard Anglin (2022) — a former adviser to Stephen Harper — termed the Free Alberta Strategy "nuttier than a squirrel's turd." Numerous journalists and constitutional experts likewise denounced Smith's proposal as vague, unworkable, unlawful, and bizarre. Several of her leadership opponents, notably Jean, Toews, and Shultz said Smith's proposal would create uncertainty and drive away investment.

Nutty or not, Smith's advocacy of sovereignty separated her from the other candidates and solidified her bona fides with the party's more extreme and angry supporters; and, further, set down a marker that pulled the party further to the right, even redefining the nature of the party's position on the political spectrum (see Acuña, Chapter 3). Smith's six leadership opponents criticized the Act. (Jason Kenney described it as a "full-frontal attack on the rule of law," as well as a step towards separation and a "banana republic.") Quickly, however, several candidates came out with their own proposals for greater Alberta independence. Jean proposed an Autonomy for Albertans Act that would "enhance Alberta's autonomy within Canada." Shultz announced her own "100 Day Provincial Rights Action Plan to fight, partner, and strengthen Alberta's position in Canada." Loewen called for Alberta to create its own Constitution. Loyalty tests became *de rigeur*.

Most of the candidates — and all of the leading ones — thus staked out positions as "Alberta Firsters."[12] In the eyes of many UCP's supporters, the departing premier had lacked the intestinal fortitude to "take on" Alberta's enemies. Of all the contenders for the UCP leadership, Smith presented as the street brawler best up for this task.

The other candidates did not go down without a fight: Each sent out repeated warnings to the membership about Smith's past mistakes. Likewise, Kenney — fearful of what a Smith leadership might mean for the party — became increasingly aggressive in warning against a Smith takeover of the party he had welded together. A Political Action Committee (PAC), Shaping Alberta's Future, formed originally to promote

Kenney and the party, began posting in mid-September a series of ads on Facebook, Instagram, and Google questioning Smith's qualifications and warning that her victory would likely result in an NDP win come next election (Tait, 2022).

The Sovereignty Act remained her seminal attraction. She told the audience during the party's final official leadership debate that,

> We might be facing mandatory vaccination; we will say we will not enforce that.... If there's an emergencies act that wants to jail our citizens or freeze their accounts, we will say we will not enforce that. Arbitrary fertilizer cuts, arbitrary phaseout of our natural gas for electricity and power. Arbitrary caps on our energy industry and perhaps even a federal digital ID. (If) we have the Alberta Sovereignty Act, we will not enforce that. We'll put Alberta first (quoted in Thomson, 2022a).

In other times and places, Smith's conjuring of threats that simply did not exist might have disqualified her as a serious candidate. But nothing — neither her opponent's criticisms nor her own inflammatory rhetoric — derailed Smith's quest; indeed, it cemented her reputation among UCP members as a "fighter."

Though it took six ballots, on October 6, Smith claimed the UCP's leadership prize. The final tally saw her defeat her main rival, Toews, with 42,423 votes (53.77 per cent) to his 36,480 votes (46.23 per cent), a winning percentage of support only two per cent more than Kenney had received in stepping down. The number of votes cast (84,593) represented only 69 per cent of the party's membership (123,915). In turn, the vote for Smith meant that 1.5 per cent of Alberta's electorate (roughly 2.8 million voters) had now put in charge of the province an unelected individual whose political past, in the eyes of many, is checkered. No matter; with Smith's victory, Alberta's Wildrose faction had secured the outcome denied it in 2012.

How did Smith win? As Chapman and Epp point out (Chapter 19), it is too simplistic to describe Alberta's electorate as divided between rural and urban constituents. Better than the other candidates, however, Smith succeeded in playing to the feelings of anger, fear, and disempowerment felt by the narrow base of UCP members. A CBC analysis (Markusoff, 2022) of UCP members, who make up only 3.5 per cent of Alberta's population, showed a large number come from a small number of ridings south of Red Deer, and only 41 per cent from the big cities of Edmonton and Calgary. Smith appealed to these members' fears and anger, particularly around the COVID-19 mandates and resultant protests. Another study shows that, while 61 per cent of Albertans disagreed with the Freedom Convoy's goals, and 67 per cent with its methods, 56 per cent of UCP respondents supported its goals and 48 per cent its methods (Duhatschek, 2022).

Smith's victory speech on the evening of October 6 echoed in many ways Kenney's from election night in 2019. Beginning with an exuberant "I'm back!" she thanked her opponents, as well as Kenney, while encouraging party members to remain united and strong. Alberta was about to write "a new chapter" in its story, she said. "It's time for Alberta to take its place as a senior partner to build a strong and united Canada." But, "No longer will Alberta ask for permission from Ottawa to be prosperous and free," continuing:

> We will not have our voices silenced and censored. We will not be told what we must put in our bodies in order to work or to travel. We will not have our resources landlocked or our energy phased out of existence by virtue-signalling prime ministers. Albertans, not Ottawa, will chart our own destiny on our own terms. ... (quotes in Climenhaga, 2022a).

Jason Kenney is a proud man. At that moment, he must surely have wondered about his legacy — or, perhaps, competing legacies, neither of them to his liking. In the immediate term, Danielle Smith is a renegade libertarian who could well destroy the party he founded; in the longer term, Rachel Notley and the NDP may return to office, an outcome which he had come to Alberta to prevent. Other conservatives likewise observed warily what had transpired and the future lying ahead. In the words of long-time advisor Ken Boessenkool, "Premier Smith is a kamikaze mission aimed at the UCP, conservatism and Alberta" (quoted in Ascah, 2022).

On November 8, Smith won a byelection in Brooks-Medicine Hat, taking 55 per cent of the vote, compared to 27 per cent for her NDP and 17 per cent for her Alberta party opponents. It was not an overwhelming victory, but it was enough for Smith to take a seat in the legislature and to launch her libertarian agenda.

Who is Danielle Smith?

Basic biographical information regarding Smith is readily available (see Markusoff, 2022; Canadian Press, 2022). She was born in Calgary on April 1, 1971, the second of five children. Her paternal great-grandfather was Philipus Kolodnicki, a Ukrainian immigrant who changed his name to Philip Smith upon arriving in Canada in 1915. Smith has also claimed Indigenous ancestry, but no evidence exists in support of this contention. Her parents worked in the oil patch. The family lived for a time in subsidized housing.

She completed a B.A. (English) at the University of Calgary, where she met, married, and later divorced Sean McKinsley, and also met such arch-conservatives as Ezra Levant (founder of Rebel News) and Rob Anders (later, a Conservative MP). She subsequently also completed a B.A. in economics, during which time she met political scientist Tom Flanagan who became a kind of mentor to her. He recommended Smith for a one-year internship with the Fraser Institute. By now, she was well on her way to becoming, as she defines herself, a "libertarian populist" whose primary intellectual influences include Friedrich Hayek, Adam Smith, John Locke, and Ayn Rand (see Harrison, Chapter 5).

In 1998, Smith was elected a trustee of the Calgary Board of Education. A year later, however, then Minister of Learning Lyle Oberg dismissed the entire board as it had become dysfunctional, of which by her own admission, Smith was a chief cause. Subsequently, Smith worked for the Alberta Property Rights Initiative and the Canadian Property Rights Research Institute, before joining the Calgary Herald as a columnist with the editorial board (where she notoriously did not hesitate to cross the picket line during the 1999-2000 strike at the newspaper). She later succeeded Charles Adler on a Global Television interview show.

She married David Moretta, a former executive with Sun Media, in 2006. That same year, the Canadian Federation of Independent Business hired her as its provincial director for Alberta. Disenchanted with the premiership of Ed Stelmach, she left the PCs and joined the Wildrose Alliance party, becoming its leader in October 2009. The ensuing events — the 2012 election, the floor crossing in 2014 — were recounted earlier and are not repeated.

Smith was defeated in her bid for the PC nomination in Highwood on March 28, 2015. Any formal political future for Smith seemed dim. By now, she and her husband had moved to High River, a town south of Calgary. They survived the infamous floods of 2013 and in 2018 opened a restaurant, the Dining Car at High River Station (formerly, the Whistle Stop Café). She became a popular talk radio host on Calgary's QR77 in 2015, a job she held for the next several years, but left in early 2021 citing personal attacks on Twitter.

In 2019, while still a talk show host, she registered as a lobbyist for the Alberta Enterprise Group, an association of which she was also president. The Calgary-based association represents 100 companies involved in such areas as health care, transportation, construction, energy, law and finance. Many of the things Smith has lobbied for in the past — for example, health spending accounts and royalty breaks for energy companies that clean up abandoned wells — she now promotes in her formal political role (story in Weber, 2022).

Smith's occupational career can best be described as that of a serial lobbyist and media personality. Her chief ability seems that of convincing others of her abilities. In politics, however, her ambitions and promise have often fallen short of actual performance. Ideologically, she is a committed right-wing libertarian, for whom freedom trumps equality, markets trump politics, and democracy is little more than an exercise in populist agitation and manipulation. Her actions since becoming premier, in centralizing power within her office, suggests an authoritarian streak.

Smith is a clever wordsmith, and described by many as intelligent, but she has not shown herself to be a critical thinker; instead, she seems wedded to novelty for novelty's sake. She is generally dismissive of "experts," except when their ideas validate what she already believes. In the words of journalist Graham Thomson (2022b), she is "noted for constructing a world view based on anecdotal evidence, confirmation bias and bad choices."

Befitting a talk radio host, Smith has a lot of opinions. (In an Ask Me Anything broadcast in June 2021, Smith said, "I literally have an opinion on everything.") But her opinions often lack evidence. The examples are multitude. In 2003, while a columnist for the Calgary Herald, she cited tobacco-funded research that "smokers of just three to four cigarettes a day have no increased risk of lung cancer, coronary heart disease, bronchitis or emphysema." Her particular focus on cancer was repeated during the 2022 leadership race when she implied that everything, up to "stage four and that diagnosis" is completely within an individual's capacity to control.

In the midst of a massive beef recall due to E. coli contamination in 2012, Smith — at the time, still Wildrose leader — claimed that thoroughly cooking the meat would kill the bacteria, and that it could then be fed to those in need (Rieger, 2020). During the COVID-19 pandemic, she used her radio show, newsletters, and podcasts to criticize health restrictions, and the science behind them, and to promote debunked treatments such as hydroxychloroquine and ivermectin[13], for which the radio station disciplined her.

Smith's over-the-top rhetoric continued even after winning the UCP leadership. During a media scrum immediately after her swearing-in ceremony as premier, Smith declared the unvaccinated were "the most discriminated against group that I've ever witnessed in my lifetime," having faced "restrictions on their freedoms" based on having made a "medical choice" (Climenhaga, 2022b). (Critics noted the statement's ignoring of a host of other groups historically persecuted, imprisoned, and even murdered, and the absence of any consideration of the rights of individuals to safe working environments.)

Smith later said she would amend the *Alberta Human Rights Act* to protect the rights of those refusing to be vaccinated against COVID-19, and mused about a possible "blanket amnesty" for anyone charged with violating public health restrictions. She also accused Alberta Health Services of "manufacturing" staffing shortages and of being in cahoots with the World Economic Forum.

Smith said during the leadership race and after that she would be a "unifier," a necessary trait given the party's difficulties during Kenney's time as premier. Her track record speaks otherwise, however. While she may succeed, where Kenney failed, in unifying the party, the contrary evidence points to a party and a province facing further division.

Conclusion

And so it is that Alberta has as its premier a right-wing libertarian and conspiracist who, despite lacking a personal mandate from the people, is in position, as we write, to attempt remaking the province according to her own fantasy vision of what Alberta is and should be. Will Danielle Smith achieve that mandate in 2023 — or sometime thereafter?

Elections are never entirely predictable. The NDP has held a solid lead in the polls for most of the last two years, repeated in polls conducted by Janet Brown and Associates (see Chapter 2) in the fall of 2022. But the UCP cannot be counted out. Resource revenues have again filled Alberta's coffers, money with which governments can reward friends, make amends to others, and shore up support in key ridings. The habit of voting for conservatives — whatever that term actually means — remains entrenched in Alberta's political culture; there remain scores of angry and alienated voters for whom the UCP message resonates, especially during a crisis of inflation and affordability that many Alberta have faced throughout 2022, and which Smith has continuously laid at the feet of the federal Liberals and NDP.

Yet, should Smith win, at least one prediction seems safe: That she will — perhaps sooner, perhaps later — disappoint her followers and face a party revolt that will force her from office. The reason is simple: Like all recent conservative premiers, she has promised more than she can deliver.

Kenney, Smith, and the UCP are the symptom of a failure of Alberta's entrenched political class to deal with the province's deeper problems. This failure takes the form of demands that the Alberta state be given more power; that is, that those who have held power in the province going on forty years be given even more power. But nothing ever changes. Their fantasy solutions always crash and burn against the political, environmental, social, and economic realities of our time. The years of UCP government represent the thrashings of this old order; the railing of anger and despair against the light.

What Abraham Rotstein said about Canada in 1964 — "Much will have to change in Canada if the country is to stay the same" — could be applied Alberta. It is a great province; but it could be better — it must be better. Abandoning the politics of anger and fear would be a good start.

References

Acuña, R. (2019, May). What they did: The Notley NDP government record. *Alberta Views*, 24-28.

Aldrich, J. (2022, Sept. 2). Support for independent Alberta drops to 23 per cent: poll. *Calgary Herald*.

Anglin, H. (2022, July 13). The Alberta Sovereignty Act is nothing but a sideshow scam. *The Hub*. https://thehub.ca/2022-07-13/howard-anglin-the-alberta-sovereignty-act-is-nothing-but-a-sideshow-scam/?utm_medium=paid%20social&utm_source=facebook&utm_campaign=boost&fbclid=IwAR2CpmqnBUKBLsgPh70RqYDTGF1d8yg3arL3B132m0KO5nOHIKmg0F1CHJM

Ascah, R. (2022, Oct. 10). Premier Danielle Smith- buckle-up. http://abpolecon.ca/2022/10/10/premier-danielle-smith-buckle-up/

Braid, D. (2019, Dec. 6). Legislature session completes UCP teardown of NDP era. *Calgary Herald*.

Braid, D. (2021, May 13). UCP caucus expels the latest dissident, but the fight may not be over. *Calgary Herald*.

Braid, D. (2022, June 1). UCP erupts again over Kenney's claims about anti-vaxxers. *Calgary Herald*.

Bratt, D., Brownsey, K., Sutherland, R., & Taras, D. (Eds.). (2019). *Orange Chinook: Politics in the New Alberta*. Calgary: University of Calgary Press.

Canadian Press. (2019, May 1). A look at the cabinet members in Alberta's new UCP government.

Canadian Press. (2022, Oct. 6). Redemption: Danielle Smith aims to be "force of unity" as new Alberta premier.

Climenhaga, D. (2022a, Oct. 7). Danielle Smith is narrowly chosen to lead a divided UCP. *Rabble*. https://rabble.ca/politics/canadian-politics/danielle-smith-is-narrowly-chosen-to-lead-a-divided-ucp/

Climenhaga, D. (2022b, Oct. 12). Danielle Smith's first news conference as premier: Sovereignty, health care changes, succour for the Great Unvaxxed! Blog.

Conway, J. (2014). *The Rise of the New West: The History of a Region in Confederation*. Toronto: Lorimer.

Duhatschek, P. (2022, Sept. 2). Most Albertans did not support Freedom Convoy, University of Alberta survey suggests. *CBC News*.

Finkel, A. (1989). *The Social Credit Phenomenon in Alberta*. Toronto: University of Toronto Press.

Fukuyama, F. (1992). *The End of History and the Last Man*. New York: The Free Press.

Gramsci, A. (1971). *Selections from the Prison Notebooks*. Originally written in 1930. New York: International.

Hansen, H. (2001). *The Civil War: A History*. New York: Signet Classics.

Harrison, T. (1995). *Of Passionate Intensity: Right-Wing Populism and the Reform Party of Canada*. Toronto: University of Toronto Press.

Harrison, T. (2002). *The Great Right Hope: Stockwell Day and Image Politics*. Montreal: Black Rose Books.

Harrison, T. (ed.). (2005). *The return of the trojan horse: Alberta and the new world (dis)order*. Montreal: Black Rose Books.

Harrison, T. (2019). Morbid symptoms and Alberta's yellow vest movement. *Canadian Dimension* 52(4), Winter, pp. 11-12.

Harrison, T. & Krahn, H. (Unpublished). *Provincial versus national identity in Alberta*.

Hiller, H. (1977). Internal problem resolution and third party emergence. *Canadian Journal of Sociology* 2(1), 55-75.

Johnson, L. (2021, Apr. 7). Sixteen UCP MLAs say Kenney's latest COVID-19 restrictions move Alberta "backwards." *Calgary Herald*.

Julie, A. (2017, Sept. 8). Brian Jean more popular than Jason Kenney amongst Alberta voters: ThinkHQ. *Global News*. https://globalnews.ca/news/3726632/brian-jean-more-popular-than-jason-kenney-amongst-alberta-voters-thinkhq/

Kaiser, Saif. (2022, Apr. 9). Alberta premier Jason Kenney gives united conservatives ultimatum. *City News*.

Lambert, T. (2022, Jan. 17). "This is 'Kenneyism.'" *The Tyee*.

Laxer, G., & Harrison, T. (1995). *The Trojan Horse: Alberta and the Future of Canada*. Montreal: Black Rose Books.

Lisac, M. (1995). *The Klein Revolution*. Edmonton: NeWest Press.

Lisac, M. (2005). *Alberta Politics Uncovered: Taking Back Our Province*. Edmonton: NeWest Press.

Martin, D. (2016, June 23). Jason Kenney's mission unlikely, but not impossible. *CTV News*.

Macpherson, C. B. (1953). *Democracy in Alberta: TThe Theory and Practice of a Quasi-Party System*. Toronto: University of Toronto Press.

Markusoff, J. (2022, Aug. 25). Why choosing Alberta's next premier largely lies in the hands of folks in Rimbey, Strathmore and Three Hills. *CBC News*.

Richards, J., & Pratt, L. (1979). *Prairie Capitalism: Power and Influence in the New West*. Toronto: McClelland & Stewart.

Rieger, S. (2020, March 22). Alberta talk radio host deletes tweet with false claim that there's a 100% cure for coronavirus. *CBC News*. https://www.cbc.ca/news/canada/calgary/coronavirus-cure-claim-1.5506187

Tait, C. (2022, Sept. 21). Political advertiser spent thousands campaigning against UCP leadership candidate Danielle Smith. *Globe and Mail*.

Thomson, G. (2022a, Aug. 31). Ignore the facts and pump the rhetoric: Welcome to the alternate reality of those vying to become Alberta's next premier. *The Star Calgary*.

Thomson, G. (2022b, Oct. 26). Why Alberta's election pitting Danielle Smith against Rachel Notley will be a ferocious but uneven fight. *Star Calgary*.

von Scheel, E. (2021, Aug. 26). Albertans got more federal COVID support per capita than people in any other province, report says. *CBC News*.

von Scheel, E. (2022, March 24). "I don't need this job": Kenney says he has to stay to keep "lunatics" from "trying to take over the asylum." *CBC News*.

Weber, B. (2022, Nov. 10). Danielle Smith's lobbying record holds clues to her governing agenda, observers say. *The Star Calgary*.

Wesley, J., and Young, L. (2022, July 15). What the spectre of Alberta separatism means for Canada. *Lethbridge Herald*.

Wright, S. (2019, December). Kenney's first 100 days. *Alberta Views*, pp. 24-28.

Zimonjic, P. (2022, Sept. 3). World economic forum official says Canada has bigger issues to discuss than conspiracy theories. *CBC News*.

NOTES

1 Morton is also a former professor of political science at the University of Calgary, part of the so-called Calgary School of conservative academics.

2 A few examples: Social Credit in the 1930s, the Reform party in the 1980s, the separatist Western Canada Concept, also in the 1980s, and the more recent Wildrose Independence party.

3 Jean was, and remains, a sympathetic figure. His son passed away from cancer in March 2015 in the midst of Jean's bid for the Wildrose leadership. The following spring, he lost his house to the Fort McMurray fires.

4 The flat rate tax, unusual in Canada, had been introduced by then Finance Minister Stockwell Day in 1999, at a significant cost to the Alberta treasury.

5 The UCP subsequently cancelled the NDP initiative, though grudgingly it later signed on to a $10-per-day program initiated by the federal Liberals.

6 The accusations of dirty tricks were reminiscent of several shady undertakings during the years of the Harper Conservatives, both before and after they attained government. These included an alleged financial inducement offered to a dying MP, Chuck Cadman, in 2005 to help defeat Paul Martin's minority Liberal government on a confidence vote; the Mike Duffy Senate scandal in 2013; and the Robocalls incident during the 2011 election.

7 Danielle Smith's first cabinet, after becoming premier in October 2022, went even further. Despite appointing a record 38 people to cabinet posts, the separate Ministries for the Status of Women, Labour, and Environment were eliminated entirely.

8 The letter's original signers were House Speaker Nathan Cooper, Tracy Allard, Michaela Glasgo, Miranda Rosin, Todd Loewen, Angela Pitt, Drew Barnes, Jason Stephan, Roger Reid, Nate Horner, Glenn van Dijken, Ron Orr, Dave Hanson, RJ Sigurdson and Mark Smith (Johnson, 2021). Garth Rowswell later added his name.

9 There are good and legitimate criticisms of the World Economic Forum, but arguing it is part of a global communist plot is not one of them. For a good analysis of the etiology of this current hysteria, see Zimonjic (2022).

10 Giving individual states the ability to prohibit abortion is a favourite position of right-wing conservatives in the U.S., but as yet has gained little traction among most conservatives in Canada (see, Steward, Chapter 4).

11 President James Madison, upon hearing this assertion, is said to have remarked, “For this preposterous and anarchical pretension there is not a shadow of countenance in the Constitution” (quoted in Hansen, 2001, p. 15).

12 It is important to note the vast majority of Albertans identify themselves as Canadians first. The percentage of Albertans identifying themselves first as Albertans and second as Canadians is about 20 per cent, with only about one per cent identifying as only Albertan. Only a small percentage would vote to separate (Harrison and Krahn, unpublished; Wesley and Young, 2022; Aldrich, 2022).

13 Smith’s support for unproven and dangerous remedies mirrored similar claims by right-wing conservatives in Canada such as former Conservative MP Derek Sloan and Calgary mayoral candidate Kevin J. Johnston, as well as U.S. President Donald Trump

POLITICS

CHAPTER 2

SORRY, NOT SORRY: THE NASTY, BRUTISH, AND SHORT PREMIERSHIP OF JASON KENNEY

Janet Brown and Brooks DeCillia

> "We shall fight on the beaches, we shall fight on the landing-grounds, we shall fight in the fields and in the streets, we shall fight in the hills. We shall never surrender!"
>
> — Winston Churchill, June 1940, House of Commons, following the evacuation of British and French armies from Dunkirk as the German tide swept through France.

JASON KENNEY'S political hero famously vowed during the darkest hours of the Second World War to "never surrender." And Kenney's premiership certainly did "not flag nor fail" to fight on many fronts and up until the very end of his time in office, seemingly adopting the wartime words of British Prime Minister Winston Churchill as a political battle cry to "never surrender" — even in the face of withering criticism, plummeting public opinion, and perilous political infighting that, arguably, necessitated giving in, or, at the very least, being more flexible (Keller, 2019). Kenney's tenure as United Conservative Party (UCP) leader and Alberta premier began with a promise to "unapologetically" fight for Alberta. This pugilistic posture, arguably helped him unite the once rival Progressive Conservatives and Wildrose Party and win power in 2019. The former prominent federal cabinet minister in Stephen Harper's government was, in fact, praised at the start of his provincial political career for his intellect, steadfastness, shrewd political instincts and political communications, but the UCP leader's uncompromising character eventually backfired when, in May 2022, with one year remaining in his mandate, Kenney was forced to step down as premier after receiving only 51.4 per cent support from UCP members in a leadership review (Amato, 2022).

Throughout his time as premier, Kenney seemingly could not admit mistakes and bristled frequently when pressed for accountability by journalists, derisively earning him a reputation as an "I reject the entire premise of your question" politician (Climenhaga, 2021). Kenney also demonstrated an inability to bend or soften his conservative ideology in the face of an increasingly centrist Alberta population (DeCillia, 2022; DeCillia, 2023). Additionally, he often failed to read the politics of his precarious positions, and some have even questioned if Kenney really understood the province that he spent so much time away from while serving in federal politics (Bratt, 2021; Epp, 2023).

While Churchill's now famous "We shall fight on the beaches" speech is best remembered for its commitment to never capitulate, much of the iconic address to the House of Commons candidly reflects on the Allied military losses, while also offering a frank assessment of Great Britain's precarious position and potential imminent invasion by Germany. Churchill aimed to dampen public joy about the evacuation of Dunkirk and reminded Britons, "We must be very careful not to assign to this deliverance the attributes of a victory" (Churchill, 1940). Kenney's political hero counterbalanced his warning of a potential Nazi invasion with a need to buoy public support of the war. Kenney's communication about COVID-19 often lacked this symmetry, infamously insisting prematurely in June 2021 the pandemic was ending and the "best summer ever" laid ahead for a beleaguered province. By the fall, though, a surging fourth wave of the pandemic threatened to collapse the province's health system (Cecco, 2021). The pandemic also exposed the ideological tensions in the governing party with United in its name (Bratt, Sutherland, and Young, 2023).

It is as if Kenney hoped his bravado in June 2021 would bring an end to the pandemic. He argued the vaccines offered Alberta "a superpower" to fight the virus (CBC News, 2021c). His party even sold "best summer ever" ballcaps as a fundraiser. But as experts warned and his critics stressed, Kenney's rhetoric proved wishful thinking (Anderson, 2021). Over the summer, while the premier vacationed, COVID-19 cases soared, the province's hospitals filled with sick people, and public anger swelled. When he returned from holidays, Kenney was forced to impose strict public health measures and backtrack on a promise to never introduce a vaccination passport.

Throughout his time in office, Kenney rarely — or only begrudgingly — apologized for his actions. When he did say sorry, Kenney was slow and often backhanded (CTV News, 2021). He initially refused to offer an apology on behalf of MLAs and other members of the UCP who travelled over the 2020-21 Christmas break, despite official provincial government guidelines asking Albertans to avoid non-essential travel. Days later he accepted the resignation of his minister of municipal affairs and chief of staff, who had both travelled internationally over the holidays (Hudes, 2021).

Always quick to blame others — Ottawa, the former NDP government, to name some of his favourite targets — Kenney frequently deflected criticism of his government's missteps. Yet, amidst surging COVID-19 cases in the spring of 2021, Kenney sidestepped a reporter's questions about his handling of the pandemic by arguing it was regrettable that the pandemic had triggered a "tendency to politicize [COVID-19] and turn it into a blame game"(Climenhaga, 2021). Kenney also appeared to play both sides in the debate about pandemic restrictions, attempting to appease his rural and culturally conservative bases. The majority of Albertans — especially those living in urban areas — supported pandemic precautions, according to numerous polls (Brown, 2022). The UCP premier, for instance, initially resisted mask mandates, arguing rural people would not comply (Tait, 2020). In the summer of 2021, Kenney vowed to never introduce vaccine mandates, but by September of that year reality crashed into his rhetoric, and Kenney brought in a proof of vaccination program, saying that "we regrettably came to the conclusion that the only way to cut viral transmission, without destroying businesses, people's livelihoods, was a program like this" (Dawson, 2021).

He did manage to muster an apology after photos of him and other cabinet ministers and senior aides surfaced showing them flaunting pandemic rules and dining on a rooftop of the government office suite nicknamed "Sky Palace" in the aftermath of a spending imbroglio involving former premier Alison Redford. Initially, Kenney insisted all public health regulations were followed at the outdoor gathering (CBC News, 2021a). Kenney eventually apologized to his colleagues and all Albertans, conceding he had let the people of the prairie province "down for not being more careful" (CBC News, 2021a). But his apology came a week after the controversial photo first surfaced. Amidst a public outcry and criticism from fellow cabinet ministers and other members of his caucus that labelled his behavior as inappropriate, Kenny told a news conference that while he tried to "lead by example," he admitted he had not "always done that perfectly" (von Scheel, 2021a).

Kenney's approach to apologies differed dramatically from former PC premier Ralph Klein. Klein — a former broadcast journalist turned politician — knew there was real power in saying sorry. The plain-spoken beer-drinking, smoking, working-class champion often got out of the doghouse with voters by admitting he was wrong or had messed up. It made him relatable. Kenney, during his time as premier, on the other hand, rarely offered apologies (and mostly meagre ones), and then only after he was caught doing something wrong or sustained internal or public criticism.

It took until late May 2022 — two weeks after Kenney had announced he'd be stepping down as premier — for him to show a rare act of contrition. During opening remarks at a news conference, Kenney said he regretted earlier actions such as announcing plans to stop mass testing for the coronavirus. But moments later, when taking questions from reporters, he reversed himself, pointedly abrogating his responsibility to relax public health restrictions at the start of summer, emphasizing that "you can't sustain serious intrusions into people's lives permanently. And so no, I don't apologize for this decision" (Amato, 2022). In short, Kenney was sorry, but not sorry.

The result of Kenney's intransigence was harsh criticism from all sides. Kenney was blasted for both not doing more sooner and for doing too much. And with pro-lockdown and anti-lockdown sides agreeing Kenney's actions were undermining public trust, polling numbers for Kenney himself and his party plummeted. A CBC News poll conducted in October 2022 found half of Albertans, in retrospect, were fine with the restrictions placed on businesses and gatherings throughout the pandemic. Nearly two in ten Albertans (18 per cent) thought the COVID-19 precautions were not strict enough. Only a minority — 30 per cent — thought the measures were too strict (Brown, 2022). As political scientist Duane Bratt argued, in his interpretation of CBC News' polling data, "most politicians, when faced with a 70-30 question, would easily choose the majority" (Bratt, 2022). Evoking Klein, Bratt noted the former premier often quipped that his political strategy was to figure out what way the public opinion parade was headed and get out in front of it. Throughout his time as Alberta premier, Kenney failed to demonstrate the political acumen and strategic thinking he was known for as a federal cabinet minister in Stephen Harper's government. The following pages argue there are four reasons for Kenney's precipitous drop in the polls: (1) His conservative ideology put him at odds with most Albertans who are centrist; (2) His failure to position himself and his party in the mainstream of public opinion; (3) His government's long list of gaffes and scandals; and (4) His intransigence (or inability to say sorry) when he or his government made a misstepped.

Polling data from the *Wild Ride Update*, a subscription-based polling report published by Janet Brown and journalist Paul McLoughlin, documents Kenney's dramatic fall from grace. His UCP party won the 2019 election with an overwhelming 55 per cent of the popular vote. But by October 2021, following Alberta's disastrous "Best Summer Ever," UCP support had fallen to 29 per cent. The premier's own popularity peaked in September 2019, when 53 per cent of Albertans said they approved on the job his was doing. By October 2021 his approval rating had fallen to a dismal 19 per cent — the lowest Brown had ever recorded in her 30-year history of conducting polling in Alberta.

Comparing polling results from September 2019, when Kenney was his most popular, to October 2021, when Kenney was his least popular, shows those indicating they would vote for the UCP if an election were held today fell a breathtaking 25 points from 55 per cent to 29 per cent. But looking closer at the data, the sharpest decline in support was among those with an annual household income over $120,000 (down 37 points from 64 per cent to 27 per cent), Calgarians (down 35 points from 64 per cent to 29 per cent), those working in the private sector (down 30 points from 63 per cent to 33 per cent), men (down 30 points from 60 per cent to 30 per cent), and those with children under 18 in the home (down 30 points from 55 per cent to 25 per cent) (see Figure 2.1 below).

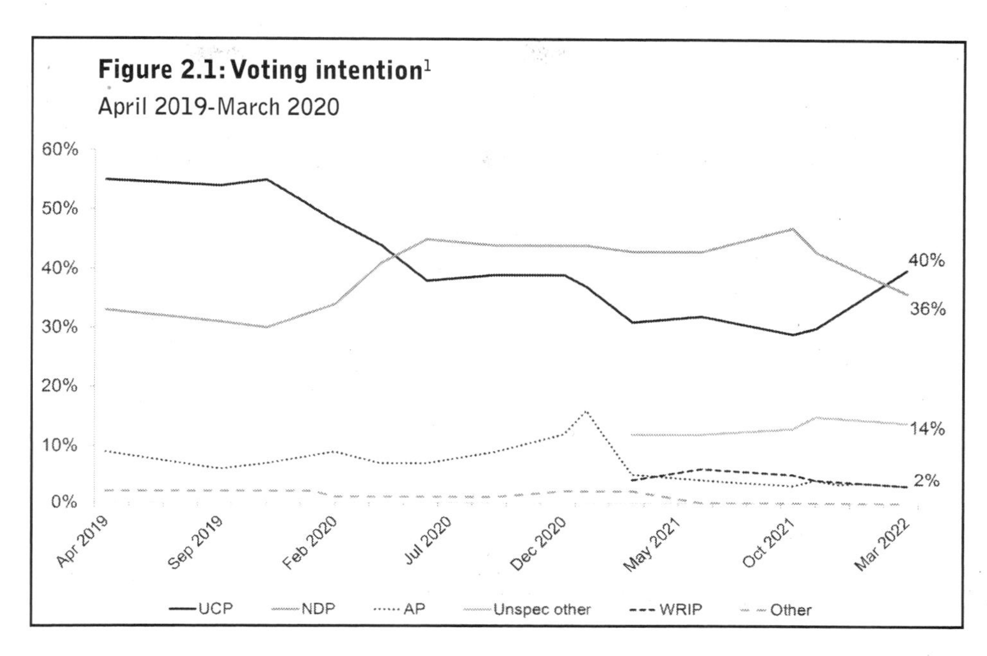

Figure 2.1: Voting intention[1]
April 2019-March 2020

The pattern was even more striking for Kenney's personal approval, which fell from 53 per cent in September 2019 to 19 per cent in October 2021. The greatest declines in approval were among those with an annual household income over $120,000 (down 47 point from 64 per cent to 17 per cent), Calgarians (down 43 points from 63 per cent to 20 per cent), those with children under 18 in the home (down 40 points from 56 per cent to 16 per cent), those working the private sector (down 39 points from 60 per cent to 21 per cent), and men (down 39 points from 61 per cent to 22 per cent).

CBC News polling data from the fall of 2022 highlighted how the UCP's policy positions were disconnected from the more moderate values of most Albertans — especially in urban areas — on issues ranging from the environment to the handling of the pandemic (DeCillia, 2022). Arguably, Danielle Smith's elevation as UCP premier — combining Kenney's strident rhetoric about more autonomy for Alberta with her own outside-the-mainstream views on COVID-19 — likely intensified this misconnection. Polling data suggests Alberta voters were becoming more moderate just as the governing party was becoming more militant.

The NDP under the leadership of Rachel Notley was able to capitalize on Kenney's unpopularity. After winning 33 per cent of the popular vote in the 2019 election, they overtook the UCP in support in June 2020 and maintained that lead until the end of 2021. Throughout 2020 and 2021, Notley's personal approval hovered between 48 and 57 per cent.

Voting Intention

A note about how voting intention is measured:

Prior to March 2021, Brown posed a voting intention question that was like most other pollsters. It asked respondents how they would vote if an election were held today. If respondents were undecided, they were asked if they were leaning toward a party. The decided and leaning vote is added together, and the undecideds are excluded.

In early January 2021, Brown recorded the Alberta Party at 16 per cent support, despite not having a leader or any real presence on the political scene. Suspecting people were using the Alberta Party as a "none of the above" category, Brown changed the way she asks about voting intention, starting in March 2021. She now asks a three-part question: (1) would you vote UCP, NDP or for another party if an election were held today, (2) if another party, which party, (3) if undecided, are you leaning toward a party. Those who initially say "another party" but do not name a party in the follow-up question are classified as "orphan" voters. Orphan voters are not the same as truly undecided voters, as they know who they don't want to vote for. They just don't know yet who they will vote for.

Leader Approval

Kenney jumped from federal to provincial politics with the big ambition of uniting Alberta's two conservative parties. In the summer of 2016, he sped westward from Ottawa to Alberta in a blue Dodge Ram 1500 — complete with four oversized wheels — pledging to unite Alberta's conservative parties. After 20 years in federal politics, Kenney crisscrossed the prairie province in his so-called "Cowboy Cadillac," hammering the governing NDP's handling of the sluggish economy and promising better days ahead if free enterprise voters united under one political banner to fight the 2019 election. Initially, Kenney, the prime architect of the unite the right movement, racked up win after win, easily winning the Progressive Conservative Association leadership in the winter of 2017 and later that same year capturing the leadership of the newly formed United Conservative Party, a merger of the former PC and Wildrose parties. Described by one commentator as a "political colossus who'd make Alberta muscular again," Kenney initially "dazzled [Alberta voters] with his smarts, steadfastness, tactical cunning and communications savvy" (Markusoff, 2021). Kenney's propensity for winning continued.

Although it did not seem to have a negative impact on his polling numbers during the 2019 election, Albertans got a glimpse into Kenney's compunction to never concede or offer genuine contrition. During an interview just two weeks before election day, Charles Adler — the conservative radio host and supposed friend of the UCP leader — couldn't pull an apology out of Kenney for his efforts in the 1980s in

San Francisco to deprive dying AIDS patients from visiting with partners. Kenney would not budge, only conceding that he "regret[ed] many things he did as a young man." The interview with the veteran broadcaster came in the wake of Kenney's defense of one of his party's candidates in the 2019 general election who had faced a barrage of criticism for past remarks that compared homosexuality to pedophilia (As It Happens, 2019). Speaking about the interview with CBC Radio's *As It Happens*, Adler said he felt an obligation to hold his friend's "feet to the fire" during the often-heated exchange on his *Global News Radio* program. "He knew," Adler told CBC, "that I was searching for his moral compass" (As It Happens, 2019). Adler challenged — even implored — the UCP leader to say sorry, stressing it was time to offer a "genuine, remorseful apology for the many people that you and your colleagues hurt" (Bartko, 2019). Kenney was immoveable during the interview. He continually repeated his talking point about regretting his past position, but never uttering an apology, foreshadowing what was to come when he became premier.

Kenney won a commanding electoral victory in the April 16, 2019, provincial election. His United Conservative Party captured 55 per cent of the popular vote, ousting Notley's NDP after a single term. Kenney's performance, in fact, bested Ed Stelmach's landslide victory in 2008 that captured 53 per cent of all votes cast. As some analysts predicted when Kenney became Alberta's 18th premier, his honeymoon with voters would be short (DeCillia, 2019). Even before Kenney — who had a reputation for never apologizing or admitting mistakes — took control of the prairie province's government, he was not the most popular politician. A public opinion poll by Brown released during the 2019 provincial election campaign, in fact, found that Notley, by comparison, was more popular (47 per cent) with Albertans than Kenney (44 per cent) (Brown, 2019). So, even before Kenney became premier — and made any unpopular decisions — he was less popular with Alberta voters than a premier who had governed through hard economic times for four years (see Figure 2.2 below).

On election night, a jubilant Kenney declared "help is on the way and hope is on the horizon," echoing his campaign promise to breathe life into Alberta's moribund economy by focusing on "creating good jobs, growing the economy and building pipelines" (DeCillia, 2019). Without a doubt, Kenney had big plans when he grabbed the levers of government. He took real pleasure in touting his new government's accomplishments, even creating a "Promises Kept" webpage on the UCP's website that detailed the government's efforts to follow through on its 2019 platform pledges (United Conservative Party, 2022).

On his first day on the job, Kenney vowed to cut taxes, streamline regulation and attract new investment to the province. The UCP made good on many of its campaign pledges, including killing the provincial carbon tax, slashing the province's corporate

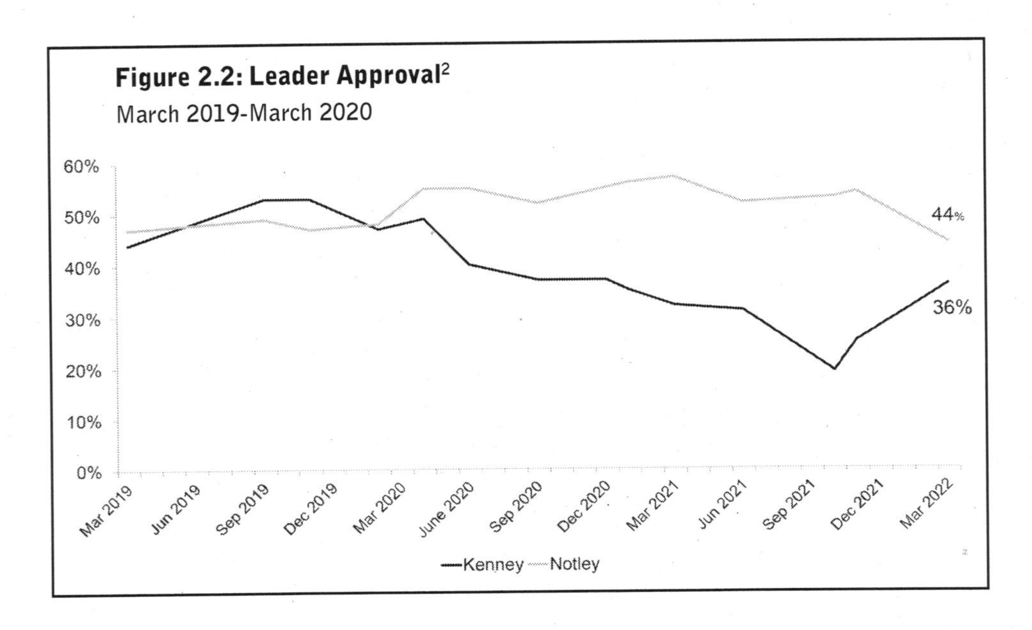

tax rate from 12 to eight per cent. Kenney also dialled up the anti-Ottawa rhetoric, even warning of a "growing crisis of national unity" if the federal Liberals did not reverse plans to overhaul federal environmental assessment legislation. Kenney also made good on his promise to set up a $2.5-million public inquiry, called a witch hunt by its critics, to investigate alleged foreign-funded attacks on Alberta's oil industry (Anderson, 2019). The UCP also tasked former Saskatchewan NDP finance minister Janice MacKinnon to come up with a roadmap for slaying the province's ballooning deficit, without raising taxes (Bellefontaine, 2019). Part of the UCP's belt tightening in its first year in office involved asking public sector workers — nurses, doctors, teachers, civil servants — for wage freezes and, in some cases, wage rollbacks, putting the government at odds with powerful public sector unions in Alberta.

By the fall of 2019 — about half a year into the UCP's mandate — the glow of winning was beginning to dim as the tough, and sometimes unpopular, decisions of governing clouded the party's blazing electoral win in the spring. Around this time, the first self-inflicted wounds in what would become a long list of scandals and gaffes emerged, planting the seeds of doubt, arguably, in Alberta voters' minds about Kenney and the UCP's competence.

The UCP government's much vaunted $30-million Canadian Energy Centre (CEC) — commonly known as the War Room — failed to impress many with its amateur and blunder-filled launch in the fall of 2019. The centre's clumsy beginning is emblematic of the UCP fumbles to come. The Kenney government created the controversial organization to combat what it called misinformation about Alberta's energy industry (Clark, 2023). Headed by a former UCP candidate, the private corporation — which is immune from public scrutiny, including freedom of information laws — was branded a propaganda operation by its critics when it opened. The Canadian Association of Journalists, in fact, condemned the CEC for misleading the public and describing its public relations professionals as "reporters" to people they interviewed for content published by the organization (Bruch, 2019). The War Room also took heavy fire in its first week in operation when it had to change its logo after social media posts revealed the symbol was already being used by an American tech company. In the same week, an official with the CEC seemingly ordered a local newspaper in Medicine Hat to print a CEC rebuttal to a critical column. And again, in the same week, the head of the CEC, Tom Olsen, raised eyebrows when he told a Calgary TV news show that "We are not about attacking, we are about disproving true facts" (Braid, 2019). It was an inauspicious beginning to an endeavour that continued to spark controversy and recrimination for the UCP government.

On top of the gaffes, the economy did not magically rebound when the UCP came to power. Investment did not pour into the province when the new government cut corporate taxes. The UCP cut the provincial carbon tax, but Albertans didn't get any respite because a federal one quicky replaced it. Despite all of Kenney's economic boosting rhetoric, Alberta's economy ended 2019 on a sour note with more job losses and a mild recession (Fletcher, 2020a). The year was, in fact, marked by continued high unemployment, low consumer confidence and sluggish activity in the energy, housing and construction sectors in Alberta (Varcoe, 2019). From an election night high of 55 per cent, public support for the UCP took its first noticeable drop in February 2020 when 48 per cent of Albertans said they would vote for the UCP if an election were held at that time.

Fundraising

The party's fundraising in late 2019 also dropped considerably. Even before anyone in Alberta had heard of the novel coronavirus or knew they would soon be overtaken by a global pandemic, UCP fortunes were starting to sputter.

The first Canadian case of COVID-19 was reported on Jan. 25, 2020. The first case in Alberta was reported on March 5. By Friday the 13th, most Albertans were departing their workplaces and schools uncertain about when they would be returning. In the early stages of the pandemic, Alberta was in an enviable position compared to other provinces, as it had lower rates of infection, hospitalizations, ICU admissions and deaths. Data collected in May 2020 by Brown as part of a Road Ahead Poll for CBC Calgary showed that more than six in 10 Albertans (62 per cent) approved of how the federal government was handling the pandemic, either strongly (23 per cent) or somewhat (39 per cent) (Fletcher, 2020b). An even higher proportion (70 per cent)

Table 2.1: Quarterly UCP and NDP Donations. By quarter since 2017-2022

Year	NDP Donation Amount	UCP Donation Amount
Quarter 3 2017	$439,101.08	$396,962.35
Quarter 4 2017	$335,546.59	$240,098.60
Quarter 1 2018	$329,416.93	$293,147.61
Quarter 2 2018	$853,357.15	$1,159,071.57
Quarter 3 2018	$678,233.20	$1,123,149.58
Quarter 4 2018	$1,544,193.65	$2,755,730.08
Quarter 1 2019	$2,675,416.60	$2,862,177.10
Quarter 2 2019	$1,451,532.28	$1,578,741.07
Quarter 3 2019	$385,328.50	$793,941.70
Quarter 4 2019	$1,035,896.46	$1,144,280.16
Quarter 1 2020	$581,195.41	$1,306,949.79
Quarter 2 2020	$ 1,031,316.85	$644,853.09
Quarter 3 2020	$1,124,575.02	$1,223,889.33
Quarter 4 2020	$2,322,450.38	$1,898,494.18
Quarter 1 2021	$1,183,500.03	$592,039.46
Quarter 2 2021	$1,510,659.87	$771,608.48
Quarter 3 2021	$1,367,100.50	$1,234,819.95
Quarter 4 2021	$2,090,743.53	$1,208,799.46
Quarter 1 2022	$1,037,511.32	$887,974.49

approved of the way the provincial government was handling the pandemic, either strongly (22 per cent) or somewhat (48 per cent).

Arguably, the merger of the former Wildrose Party and Progressive Conservative Association merely papered over the ideological, political, and personal divisions between the two factions. As Stewart and Sayers (2023) contend, the pandemic did not trigger the internal division in the supposedly big tent united conservative governing party; instead, it dialled up the real internal conflict within the political enterprise aimed at denying the New Democrats power. At its heart, opined Gary Mason in the *Globe and Mai*, the party "is an amalgam of two political philosophies, two ideological forces. They are often at odds." "To put it another way," he added, "the old Wildrose forces often disagree and resent the old Progressive Conservative types. Their interests aren't aligned. They don't like one another. Old war wounds have not healed and may never heal" (Mason, 2022).

These tensions played out in public, arguably, leading the public to question Kenney and the UCP's judgement and competent management of the government during a global crisis. (Table 2.2 below highlights the key moments in Kenney's leadership that, arguably, planted seeds of concern or discontent in Alberta voters' minds.)

The first cracks in the United Conservative Party caucus began to show a few months after the governing party's polling numbers began to slip. In June of 2020, the UCP government released its so-called 68-page "fair deal" report examining how the federal government treats Alberta. The panel, which travelled the province consulting thousands of Albertans in townhalls, made 25 recommendations on issues ranging from establishing a province police force to an Alberta pension plan. UCP Cypress-Medicine Hat MLA Drew Barnes, a panel member, surprised many when he denounced the panels' consensus report, saying it didn't go far enough (Dryden, 2020). Barnes also did not rule out the idea of Alberta breaking away from Canada. In his open letter criticizing the "fair deal" panel report, Barnes said Albertans should be given the opportunity to vote on their independence if their demands were not met.

A bigger crack in the UCP ranks came months later with the scandal that became known as Aloha-gate. With the promise of vaccines coming in the new year, most Albertans hunkered down in their homes for the 2020 Christmas holiday, forgoing the usual travel and large gatherings. While the government warned Albertans to "avoid non-essential travel outside Canada until further notice," several UCP MLAs,

Table 2.2: Key Moments of Controversy in Kenney's Premiership

Date	Seeds of Discontent
December 2019	Rocky start for Energy "War Room"
June 2020	Dissent over "Fair Deal Panel"
January 2021	Aloha-gate
January 2021	Speaker Nathan Copper and other denounce Kenney
March 2021	Rumblings of discontent with Kenney's leadership
April 2021	18 UCP MLAs say it's time to rethink COVID-19 strategy
May 2021	Todd Loewen and Drew Barnes booted from caucus
June 2021	Kenney apologizes for "Sky Palace" dinner
June 2021	Kenney promises "best summer ever"
Summer 2021	COVID-19 rates rise dramatically
September 2021	Kenney imposes vaccine passports
September 2021	Kenney puts down caucus revolt
October 2021	Agriculture and Forestry Minister Devin Dreeshen resigns
November 2021	Quarter of UCP constituency assoc. call for leadership review

political staffers and one cabinet minister escaped to foreign locations (Pasiuk, 2021). When the story broke, Kenney, in a rare New Year's Day news conference, resisted punishing his minister and others in the UCP government who travelled, saying the caucus members and other staff who travelled were acting in "what they believed to be good faith" (Pasiuk, 2021). The public and pundits erupted in rage. Calgary Sun columnist Rick Bell's usually sympathetic commentary this time was titled "Kenney's COVID-19 slap in the face to Albertans." Bell blasted Kenney with the opening line: "I am gobsmacked. Disgusted. Sickened" (Bell 2021). "There is no hiding," Bell added, "the anger I feel towards the United Conservative government of Premier Jason Kenney this New Year's Day" (Bell, 2021). Three days later, Kenney accepted the resignation of his minister of municipal affairs, Tracy Allard, and chief of staff, Jamie Huckabay. Kenney also demoted five other UCP MLAs who left the country over the holiday season (Bellefontaine, 2021a). As an editorial in the Edmonton Journal around this time stressed, "many Albertans feel betrayed" (Edmonton Journal, 2021). The editorial board argued: "The moral authority that the Kenney government must wield in convincing Albertans to obey public-health recommendations is now severely diminished by the apparent double-standard" (Edmonton Journal, 2021).

The so-called Aloha-gate scandal seemingly opened the floodgates of the simmering tension within the UCP government. A week later, both the speaker and deputy speaker of the Alberta Legislature denounced the premier's slow action to punish UCP caucus members and officials for vacationing abroad (von Scheel, 2021d). Nathan Cooper, the speaker of Alberta's Legislative Assembly, pointed out what he labelled the "hypocrisy" to his Olds-Didsbury-Three Hills constituents in an email, calling the scandal a "great embarrassment to the government, especially Premier Jason Kenney, who chose not to sanction these senior officials and staff members until he was prompted to do so by widespread public outrage" (von Scheel, 2021d). Other ministers and MLAs echoed Cooper's sentiments, stressing the actions of several people connected to the government had undermined the government's moral authority. A couple months later, in March of 2021, rumblings began to emerge of UCP constituency associations talking about forcing a leadership review of Kenney (von Scheel, 2021c). Shortly after these stirrings were made public by CBC News, the party's leadership announced it would, in fact, hold a leadership review in the fall of 2022, about eight months before the expected 2023 May election (von Scheel, 2021f).

Amidst the questions about Kenney's leadership, opposition to pandemic restrictions began to grow, particularly in the rural powerbase of the United Conservative Party. In April of 2021, 18 UCP MLAs voiced their opposition to the government's public health restrictions aimed at slowing spiking COVID-19 cases (French, 2021). The MLAs — including the speaker, Nathan Cooper, the deputy speaker, Angela Pitt, and former municipal affairs minister, Tracy Allard — published a statement on Facebook asking their government to rethink public health measures. The next day, Kenney pushed back against the more than a dozen backbenchers openly criticizing the government, warning them that they could face discipline and even expulsion if they broke health rules (Bennett, 2021a).

A month later, simmering internal discontent within the UCP caucus about their leader boiled over in an open challenge to Kenney's leadership. On May 13, UCP caucus chair Todd Loewen, in another open letter on Facebook, demanded that Kenney quit, telling the premier "many Albertans, including myself, no longer have confidence in your leadership" (Bennett, 2021b). The Central Peace-Notley MLA became the first government caucus member to openly call on Kenney to step down. He would not be the last. Kenney expelled Loewen and fellow frequent critic of the premier, Drew Barnes, from the party after a lengthy caucus meeting (Bellefontaine, 2021b). A few weeks later, Kenney faced more public and internal party criticism when the premier and three of his ministers skirted pandemic rules by dining on an outdoor patio. As discussed above, Kenney, at first, seemed to push back on whether he broke any rules, but later, in the face of growing criticism from the public and his own caucus, conceded he had let Albertans down by "not being more careful" (CBC News, 2021a). Kenney initially insisted that all public health regulations were followed at the outdoor gathering (von Scheel, 2021a). In the aftermath, UCP MLA Richard Gotfried resigned as the government's Calgary caucus chair over the incident, stressing that elected officials needed to "show leadership, to act responsibly and to avoid the hypocrisy that makes a mockery of the tough decisions we have to make" surrounding pandemic precautions (Kanygin, 2021). The UCP minister of culture, multiculturalism and the status of women also publicly chastised Kenney for the optics of the dinner (CBC, 2021). A month after her public admonishment, Leela Aheer was out of cabinet (Heintz, 2021).

Open revolt to Kenney's premiership in the UCP caucus took until the fall of 2021. At a caucus meeting in Calgary in late September, Kenney faced down a rebellion with a promise for a leadership review in the spring of 2022 (Anderson and von Scheel 2021). Discontent amidst the UCP caucus ranks rose all summer long alongside climbing — and alarming — COVID-19 rates. Kenney, remember, promised Albertans in June the "best summer ever" and vowed that the province, which had shuttered to weather the pandemic in the early days of COVID-19, would remain open "for good" (Anderson, 2021). By September, Alberta voters and Kenney's caucus were angry. Some of his internal critics, in fact, called on him to publicly apologize for his vacation and handling of the fourth wave, saying he "must show some humility and admit he was wrong" (CBC News, 2021b).

The UCP government also faced embarrassment for allegations of alleged sexual harassment and a toxic workplace in the fall of 2021 (von Scheel, 2021e). A few weeks after CBC News' investigation broke, Agriculture and Forestry Minister Devin Dreeshen resigned from cabinet to focus on dealing with problems related to alcohol (von Scheel, 2021b). By mid-November, more than a quarter of the United Conservative Party's constituency associations, fed up with Kenney's leadership, passed a special motion they hoped would trigger a leadership review of Kenney within three months (CBC News, 2021d).

After months of controversy and defiance, approval for how the Kenney government managed the pandemic had collapsed alongside the intention to vote UCP and Kenney's personal approval. A mere 21 per cent of Albertans approved of the provincial government's handling of the pandemic to that point (including three per cent who strongly approved and 18 per cent who somewhat approved) (French, 2022). While approval of the federal government's handling of the pandemic had softened among Albertans (down to 52 per cent, including 10 per cent who strongly approved and 42 per cent who somewhat approved), it was substantially higher than the UCP government's approval rating. The pandemic, as Young (2023: 435) contends, "ended Jason Kenney's political career." COVID-19, "exacerbated cleavages" within Alberta politics, adding oxygen to a "brewing libertarian populist movement" (Young, 2023, p. 462).

Kenney, by one account, tried to ride two horses during the pandemic (Herle, 2022). He attempted to appeal to his libertarian and rural bases while also placating most Albertans who were fine with public health restrictions recommended by health experts to combat COVID-19. As emphasized earlier, seven in 10 Albertans, in retrospect, were fine with pandemic precautions or wanted more; less than a third of Albertans called the rules on vaccine requirements, masking, and limited social gathering too strict (Bratt, 2022). Kenney, however, expressed sympathy with the truck convoy that blockaded a border crossing in southern Alberta and occupied the nation's capital for weeks (Climenhaga 2022). Yet a CBC News poll in the fall of 2022 found that nearly six in 10 (59 per cent) of Albertans had little or no sympathy for the convoy protest (Bratt, 2022), suggesting, once again, Kenney misread the politics, leaving the public questioning his judgement.

Fundamentally, as polling data in the fall of 2022 suggests, most Albertans hold moderate or centrist political values. The Progressive Conservative dynasty, arguably, remained in power for decades because it moderated its public policy stances and rhetoric to public opinion. That is, the party never got too far ahead or behind public opinion. The PCs were traditionally the "big tent party" that moved with mainstream public opinion, while the NDP's policies were more guided by its core progressive ideology. Arguably, Alberta's current New Democratic Party is attempting to broaden its appeal to align its policies and political communication to mainstream public opinion.

Government Handling of COVID-19

By the beginning of 2022, we argue the die was cast for Kenney. Low public opinion polls, dissension in the UCP caucus, and a surge of UCP membership sales among the anti-vaccine advocates set the stage for the end of Kenney's reign as premier. A planned in-person leadership review in Red Deer on April 9 was postponed and turned into a mail-in leadership vote on May 18.

Just as political forces were conspiring against Kenney, the public mood was starting to thaw. By March 2022, COVID-19 case numbers had subsided greatly, and the provincial government celebrated the release of its first balanced budget in eight years. Just as many of Kenney's advocates had predicted, once COVID-19 concerns lessened and good economic news emerged, Kenney's numbers rebounded. A *Wild Ride* poll conducted immediately after the release of the provincial budget in March showed the UCP with 40 per cent support and the NDP with 36 per cent support. In that same poll, Kenney's approval score, though still low, had rebounded to a more respectable 36 per cent.

In a last-ditch attempt to convince party members to let him lead the UCP into the next election, Kenney urged them not to compare him "to the Almighty, but to the alternative" (Frew, 2022). Kenney was hinting that Albertans might come to regret his departure. And you could not blame Kenney if he felt a small amount of vindication when CBC Calgary released polling data shortly after Smith was sworn into as Alberta's 19th premier that showed she was off to a rough start and that approval for

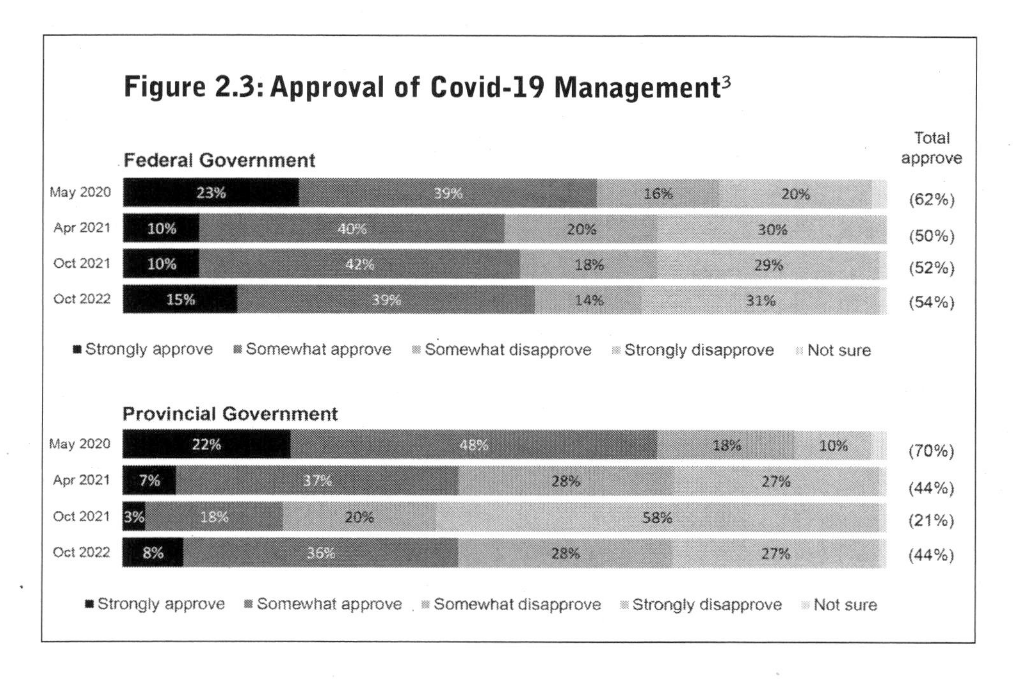

the way his government had handled COVID-19 had rebounded to 44 per cent (Dryden and Markusoff, 2022).

During her first few weeks as premier, Smith made it clear she would take a different tack than Kenney. At her first news conference after being sworn in as premier, Smith referred to people who chose not to receive a COVID-19 vaccine as the "most discriminated against group that I've ever witnessed in my lifetime." The very next day, she responded to the outrage expressed toward this comment with a clarification, saying, "I want to be clear that I did not intend to trivialize in any way the discrimination faced by minority communities and other persecuted groups both here in Canada and around the world or to create any false equivalencies to the terrible historical discrimination and persecution suffered by so many minority groups over the last decades and centuries" (von Scheel, 2022).

One week after her swearing in, Smith was apologizing again, this time for remarks suggesting Ukraine is partially responsible for Russia's invasion. She issued a statement saying she "categorically condemn(s) the Russian invasion of Ukraine", saying her "knowledge and opinion of this matter have drastically evolved since that time" (Black 2022).

Sensing her history as a media personality and radio talk show host was going to continue to dog her premiership, Smith used her first televised address to Albertans to acknowledge that, "Having spent decades in media and hosting talk shows, I discussed hundreds of different topics, and sometimes took controversial positions, many of which have evolved or changed as I have grown and learned from listening to you" (Markusoff, 2022).

In an attempt to get Albertans to disregard controversial statements she has made in the past, Smith harkened back to Klein's approach to apologizing — "admit to it, learn from it, and get back to work" (Markusoff, 2022). For his part, Kenney continued to defiantly offer no regrets or apologies as his premiership drew to a close.

At a news conference shortly after announcing he would be stepping down as premier, reporters asked if he would have done anything differently. Kenney offered punditry about the accuracy of Alberta public opinion polls and prognosticated about "people angry about vaccines" precipitating his political downfall (Bell, 2022). He put the blame on his downfall on "a small but highly motivated, well-organized and very angry group of people who believe that [he] and the government have been promoting a part of some globalist agenda, and vaccines are at the heart of that" (Braid, 2022). Even after signalling his exit from politics — a moment where many politicians offer magnanimous pronouncements about politics, their opponents, and the role of the news media in democracy — Kenney stuck with his political mantra to never surrender.

References

Amato, S. (2022, May 31). *"He does not apologize": Kenney points fingers when asked if he has regrets.* CTV News. https://edmonton.ctvnews.ca/he-does-not-apologize-kenney-points-fingers-when-asked-if-he-has-regrets-1.5926799

Anderson, D. (2019, July 4). *Alberta to spend $2.5M on public inquiry into "foreign-funded special interests."* CBC News. https://www.cbc.ca/news/canada/calgary/alberta-war-room-public-inquiry-1.5200549

Anderson, D. (2021, Sept. 15). *A reality check on Alberta's path to the devastating 4th wave of COVID.* CBC News. https://www.cbc.ca/news/canada/calgary/covid-fourth-wave-alberta-reality-check-1.6175564

As it Happens. (2019, April 5). *Radio host Charles Adler on his fiery interview with Alberta UCP Leader Jason Kenney.* CBC. https://www.cbc.ca/radio/asithappens/as-it-happens-friday-edition-1.5086498/radio-host-charles-adler-on-his-fiery-interview-with-alberta-ucp-leader-jason-kenney-1.5086938

Bartko, K. (2019, April 4). *Transcript: Charles Adler's fiery interview with UCP Leader Jason Kenney.* Global News. https://globalnews.ca/news/5129721/alberta-election-charles-adler-jason-kenney-transcript/

Bell, R. (2022, June 1). Bell: Jason Kenney is blameless, just ask him. *Calgary Sun.* https://calgarysun.com/opinion/columnists/bell-jason-kenney-is-blameless-just-ask-him

Bell, R. (2921, Jan. 1). Bell: Kenney's COVID-19 slap in the face to Albertans. *Calgary Sun.* https://calgarysun.com/opinion/columnists/bell-kenneys-covid-19-slap-in-the-face-to-albertans

Bellefontaine, M. (2019, May 7). *Alberta panel asked to recommend path to balance without raising taxes.* CBC News. https://www.cbc.ca/news/canada/edmonton/alberta-jason-kenney-panel-1.5126458

Bellefontaine, M. (2021a, Jan. 4). *Alberta cabinet minister, premier's chief of staff resign over holiday travel, other MLAs demoted.* CBC News. https://www.cbc.ca/news/canada/edmonton/alberta-cabinet-minister-chief-resign-1.5860869

Bellefontaine, M. (2021b, May 13). *Alberta MLAs Todd Loewen, Drew Barnes booted from UCP caucus.* CBC News. https://www.cbc.ca/news/canada/edmonton/barnes-loewen-removed-ucp-caucus-1.6025897

Bennett, D. (2021a, April 8). *Caucus dissent over COVID OK, breaking health rules means expulsion: Alberta premier.* CBC News. https://www.cbc.ca/news/canada/edmonton/caucus-dissent-over-covid-ok-breaking-health-rules-means-expulsion-alberta-premier-1.5980893

Bennett, D. (2021b, May 13). *Internal dissent boils over into open call for Alberta Premier Jason Kenney to resign.* CBC News. https://www.cbc.ca/news/canada/edmonton/alta-kenney-caucus-todd-loewen-1.6024857

Black, M. (2022, Oct. 18). Alberta Premier Danielle Smith apologizes for past comments on Russian invasion of Ukraine. *Edmonton Journal*. https://edmontonjournal.com/news/local-news/smith-apologizes-for-past-comments-on-russian-invasion-of-ukraine

Braid, D. (2019, Dec. 19). Alberta's Canadian Energy Centre off to rocky start thanks to trio of blunders. *Calgary Herald*. https://calgarysun.com/opinion/columnists/braid-the-first-war-room-battle-is-over-a-logo/wcm/cdad42bf-cbf2-4bf4-9be1-25ea8ba7275d

Braid, D. (2022, June 1). Braid: UCP erupts again over Kenney's claims about anti-vaxxers. *Calgary Herald*. https://calgaryherald.com/opinion/columnists/braid-ucp-erupts-again-over-kenneys-claims-about-anti-vaxxers

Bratt, D. (2021, Sept. 8). *How Kenney's political ideology is out of touch with Alberta values*. CBC News. https://www.cbc.ca/news/canada/calgary/kenney-political-ideology-1.6167303

Bratt, D. (2022, Nov. 12). *Don't assume Danielle Smith will stop re-litigating COVID if that's what Albertans want*. CBC News. https://www.cbc.ca/news/canada/calgary/opinion-poll-danielle-smith-covid-1.6646784

Bratt, D., Sutherland, R., & Young, L. (2023). Introduction: Jason Kenney and the perfect storm. In D. Bratt, R. Sutherland, & D. Taras (Eds.), *Blue Storm: The Rise and Fall of Jason Kenney* (pp. 1-16). Calgary: University of Calgary Press.

Brown, J. (2019). *Global Petroleum Show: Provincial Election Survey*. Retrieved from Calgary, CA: http://planetjanet.ca/wp-content/uploads/2019/04/2019-04-02-NWPA-March-2019-Election-Poll-Report.pdf

Brown, J. (2022). *Supporting data for Smith relitigation story by Duane Bratt*. CBC Calgary: The Road Ahead. Retrieved from http://planetjanet.ca/wp-content/uploads/2022/11/2022-11-12-COVID-19-supporting-data.pdf

Bruch, T. (2019, Dec. 26). *"Spin doctors": Alberta war room draws criticism over writer titles*. CTV News. https://calgary.ctvnews.ca/spin-doctors-alberta-war-room-draws-criticism-over-writer-titles-1.4744478

CBC News. (2021a, June 7). *Alberta premier apologizes for controversial sky palace dinner.*

CBC News. https://www.cbc.ca/news/canada/calgary/kenney-apology-sky-palace-dinner-cancel-culture-1.6056177

CBC News. (2021b, Sept. 15). *Backbencher urges Alberta premier to admit he botched COVID-19 response, take action*. CBC News. https://www.cbc.ca/news/canada/calgary/ucp-covid19-kenney-alberta-1.6177212

CBC News. (2021, Sept. 16). *"I apologize": Kenney says Alta. wrong for COVID-19 pandemic to endemic shift, not sorry for open for summer plan*. CTV News. https://edmonton.ctvnews.ca/i-apologize-kenney-says-alta-wrong-for-covid-19-pandemic-to-endemic-shift-not-sorry-for-open-for-summer-plan-1.5587497

CBC News. (2021c, June 18). *Ready to reopen: Alberta set to lift almost all health restrictions by July 1*. CBC News. https://www.cbc.ca/news/canada/edmonton/jason-kenney-deena-hinshaw-covid-reopening-update-1.6070925

CBC News. (2021d, Nov. 15). *UCP constituency associations say they passed vote triggering early Kenney leadership review*. CBC News. cbc.ca/news/canada/edmonton/ucp-trigger-early-kenney-leadership-review-1.6249304

Cecco, L. (2021, Sept. 15). Canada: Alberta healthcare system on verge of collapse as Covid cases and anti-vax sentiments rise. *The Guardian*. https://www.theguardian.com/world/2021/sep/15/canada-alberta-healthcare-system-covid-cases-rise

Churchill, W. (1940). *We Shall Fight on The Beeches*. Retrieved from https://winstonchurchill.org/resources/speeches/1940-the-finest-hour/we-shall-fight-on-the-beaches/

Clark, B. (2023). Just our facts: The energy war room's adventures in branded content. In D. Bratt, R. Sutherland, & D. Taras (Eds.), *Blue Storm: The Rise and Fall of Jason Kenney* (pp. 233-256). Calgary: University of Calgary Press.

Climenhaga, D. (2022, Jan. 31). How Kenney's convoy support backfired. *The Tyee*. https://thetyee.ca/Analysis/2022/01/31/Kenney-Convoy-Support-Backfire/

Climenhaga, D. J. (2021, May 4). *Jason Kenney edges closer to blaming Albertans for the province's COVID-19 predicament*. rabble.ca. https://rabble.ca/health/jason-kenney-edges-closer-blaming-albertans-provinces-covid-19/

Dawson, T. (2021, Sept. 17). Jason Kenney on why he brought in vaccine passports for Alberta: "We just ran out of options." *National Post*. https://nationalpost.com/news/canada/hit-a-wall-kenney-relents-on-vaccine-passports-to-mitigate-alberta-hospitals-grave-collapse

DeCillia, B. (2019, May 20). *Why Kenney's honeymoon popularity may last less than 9 months*. CBC News. https://www.cbc.ca/news/canada/calgary/kenney-honeymoon-decillia-brown-taras-analysis-1.5138644

DeCillia, B. (2022, Nov. 15). *Danielle Smith is losing Alberta's moderate middle. Now she'll try to reclaim them*. CBC News. https://www.cbc.ca/news/canada/calgary/danielle-smith-alberta-moderate-middle-ucp-ndp-poll-1.6651460

DeCillia, B. (2023). Standard error: The polls in the 2019 Alberta election and beyond. In D. Bratt, R. Sutherland, & D. Taras (Eds.), *Blue Storm: The Rise and Fall of Jason Kenney* (pp. 59-79). Calgary: University of Calgary Press.

Dryden, J. (2020, June 18). *2 "fair deal" panel members at loggerheads after report's release*. CBC News. https://www.cbc.ca/news/canada/calgary/fair-deal-drew-barnes-donna-kennedy-glans-alberta-at-noon-1.5618178

Dryden, J., & Markusoff, J. (2022, Nov. 3). *Danielle Smith's rough first impression puts Alberta NDP in likely majority territory: new poll*. CBC News.

https://www.cbc.ca/news/canada/calgary/danielle-smith-rachel-notley-ucp-ndp-alberta-janet-brown-1.6638402

Edmonton Journal. (2021, Jan. 4). Editorial: Albertans won't forget Aloha-gate. *Edmonton Journal.* https://edmontonjournal.com/opinion/editorials/editorial-albertans-wont-soon-forget-aloha-gate

Epp, R. (2023). Always more than it seems: Rural Alberta and the politics of decline. In D. Bratt, R. Sutherland, & D. Taras (Eds.), *Blue Storm: The Rise and Fall of Jason Kenney* (pp. 281-299). Calgary: University of Calgary Press.

Fletcher, R. (2020a, Feb. 27). *Alberta ended 2019 with more job losses and "mild recession": reports.* CBC News. https://www.cbc.ca/news/canada/calgary/alberta-conference-board-2020-winter-and-seph-data-1.5478183

Fletcher, R. (2020b, June 15). *Most Albertans approve of government responses to COVID-19, but "political lens" skews our views.* CBC News. https://www.cbc.ca/news/canada/calgary/alberta-poll-covid-government-handling-pandemic-approval-1.5608611

French, J. (2021, April 7). *Sixteen government MLAs speak out against latest Alberta public health restrictions.* CBC News. https://www.cbc.ca/news/canada/edmonton/alberta-mlas-public-health-restrictions-1.5978864

French, J. (2022, March 24). *Alberta legislature approves $62-billion budget for 2022-23.* CBC News. https://www.cbc.ca/news/canada/edmonton/alberta-legislature-approves-62-billion-budget-for-2022-23-1.6397059

Frew, N. (2022, April 9). *Kenney pleads for UCP members not to compare him "to the Almighty, but to the alternative."* CBC News. https://www.cbc.ca/news/canada/edmonton/ucp-jason-kenney-review-1.6414800

Heintz, L. (2021, July 16). Leela Aheer out as minister after Kenney shuffles cabinet. *Airdrie Today.* https://www.airdrietoday.com/rocky-view-news/leela-aheer-out-as-minister-after-kenney-shuffles-cabinet-3955326

Herle Burly, The. (2022, Nov. 17). Alberta politics and polling with Brown & Petty. *Podcast.* https://www.youtube.com/watch?v=84xWH6vDDVg

Hudes, S. (2021, Jan. 4). Allard resigns from cabinet, Kenney's chief of staff steps down following international travel. *Calgary Herald.* https://calgaryherald.com/news/politics/allard-resigns-from-cabinet-kenneys-chief-of-staff-steps-down-following-international-travel

Kanygin, J. (2021, June 7). *MLA resigns as UCP Calgary caucus chair, calls on leadership to "avoid the hypocrisy."* CTV News. https://calgary.ctvnews.ca/mla-resigns-as-ucp-calgary-caucus-chair-calls-on-leadership-to-avoid-the-hypocrisy-1.5459716

Keller, J. (2019, Dec. 20). Alberta's "energy war room": Reframing the energy debate or attempt to mimic legitimate journalism? *Globe and Mail.* https://www.theglobeandmail.com/canada/alberta/article-albertas-energy-war-room-reframing-the-energy-debate-or-pretending/

Markasoff, J. (2021, Nov. 1). Jason Kenney is sinking. How it all went wrong for him. The premier fumbled Alberta's pandemic response so badly that his job is now in jeopardy. *Maclean's.* https://www.macleans.ca/longforms/jason-kenney-is-sinking-how-it-all-went-wrong-for-him/

Markusoff, J. (2022, Nov. 24). *Danielle Smith, the pundit turned premier, wants to self-immunize from her opinionated past.* CBC News. https://www.cbc.ca/news/canada/calgary/danielle-smith-makes-mistakes-from-media-pundit-to-alberta-premier-1.6662243

Mason, G. (2022, May 20). The spectacular fall of Jason Kenney. *Globe and Mail.* https://www.theglobeandmail.com/opinion/article-the-spectacular-fall-of-jason-kenney/

Pasiuk, E. (2021, Jan. 1). *Kenney orders MLAs not to leave Canada unless on government business after minister's vacation.* CBC News. https://www.cbc.ca/news/canada/edmonton/jason-kenney-conference-1.5859278

Tait, C. (2020, July 23). Jason Kenney gives tacit approval to Calgary's mandatory mask bylaw. *The Globe and Mail.* https://www.theglobeandmail.com/canada/alberta/article-jason-kenney-gives-tacit-approval-to-calgarys-incoming-mask-mandate/

United Conservative Party. (2022). *United Conservatives: Promises Kept.* https://keepingourpromises.ca/

Varcoe, C. (2019, Dec. 31). A no-growth economy is Alberta's top business story of 2019. *Calgary Herald.* https://calgaryherald.com/business/varcoe-a-no-growth-economy-is-albertas-top-business-story-of-2019

von Scheel, E. (2021a, June 5). *2 Alberta cabinet ministers push back against Sky Palace dinner.* CBC News. https://www.cbc.ca/news/canada/calgary/leela-aheer-asks-kenny-for-apology-1.6055143

von Scheel, E. (2021b, Nov. 5). *Alberta minister resigns after allegations of frequent drinking in legislature.* CBC News. https://www.cbc.ca/news/canada/calgary/devin-dreeshen-drinking-allegations-alberta-ucp-1.6237656

von Scheel, E. (2021c, March 13). *Alberta Premier Jason Kenney facing party leadership review next year.* CBC News. https://www.cbc.ca/news/canada/calgary/kenney-leadership-review-ucp-alberta-covid-1.5949069 von Scheel, E. (2021d, Jan. 11). *Alberta Speaker denounces government's "hypocrisy" as MLAs slam premier, UCP colleagues over travel scandal.* CBC News. https://www.cbc.ca/news/canada/calgary/alberta-travel-hawaii-kenney-ucp-cooper-criticize-1.5868603

von Scheel, E. (2021e, Oct. 27). *Fired staffer alleges sexual harassment, intoxication in suit against Alberta premier's office.* CBC News. https://www.cbc.ca/news/canada/calgary/kenney-government-lawsuit-sexual-harassment-alberta-1.6225606

von Scheel, E. (2021f, March 8). *Kenney's close call: How the conservative grassroots put Alberta's premier on notice.* CBC News. https://www.cbc.ca/news/canada/calgary/kenney-leadership-review-ucp-alberta-covid-1.5935730

von Scheel, E. (2022, Oct. 12). *Smith didn't mean to trivialize discrimination of others in remarks about unvaccinated, she says.* CBC News. https://www.cbc.ca/news/canada/calgary/smith-covid-vaccines-discrimination-minority-communities-1.6613920

Young, L. (2023). "With comorbidities": The politics of COVID-19 and the Kenney government. In D. Bratt, R. Sutherland, & D. Taras (Eds.), *Blue Storm: The Rise and Fall of Jason Kenney* (pp. 435-466). Calgary: University of Calgary Press.

NOTES

1 If a provincial election were held in Alberta today, which one of the following parties would you vote for?
IF ANOTHER PARTY, ASK … Which other party would you vote for?
IF "NO PREFERENCE" OR DON'T KNOW... Perhaps you have not yet made up your mind. Is there nevertheless a party you might be inclined to support?

2 Please indicate whether you strongly approve, somewhat approve, somewhat disapprove, or strongly disapprove of the following.
a) The performance of Jason Kenney, as Premier of Alberta
b) The performance of Rachel Notley, as leader of the Alberta New Democratic Party

3 Do you strongly approve, somewhat approve, somewhat disapprove, or strongly disapprove of the way the following have managed the COVID-19 pandemic?

CHAPTER 3

A WINDOW ON JASON KENNEY'S FALL FROM GRACE

Ricardo Acuña

"Don't follow the crowd, let the crowd follow you."

— Margaret Thatcher

SINCE FIRST ELECTED as a Member of Parliament in 1997, Jason Kenney built a reputation as a cunning strategist and brilliant politician. Seen early on as a rising star, he held increasingly important portfolios in the Reform and Canadian Alliance caucuses, and then in Stephen Harper's Conservative Government following the 2006 election. Described as smart, loyal and hard-working, Kenney was broadly credited during his tenure as Minister of Citizenship, Immigration, and Multiculturalism with building strong connections between various ethnic communities and the Conservative Party.

His reputation garnered further lustre when, within 18 months of announcing his intention to do so, he returned to Alberta, became leader of the Progressive Conservative (PC) Party, merged that party with the Wildrose Party, got elected leader of the new party, and sat in the Alberta Legislature as Leader of the Opposition. In the words of columnist Margaret Wente (2017), he had "achieved the near impossible: engineered a brand-new party from scratch, steamrollered his opponents and won the leadership by a decisive margin." Wente went on to praise Kenney's political acumen and work ethic:

> In the eyes of many, Mr. Kenney is a brilliant politician. Like Ralph Klein, he has a populist's ability to connect with ordinary people and articulate what they're thinking. He's far more comfortable with Main Street than with Bay Street. Unlike Mr. Klein, he is personally abstemious, disciplined and focused, with a track record of getting things done.

Yet, as he approached the end of his third year in office Kenney found himself with the second lowest approval rating among Canada's premiers, the lowest rating for COVID-19 handling, and an internal caucus and party revolt in advance of an April 2022 leadership review (Angus Reid Institute, 2022). A month later, after receiving only 51 per cent support from members in the leadership review, Kenney announced his resignation as UCP leader, and called on the party to launch a leadership race as soon as possible. It is hard to square the Jason Kenney who resigned that day with the person who had adeptly risen to power just three years earlier. By way of explanation, this chapter argues the very skill set and leadership style that had facilitated Kenney's meteoric rise to power and prominence were responsible for his struggles once he became premier. He was unable to understand that his survival as premier depended entirely upon his ability to understand, and govern within, the province's dominant paradigm, rather than influencing and moving the paradigm to the right as he had done throughout most of his career. Instead of listening to Albertans and adjusting his policies to meet the majority's wishes, Kenney spent his term in office pushing his own ideological preferences and imperatives (see Harrison, Chapter 5). His focus on ideology ultimately cost him the support of Alberta's centre-right, while his vacillating actions on COVID-19 measures eventually cost him support of th UCP's more radical far right.

The Overton Window

Former Alberta Premier Ralph Klein liked to say that his job as a politician was simply to determine which way a parade was headed and then jump in front of it. What Klein was articulating is the reality that, except for very rare circumstances, politicians do not lead, they follow. The most adept politicians are the ones able to identify the outer boundaries of what has been termed "the Overton Window," and try to shape their platforms and possibilities for action within those boundaries.

Joe Overton introduced the concept while vice-president of the Mackinac Center for Public Policy, a right-wing think tank based in Michigan. Overton's theory is useful in explaining how policy agendas emerge as dominant paradigms, and how political contexts shift over time. The essence of the Overton window is this: if all of the possible policy choices for any issue are placed on a spectrum, there will always be a small window along that spectrum where those policies that are considered acceptable, sensible, and popular reside. At the very centre of that window is where public policy happens. As one moves away from the window's centre to the left or right of the spectrum — however these terms are defined at the time — one finds policies considered radical and unthinkable (Mackinac Centre for Public Policy, 2022).

Overton asserted that politicians will only act to implement policy options that fall within the boundaries of the window, as options that fall outside the window do not have enough public support to be viable. He argued politicians cannot set or move the window, as it depends on public perception of what is acceptable and sensible at any given point.

After Overton's death in 2003, his Makinac Centre colleague Joseph Lehman expanded the analysis to demonstrate how think tanks and media influencers can move or expand the window by flooding the public discourse with radical and previously unthinkable ideas. The more people hear these ideas and see them represented in the media, the more likely they will begin seeing them as acceptable and sensible; the ideas become normalized. Once the window has moved to a new point on the spectrum or its boundaries have expanded to include new ideas, politicians will follow, opting for the politically safe options inside the window rather than the electorally riskier options outside of it. Indeed, they will rush to occupy the Overton window (Astor, 2019). Pragmatic politicians will have their pragmatism rewarded with popular support, but those driven exclusively by ideology will find little success.

Overton and Lehman failed to articulate explicitly, or chose to ignore, the role that wealth, power and privilege can play in shifting the Overton window. The more a think tank is backed by significant corporate wealth and easy media access, the more success it will have at flooding the public airwaves and discourse with their radical ideas. This concept was perhaps best articulated by Lewis F. Powell, a high-profile lawyer and future Associate Justice of the U.S. Supreme Court, in a memo he wrote in 1971 to the director of the US Chamber of Commerce. In the memo, since termed the Powell Manifesto, Powell urges "pro-enterprise" conservatives to take control of public discourse by funding think tanks, using their influence in the media, reshaping universities, and using the judiciary (Powell [1971], 2022). Powell's memo is largely credited for sparking the birth of corporate-funded right-wing think tanks in the United States. Since that time, the media profile and visibility of these think tanks has increased dramatically due to support from "some very wealthy people" (Lakoff, 2004, pp. 15–16). The media's privileging of right-wing ideas resulted in the mainstreaming of radical neoliberalism, beginning in the late 1970s. The trajectory of these ideas in the US was mirrored in Canada by the birth and stellar growth of counterparts like the Canada West Foundation, the C. D. Howe Institute, and the Fraser Institute, all with the same mission of moving the Overton window to fully encompass the neoliberal suite of economic and social policies (Abelson, 2000, p. 18).

Historically, in times of war, global financial crises, big natural disasters, and pandemics, the public paradigm tends to shift toward collective responses led by government. As Seth Klein (2020) points out in his book *A Good War*, this was the case during World War II, as the people and governments of the world came together to fund, mount, and carry out an unprecedented war effort to defeat the Nazis. That paradigm of concerted collective action held for 30 years after 1945 as countries worked to rebuild their economies and their communities. It was that broad and persistent consensus around activist government and the expansion of government and social supports that Powell encouraged corporations and right-wing ideologues to rebel against and shift public discourse toward that extreme brand of anti-government, pro-market capitalism often referred to as neoliberalism.

Alberta and the Window

Alberta has been ground zero within Canada for the efforts of corporate-funded neoliberal think tanks and organizations to move the Overton window to the right.

Peter Lougheed became Premier in 1971 by unseating a Social Credit Party that had governed the province for 36 years. He did so by realizing what the Socreds had not: that Alberta had become a more urban province and that its young and growing population wanted the province's energy resources developed responsibly and in the long-term interest of Albertans (Richardson, 2012, 36). Albertans were increasingly concerned about over-dependence on a single resource, but they were equally interested in seeing the whole province benefit from its oil wealth through investments in quality public services, arts and culture, and a more metropolitan future. These demands were consistent with the post-war consensus in favour of bigger, activist governments supporting post-war recovery and province-building. While the Socreds proposed maintaining their hands-off approach to the energy industry and the economy in general, Lougheed ran on a much more interventionist platform calling for higher royalties, investment in secondary and tertiary value-added industries, and greater diversification. Lougheed had successfully read the shifting paradigm in the province, and although the popular vote was close, his strong showing in the cities gave him a significant majority of legislature seats. Lougheed and his government moved quickly to implement his platform. They tripled royalty rates, created the Heritage Savings Trust Fund in order to save 30 per cent of oil revenues every year, invested in value-added industries, and invested heavily in health care, infrastructure, and K-12 and post-secondary education. Lougheed's government also maintained and increased ownership of corporations in telecommunications, air transport, banking, and in the oil, gas, and petrochemical industries.

It was precisely against these types of interventionist, high-tax, high-royalty, large public sector policies that Canada's think tank sector emerged in the 1970s. The slightly right-of-centre consensus was working well, but the country's extractivist elites knew moving the window even further to the right — even smaller government, lower taxes, privatization, and voluntary self-regulation — could greatly increase their power and wealth. Donations to think tanks by Canada's corporate elite thus grew significantly in the 70s and early 80s. In keeping with the strategy outlined in the Powell Manifesto, their ideas began gaining greater privilege in the country's mainstream media. In turn, the greater visibility of their ideas led to right-wing think tanks garnering further fundraising success (Fraser Institute, 1999, 29).

In Alberta, the growth of the far right think tanks, and acceptance of their policy proposals, was fueled by high interest rates and the OPEC-driven global recession of the early 1980s. By the time the international price of oil bottomed out at $10/barrel in 1986, unemployment had more than doubled, people were leaving the province in droves, and foreclosures and bankruptcies were the order of the day. Alberta's energy companies began contributing significantly to the Fraser Institute and other far right think tanks and advocacy groups, who leveraged people's experiences of the recession to spread their message that high taxes, royalties, regulation, and government intervention in the market (through such things as the National Energy Program) were making the recession worse. The Fraser Institute alone saw its revenue base grow six-fold between 1988 and 1993. This time period also saw the birth of the Canadian Taxpayers Federation (CTF), a new national advocacy organization dedicated to "lower taxes, less waste and accountable government" (Canadian Taxpayers Federation, 2022).

These think tanks and advocacy groups successfully leveraged Albertans' trauma over the collapsing economy to move the Overton window towards acceptance of proposals for small government, lower taxes, privatization, and less regulation. Alberta's 1993 general election provides a textbook study of politicians rushing into the centre of the window in search of votes and popular support. The first to do so publicly was Laurence Decore, leader of the Alberta Liberals, the third party in the provincial legislature with just eight elected members. After the 1989 election, Decore began focusing his attacks in and out of the legislature on the government's growing debt and spending levels, undue intervention in the market, and the need to get Alberta's fiscal house in order (Kheiriddin and Hennig, 2010). Ralph Klein, Alberta's Environment Minister at the time, saw Decore's views gaining traction and made them central in his successful 1992 run for the Progressive Conservative Party's leadership.

The 1993 election featured a choice between Decore and Klein, whose respective campaigns were both heavily financed by oil and corporate money, and who proposed virtually identical policy platforms of cuts and austerity. The New Democrats, Alberta's Official Opposition party going into the election, ran on a traditional left-of-centre platform well outside the parameters of the Overton window. In the end, Klein's Conservatives won the election with 44.5 per cent of the popular vote, the Liberals became the Official Opposition with 39.7 per cent, and the New Democrats were completely shut out of the Legislature. In effect, 84 per cent of Albertans supported the broad platform advanced by both the PCs and Liberals.

Ralph Klein thus began his time as premier fully confident his government's policy agenda was well within the province's Overton window. The 1993 election result highlights the success of Klein's strategy of seeing where the parade was going and then jumping in front of it — a strategy that served him well throughout his tenure as premier, and that Jason Kenney might have benefited from studying.

After the 1993 election Ralph Klein moved quickly to implement his corporate neoliberal platform. He slashed public spending, eliminated thousands of public sector jobs, and downloaded social service delivery to non-profits who would use donations to make up for lost government funding and pay their staff significantly less. Those who got to keep their jobs in the public sector were pressured to — and eventually did take — five per cent salary rollbacks (Martin, 2013). The government also privatized liquor stores, registries, and highway maintenance, sold off its remaining stock in Alberta Energy Company (AEC), and began the process of fully deregulating the province's energy markets. The proceeds of the AEC sale did not go into savings or a special fund, but instead were used to eliminate that year's fiscal deficit.

Reflecting how much the Overton window had shifted in twenty years, the Klein government proceeded to implement significant changes to Alberta's royalty regime, introducing inflation indexing and new tiers of royalties to reduce rates and establishing a royalty rate of just one per cent for bitumen sands operations that had not yet recouped their initial capital expenditures. The Conservatives reduced Lougheed's target of what percentage of oil and gas revenue would go to government from 35 per cent to 25 per cent, a target that they never met (Campanella, 2012, pp. 9-10).

Success breeds success. The actual implementation of their neoliberal economic project grew and emboldened further the voice and influence of far right think tanks and advocacy groups. Being able to draw a direct line from their work to public opinion to government policy made corporate fundraising far easier. The increased corporate consolidation of mainstream media served to amplify their ideas and policy proposals even further.

The increased flood of ideas and money continued to move the window to the right, encompassing even more ideas that just twenty years previous would have been considered unthinkable. In his second term in office, from 1997 to 2001, Klein reduced corporate taxes and introduced the ten per cent single-rate income tax regime, often called the flat tax. These changes to the province's revenue regime had the double effect of drastically reducing the province's tax revenues and undoing all of Lougheed's work to significantly reduce economic dependence on volatile fossil fuel revenues.

Despite these policy initiatives, however, Ralph Klein also had an intuitive understanding of what was politically possible and what was not — the Overton window. This was most on display in the area of health care. Right-wing think tanks and advocacy groups wanted the introduction of a full-on two-tier system with private hospitals, clinics, and insurance companies being allowed to compete directly with the public system. Klein sensed early on that Alberta's window for radical health care changes had not yet moved that far. His approach was to introduce fairly aggressive legislation and then watch for the public's reaction. In 2000, he introduced Bill 11, which would have permitted for-profit corporations far greater access to the public health care system. But the reaction of Albertans, concerned it would open the door for American style two-tier health care, was strongly negative, so Klein modified the bill slightly to take it out of the realm of what a majority of Albertans would consider radical and unthinkable. The bill still permitted for-profit corporations to operate "day surgery" clinics, including for overnight stays, that charged the public system for procedures, but did not go as far as the right-wing think tanks and ideologues wanted.

In July 2005, Klein looked again at the possibility of more radically altering the balance between private and public in Alberta's health care system. This time, however, he set out to determine the public appetite for change before introducing any legislation. He launched the "Third Way" discussion paper to start the process. Nine months later, despite extensive efforts by far right think tanks and advocacy groups, Klein shelved the legislation. It was clear to him the ideas in the paper were still well outside the realm of political possibility in the province (CBC News, 2006).

Ralph Klein's ability and willingness to talk with people, to float trial balloons for potentially controversial policies, and to back-off when necessary, reflected his intuitive understanding of the Overton window, an understanding that facilitated his success and long-term survival as premier. He intuitively understood the difference between influencer and politician, and although he was very good at packaging and selling policy proposals to Albertans, he knew that those were fundamentally public relations and promotional exercises rather than attempts to fundamentally shift the public paradigm.

Throughout the 1980s and the early 1990s, Alberta's and Canada's right wing think tanks, advocacy groups, and corporate leaders had fostered in the public a sense of crisis about government spending, deficits, debt, and taxation levels. Klein's ability to leverage that sense of crisis and respond to it served him well for over a decade. By the mid 2000s, however, extreme oil revenues and record surpluses made it impossible to continue even pretending there was a crisis. Albertans began looking for something more from their government. A return to economic prosperity had moved the window to a place where Albertans were ready for a genuine investment in their province and communities, and Klein was unable to offer anything more substantial or long-term than $400 prosperity cheques. Above all, Albertans wanted a government and a premier that was visionary, but this was not Klein, a man once described by political science professor David Taras as "a politician without 'nuance,' who couldn't look beyond the immediate target to a long-term goal and who was a 'great failure' as a visionary" (Martin, 2013). He could read the window and respond in the short term, but unlike Lougheed, Klein was unable to build a bridge from the paradigm of the day to one leading to the future. That was his ultimate undoing as premier.

Jason Kenney's Rise to Prominence

Among the groups born on the prairies in the late 1980s to help disseminate the ideas of far right think tanks were the Association of Saskatchewan Taxpayers and the "Resolution One Association" of Alberta. In 1990, those two groups joined forces to create the Canadian Taxpayers Federation, whose founding mission was to push back against government spending, public deficits and debt, and high taxes. Founder Kevin Avram selected Jason Kenney, a recent university drop-out and former Saskatchewan Liberal Party activist, to head the Alberta chapter. Three years later, at age 25, Kenney became CTF President and Chief Executive Officer.

According to journalist Susan Delacourt, Kenney learned early on at the CTF that a highly visible stunt was far more valuable than a well-researched report or study in influencing public anger and outrage. She recalls how a major CTF paper on public sector pensions and defined contributions yielded next to zero public attention, while a stunt a few weeks later that saw Kenney set up a couple hundred plastic-pig lawn ornaments on Parliament Hill to symbolize "pension hog" MPs generated huge media and public interest. Delacourt quotes Kenney as saying afterward that, "After the dozens of hours and thousands of dollars we'd spent on the policy paper, the three hundred bucks I'd spent on plastic pigs was a major news story" (Delacourt, 2016, pp. 119-120).

This became the CTF's mode of operating. There was no need to do research. The CTF could spark public anger and outrage, and extensive media coverage, simply by triggering frames already well established by far right think tanks like the Fraser Institute, the CD Howe Institute, and numerous others in the United States. Within two years, Kenney's skill at this style of advocacy resulted in the CTF having 83,000 supporters and significant political influence. Media outlets across the country soon regularly published CTF releases as news stories with few or no edits or changes. The front page of *Maclean's* magazine in 1995 featured Kenney with the heading "The Tax Fighter."

By focusing his energy on tangible issues like lucrative political pensions, public sector defined benefit pensions, and government boondoggles in the market, Kenney was able to effectively trigger well-entrenched right-wing frames of government largesse, overpaid and underworked public servants, government inefficiency and ineffectiveness, and taxpayer funded "waste." His campaigns and advocacy were specifically designed to keep moving the window in Alberta (and Canada) to the right, and hopefully generate some policy wins along the way.

Under Kenney, the CTF chalked up several key policy wins in Alberta, including the Klein government's balanced budget legislation and a commitment to stop direct government investment in and support for specific industries and businesses. But the CTF's greatest win — one many commentators and analysts specifically credit to Kenney — was the government's elimination of the Alberta MLA's pension plan.

In a 2016 article for CTV News, reporter Don Martin tells of a heated exchange in the basement of the Alberta Legislature between Ralph Klein and Jason Kenney about the plan. Kenney had been publicly trying to shame Klein and his government about the pension plan, and rattled off CTF talking points about government largesse and entitled politicians. Klein was reportedly furious, but a few days later, realizing that Kenney and the CTF had successfully moved the window on this issue, made a big show of abolishing the pension plan. In making the announcement, Klein used many of Kenney's talking points. Klein credited the move with helping him win the election later that year (Martin, 2016).

This story highlights what Joe Overton and Joseph Lehman had in mind. Well-funded think tanks, lobbyists, and their media colleagues echo and amplify a set of ideas repeatedly until a majority of the population comes to see as reasonable the adoption of these ideas into policy. In turn, having seen the window move, politicians legislate those policies. This is the political operating style that Jason Kenney mastered and excelled at during his time at the CTF.

Federal Kenney

Kenney's profile and influence did not go unnoticed in western Canadian conservative circles. Soon, Preston Manning's Reform Party team reached out to him about becoming a candidate. Following an easy win in the Calgary Southeast seat in 1997, he became revenue critic for the Reform Caucus, and played important roles in the merger of the Progressive Conservative and Reform parties and Stockwell Day's successful bid to lead the resultant Canadian Alliance in 2000. By then, Kenney had become a strong voice for the party's religious and socially conservative right wing. As such, he was well-positioned to help bring Stephen Harper and a united Conservative Party to power in 2006.

Throughout his first nine years in federal politics, Kenney's role was much the same as it had been at the CTF. His ideologically conservative credentials helped establish and solidify the party's right-wing base, while also broadening the party's appeal through his ability to articulate right-wing messaging and to leverage frames by then already well entrenched in the Overton window.

Kenney rose quickly through the caucus ranks after the 2006 election and in 2007 was named secretary of state for multiculturalism and Canadian identity. Leveraging his own social and religious conservatism, and ability to trigger established conservative frames, Kenney developed a plan to reach out to immigrant voters and ethnic communities — especially those coming from Africa and Asia — by showing the Conservative Party as the only one in Canada that could properly reflect their own "natural values and aspirations." If successful, the plan would help expand the party and bring in votes and donations, while also breaking a long-standing connection between immigrant communities and the Liberal party. By most accounts the strategy worked well in a number of specifically targeted ridings in 2011, as Conservative vote totals and seats won increased, though there's no evidence the plan increased ethnic support for the party broadly across the country (Lum, 2018). Still, Harper and the party were sufficiently pleased with the plan that, despite serving as minister in a number of other ministries during his time in the federal government, Kenney always retained responsibility for multiculturalism and Canadian identity.

In government, Kenney's approach to multicultural communities was no different from what he had done at the CTF. He worked incredibly hard attending events across the country, repeating conservative talking points reflecting his own value and belief system. He didn't have to read the rooms he was in; he didn't have to identify where folks stood on numerous complex and nuanced issues. He was there to shake hands and connect with people exclusively on their mutually shared religious and socially conservative values.

As Minister of Immigration and Citizenship, Kenney's style didn't change. He overhauled the rules, making it harder for refugees to come to Canada, and continued to echo talking points from the far right, even after the 2011 election as these inched toward unabashed racism and xenophobia. Some of Kenney's talking points during this period revolved around "barbaric cultural practices," "bogus" and fraudulent refugees, limiting the number of skilled immigrant applications, and expressing offense at women wearing the niqab during citizenship ceremonies. While these ideas were popular among certain segments of the Conservative Party's ultra-right-wing base, they had not yet moved into the Overton window, nor were they front of mind for the majority of Canadians across the country. This was especially the case for the immigrant communities Kenney had worked so hard to court in advance of the 2011 election, but now faced increased racism and Islamophobia in Canada's major centres.

During the 2015 election, as the party sought to revive a lack-lustre campaign, the Conservatives pledged to set up a "barbaric cultural practices" tip line and ban the niqab during citizenship ceremonies. The polling reaction to those promises, long promoted and championed by Kenney and others in the party, demonstrated they were far outside the Overton window. A news report that Canada's new refugee rules had resulted in the denial of a claim by the family of Alan Kurdi — the drowned three-year-old Syrian refugee whose image made global news — put even more distance between Kenney's policies and the Overton window. Although the family later clarified that the application filed had been for the boy's uncle, not his father, the clarification did little to change the public perception of the new rules, or reverse the electoral tide moving in the Liberal's direction. Despite Kenney's past role in helping to move the window on refugee issues, his failure to read the window correctly during the 2015 election was, in large part, responsible for the Conservative loss.

Kenney's penchant for media stunts and desire to generate public outrage had served him well at the CTF, but often failed him during his time in Ottawa. Two instances stand out in particular. The first was a fake citizenship ceremony staged by Kenney's department and broadcast on Sun News as the real thing. Kenney refused to apologize and deflected blame for the event onto Sun News and his own staffers, but the fact he had staged a fake ceremony while regularly railing against fake refugees was not lost on many in the media and the public (Cohen, 2012).

The second happened in March 2015 when Kenney sought to stir public outrage and build support for parliament's authorization of air strikes against Islamic State targets in Syria. On International Women's Day, he tweeted pictures allegedly representing the enslavement of women by ISIS, and included the message, "On #IWD2015, thank-you to the @CanadianForces for joining the fight against #ISIL's campaign to enslave women & girls." It was later revealed that none of the images actually represented what Kenney implied they did. One of the images was staged by Kurdish protestors; another was a photo from a ceremonial procession representing the taking of Muhammad's grandson to Damascus in chains. All the pictures had been circulating for some time on racist and Islamophobic social media accounts and platforms, and had been debunked numerous times (McGregor, 2015). The tweeting of fake pictures to stir public outrage is a tactic we will see Kenney use again as Alberta premier.

Opposition Kenney

Jason Kenney handily won his seat in the 2015 federal election. The Conservative Party as a whole did not fare as well, however, losing 60 seats and being downgraded to Official Opposition status. After Stephen Harper's election night resignation as leader, Kenney was widely considered a contender to become the party's new leader. But another path to power was emerging.

The 2014-2015 oil price crash and resulting loss of jobs in Alberta were taking a toll on the fortunes of Rachel Notley's New Democratic (NDP) in Alberta — the first one in history, that in 2015 had put an end to 43 years of Conservative rule. Alberta's conservatives saw an opening for a return to power, but knew it would only be possible if Alberta's two right-wing parties — the Progressive Conservatives and Wildrose —

were unified under a strong and high-profile leader. Conservatives began imagining Jason Kenney as that leader even before Kenney himself mused publicly about the possibility.

A story in the *National Post* from early 2016 quoted an anonymous conservative activist as saying, "Kenney is sort of the last giant standing in Alberta politics on the right. If he decided to make some kind of move, it would be seismic." The same article quoted long-time conservative insider Ken Boessenkool as asserting that, "Jason Kenney has earned the right over the last 10 years to do whatever in politics he thinks he should do. I think he would have broad support whatever he decides to do." Amidst ongoing speculation, Kenney officially resigned his parliamentary seat in September 2016 with the mission to unite Alberta's right-wing parties and defeat the NDP.

His decision to start the process by first running for the leadership of the PC Party was as much a function of convenience (the party was without a leader at the time) as it was strategic. There is no question Kenney's own views were much closer (if not identical) to those of the far right Wildrose Party than those of the more moderate PC Party. Strategically, therefore, having him as leader of the PCs would greatly increase the likelihood of ideological Wildrosers wanting to unite. His profile, and the deep understanding by many PCs that uniting was a political necessity, gave him a strong advantage in the PC leadership race. Not being one to leave things to chance, however, Kenney put together an aggressive campaign team whose merciless strategies and tactics could take advantage of the delegated convention format (rather than one-member-one-vote) to ensure winning the leadership. Once that was done, unification under a new party was a certainty.

Once elected to the legislature in December 2017, and assuming the role as leader of the opposition under the new United Conservative Party (UCP) banner, Jason Kenney continued playing to his strengths. Once again, he had no need for concrete policy proposals: he just had to build a loud and aggressive critique of the Notley government based almost exclusively on talking points from the likes of the CTF, the Fraser Institute, the Canadian Federation of Independent Business (CFIB), and Rebel Media. Amidst the oil patch's continued struggles and lagging private sector job growth, Kenney was able to effectively trigger right wing frames by demonizing the carbon tax, environmental policy, public sector unions, and government spending.

Despite Notley's hard work during the first half of her government's term positioning the NDP in the middle of the Overton window and building a big tent there, Kenney was able to effectively brand her as a socialist, a radical environmentalist, a puppet of the province's unions, and part of an anti-oil, anti-Alberta allegiance with Justin Trudeau. This latter point had the added advantage of playing to Albertans' long-entrenched alienation and antagonism to the federal government, most especially the Liberals and any prime minister named Trudeau. It didn't matter that many unionists, environmentalists, and progressive activists felt Notley's policies were too timid (Acuña 2019a). After 40 years of corporate-funded think tanks and advocacy groups successfully moving the window rightward in Alberta, the idea that the province's economy was suffering because of high taxes, environmental policies, government spending, and the policies of the federal Liberal government was an easy sell, especially in rural Alberta and Calgary where those frames are the most entrenched. It further helped that Kenney, through his past work in Ottawa and before that with the CTF, had established a strong public connection between himself and those frames. Although Notley had moved to the right in search of the window on issues like pipelines, royalties, and public sector salaries, Kenney owned the frames in a way that Notley never could.

It is important to reinforce that Notley spent her entire term seeking out and working within the established parameters of the Overton window in Alberta. In the lead-up to the 2015 election, substantial space had opened up in Alberta's cities for a discussion around the environment, a carbon tax, the need for revenue reform, and a more thoughtful approach to energy development. Albertans, especially those in the major cities, had become significantly less socially conservative by 2015, and a significant percentage supported the notion that the government has some responsibility for ensuring a decent standard of living for all. And, as demonstrated during Premier Klein's tenure, Albertans have also always been incredibly protective of their public health care system (Sayers and Stewart, 2019). Rachel Notley was very comfortable operating within all of these established frames, and effectively leveraged them during the 2015 election by painting the PCs, on the one hand, as entitled and out of touch and the Wildrose Party, on the other, as too socially conservative for Albertans.

During the 2019 Alberta election, Jason Kenney stayed true to form, playing from his strengths. The cornerstone of the campaign was the UCP's platform document titled Alberta Strong & Free: Getting Alberta Back to Work. At 114 pages in length, the platform document was structured around an extensive narrative that often read more like a Fraser Institute or CFIB report than a traditional election platform. It hit all of the standard talking points used by right-wing think tanks in all their reports and press releases, and featured extensive quotes from the CTF, economist Jack Mintz, the CFIB, the Alberta Chamber of Commerce, the Canadian Association of Petroleum Producers, conspiracy blogger Vivian Krause, columnist Licia Corbella, and numerous others — many of them veterans since the 1980s in efforts to move the Overton window in Alberta to the right. The UCP's platform was all about triggering the frames these groups and individuals had previously established.

The document frequently references Alberta's "tax burden," the "cash grab" that is the carbon tax, the belief that low corporate taxes will increase jobs and investment, environmentalists' war against Alberta oil, over-spending, the debt and the deficit, and how regulation and high wages kill jobs. Trudeau's name appears 14 times, Notley's nine, but Kenney's only seven times. The acronym NDP appears 103 times while UCP only appears 20 times (United Conservative Party, 2019). Just as Ralph Klein had done in 1993, the UCP was clearly seeking to leverage Albertans' historic fears and distrust of the NDP and federal Liberals to blame them for the economic downturn.

As he had when criss-crossing Canada delivering the CTF's message, and again when he was tirelessly hopped from one multicultural event to another on behalf of Stephen Harper's Conservatives, Kenney was in his element making stump speeches and announcements across the Alberta. And just as he had when he was part of the CTF, his focus on simple messages and asks stemming from well-established right-wing frames was incredibly successful. Notley and the NDP tried very hard to brand Kenney's proposals for tax cuts and spending cuts as radical and irresponsible, but forty years of work by right-wing think tanks and lobbyists ensured Albertans would see these proposals as normal and unobjectionable.

Once again, Kenney's think tank and advocacy playbook won the day and the 2019 provincial election. The UCP increased the combined 2015 popular vote of the PC and Wildrose parties by just 2.87 per cent, but that was enough to secure a comfortable majority and mandate in the provincial legislature. The NDP, on the other hand, lost about 8 percentage points off its popular vote from the 2015 election, but because most of that loss came in Calgary and rural Alberta, the party still managed to win every seat but one in Edmonton. The 2019 result was the Alberta NDP's second highest popular vote percentage ever, but because of a record voter turnout, was also the highest number of individual votes ever for the NDP. Despite losing the election, Notley and her team had successfully identified the parameters of the Overton window on a number of issues, and proposed policy clearly within those parameters.

Premier Kenney

Having received a strong electoral mandate, Kenney moved quickly on a number of campaign promises. Only eight months later, however, his approval rating had dropped by 15 points to 40 per cent — the largest drop, and the third lowest approval rating, among all of Canada's premiers (DART, 2019). This drop in approval, coming on the heels of the UCP's first provincial budget in October of that year, suggests that while Albertans may have been comfortable with the ideas put forth in the platform, they were not yet ready to accept the policies and budget decisions based on those ideological rhetorical flourishes.

One specific example is the UCP promise, clearly stated in the platform document, to cut the corporate tax rate by four percentage points (one third) over a four-year period. This commitment was presented in the platform with the standard right-wing trickle-down talking points that reducing corporate taxes would result in increased investment and greater job creation. A large number of Albertans accepted this premise in principle, given its entrenchment in messaging by right-wing advocacy groups and think tanks over several years (Acuña, 2015). They were not yet in a place, however, to accept such a radical policy based on that principle. The Overton window relies not only on whether an idea has moved from unthinkable to feasible, but more importantly whether policy based on that idea has moved from the unthinkable to the acceptable, sensible, and popular. The latter requires a politician able to, and interested in, determining the public context, mood, and appetite for specific policies.

The context in which the first cut to Alberta's corporate tax rate was done shows why many Albertans viewed it unfavourably. The cut, from 12 per cent to 11 per cent, was made when Alberta was already tied for the lowest corporate rate in Western Canada, and one of the lowest in North America. Employment in the oil patch had still not recovered, and many Albertans were struggling financially while major oil corporations continued to make significant profits.

At the same time, budget 2019 removed indexing of the basic tax exemption, meaning that every Albertan making more than roughly $19,000/year would pay more in taxes. The budget also included increases on tobacco taxes and registry user fees for things like motor vehicles and land titles. The budget also heralded unprecedented cuts to funding for post-secondary educational institutions, with the University of Alberta receiving the lion's share, and the elimination of tax credits for tuition and education costs (McIntosh and Hussey, 2019).

In short, Albertans were being asked to support a corporate tax cut while their own taxes and service fees were being increased; all of this, additionally, coming at a time when 51 per cent of Albertans polled believed their household's financial situation was worse than it had been the previous year (Santos, 2020).

Even the most cursory effort at determining how Albertans might view the corporate tax rate should have told Kenney to back off, but he was either unable to read the room or unwilling to pivot from his belief that he could still influence and move the window simply by repeating an ideological assertion over and over. Given the choice between continuing to try to influence the window as he did at CTF and being influenced by it as a political leader, he chose the former.

The optics of Kenney's position became even worse when, within weeks of the provincial budget's formalization of the tax cut, major energy companies' quarterly financial statements showed them reaping large profits based on the cut, while continuing to cut jobs and move investment out of Alberta. For example, Husky

Energy, who saw an immediate benefit of $223 million from the tax cut, announced the elimination of hundreds of Alberta jobs and asserted that it would be increasing investment in Saskatchewan rather than in Alberta (Bakx, 2019).

Kenney's commitment to making policy based on his ideological beliefs, rather than the position of a majority of Albertans, became even more clear when, a year later, he put forward a bill to accelerate the pace of the corporate tax cut, reducing it to eight per cent retroactively rather than in 2023 as initially promised. Accelerating the tax cut in the middle of a pandemic-driven recession, when the first step of the cut had yielded no positive results, and there was no evidence to show this step would fare better, was certain to land poorly with a majority of Albertans (Johnson, 2020). It seems impossible that Kenney did not know the policy lay far outside the Overton window. But in a situation where Ralph Klein might have floated a trial balloon to test the public mood, Kenney didn't seem to care.

Kenney's predisposition to think like an advocacy influencer rather than a government leader was also evidenced in other ways. Over the years, right-wing think tanks and advocacy groups have moved the Overton window through the technique of having an individual or committee, with some academic credentials, write a report or brief on a particular issue such as taxation levels, overregulation, or government spending. The reports' conclusions always precede the "evidence," which often includes only cherry-picked data and examples. The subsequent dissemination of a report's "findings" is based on using the established frames and asserting that the report reinforces ideas the far right has already promoted. The report doesn't change the talking points; it simply provides legitimacy to them as they are repeated, and an opportunity to move the ideas a bit further from the realm of the unthinkable to the realm of the sensible; to normalize them.

In keeping with his partiality to think tank and advocacy tactics, Jason Kenney's UCP government, immediately following the 2019 election, contracted Janice McKinnon (a former Saskatchewan finance minister and perennial favourite among the far-right advocacy crowd) to convene a "Blue Ribbon Panel" to write a definitive report outlining everything that is wrong with Alberta's finances, and how to fix it. The government then used those findings to justify all of its economic policies going forward.

Predictably, McKinnon's report cherry-picked data and analyzed it so as to validate the UCP's election platform and reinforce Kenney's far right talking points (Ascah, Harrison and Mueller, 2019). Having been clearly instructed as part of its mandate to not consider any changes to provincial revenue, the panel's recommendations revolved entirely around cutting government spending, reducing public sector salaries, increasing private delivery of services, and shifting post-secondary costs from government grants to the public through things like increased tuition. Notably, the report recommended a complete overhaul of the public health care system to facilitate better service at reduced costs (Blue Ribbon Panel on Alberta's Finances 2019). Given Kenney's reasons for commissioning the report, many saw this as an echo of the far right's long-standing call to privatize the health care system.

The McKinnon report's section on health care argued the cost of physician services was both exceedingly high and rapidly increasing. To deal with this, the report recommended the government move quickly to restructure how physicians are paid and to reduce those costs. Kenney saw this as an opportunity to simultaneously leverage some of his favourite frames, namely that public health care is unaffordable, that health care workers make too much money, and that health workers, especially doctors, are self-serving and out-of-touch elites out to get rich off the system. According to journalist Graham Thomson (2020), "[d]uring debate in the legislature, Kenney has tried to undermine the credibility of doctors by referring to them as affluent "members of the one per cent club" who should share in the province's economic pain." Kenney's then health minister, Tyler Shandro, also repeatedly sought to trigger those same frames publicly through the use of similarly aggressive rhetoric, often highlighting that physician pay and compensation made up the largest portion of the health budget.

Kenney's and Shandro's efforts at the bargaining table to cut costs and overhaul how doctors' compensation was met with predictable resistance from the Alberta Medical Association (AMA). Contract negotiations broke down completely in late February of 2020, whereupon Minister Shandro announced the government was ripping up its funding contract with doctors and would arbitrarily impose the changes being sought to physician rules and fees.

The collapse of negotiations shifted Shandro's and Kenney's persistent right-wing messaging into overdrive: physicians were unwilling to offer proposals or solutions for the public good; the government was simply responding to the research and trying to bring health costs under control; Alberta physicians were the highest paid in the country; and there should be a sunshine list for "taxpayer-funded" doctors.

The references to research and the public good, and the use of the phrase "taxpayer-funded" are all key tactics from Kenney's days at the Canadian Taxpayers Federation. Once again, however, Kenney embraced these tactics while seemingly having no idea of, or interest in, how Albertans stood with regard to doctors and their pay. He simply believed that repeating the same talking points about taxes, big government, and public servants was all that was needed to win the day. A comparison with Klein is once again in order: where he had learned the hard way how deeply Albertans care about their public health care system, Kenney simply didn't think it mattered.

Kenney's flawed strategy soon faced a larger challenge from the doctors, one that he was not expecting. First, the doctors fought back publicly. They highlighted that they had offered to take a five per cent cut in fees. They released figures showing the government's proposed changes would force rural doctors to close practices. They pointed out the government was in violation of doctors' Charter rights by denying the right to third-party arbitration. The AMA sued the province over the changes. Later on, they ran full-page ads in Alberta newspapers saying the AMA was willing to hold physician pay at current levels. They released the results of a survey showing over 40 per cent of physicians were looking to move away from Alberta. Three weeks later, they released the results of a referendum showing that 98 per cent of physicians, residents, and medical students had lost faith in the health minister (*Edmonton Journal*, 2020).

Despite growing public support for doctors, and fears in rural communities — the heartland of UCP support — about the possibility of losing rural services, Kenney and Shandro were undeterred. The government passed legislation that would make it illegal for physicians to leave their practices, and asserted repeatedly that there were physicians from across the country who would love to move to Alberta. As late as August 2020, Kenney continued to state publicly that he was open to meeting with physicians, but that the dispute continued to be about money and not principles.

That month, the government and the AMA returned to the bargaining table, arriving at a tentative agreement that was presented to AMA members for a vote in March 2021, but the agreement was rejected by a margin of 53 per cent to 47 per cent. Data for 2021 from the Alberta College of Physicians and Surgeons show that 568 physicians left the profession in Alberta in 2021, compared to 87 in 2020 and 54 in 2019. The 2021 number includes 140 doctors who left the province, and likely many more who moved away to work elsewhere but had simply not yet given up their Alberta license (Short, 2022). Once again, Kenney's insistence on working to move the window rather than working within it resulted in a failed strategy, both politically and in bargaining. But the failure highlights, even more, how an historical crisis can abruptly shift the Overton window.

Kenney and Shandro waged their initial campaign against doctors during the onset and first wave of the COVID-19 pandemic. At a time when Albertans were more aware than ever before of the value of well-paid health professionals and public health services in their communities, Kenney stayed the course on trying to demonize doctors, other health workers, and the health system as a whole. Once again, he showed either a complete inability, or unwillingness, to read the room in the pursuit of his ideological goals. The episode also highlights the degree to which his other persistent strategy, of identifying a public enemy and picking a fight with them, was blind to whether public sentiment favoured that enemy or not. Trudeau, Ottawa, and environmentalists are always a safe bet as enemies in Alberta (Acuna, 2019b), but clearly doctors and nurses are not — especially in the midst of a health care crisis. It is hard to believe that neither Kenney, nor anyone on his team, acknowledged that reality.

Kenney's desire to lead with far-right ideology while ignoring the political need to read the room was most evident throughout the UCP government's response to the COVID-19 pandemic. From the very beginning, and every step along the way, Kenney's default and bias has been toward his own libertarian and anti-government beliefs and values (see Harrison, Chapter 5) — regardless of advice from doctors and scientists, and the desires of a majority of Albertans.

Kenney initially resisted implementing any types of restrictions, opting instead for a list of guidelines and voluntary measures that people were encouraged to follow. Even after the World Health Organization officially declared a pandemic on March 11, 2020, Jason Kenney waited almost a week before declaring a public state of emergency in Alberta and instituting any restrictions or limits on gatherings.

From the pandemic's beginning, one of the libertarian right's biggest pushbacks across North America was against mask mandates as unnecessary, a violation of basic freedoms, and, according to numerous conspiracy web sites, dangerous. It is not at all surprising, therefore, that Kenney's March 17, 2020 restrictions did not include a province-wide masking mandate. Instead, Kenney asserted this was something municipalities could implement if they so desired, ignoring the calls from a growing majority of Albertans for a standardized province-wide set of restrictions. In fact, Alberta did not implement a provincial mask mandate, along with other significant public health measures, until December 8, 2020, at the height of the second wave. When he introduced the extended measures on December 8, Kenney made sure Albertans understood his preference is always to rely on personal responsibility and personal choice, making clear the restrictions he was imposing went against his own core values (Bratt, 2021).

A similar dynamic played out in the summer of 2021 as jurisdictions across North America began mandating vaccines and implementing vaccine passports to verify vaccination status for access to public venues and events. The government had introduced in early 2021 a step-by-step framework to ease restrictions, based on hospitalization rates. By early April, as the third wave of the dynamic started to hit the province and hospitalizations spiked, Kenney had no choice but to step back on the easing of restrictions. Instead of citing the earlier weakening of public health measures, Kenney insisted the spike in infections was the result of some Albertans breaking the rules. The April tightening of the rules was met with pushback from the far right, as fifteen UCP MLAs released a public letter criticizing Kenney and the new measures, though they were still by far the weakest measures nearly anywhere in

Canada. Although Kenney had reluctantly embraced a collective response to the first wave, as the pandemic wore on, he became more conflicted between the need to keep the health care system from collapsing and his own libertarian beliefs, now being echoed from within his caucus.

On May 4, 2021, at the height of the third wave, after months of resisting the kinds of lockdown measures seen in Ontario and Quebec, and after months of dismissing warnings from health experts and professionals, Kenney finally relented and introduced a series of new restrictions. These included online learning in K-12 schools and post-secondary institutions, closing dining establishments for indoor service, and new limits on gatherings, but the measures were too late — Alberta's third wave resulted in some of the highest case rates in North America (Boyd and Quan, 2021).

Just three weeks later, while the third wave was still at its peak, Kenney again reverted to his libertarian talking points around freedom, taking personal responsibilities, and the need for businesses to re-open and so thrive. On May 26, 2021, he announced a new plan that would see all provincial restrictions dropped on July 1, in time for the Calgary Stampede. He told Albertans, "don't live in fear" and promised "the best summer ever."

The day came and, true his word, Alberta became the first jurisdiction in Canada to drop all restrictions. Kenney and his communications team chided the NDP opposition for warning it was still too risky to eliminate all measures. His director of issues management, Matt Wolf, publicly tweeted, "The pandemic is ending. Accept it."

A week later, as Kenney hosted the annual premier's Calgary Stampede Breakfast, he asserted to reporters that vaccine passports were likely a violation of information privacy legislation, and an affront to freedom generally. He promised Alberta would, under no circumstances, follow the lead of Quebec and Manitoba in implementing a vaccine passport, and added that he would fight any effort by the federal government to do so. "These folks who are concerned about mandatory vaccines have nothing to be concerned about," he told reporters (The Canadian Press, 2021).

Early in August, infection rates of the Delta variant started to spike. By mid-month, it was clear the province was well into a fourth wave that, by all accounts, stood to be the worst one yet in terms of overloading the province's health care system. After weeks of pressure from local politicians, physicians, and a large majority of Albertans, the government agreed to pause the elimination of the few remaining public health provisions, including isolation requirements for those who tested positive. Though Alberta again had the highest case count in the country, the government refused to reintroduce any of the previously cancelled measures and health protections.

By early September, Alberta Health Services was overwhelmed: ICUs were operating over capacity, and elective surgeries and outpatient procedures were being postponed. A dozen mayors from around the province sent a letter pleading with the provincial government to implement a vaccine passport. Kenney continued to insist, on ideological terms, that vaccine passports would never be introduced in Alberta. Instead, he offered up $100 gift cards as an incentive to the unvaccinated to take the shot. On September 13, Alberta's Chief Medical Officer of Health admitted publicly it had been a mistake to lift all public health restrictions in July. Two days later, Kenney declared a state of public health emergency and brought in a number of new restrictions and measures, including a vaccine passport, albeit under a different name ("the restriction exemption program").

A similar situation played out in early February 2022, when within one week Kenney went from asking people to be a little more patient about the easing of restriction, as some urban hospitals were still facing significant pressure, to announcing that passports and food and beverage restrictions were being cancelled immediately, and that all remaining restrictions would be lifted on March 1. The abrupt change coincided with the Freedom Convoy's occupation of Ottawa and the blockade at the Coutts border crossing — a constellation of protests by Kenney's ideological base against vaccine mandates and other restrictions that Kenney himself had tacitly encouraged (see Harrison, Chapter 5). The lifting of restrictions also coincided conveniently with the lead-up to the UCP leadership review, required by party bylaws at every third annual general meeting, then scheduled for Red Deer on April 9.

At virtually every decision point throughout the pandemic, with the possible exception of the first wave where he acted fairly quickly and called for strong collective action, Jason Kenney chose to focus on far-right ideology and talking points rather than the wishes of the majority of the public, the warnings of experts, or the broad public good. This approach, predictably, resulted in two years of decisions that served neither the public interest nor his political career.

Conclusion

In the weeks before Kenney's February announcement of his final plan to cancel restrictions, his public rhetoric took a noticeable turn toward the far right, demonizing environmental radicals, calling the NDP opposition socialists, and renewing his over-the-top rhetoric against the federal Liberals. In a stunt typical of Kenney's past efforts to rile up public anger and outrage, in the days leading up to the Ottawa convoy and Coutts blockade he tweeted pictures that were ostensibly of empty grocery store shelves in Alberta, brought about allegedly by restrictions on truckers. It was a clear attempt to deflect anger about restrictions away from himself and his government and toward Justin Trudeau. Of course, neither he nor his team were able to provide any specifics about where the pictures were from, when they were taken, and why the shelves were empty. Hundreds of Albertans responded to this failed social media stunt by taking and posting pictures of grocery store shelves full of items. Once again, the Kenney's penchant for CTF style publicity stunts had damaged his reputation among everyone but Alberta's far-right fringe.

The incident did not lessen his far-right rhetoric and aggressive social media name-calling, however. Indeed, it intensified. He refused to discipline MLAs who had attended and publicly supported the illegal Coutts blockade; he passed legislation making it impossible for municipalities to implement mask mandates; and began essentially blaming Canadian environmental activists for the war in the Ukraine. As the April 9 leadership review in Red Deer approached, and sensing his leadership increasingly threatened, Kenney decided to court the UCP's far right who had become increasingly vocal and organized in opposition to COVID-19 public health measures.

After Brian Jean's by-election victory in February 2022, and following the UCP's decision to change rules for the leadership review to make it a mail-in vote by all members, Kenney made one more desperate, last-minute change in his rhetoric. Suddenly, he began asserting that only he could build a big conservative tent; that only he could keep the "lunatics" from "trying to take over the asylum"; and that he would "not let this mainstream conservative party become a, uh, an agent for extreme, hateful, intolerant, bigoted and crazy views" (Tran, 2022). But it was those same "lunatics" that Kenney had been courting with his social media posts just a few days prior.

Kenney's problems within the UCP mirrored much the same dynamic as had seen his fall in public support and popularity among most Albertans. He tried to govern in the same way that he worked when at the CTF. He demonstrated no interest whatsoever in identifying where the Overton window — the majority of Albertans — was on any one issue, opting instead to act entirely with language, speaking points, and policy alternatives based on his own far right ideology. On those occasions when he did ultimately come around and listen to the majority, it was always too late. Moreover, it was clear that he did not believe in what he was ultimately doing, and would seek to undo it at the first available opportunity.

Jason Kenney's inability and unwillingness to read the room cost him the support of the very mainstream conservatives that had helped him become Premier, while his eventual and reluctant adoption of public health measures during the pandemic also eventually cost him the support of his far-right base. It is telling, though, that in the lead-up to the leadership review his initial instinct was to start by winning back his base rather than the mainstream majority. That move, perhaps more than any other, highlights how he went from being the golden child of Alberta's unified right, and eventual premier of the province's first UCP government, to resigning as premier just three years later.

Forty years of effort by right wing think tanks and advocacy groups have successfully moved the Overton window in Alberta to the right by constantly repeating far right talking points and ideas. The result has been a public policy environment open to, and embracing of, the unregulated free market, privatization, and anti-union ideas, hence promoted by the province's conservative parties. But Jason Kenney took that openness for granted. He never bothered to check just how far right the majority had actually moved, and never considered the possibility that, especially during a pandemic, the Overton window may have moved in the opposite direction. He either forgot, or didn't care, that he would not have become premier without the support of the province's moderate right — those that had previously filled the ranks of the PC party. He didn't notice that as he continued blindly to govern from the far right, the NDP had continued seeking out the centre-right of the Overton window, providing an alternative big-tent home for more moderate conservatives. He arrogantly assumed that none of that mattered; that all he needed to do was continue repeating his far-right rhetoric and ideas, and Albertans, especially conservative Albertans, would continue to follow him blindly. They didn't. His lack of understanding of his role vis a vis the Overton window, and his hubris vis a vis the views of Albertans cost him his job. In the process, conservatives in Alberta today are arguably more divided than in 2015, and facing an NDP opposition that has set up a very large tent on the centre and centre-right of the political spectrum.

References

Abelson, D. E. (2000). Do think tanks matter? Opportunities, constraints, and incentives for think tanks in Canada and the United States. *Global Society* 14(2), 213–36.

Acuña, R. (2015). A window on power and influence in Alberta politics. In M. Shrivastava and L. Stefanick (Eds.), *Alberta Oil and the Decline of Democracy in Canada.* Edmonton: AU Press

Acuña, R. (2019a, May). What they did: The Notley NDP government record. *Alberta Views*, 24-28.

Acuña, R. (2019b, July/August). Kenney's enemies: A long-standing Alberta tradition of playing the victim is taken to new extremes. *CCPA Monitor*, 19-20.

Angus Reid Institute. (2022, Jan. 17). *Premiers' performance: Ford continues to fall in approval, Houston rides high on strength of COVID-19 handling.* https://angusreid.org/premiers-performance-january-2022/

Ascah, B., Harrison, T., and Mueller. R. E. (2019). *Cutting Through the Blue Ribbon: A Balanced Look at Alberta's Finances.* Edmonton: Parkland Institute. Sept. 9. https://www.parklandinstitute.ca/cutting_through_the_blue_ribbon

Assaly, R. (2021, Sept. 16). How we got here: A timeline of Alberta's response to the COVID-19 pandemic. *Toronto Star*. https://www.thestar.com/news/canada/2021/09/16/how-we-got-here-a-timeline-of-albertas-response-to-the-covid-19-pandemic.html

Astor, M. (2019, Feb. 26). How the politically unthinkable can become mainstream. *The New York Times*. https://www.nytimes.com/2019/02/26/us/politics/overton-window-democrats.html

Bakx, K. (2019, Oct. 26). *Why Alberta's corporate tax cut might not keep investment at home*. CBC News. https://www.cbc.ca/news/business/husky-kenney-sask-nfld-alta-alberta-1.5335823

Blue Ribbon Panel on Alberta's Finances (Janice MacKinnon). (2019). *Report and Recommendations*. https://open.alberta.ca/dataset/081ba74d-95c8-43ab-9097-cef17a9fb59c/resource/257f040a-2645-49e7-b40b-462e4b5c059c/download/blue-ribbon-panel-report.pdf

Boyd, A., & Quan, D. (2021, May 2). It's got the highest COVID-19 rates in Canada – again. Is Alberta headed for disaster? *Toronto Star*. https://www.thestar.com/news/canada/2021/05/02/its-got-the-highest-covid-19-rates-in-canada-again-is-alberta-headed-for-disaster.html

Bratt, D. (2021, Sept. 8). *How Kenney's political ideology is out of touch with Alberta values*. CBC News. https://www.cbc.ca/news/canada/calgary/kenney-political-ideology-1.6167303

Campanella, D. (2012). *Misplaced Generosity Update 2012: Extraordinary Profits in Alberta's Oil and Gas Industry*. Edmonton, AB: Parkland Institute. March 15. https://www.parklandinstitute.ca/misplaced_generosity_update_2012https://www.parklandinstitute.ca/misplaced_generosity_update_2012

Canadian Press, The. (2021, July 12). *Alberta will not bring in vaccine passports, premier says*. CBC News. https://www.cbc.ca/news/canada/calgary/covid-jason-kenney-vaccine-passports-1.6098986

Canadian Taxpayers Federation. (2022). *Who we are*. https://www.taxpayer.com/about/

CBC News. (2006, April 20). *Alberta backs away from 'Third Way' health reforms*. CBC News. https://www.cbc.ca/news/canada/alberta-backs-away-from-third-way-health-reforms-1.598989.

Cohen, T. (2012, Feb. 2). Kenney refused to apologize for fake citizenship broadcast, blames bureaucrats. *National Post*. https://nationalpost.com/news/canada/kenney-refuses-to-apologize-for-fake-citizenship-broadcast-blames-bureaucrats

DART C-Suite Communicators. (2019, Dec. 12). *Quarterly approval rating and rankings of Canada's premiers December 2019*. https://dartincom.ca/poll/quarterly-approval-rating-and-rankings-of-canadas-premiers-december-2019/

Delacourt, S. (2016). *Shopping for Votes: How Politicians Choose Us and We Choose Them.* Madeira Park: Douglas & McIntyre.

Edmonton Journal, The. (2020, Sept. 9). A timeline of the UCP's fight with doctors over pay: From budget cuts to the COVID-19 pandemic. https://edmontonjournal.com/news/politics/a-timeline-of-the-ucps-fight-with-doctors-over-pay-from-budget-cuts-to-the-covid-19-pandemic

Fraser Institute. (1999). *Challenging Perceptions: Twenty-Five Years of Influential Ideas.* Vancouver: Fraser Institute.

Johnson, L. (2020, Nov. 3). NDP calls on UCP government to release fiscal analysis of corporate tax cut. *Edmonton Journal.* https://edmontonjournal.com/news/politics/ndp-calls-on-ucp-government-to-release-fiscal-analysis-of-corporate-tax-cut

Kheiriddin, T., & Hennig, S. (2010, April 6). Alberta's 'miracle on the prairie.' *National Post.* https://nationalpost.com/full-comment/tasha-kheiriddin-and-scott-hennig-albertas-miracle-on-the-prairie

Klein, S. (2020). *A Good War: Mobilizing Canada for the Climate Emergency.* Toronto: ECW Press.

Lum, Z-A. (2018, May 25). *Jason Kenney: "Pulp fiction" argument with Harper led to ethnic vote strategy.* HuffPost Canada. https://www.huffpost.com/archive/ca/entry/jason-kenney-pulp-fiction-harper-ethnic-vote-conservatives_a_23443144

The Mackinac Center for Public Policy. 2022. *The Overton Window.* https://www.mackinac.org/ OvertonWindow

Martin, D. (2016, June 23). *Jason Kenney's mission unlikely, but not impossible.* CTV News. https://www.ctvnews.ca/politics/don-martin-s-blog/jason-kenney-s-mission-unlikely-but-not-impossible-1.2958924

Martin, S. (2013, March 29). Obituary: Ralph Klein, 70: The man who ruled Alberta. *The Globe and Mail.* https://www.theglobeandmail.com/news/national/ralph-klein-70-the-man-who-ruled-alberta/article10569210/

McGregor, G. (2015, March 10). The gargoyle – Kenney tweets misleading photos of muslim women in chains. *Ottawa Citizen.* https://ottawacitizen.com/news/politics/the-gargoyle-kenney-tweets-misleading-photos-of-muslim-women-in-chains

McIntosh, A., & Hussey, I. (2019). *What You Need to Know About Alberta Budget 2019.* Edmonton, AB: Parkland Institute. Oct. 25. shttps://www.parklandinstitute.ca/what_you_need_to_know_about_alberta_budget_2019

National Post, The. (2016, Jan. 12). Alberta conservatives see Jason Kenney as best chance to unite-the-right and beat NDP in next election. https://nationalpost.com/news/politics/alberta-conservatives-see-jason-kenney-as-best-chance-to-unite-the-right-and-beat-ndp-in-next-election

Powell, L. F., Jr. (1971/2022). *Confidential memorandum: Attack of American free enterprise system*. Reclaim Democracy! http://reclaimdemocracy.org/powell_memo_lewis/

Richardson, L. (2012, June-July). Lougheed: Building a dynasty and a modern Alberta from the ground up. *Policy Options*, 32–37.

Santos, J. (2020, April 4). *CBC News poll: Albertans were hurting financially even before COVID-19*. CBC News. https://www.cbc.ca/news/canada/calgary/cbc-news-poll-alberta-economy-covid-19-personal-finances-1.5519085

Sayers, A. M., & Stewart, D. K. (2019). Out of the blue: Goodbye tories, hello Jason Kenney. In D. Bratt, K. Brownsey, R. Sutherland, and D. Taras (Eds.), *Orange Chinook: Politics in the New Alberta*. Calgary: University of Calgary Press.

Short, D. (2022, March 23). Number of doctors who left Alberta nearly tripled in 2021 compared to pre-pandemic times. *Edmonton Journal*. https://calgaryherald.com/news/local-news/number-of-doctors-who-left-alberta-nearly-tripled-in-2021-compared-to-pre-pandemic-times

Thomson, G. (2020, July 17). *OPINION: Doctors' heated dispute with Kenney government a creature of politics and history*. CBC News. https://www.cbc.ca/news/canada/edmonton/opinion-doctors-heated-dispute-with-kenney-government-a-creature-of-politics-and-history-1.5652934

Tran, P. (2022, March 25). *LISTEN: Extended transcript of Alberta Premier Jason Kenney's secretly recorded meeting with UCP caucus staff*. Global News. https://globalnews.ca/news/8710043/listen-extended-transcript-of-alberta-premier-jason-kenneys-secretly-recorded-meeting-with-ucp-caucus-staff/

United Conservative Party. (2019, March 29). *Alberta strong & free: Getting Alberta back to work*.

Wente, M. (2017, Oct. 30). Who's afraid of Jason Kenney. *The Globe and Mail*. https://www.theglobeandmail.com/opinion/whos-afraid-of-jason-kenney/article36767190/

CHAPTER 4

THE RELIGIOUS ROOTS OF SOCIAL CONSERVATISM IN ALBERTA

Gillian Steward

"I believe I could tell better by the praying people in this province than we could by the folks that are shouting too much for Social Credit. Let us pray and shout both. I don't want you to stop shouting but I do want every last man of you to join in a Bible study and a little prayer meeting."

— William Aberhart (1935)

WHENEVER FORMER premier Jason Kenney took to a podium during 2020-21 to speak about the COVID-19 pandemic, it wasn't long before he started talking about "freedom." Just as he did in spring 2021 when the second wave of COVID-19 hit the province and physicians and epidemiologists were urging him to impose a province-wide lockdown. Kenney refused, saying it would be "an unprecedented violation of fundamental, constitutionally protected rights and freedoms" (Fawcett, 2020).

While other provinces weren't immune to angry anti-vax protestors, church congregations ignoring public health restrictions, or restaurants defying orders to close, the freedom rhetoric that emerged during the pandemic has a long and deep history in Alberta. For several Alberta political leaders, individual freedom was the basis of their ideology and their public appeal. Its prevalence, however, must also be understood as arising from evangelical/fundamentalist Christian beliefs.

Kenney's touting of freedom over safety during the pandemic followed a long line of Christian politicians — William Aberhart, Ernest Manning, Preston Manning and Stockwell Day being the most prominent — who equated personal freedom with personal salvation. For them, individuals can only authentically choose to be reborn in Christ, and thus be saved from eternal damnation, if they are free from state interference. As they saw it, their job as politicians was to install policies and programs that would ensure that personal freedom as much as possible (Banack, 2016).

During Kenney's tenure, these religious beliefs, based on the Bible's authority, didn't just affect pandemic policies and regulations. They manifested in various public policies and legislation the UCP government enacted following the 2019 election. Its legislative agenda has included restrictive legislation regarding gay straight alliances in schools; the promotion of private, charter, and home schooling as a way to enhance parental rights for those who prefer a more Christian-based education; the revamping of school curricula to encourage a more traditional way of perceiving and acting in society; and the promotion of conscience rights for health-care practitioners who object to abortion or LGBTQ health provision. On the rhetorical level, the UCP has refused to commit to abortion rights, advocated for personal freedom over restrictions as a means of controlling the pandemic, and engaged in the quasi-denial of climate change as a serious threat to humanity.

Like Alberta politicians before him, Kenney cast his lot with those adhering to the U.S evangelical/fundamentalist Protestant tradition of prizing individualism and protest compared to the more hierarchical and collective traditions of the Anglican and Catholic churches. And since the former also believe the Bible to be the fundamental bedrock of authority, there is little tolerance for anyone, be they Christian or otherwise, who deviates from the strict code of conduct as revealed in that text's Old and New testaments (Banack, 2016).

Evangelical/fundamentalist Christianity has deep roots in the pre-revolutionary United States, where it became synonymous with a radical individualism that challenged traditional church hierarchies, placing authority instead in the congregation's hands. Its adherents viewed the Bible as the literal word of God, and thus to be heeded; and, further, if pastors spoke in plain language the common sense of the common man would prevail — a radically egalitarian belief opposed to the elitist, intellectual ideas proffered by other Christian denominations (Banack, 2016). These beliefs stretch even further back to Martin Luther's rebellion against the authority of the Catholic Church as personified in the pope, Catholic cardinals, and bishops. Luther believed each person could maintain a relationship with God through his or her faith, the clergy were not needed as an intermediary, and the Bible conveyed the only truths Christians need heed.

Individual freedom and faith in the common sense of the common man became touchstones of Alberta's political leaders and their followers in the 20th century, marrying politics and religion to policy agendas synonymous with Alberta's brand of conservatism. In the 1930s, the Reverend William Aberhart, a popular radio preacher, high school principal and political gadfly became convinced that an unorthodox monetary theory — Social Credit — would not only make it easier for Albertans to endure the Depression's hardships, but was in fact a solution sent by God. Aberhart garnered thousands of followers through his radio preaching and the Prophetic Bible Institute in Calgary, which became a church, social centre and political training school all rolled into one. As his movement grew, Aberhart's political speeches were so full of biblical and religious language it was hard to tell if he was trying to convert people to Social Credit or to his version of Christianity. Aberhart suggested during the 1935 provincial election campaign that prayer should be an integral part of the Social Credit campaign, telling his followers (quoted in Elliott, 1987, p. 191):

> I believe I could tell better by the praying people in this province than we could by the folks that are shouting too much for Social Credit. Let us pray and shout both. I don't want you to stop shouting but I do want every last man of you to join in a Bible study and a little prayer meeting.

To even his surprise, Social Credit candidates garnered more seats than the previously governing United Farmers of Alberta (UFA), and Aberhart was named premier by his caucus. He also served as minister of education and within a year that department produced a booklet — Bible Readings for Schools — under Aberhart's name (Alberta Department of Education, 1940). It was reissued in 1967 by the Social Credit government of the day. Just like Kenney, Aberhart challenged the jurisdiction of the federal government when it came to provincial matters. After his government introduced legislation severely restricting bank operations, the federal government soon vetoed it. Aberhart was furious and soon suggested replacing the RCMP with a provincial police force. When the federal government established a royal commission to look into federal-provincial relations, the Alberta government refused to participate. Just like many Alberta politicians to follow, Aberhart stoked the idea of the federal government as the enemy as a way to rally support from voters (Elliott, 1987).

Aberhart's eager student at the Prophetic Bible Institute, and his eventual successor in the premier's office, was Ernest Manning. After he took over as premier in 1943, following Aberhart's untimely death, Manning followed in his mentor's footsteps with his own Sunday gospel radio program, the Back to The Bible Hour. Given the predominant role of the Christian religion in western society at the time, it was not unusual for politicians' faith to inspire their various legislation or reforms. But Manning made it clear, through his political ideology and his radio broadcasts, he saw his role as being to convert people to a particular kind of Christianity. He was not subtle about this, constantly encouraging people during the Back to the Bible Hour broadcasts to be "born again" before the immanent Second Coming of Christ. Noting the resultant blurring of lines, Marshall (2016, p. 14) comments, "It was one thing for clergy to advocate for Christian-inspired reform but quite another for a premier to engage with the public in the most private of matters by making appeals to their personal faith and inquiring into the state of their souls." By then, however, the link between politics and Christian fundamentalism had been so firmly established in Alberta that, although faith in Social Credit monetary theories had faded, faith in quasi-religious government had not.

Manning stepped down as Alberta premier in 1968. Three years later the more secular Peter Lougheed and his Progressive Conservatives defeated Social Credit and religion slipped into the background. But religion rebounded when Manning's son, Preston, founded and became leader of the federal Reform Party in 1987. While the Reform Party was not as openly religious as the Christian Heritage Party, founded in that same year, the younger Manning retained many of the political and religious beliefs his father held. Preston believed contemporary society had grown distant from God and only through individual effort could citizens re-establish a relationship with God. While a truly perfect society was beyond humanity's reach, the state could play a divine role by guaranteeing individuals the personal freedom necessary to allow this relationship with God to flourish (Banack, 2016).

Manning's quest for a smaller, more fiscally conservative state more responsive to the wishes of the "common people" was derived from a pre-millennial Christian perspective he shared with his father and Aberhart. Pre-millennialism applies a literal interpretation to the New Testament chapter of Revelations in which Christ physically returns to Earth, after the Anti-Christ's defeat, to save those "reborn" through their Christian faith. Although the younger Manning was usually tagged as a neoconservative in the vein of Ronald Reagan and Margaret Thatcher, his political ideology is better understood as rooted in an anti-collectivist (anti-Communist, anti-socialist, anti-union) attitude originating with the belief in the state's divine, and minimalist, purpose of ensuring individual freedom, especially the freedom to be "reborn" (Banack, 2016). Both Ernest and Preston Manning's quest for ultimate freedom was directed at the prevailing political parties, including then prime minister Brian Mulroney's Progressive Conservatives, and the elites of central Canada who they saw as inhibiting western Canada's independence and prosperity with such "socialist" policies as national Medicare and the Liberal's National Energy Program.

As Manning crisscrossed Western Canada to promote the Reform Party in the late 1980s, he gained support from Ted Byfield, publisher of the popular *Alberta Report* magazine and an ardent conservative Christian. Through his magazine, Byfield promoted Reform's populist and religiously-inspired ideology to his growing number of readers. Manning and Byfield's ideas found fallow ground in Alberta and the other western provinces. Reform's success was soon manifested in overwhelming electoral success at the expense of the Progressive Conservatives. In 1993, the Reform Party won 52 seats, falling just short of becoming the official opposition to the Liberal government of Jean Chretien, while the Progressive Conservatives were reduced to two seats (Harrison, 1995). The subsequent election four years later saw Reform become Canada's Official Opposition.

The Reform Party was a natural religious and political fit for Kenney. In 1997, when he was only 29, he was elected as the Reform MP for a Calgary riding and joined the ranks of the official opposition led by Manning. At the time, Kenney was a staunch conservative Catholic firmly opposed to birth control, abortion, gay rights, sex education in schools, children's rights, and medically assisted death. He converted from Anglicanism to Catholicism in the 1980s while attending the St. Ignatius Institute, an undergraduate program, housed at the University of San Francisco which was founded by Jesuit priests. Most of the faculty and students at the university were quite liberally minded compared to the St. Ignatius Institute, which was overseen by ultra-conservative Jesuits who believed the Church had become too liberal and should return to its more traditional roots. They strongly opposed the liberalization of Catholic dogma and religious rituals that had followed the Vatican Two Council (1962-1965) led by Pope John XXIII and supported the more conservative agenda of Pope John Paul II, who was elected head of the Catholic Church in 1978 (Hudson, 2001; Lambert, 2019). The St. Ignatius disciples were firmly opposed to abortion. So much so that when the USF Women Law Students Association undertook a pro-choice campaign and began collecting signatures for a petition on campus, Kenney and other male students from St. Ignatius resorted to physical intimidation in an effort to stop them. The conflict eventually led Kenney to organize a petition to the hierarchy of the Catholic Church to decree USF no longer had the right to call itself a Catholic institution (Lambert, 2019). Kenney was also active in campaigns opposing spousal rights for gay couples afflicted by AIDS.

By the time he became a Reform MP, this agenda was resonating with evangelical/fundamentalist Protestant congregations in both Canada and the United States. Some Catholic and Evangelical leaders had begun meeting in 1985 to discuss ways they could achieve common ground on political and social issues. In 1994, they issued a statement — Evangelicals and Catholics Together: The Christian Mission in the Third Millennium — which, while highlighting their theological differences, nonetheless established a common stand against abortion and physician-assisted death. The documents also supported parents' rights to educate their children with a Christian-based curriculum and called for government and legislative support for the family.

Following the 1997 election, the Reform Party underwent a renovation as it attempted to draw in Progressive Conservative voters and expand its base nationally. The result was the creation of the Canadian Alliance party, led by Stockwell Day. Like Manning, Day was a prominent Alberta politician who was not shy about publicly expounding on his fundamentalist Christian beliefs. He believed in creationism, the

Bible's literal account of how God created the world, and suggested it should be taught in schools alongside theories of evolution. Though he came to religion late, by the late 1980s Day, like Aberhart and the Mannings, had forged deep religious roots and was the pastor and principal of a private Christian school in Bentley, Alta. He was first elected to public office in 1986 as a Progressive Conservative MLA and later served in Ralph Klein's cabinet as minister of labour, social services and treasurer (Harrison, 2002). Although Day was instrumental in the government of the day, he believed government should be restrained from interfering in the life of the individual whether it be through public education, public health care, or social welfare programs. People should be free to make their own way and rely on God's help when they needed it.

After Day left provincial politics to run for the leadership of the Canadian Alliance, Kenney stepped up to lead his campaign. Now a Canadian Alliance MP, and — like Manning and Day — still holding firm religious beliefs, Kenney told 300 attendees at a Catholic Home School conference in Edmonton while campaigning for Day that he went to San Francisco in the 1980s because he felt it was his vocation to be engaged in the "culture wars" then raging in California. "I wanted to know firsthand the forces that are undermining human life," he said. Thirty-two years old and single, Kenney's conference speech referenced the liberalization of birth control, abortion and gay rights as a "death culture" and a "pernicious agenda" that sought to destroy the nuclear family and parental rights. He also condemned the United Nations' support for birth control and abortion in developing nations as a plot to reduce the world's number of "brown people." Kenney's reference to a "culture of death" was very much in line with the dogma of Pope John Paul II, who had become head of the Catholic Church in 1978. But despite Kenney's insistence that birth control and abortion were morally wrong, he never once throughout the hour-long address mentioned the fact it is women who birth children and assume most of the burden of rearing them in their younger years. He never mentioned that women denied birth control or abortion in developing nations are often consigned to raising several children on their own (St. Joseph Communications Canada, 1999).

Kenney belongs to a branch of the Catholicism known as the Ordinariate. It was established by Pope Benedict XVI in 2009 for Anglican clergy and laity who wanted to return to the Catholic Church because they opposed trends in the Anglican Church that supported the ordination of women, gay clergy and same sex marriage. According to a notice in an Ordinariate publication which extended congratulations to Kenney when the UCP won the 2019 election, he and Bishop Steven J. Lopes, head of the Ordinariate, both attended the St. Ignatius Institute at the University of San Francisco for their undergraduate studies. Kenney is a parishioner of the St. John the Evangelist Catholic Church in Calgary, which was once a parish of the Anglican diocese but now counts itself as a Catholic parish of the Ordinariate tradition. Kenney is regarded as such a staunch traditional Catholic that, in 2012, then prime minister Stephen Harper appointed him to the Canadian delegation that accompanied the archbishop of Toronto and Cardinal-Designate Thomas C. Collins to the Vatican for the pomp and ceremony that is part of being appointed a Prince of the Church.

Kenney Unites Alberta's Fractious Conservatives

The early 2000s saw the continued recombination of conservative parties, both federally and in Alberta. In 2003, the same year the Canadian Alliance led by Harper and the Progressive Conservative party merged to form the Conservative Party of Canada, the Alberta Alliance held its founding convention in Red Deer, Alta. Its first leader was Randy Thorsteinson, a former leader of the Alberta Social Credit party. Though never formalized, the party billed itself as the Alberta wing of the Canadian Alliance party; indeed, Alberta's only truly conservative party. By 2004, the Canadian Alliance and the Progressive Conservatives had merged to form the Conservative Party of Canada and elected Harper as leader. Four years later, the Alberta Alliance merged with another upstart provincial conservative party — the Wildrose Party of Alberta — to create the Wildrose Alliance. When then premier Ed Stelmach, a Progressive Conservative, initiated a review of oil royalties with hopes of bringing in more revenue for the provincial treasury, the Wildrose Alliance — with the significant financial help of oil and gas producers — mounted such strong opposition that Stelmach eventually backed down (see Chapter 1). The Wildrose Alliance went on to become simply the Wildrose Party and by 2012, under the leadership of Danielle Smith, had elected enough MLAs to become the official opposition. In many ways, Wildrose stayed true to its Social Credit roots, standing for more provincial freedom from federal government interference and for more individual freedom, generally, from government. When Kenney embarked on uniting the Progressive Conservatives and the Wildrose parties after Rachel Notley's NDP won the 2015 election, the Wildrose were the stronger party. While his efforts birthed the United Conservative Party (UCP), and went on to defeat the NDP in the next election, the party's Wildrose element still dominated, to which Kenney remained beholden — if not downright intimidated — during his premiership.

As the pandemic wore on through 2020-21, and Premier Kenney repeatedly established public health restrictions, then lifted them, and then imposed them again, opposition erupted from his caucus' former Wildrose flank. In early April 2021, as the third wave of the pandemic was building, 15 UCP MLAs publicly criticized the restrictions imposed. A letter posted online declared, "We have heard from our constituents, and they want us to defend their livelihoods and freedoms as Albertans."

The rebels were led by Drew Barnes, who, in 2015, ran for the leadership of the Wildrose Party while it still formed the official opposition to then premier Jim Prentice's PC government. Around the same time, people opposed to masking and other restrictions organized "freedom rallies" that quickly gained steam in Calgary, Edmonton and other parts of the province. The rallies were often attended by right-wing political figures, such as Paul Hinman, who had also once been a Wildrose MLA and who, in 2021, secured the leadership of the latest entrant into Alberta's right-wing political arena, the Wildrose Independence Party (WIP). Hinman and other ultraconservative political operators were front and centre at the Anti-Lockdown Rodeo Rally in Bowden, Alta. at the end of April. They were also on hand when Chris Scott's restaurant in Mirror, Alta. was shut down by Alberta Health Services and the RCMP for defying public health restrictions. Scott told supporters he was so angry he would seek a nomination for the Wildrose Independence Party and run in the next election. If elected, Scott said, the first bill he would bring forward would entrench personal rights and freedoms (see Harrison, Chapter 5). And what exactly does the WIP stand for? Its first priority, as stated on its website, reads, "the freedom, dignity and worth of every individual" — that freedom word again.

The JCCF's Crusade

The fiercest, most organized and best funded opposition to government laws, regulations and restrictions arising from the pandemic came from the Justice Centre for Constitutional Freedoms (JCCF). The organization is based in Calgary and led by conservative Catholic lawyer, John Carpay. Carpay has been involved in conservative politics in Alberta for decades and was once a close ally of Kenney. In 2010, he was an unsuccessful candidate for the Wildrose Party in a Calgary constituency. He once compared the pride flag to the Nazi swastika banner and actively campaigned against legislation that provided confidentiality for students who joined gay-straight alliances organized in their schools. For Carpay, parents' rights trumped the need for confidentiality. He has also been front and centre in the campaign for what conservatives call "school choice." For Carpay, school choice means private schools, charter schools, religious schools and home schools should receive government funding at the same level as Alberta's public and Catholic school systems receive (see Stirling, Chapter 12).

Carpay and the JCCF also fought hard against COVID-19 public health restrictions, which it said had turned Canada into a "police state" (Justice Centre for Constitutional Freedom, 2021). It went to court on behalf of churches that refused to abide by public health restrictions because they impeded on their right to worship together as they saw fit. Those churches, among them Fairview Baptist Church of Calgary and

GraceLife Church in Edmonton, are stridently anti-abortion, anti-gay, anti-same sex marriage, and hold to traditional, if outdated, views about the role of women in marriage and society in general. In its zealotry, the JCCF pursued one such case on behalf of churches in Manitoba by hiring a private investigator to tail the presiding judge. Carpay admitted he had made the decision to hire the investigator and explained he did so to determine whether government officials were themselves complying with public health orders. He also disclosed surveillance had been conducted on a number of public officials. Apparently, for the JCCF, freedom for those public officials was expendable. Carpay took a leave of absence after the judge revealed he had been surveilled but was reinstated six weeks later. Six JCCF board members resigned including Queen's University law professor Bruce Pardy, former British Columbia attorney general Bud Smith, and journalist Barbara Kay (Raymer, 2021).

The Mantra of School Choice

Kenney acquired most of his education at a Catholic private school — Notre Dame in Wilcox, Saskatchewan — where his father was headmaster. That makes him Alberta's first post-Depression premier to have graduated from a private school. It might explain his strong support for "school choice" and his insistence that the public school system must be cleansed of what he sees as "ideological curriculum," a belief he made quite clear in 2016 when he was attending a Conservative Party of Canada convention in Vancouver. During an interview with Rebel Media, Kenney said Conservatives have a difficult time recruiting young people because they are indoctrinated in school with leftwing ideology (Press Progress, 2016):

> I think it's the first generation to come through a schooling system where many of them have been hard-wired with collectivist ideas, with watching Michael Moore documentaries, with identity politics from their primary and secondary schools to universities. That's kind of a cultural challenge for any conservative party, any party of the centre-right, and we've got to figure out how to break that nut.

Two years later, at the UCP's inaugural policy convention in Red Deer, he lobbied the accusation straight at the NDP government. "If the NDP tries to smuggle more of their politics into the classroom through their curriculum we will put that curriculum through the shredder and go right back to the drawing board," he told the crowd of 2,000 which responded with a round of hearty applause (Braid, 2018). Kenney kept that promise when elected premier in 2019.

Under Education Minister Adriana LaGrange, the K-6 curriculum was thoroughly overhauled, even though over several years both Progressive Conservative and NDP governments had already worked to update it. In the end, parents, teachers, and education experts — not to mention the Alberta Teachers' Association (ATA) — rejected the UCP's version as out-of-date and unsuitable for primary school students. Fifty-six school boards representing about 95 per cent of students in the public system refused to test it in their classrooms. However, a high proportion of teachers in private schools agreed to test the material. In 2020 the UCP passed the Choice in Education Act. It removes the requirement for parent home schooling their children to get approval from provincial authorities for the curriculum. It also removes the requirement for those wanting to establish charter schools — one Calgary-based charter school society has 3,700 K-12 students across eight campuses — to get approval from the local school board: they can now go straight to the Minister of Education.

The UCP's vetting of school curricula and its moves to make it easier for parents to exercise choice concerning primary and secondary education echoes movements in the United States over the past 40 years. Christian conservatives in that country have also insisted that public school curricula is ideological because it indoctrinates students with a secular view of the world and not a Christian view. (That such a view is also ideological appears to escape the UCP's recognition.) In the meantime, in Alberta private and charter schools receive 70 per cent of their funding from provincial coffers, money that could be used to hire more teachers for public and Catholic schools where student-to-teacher ratio is much higher.

Parents' rights were also at the fore when the UCP government rolled back the privacy protections for students who participated in school-based Gay-Straight Alliances that the NDP government had established in legislation. Under the NDP's legislation, teachers and other school staff in public, Catholic, charter and private schools were prohibited from disclosing student participation in a GSA in case a student might face retribution at home. If a parent inquired, the principal was allowed only to confirm the school had a GSA. Critics saw the UCP's changes to legislation as designed to discourage GSAs from being established in the first place. Now, if a parent asks if their child is participating in a GSA school staff can out them. The new legislation states that, if school principals delay or refuse a student's request for a GSA, students can appeal to the school board or the Minister of Education. Critics note that most high school students would find this a daunting experience that would also mean relinquishing their privacy in the matter, but the UCP believes parents have a right to know everything about what their children are doing while at school, no matter their age.

The UCP's emphasis on schooling as key to an individual's freedom to adhere to the authority of the Bible has a long history in Alberta. As mentioned earlier, Premier

Aberhart also took on the post of Minister of Education following the 1935 election. But schooling and educational curricula are also key issues for U.S Evangelicals, and in many ways brought them into the political arena. When Jimmy Carter was president, the U.S Internal Revenue Service threatened to revoke the tax exemptions granted to church schools because they wouldn't desegregate. Although the threatened schools were only those segregationist hardliners who refused to adopt a non-discrimination policy, even on paper, Evangelical and Catholic leaders chose to demonize the government's supposedly heavy hand. Paul Weyrich, co-founder of the U.S conservative think tank the Heritage Foundation and a leader of the religious right, was insistent that the movement didn't start with abortion or equal rights but with Carter's attempt to shut down private Christian schools (Balmer, 2021). The number of private Christian schools exploded across the South in the wake of federally mandated busing as a way to desegregate schools. While enrollment in U.S. public schools declined by 13.6 per cent between 1970 and 1980, the number of students in private Christian schools doubled. By the mid-1980s, more than 17,000 such schools had been created (Hudson, 2010, p.66). An enduring political movement was created, centred on the shibboleth that the U.S federal government is the enemy of Christian religious freedom — unless it is staffed by them (Posner, 2020). This hostility to government spilled over the border, reinforcing already existing fears and the belief that parents should be the guiding hand in children's education.

Parents for Choice in Education (PCE) is an influential Alberta lobby group that believes parents should be offered more than what the public school system has to offer. According to its website, the non-sectarian non-profit advocates for an excellent, quality oriented, choice driven education system which recognizes parental authority. Its current Executive Director is John Hilton-O'Brien, who was a founder of the Wildrose party and once served as its president. The PCE's previous executive director also worked for the Wildrose Party.

Parental choice on school issues continued to motivate the UCP throughout Kenney's time as premier. In an April 2020 speech, kicking off the vote on his leadership, Kenney touted his government's successes in revamping education: "We reversed the NDP's attack on parental authority in education and ended their war on faith-based education" (G. Steward, personal communication, April 9, 2022). He also told the carefully-selected group of UCP members in attendance that the UCP had "shredded" the NDP's ideological curriculum rewrite and begun developing a modernized curriculum that gets back to basics in math and reading "with balanced content on our history and institutions." Kenney then went on to say that, "Instead of divisive woke-left ideology like critical race theory, cancel culture, and age-inappropriate sex education, we are putting kids and the authority of parents back in charge of our education system" (G. Steward, personal communication, April 9, 2022). Kenney's rhetoric repeats what most Evangelicals, Catholics and Republicans in the U.S use to attack public education and push for schools that follow Christian teachings and practices and ensure that parents have the authority to include those teachings in the curriculum if they are home schooling.

Abortion, Homosexuality, Same-sex Marriage, Transgenderism.

Social conservative principles on abortion, homosexuality, transgenderism, and conscience rights are driven by a deep belief in literal interpretations of the Bible and fears that society is being overridden by people who don't believe in its authority. In the U.S, organizations that base their principles on biblical authority have become politically powerful. One such organization, Focus on the Family, states on its website, "[W]e believe the Bible to be the inspired, only infallible, authoritative Word of God." A Focus on Family offshoot, The Family Research Council, states on its website (Family Research Council, 2022):

> Family Research Council champions marriage and family as the foundational cornerstone of civilization, the seedbed of virtue, and the wellspring of society. Properly understood, "families" are formed by ties of blood, marriage, or adoption, and "marriage" is a union of one man and one woman. The only appropriate context for sexual relations is within the marriage of a man and a woman. Moreover, we believe that because God created us "male and female" (Gen. 1:17) we have no right to recreate ourselves otherwise.

Needless to say, organizations such as these are firmly opposed to gay rights, same-sex marriage, sex outside of marriage, and abortion and have fought for these beliefs in the courts, legislatures and news media. They also fund politicians, mostly Republicans, who support their views. They are behind many of the battles in various states to restrict or outlaw abortion and have also supported businesses that refuse to cater to gay customers.

In Canada, these issues fall under federal jurisdiction and have been settled through the courts or Parliament. So, there is little provincial authorities can formally do to override those decisions. But in the past Kenney made clear his views on these issues. The story of Delwin Vriend provides an example. Vriend taught at an Edmonton religious college, but was fired because he was gay. The Alberta Human Rights Commission could not rule on the case because sexual orientation was not included in the Alberta Human Rights Act, even though in 1995 the Supreme Court had ruled that sexual orientation should be included in the Canadian Charter of Rights and

Freedoms. Though Alberta argued that forcing the province to include it in the Alberta Human Rights Act would be judicial overreach, Canada's Supreme Court ruled in 1998 that it must do so. At the time, Kenney was a Reform MP and advised Premier Ralph Klein to invoke the Charter's Notwithstanding Clause. "If the court rules to enforce gay rights, and the Alberta government rolls over, they will clearly be implicated in the decision," Kenney said during a media scrum. "If, on the other hand, they have the courage to invoke Section 33, to use the one remedy in the Charter, they will have begun the recovery of democracy" (Woodard, 1998). If Klein had done what Kenney advised, gay or lesbian people would have continued to be discriminated against in Alberta.

Kenney took his criticism even further. "When the Supreme Court invented a constitutional right to sexual orientation, a right based on sexual conduct, they opened the door for polygamists, advocates of incest and others to claim the same status as homosexuals. It was inevitable." Kenney also opposed legislation legalizing same-sex marriage in Canada. During a 40-minute interview with a Punjabi news service in 2005, translated by CanWest News, Kenney said, "Marriage is by definition about a potentially procreative relationship. As much as people of the same sex may love each other, as much as they try, they don't even have the potential to procreate or raise children" (Dawson, 2005). Whether for reasons of political expediency or a genuine change in his thinking, Kenney has since recanted his opposition to same-sex marriage. In 2016, when members of the Conservative Party of Canada voted to drop the traditional definition of marriage from its policy book, Kenney supported the move.

While provincial governments cannot over-ride laws that fall within federal jurisdiction, they can restrain the implementation of those laws, for example in the area of health. The UCP government has made it more difficult for Albertans to access gender affirming surgery. A few months after the UCP took over the reins of government, Alberta Health required people seeking that surgery to get a referral from a psychiatrist whereas previously a referral from a General Practitioner had been sufficient. Since there aren't many psychiatrists in Alberta who provide transgender care, this has meant long, often indefinite, wait times for patients. Perhaps more subtly, the requirement of a psychiatric referral implies that transgender patients are mentally unstable.

In contrast, the premier and most UCP MLAs prefer to remain inactive and silent on another controversial issue, abortion. When the NDP introduced a Bill in 2019 that would create 50 metre zones, tied to hefty fines, around abortion clinics to keep away people who wanted to harass clinic patients and staff, UCP MLAs refused to debate. When the vote was called the UCP MLAs in attendance walked out." Six months after the UCP took over the reins of government, Peace River MLA Dan Williams tabled a private member's bill that sought to affirm health care workers conscience rights and release them from having to refer patients who requested an abortion, medically assisted death or appropriate LGBTQ medical services to another practitioner. Three quarters of the UCP caucus supported moving the bill to first reading but it was later defeated. In early 2022, after a draft decision of the U.S Supreme Court was leaked suggesting the court will strike down abortion rights in that country, NDP leader Rachel Notley asked Kenney in the Legislature, "Can the premier commit that the UCP will never act to reduce access to abortion in this province? Yes or No." The premier did not answer the question and said instead that it was a matter for U.S courts. No commitments were made by any UCP MLAs either. But the UCP's lack of response is perhaps not surprising, given that just before the 2019 election Cameron Wilson, political director of the anti-abortion group, the Wilberforce Project, proclaimed, "If the UCP wins the upcoming election we will have the most pro-life legislature in decades, maybe ever" (Wilberforce Project, 2019).

Climate Change

It may seem odd to include climate change in a grouping of issues where social conservative or religious attitudes influence government policy. But various studies have shown that white evangelical Protestants are the most skeptical major religious group in the U.S when it comes to climate change (Veldman, 2020). The evangelical community is not monolithic, by any means, but a Pew Research Centre survey in 2014 showed just 28 per cent of white evangelicals accepted that the climate was changing due to human activities. This result tracks with other surveys finding that white evangelical Protestants' acceptance of climate change is consistently lower than other religious groups. Robin Globus Veldman (2020), an assistant professor at Texas A&M University, wanted to understand why this was so. Was it about theology? Politics? Or culture? In her book on the matter — *The Gospel of Climate Skepticism: Why Evangelical Christians Oppose Action on Climate Change* — she discovered through field work in evangelical communities that their stance on climate change was influenced by a combination of factors. One key factor was an "embattled mentality" held by evangelicals as opposed to other religious groups. Veldman described it as arising from beliefs that climate change science is "a competing eschatology concocted by secularists who sought to scare people into turning to government instead of God" (p. 8). Evangelicals interpret their battle against secularism as intertwined with their beliefs about education and sexual issues. Veldman also concluded that, since most political conservatives have doubts about the need for climate change action, and since most evangelicals are conservative, the issue has also become political for them. The evangelical leadership played an important role in creating these attitudes towards climate change through its wide-ranging radio and television networks. They enabled a culture in which expressing a different view could mean ostracization from the group. Being skeptical about climate change, as well as environmentalism, became an aspect of evangelical identity as someone "holding fast to one's faith in a hostile, rapidly secularising world" (p. 9).

The "embattled mentality" Veldman describes certainly fits Kenney's persona. It may even explain his ongoing battle with environmentalists and his aggressive "Fight Back" strategy which included a $3.5 million inquiry into what he described as "anti-Alberta energy campaigns" by environmentalists (see Taft, Chapter 8). Kenney's government also established a $30 million a year oil and gas War Room designed to provide rapid responses to environmentalists' criticisms of Alberta's energy industry. At one point the War Room took aim at Bigfoot, an animated movie for children, that it claimed impugned the oil and gas industry. "Climate change" is a phrase neither Kenney nor other UCP ministers often utter in public. In 2013, when parts of Calgary were deluged for days with flood waters, Kenney said there was no link between extreme weather and climate change. During the 2019 provincial election campaign, Kenney conceded he personally believed in man-made climate change, but wouldn't remove party members who denied it, saying there are a "spectrum of views" on the issue. After the UCP formed government, it eliminated Alberta's carbon tax brought in by the NDP, although the big emitters still had to pay for their emissions. Programs to encourage renewable energy and energy efficiency were canceled, and the province's stand-alone offices for climate change policy and environmental monitoring were dissolved.

In 2022, when the federal government announced its plan to hit net-zero carbon emissions by 2050, Kenney told a media scrum the plan was "nuts" and he planned to fight it "with everything we've got." In an op-ed in the *Edmonton Journal*, his influential cabinet ally, Jason Nixon, called the Liberal strategy to reduce emissions an "insane" plan. Since the UCP government likes to take on the federal government and Prime Minister Justin Trudeau at any opportunity, this stance may have more to do with politics than religious beliefs.

For Kenney and the UCP, climate change is not a problem to be solved in a way that causes the least harm to Albertans and the rest of the country; it's not a matter of finding the best solutions for both the oil and gas industry and the planet; it's not a matter of facing reality. Rather, it seems to be a battle against forces that would destroy the province as they envision it.

Conclusion

Alberta has been home to many political leaders who openly combined their religious beliefs with their policy agendas; William Aberhart, Ernest Manning, Preston Manning, and Stockwell Day. The Social Credit Party which was in power for 35 years began as a quasi-religious movement: Aberhart, its first leader, was a preacher who used radio to spread the Word as well as his political message. Ernest Manning preached on the radio every Sunday while he was premier. Was it any wonder that most Albertans of the day came to see their political leaders as men of God — literally. That tradition was broken when Peter Lougheed came to power in 1971. He and the Progressive Conservatives were much more secular and appealed to urban Albertans who were less inclined to be regular church goers than their rural counterparts. The PCs remained in power for nearly 44 years, even longer than Social Credit. But while they reigned almost unopposed, remnants of the Social Credit's religious conservatism continued fomenting discord within the party. In 2015, the party born out of that discord — Wildrose — became the official opposition in the Legislature. When the conservative vote split in 2015, Rachel Notley and the NDP won the election handily. But the PCs lost so many seats that the Wildrose party was still the official opposition. Kenney managed to unite the PCs and the Wildrose under the banner of the UCP, but Wildrose remained the dominant partner. After the UCP won the 2019 provincial election it became clear that Kenney, a conservative Catholic, and the Wildrose faction of the UCP were of like minds when it came to particular policy agendas. No premier since Manning has hewed to a social conservative agenda based on religious principles as much as Kenney. And he wasted no time implementing that agenda. Reversal of the NDP legislation that protected the privacy of students involved in Gay Straight Alliances was among the UCP's first pieces of legislation. It was touted as a win for parental rights but it was clear that this was also a win for conservative Christians who oppose gay and lesbian rights. The expansion of private, charter, and home schooling was another win for those who advocate for more parental choice and oversight of their children's education. At the same time the UCP took on Alberta's public school system by imposing a K-6 curriculum it said featured less leftist ideology and "more balanced content on our history and institutions." This too had religious overtones. As Kenney himself said: "We reversed the NDP's attack on parental authority in education and ended their war on faith-based education" (G. Steward, personal communication, April 9, 2022). The UCP have also refused to commit to never reducing access to abortion services as is happening in many U.S states. Much of the UCP's rhetoric around climate change echoes conservative Christian beliefs that climate change science is another secularist plot that aims to empower government at the expense of the individual citizen. At the core of these policies is a conservative amalgam of Christian identity, provincial rights, minimal government, majoritarian populism, and a mantra of freedom whose usage requires careful decoding.

Alberta has a long history of mixing religion and politics that became personified by leaders such as William Aberhart and Ernest Manning. Kenney is the first premier since the Social Credit party was in power who holds fast to a religiously-based social conservatism. Many of the UCP's policy preferences also stem from the evangelical Protestant movement in the U.S which has become a political powerhouse of election campaign financing and influence mostly within the Republican Party.

References

Alberta Department of Education (1940). *Biblical readings for schools.* Retrieved from https://archive.org/details/biblereadingsfor1940albe

Balmer. R. (2021, Sept. 8). There's a straight line from U.S racial segregation to the anti-abortion movement. *The Guardian.* https://www.theguardian.com/commentisfree/ 2021/sep/08/abortion-us-religious-right-racial-segregation

Banack, C. (2016). *God's Province: Evangelical Christianity, Political Thought, and Conservatism in Alberta.* McGill-Queens University Press

Braid. D. (2018, May 6). UCP policy meeting tilted hard right on social policy. *National Post.* https://nationalpost.com/news/politics/braid-ucp-meeting-tilted-hard-right-on-social-policy

Dawson, A. (2005, Feb. 13). Gays can marry but not each other: Calgary MP. *Calgary Herald.*

Elliott, D., & Miller, L. (1987). *Bible Bill: A Biography of William Aberhart.* Edmonton: Reidmore Books.

Family Research Council. (2022). *Marriage, Family and Sexuality.* Retrieved from https://www.frc.org/marriage

Fawcett, M. (2020, July 12). Jason Kenney's faith in personal responsibility is no match for Alberta's COVID-19 surge. *National Observer.* https://www.nationalobserver.com/2020/12/02/opinion/jason-kenney-personal-responsibility-alberta-covid-response

Harrison, T. (1995). *Of Passionate Intensity: Right-wing Populism and the Reform Party of Canada.* Toronto: University of Toronto Press.

Harrison, T. (2002). *Requiem for a Lightweight: Stockwell Day and Image Politics.* Montreal: Black Rose.

Hudson. D. (2010). *Onward, Christian Soldiers: The Growing Political Power of Catholics and Evangelicals in the United States.* Simon and Shuster; Threshold Editions.

Hudson. D. (2001) *Sed Contra: the Death of a Great College Program.* Retrieved from https://dealhudson.blog/2018/06/07/sed-contra-the-death-of-a-great-college-program/

Justice Centre for Constitutional Freedom. (2021). *Policing the Pandemic.* Retrieved from https://www.jccf.ca/category/videos/

Lambert, T. (2019). The young zealot: what Jason Kenney did in San Francisco. *The Sprawl*. https://www.sprawlcalgary.com/the-young-zealot-part-1

Marshall, D. (2016). Premier E.C Manning, back to the Bible hour, and fundamentalism in Canada. In M. Van Die, M. (Ed.), *Religion and Public Life in Canada: Historical and Comparative Perspectives*. Toronto: University of Toronto Press.

Posner, S. (2020). *Unholy: Why White Evangelicals Worship at the Altar of Donald Trump*. New York. Random House.

Raymer. E. (2021) Justice Centre for Constitutional Freedoms president John Carpay reinstated. *Canadian Lawyer*. https://www.canadianlawyermag.com/news/general/justice-centre-for-constitutional-freedoms-president-john-carpay-is-reinstated/359381

Religious Tolerance. (1994). *Evangelicals and Catholics together*. Retrieved from https://www.religioustolerance.org/chr_caev1.htm

St. Joseph's Communication Canada. (1999). Retrieved from https://www.youtube.com/ watch?v=9W_SRPlYRN4

Veldman, R., Wald, D. M., Mills. S. B, & Peterson. D. A. M. (2020). "Who are American Evangelical Protestants and why do they matter for US Climate Policy?" *WIREs Climate Change* 12(2): 1-21.

Veldman, R. (2020). *The Gospel of Climate Change Denialism: Why Evangelical Christians Oppose Action on Climate*. University of California Press.

Wilberforce Project, (2019, February). *Political Report*. Retrieved from https://www.thewilberforceproject.ca/political

Woodard., J. (1998, Jan. 19). Rumblings of a counter revolution: Alberta's justice minister blames politicians for a rising tide of judicial activism. *Alberta Report*, 10.

CHAPTER 5

DECODING THE UCP'S FREEDOM MANTRA

Trevor W. Harrison

"[I]f sometimes I do happen to tell the truth, I hide it among so many lies that it is hard to find."

— Niccolò Machiavelli, letter to Francesco Guicciardini, 1521

JASON KENNEY'S victory speech on the evening of April 16, 2019, following the Alberta election, closed with an homage, as he viewed it, to Alberta's identity (quoted in Walter, 2019); an identity:

> ... that values freedom: free expression, free inquiry, freedom of conscience, and freedom from intrusions by the state into our lives. We know that liberty is essential to human flourishing, and is what gives meaning to our choices. We also know that freedom requires us to govern ourselves, and to care for our neighbours.

Premier Kenney's emphasis on freedom was a piece with previous libertarian utterances he made throughout his political career. Earlier in the same speech, he likewise spoke of making Alberta's economy the "freest" in Canada. Over the course of the pandemic of 2020-22, "freedom" became the leitmotif of Canada's conservatives, but especially the United Conservative Party (UCP).

On the surface, the UCP is the most openly libertarian government ever elected in Canada. Libertarianism — or, at least, a simplified version of it — lies at the party's heart: its rhetoric and its political calculations. In 2019, Kenney successfully merged American-laced libertarianism with a more traditional Alberta political strand, populism, to win the election. But does the UCP's espousal of freedom mean what it purports to say? This chapter examines how the Kenney government's libertarian populist appeals contributed to its difficulties in dealing with the COVID-19 pandemic, the authoritarian tendencies lurking beneath the UCP's freedom rhetoric, and the curious trap the rhetoric of freedom has now imposed on the party.

What is Libertarianism?

Libertarianism has many parents and many siblings, ranging across conservatism, liberalism and socialism to anarchism and nihilism. Drawing from academic and popular literature, I focus here primarily on right-libertarianism (see Long, 2008).

Boaz (1997, p. 2) provides a working definition of libertarianism as, "the view that each person has the right to live his life in any way he chooses so long as he respects the equal rights of others." His checklist of right-libertarian criteria (pp. 16-18) includes: individualism; individual rights; a belief in spontaneous order (arising out of individuals actions); the rule of law; limited government; free markets; the virtue of production; a natural harmony of interests; and peace. Right-wing libertarians, in the main, also hold a positive view toward family and community, providing thus a bridge to traditional conservatism (see Friedman and Friedman, 1990, p. 3).

Right-libertarians differ importantly from their left-wing counterparts in emphasizing the rights of private over communal (or public) property; a privileging of freedom above equality; a valorizing of capital over workers; and an extolling of "gifted" individuals over common folk (see also Cowan 2016, p. xviii). For example, Friedrich Nietzsche's concept of the superman and Ayn Rand's protagonist Howard Roark in *The Fountainhead*. Right-libertarians hold liberty, not democracy, to be the most important political value (Boaz, 1997, p. 14), as they fear the state or a majoritarian populace might overturn the protection of private property. At best, libertarians favour minimal government and are generally wary of populism — though, as we will see, an uneasy relationship with populists is sometimes forged.

Modern libertarianism is quintessentially American. It draws its primary inspiration from John Locke and Adam Smith. From Locke, libertarianism takes the notion of "natural rights" and the idea that individuals may claim as their property anything they obtain from nature through their own labour. From Smith, libertarianism takes the idea that society emerges spontaneously out of individual self-interest.

These twin ideas find an allegorical home in the Friedmans' (1990) American origin story. *Free to Choose* relates how Europeans "driven by misery and tyranny" populated the New World, finding there "freedom and an opportunity to make the most of their talents [t]hrough hard work, ingenuity, thrift, and luck," (p. 1); and through voluntary and co-operative market exchanges, on lands that were available and "had been unproductive before" (p. 3), created prosperity. The story is notably absent any acknowledgement of Indigenous Peoples being dispossessed of their "unproductive" lands or of the use of slave labour.

Modern libertarianism contains three interweaving strands: economic, political/philosophical and religious. The economic strand draws heavily from two Austrian School economists, Ludwig von Mises and Friedrich von Hayek, who merged Smith's view of self-regulating markets with the political argument that state intervention in the economy is not only inefficient but also leads inevitably to totalitarianism. With only slight variances, the ideological path from von Mises and von Hayek leads directly to three American economists: Murray Rothbard (1985), who advocated the privatizing all state functions, including national defense; Milton Friedman (1962), who, like Hayek (2007), contended market systems are the guarantor of economic freedom, and that economic freedom, in turn, is the guarantor of political freedom; and James Buchanan (Buchanan and Tulloch, 1962), who popularized public choice theory, arguing the aims of public programs could be better achieved if individuals were allowed to exercise their choices through the marketplace, thus paving the way for such ideas as school vouchers and charter schools.

Libertarianism's political/philosophical strand includes Rand, whose essays and novels extoll the virtues of the superior individual who owes nothing to anyone else. Within academia, much of libertarianism lustre is due to Robert Nosick (1974), who argues the state's functions should be limited to protecting individuals against force, fraud and theft, and administering courts of law. Right-libertarians agree with Isaiah Berlin's concept of positive freedom ("freedom to") but not negative freedom ("freedom from") (Berlin, 1969).

While many libertarians are atheists (see Gier, n.d.), the ideology has links to some Christian faiths, especially Evangelicalism. The link is forged through an emphasis on free will and individual struggle and that an individual must find and choose their path to salvation (Schansberg, 2002). In turn, this belief merges with economic libertarianism's concepts of the self-made individual and back again to the prosperity doctrine.[1] For right-wing adherents, it also provides a direct philosophical basis for rejecting government or state authority, being answerable only to God — a point of departure for atheistic libertarians, but adaptable to traditional conservatives wishing to impose morality on others (Keckler and Rosell, 2015).

Taken together, one can already see the circumscribed nature of right-libertarianism's notion of freedom. Protecting the rights of private property requires state coercion; political and civil freedoms are secondary to economic freedoms; equality rights and democracy scarcely matter at all; and, in some cases, God's will hold's authority. While libertarianism has deep roots in political philosophy, its place in American political life is more recent.

The Spectre of Collectivism and the Rise of Libertarianism

Collectivism haunts libertarians. It is not only, or even primarily, the 1917 Russian Revolution that gives them nightmares; even more feared is the warm, swaddling effect of welfare statism. Post-war liberals saw Keynesianism as saving capitalism from itself, softening the market's hard edges, and lessening political and social unrest. For libertarians, Keynesianism introduced dangerous impurities into the market's functioning, making individuals dependent on government programs and threatening their freedom.

Prompted by these concerns, Hayek and others created the Mont Pelerin Society in the late 1940s to promote libertarian ideas. While these ideas flew under the public radar at first, the Cold War — pitting the United States as the self-described defender of western freedom and democracy against atheistic communism — gave libertarianism a boost. Gradually, it moved out from the fringe. Libertarianism gained further prominence in 1964 when Barry Goldwater, the Republican candidate, declared, "Extremism in the defense of liberty is no vice," though the Vietnam War exposed conflicts between conservative war-hawks and devout libertarians, such as Rothbard.

Right-libertarianism's avowed defense of private property received a corporate boost in the 1970s when the Trilateral Commission declared western government's capacity to act was being hindered by an "excess of democracy." Neoliberalism was on the launching pad and took off a few years later with the elections of Margaret Thatcher in the U.K. and Ronald Reagan in the U.S. Neither Thatcher nor Reagan were "pure" libertarians, but the former's dictum that "there is no such thing as society" coincided with its hyper-individualism and ethos of self-reliance, while the latter's oft-repeated quip — "the scariest words in the English language: I'm from the government and I'm here to help you" — encapsulated libertarianism's mistrust and fear of the government and state. Libertarianism was a perfect ideological companion to neoliberalism's promotion of markets over the state, with the former's emphasis on deregulation, privatization, low tax rates and free trade. Most right-libertarians would not be opposed to the New Right's dictum of wanting to shrink government small enough to be "drowned in the bathtub."

Libertarianism has grown in the U.S. since the 1980s. While few Americans describe themselves as libertarians – a 2014 survey conducted by Pew Research suggests about 11 per cent – many more Americans share some of its values (Kiley, 2014). Politically, libertarianism's appeal is outsized by the prominence of such Republican figures as Paul Ryan, and the father-son team of Rand and Ron Paul.

Libertarianism in Canada

Current sloganeering around freedom in Canada is heavily American. Its emphasis on individualism ("Life, liberty, and the pursuit of happiness") stands at odds, generally, with Canada's cultural traditions ("Peace, order, and good government") (see Lipset, 1968), which have attempted to balance freedom with equality, and the rights and responsibilities of the individual and the community. But American political culture is influential and has made inroads in Canada through media and industry.

When founded in 1973, the Libertarian Party of Canada drew scant attention and few votes. In 2018, however, following his failed bid to lead the Conservative party, Maxime Bernier created the libertarian Peoples Party of Canada (PPC). The party garnered 293,000 votes (1.6 per cent) in the 2019 election, but 841,000 (4.9 per cent) in 2021, much of its success in Western Canada, especially Alberta, but also Quebec, while cutting into federal Conservative support. Fearing the same kind of vote-splitting federally as occurred in Alberta in 2015, federal Conservatives since the 2021 election have ratchetted up their "freedom loving" bona fides to win back PPC supporters. Loudly vocalizing his support for the Freedom Convoy protesters in early 2022, Pierre Poilievre, on his way to winning the Conservative Party's leadership, once tweeted, "Canada is free and freedom is its nationality."[2] Many conservatives have endorsed his efforts to reshape Canada's political culture around an extreme notion of individual freedom devoid of a sense of social solidarity and responsibility. A key to understanding these efforts, however, lies in the appeal of libertarian populism in Alberta and the enormous ideological influence it has upon the federal Conservative party.

The UCP's Selective Use of Freedom

Neither Kenney, Danielle Smith (who succeeded him as UCP leader and premier in October 2022), nor the rest of the party's brain-trust are intellectuals; indeed, proudly so, as the UCP generally dislikes academic-types. Kenney's own brush with academia was brief and unsuccessful. Like all politicians, however, Kenney does understand the usefulness of dipping his toes in intellectual waters, enough to have a gloss of expertise. There are also specific ideas within libertarianism that, when stripped down to single slogans, have political use; for example, free enterprise and free speech, and private property and conscience rights. These terms resonate with key aspects of Alberta's political culture.

Much has been written about this culture. Stewart and Archer (2000, p. 13) describe it as predominantly "alienated, conservative, and populist." Wesley (2011, p. 260) describes it as built around an idea of freedom that embraces populism, individualism and autonomy. Historically, populism — of both the left and right — has been a dominant element within Alberta's political culture. Right-libertarianism's growing appeal is based on its ability to accommodate itself to Alberta's populist traditions, and the very real economic and social anxieties that many Albertans face.

Unlike libertarianism, populism is not an ideology. It is rather a discursive practice used by politicians to mobilize a group — "the people" — who feel threatened by external others – "an elite." Both the people and the elite are manufactured concepts.

Right-wing populism's anti-government/anti-state and pro-market solutions are adaptable to right-libertarianism. But "the people" is a collective concept, while libertarianism emphasizes the individual. Current conservative politicians and their supporters get around this potential conflict by appealing to anti-elite sentiments (the populist quotient) while also appealing to possessive individualism (the libertarian quotient). The individual exists as one of "the people" in a purely mathematical sense; as one counted in relation to his or her similarly blocked interests by various "others," be they governments, experts, special interests, etcetera. Libertarian individuals stand proudly independent, sole architects of their own success, while their failures — should these occur — are the result of "elites."

Before and after the 2019 election, Kenney regularly extolled the right-libertarian value of individual responsibility and the special qualities of "entrepreneurs" and "wealth creators." He was not alone. In a similar vein, Finance Minister Travis Toews' fall 2020 fiscal update stated the public sector "does not create jobs or generate wealth. Rather, public sector activities and spending are paid by withdrawing money from the economy, through taxes, or by taking money from future taxpayers by borrowing for deficit financing" (quoted in Joannou, 2020).

Toews' statement plays to the UCP's anti-government, conservative base. It is also in line with libertarian arguments dividing those who produce from "parasites" (Boaz, 1997). The UCP's avowed efforts at privatizing many public services — for example, provincial parks — are conventionally conservative. But its overt financial support for private, Christian universities and faith-based charter schools is also a bow to religious libertarian concerns with the secular state.

The UCP employs these right-libertarian/populist arguments to underpin their hostility to public services and the welfare state more generally. Public programs, such as education and health care, are funded through taxes.[3] Implicitly, the UCP suggests taxes are a kind of theft of private property — at the very least, "excessive" and "inefficient;" better to pay for them through market mechanisms, whereby consumers can exercise choice. But the UCP also implies government programs are an affront to individual's self-reliance; an insult to their ability to make it on their own, in turn ensnaring them in government "slavery."

The UCP and its supporters are inordinately exercised by the term "socialism," a term Kenney regularly dropped into his public utterances. In Alberta, socialism is the evil twin of freedom. It refers to liberals (and Liberals) and social democrats, as well as unionists, environmentalists, intellectuals, feminists and others; in fact, everyone and anything that one dislikes. Premier Lougheed's government is sometimes remembered, and not fondly, as socialist. In a similar vein, both Ed Stelmach and Allison Redford were called "liberals" in disguise, as has Kenney by some disgruntled UCP members.

Kenney and other conservatives understand the political appeal of calling his opponents "socialists," a group, he warns, bent on taking away the hard-earned dollars and freedoms of Albertans. Kenney was especially fond of attacking the federal Liberal government, a tactic that kept wary voters inside the UCP tent while distracting them from looking at Alberta's own home-grown problems, such as vast inequality, stagnant real incomes, and the perpetual spin-cycle of boom and bust; or such transnational and existential problems as global warming or pandemics. In turn, the illusion that Alberta's ills would all disappear if only the province was freed from Canada's yoke plays well with the UCP's angry and scared separatist supporters.

The UCP's defense of freedom is selective. On gender rights, for example, the party's stance falls much short of full support. (Kenney early in his career famously lobbied against gay rights.) On abortion rights, UCP caucus members — with the exception of Leela Aheer — stayed silent in condemning the U.S. Supreme Court's overturning of Roe vs. Wade in summer 2022 (see Stewart, Chapter 4), though during the leadership race to replace Kenney, Smith did declare herself pro-choice. When it comes to workers' rights, the UCP is firmly on the side of management; efforts to unionize face significant obstacles (see Foster et al., Chapter 10).

When the UCP won office in 2019, the path seemed clear for Kenney to use Alberta as a testing ground for right-libertarian/populist politics. But the times changed. Beginning in the early 1980s, the political right expertly navigated the policy window in favour of small government, markets and individualism (see Acuña, Chapter 3). But libertarian espousals of individual freedom ring hollow when crises require a collective response. The COVID-19 pandemic quickly revealed the contradictions and conflicts within the UCP's coalition and the limitations of libertarian populism.

Libertarian Populism Meets the Pandemic

In late 2019, a strange, new virus was detected in Wuhan, China. Hopes the novel coronavirus could be contained were soon dashed. It moved steadily out of Asia to Europe and then the rest of the world following the usual trade and tourist routes. On March 11, 2020, the World Health Organization (WHO) declared COVID-19 a pandemic. The virus proved highly contagious and deadly, especially for the elderly and those with underlying health issues. By November 2022, the number of confirmed cases around the world reached 628 million and the confirmed death toll, nearly 6.4 million. The United States led in both categories (nearly 96 million cases, more than one million deaths), though not in per capita terms. By comparison, Canada had 4.3 million cases and 46,000 deaths (all figures from WHO, 2022).

The virus challenged medical experts, scientists and health practitioners who faced enormous pressures to quickly gather and assess information, and to provide advice to political leaders who faced their own pressures from an increasingly fearful and sometimes skeptical public. Around the world, if unevenly, a series of standard health measures were enacted to deal with the pandemic: border closures, shutdowns, social distancing, preventive masks, enhanced cleaning and improved ventilation. Few sectors and few individuals escaped the pandemic's reach.

Many people at first ignored the virus' seriousness. Kenney early on compared COVID-19 to influenza. As the weeks turned into months and the months into years, the pandemic's obstinacy gave way to increased fear, anger, depression and blame-seeking. Legitimate inquiries about the virus' origins, its long-term effects, or the long-term efficacy of MRNA vaccines were sidelined in some quarters by wild conspiracy stories. Aided and abetted by social media pundits, conservative think tanks, and craven politicians (including, within Alberta, Smith and Brian Jean, who have misrepresented vaccine studies), a startling number of people rejected science-based knowledge in favour of bizarre "cures" involving household detergents or horse de-wormer medicines. For some, the absence of proof showing the efficacy of alternative medicines was evidence that governments, scientists and Big Pharma were engaged in a global conspiracy of spreading "fake news."

While conspiratorial beliefs cross all political lines, research conducted even before the pandemic found those on the political right are particularly drawn to such beliefs. A 2019 poll of 25,000 people in 24 countries by the YouGuv-Cambridge Globalism Project found that supporters of right-wing populist parties were much more likely than those in the general public to believe in conspiracies. In a stark warning of what was to come, the survey found those with strong views were twice as likely than others to believe the harm from vaccines was being hidden from the public (Lewis et al., 2019).[4]

Among some, fatalism — even a crude Darwinism — replaced faith in medical science. Misreadings of how herd immunity is defined grew. It found its apogee in the Great Barrington Declaration, a paper released in fall 2020 by three academics (Bhattacharya, Gupta, and Kulldorff, 2020). Published by the American Institute for Economic Research, a libertarian think tank, the paper argued that, while those at high risk should be protected, the remainder of people in the public should be "allowed to resume life as normal."

Numerous medical experts, including Alberta's Chief Medical Officer of Health Dr. Deena Hinshaw, immediately rejected the declaration, noting it misconstrued the concept of herd protection and minimized the problem of sorting out at-risk individuals from others; indeed, individual's limited capacities to evaluate their own risk (see Hinshaw, 2020; and Archer, 2020). In a word, the declaration asserted an impossible libertarian "solution" to a problem that required a collective response. Despite the evidence, however, popular talk show host Smith, along with several UCP members of the legislature, voiced support for the declaration. Why?

Far from a sincere, if misguided, belief in alternative medical approaches, the problem for libertarian-minded conservatives was that it steered public policy away from individual, market-based solutions toward a return to big government programs and away, as they viewed it, from individual freedom. Canada's Liberal government, like those elsewhere in the world, introduced a mix of health-care measures and financial supports to people who had lost their jobs; shades of welfare state programs that emerged after the Second World War.

Canada managed the pandemic better than several other countries. Within the country, some provinces also did better than others. But no province experienced the roller-coaster ride as did Alberta, where libertarian and anti-government (e.g., anti-Liberal, anti-Trudeau) rhetoric dominated talk radio and print media. After some first-wave success, the next four successive waves of 2020 and 2021 in Alberta saw disproportionately high rates of infection and death and an overwhelmed health-care system. The problem of dealing with the crisis was only partially medical; the larger problem was political.

In the fall of 2020, as Alberta experienced a surge of COVID-19 cases, Kenney repeatedly resisted recommendations from medical and public health experts that strong public health measures were needed to break the pandemic's transmission and protect the health-care system from being overwhelmed. Under increased pressure, he grudgingly brought in new, tougher restrictions on private social gatherings, including religious, business and other activities. A regretful Kenney claimed the measures meant "constitutionally protected rights and freedoms ... were being suspended."

The Justice Centre for Constitutional Freedoms (JCCF) supported Kenney's reference to charter rights. The libertarian JCCF was founded in 2010 by an Alberta lawyer, John Carpay. Carpay, like Kenney, has roots in the Canadian Taxpayers Federation (as director, 2001-05). The centre was prominent throughout the pandemic in defending the rights of individuals, businesses and churches to refuse vaccinations or to implement health measures.[5]

Constitutional experts disputed Kenney's — and the Centre's — interpretation of the Charter of Rights and Freedoms. Adams (2020), for example, agreed health measures restricted certain charter freedoms, but noted the Supreme Court's rulings that individual rights are not absolute and may have to be limited to protect and promote other important rights and freedoms. In Adams' words, "Government restrictions which justifiably limit certain civil liberties are not an abandonment of charter rights, and certainly not a suspension of them."

Nonetheless, Kenney's defense of freedom found a ready audience among much of the UCP libertarian-populist base. Many did not believe there was a health-care crisis;

that it was a Liberal or otherwise Big Government/Big Pharma hoax; or a test from God; or that vaccines were more harmful than COVID-19; or that they, themselves, had natural immunity; and that, in any case, people had the right to decide individually whether to follow health restrictions or get vaccinated. For whatever reason, a large contingent of supporters — as well as several UCP MLAs — opposed lockdowns and other restrictive health-care measures. As time went on, opposition to the measures became more pronounced. Spring 2021 saw anti-mask and anti-lockdown protests in Edmonton and Calgary by individuals denouncing "fake science" and "fake news." Some drove trucks, while others marched, waved flags and shouted calls to be released from the "tyranny" of lockdowns and masks. Sometimes the protests took on more sinister tones: neo-Nazi groups, such as the Proud Boys, and individuals carrying tiki torches and Confederate flags joined in.

Some lone restaurants and gyms refused to obey health measures. More prominent were several right-wing Evangelical churches and their pastors, many part of Liberty Coalition Canada, an alliance formed in 2021 to defend religious freedoms they argued were threatened by government actions to combat the pandemic (Leedham, 2022a), (see Steward, Chapter 4). GraceLife Church, located just west of Edmonton, was especially prominent in questioning the science behind the public health measures and refusing to limit the number of parishioners on its premises or to enforce masking. The church's pastor James Coates quoted the Bible as requiring Sabbath gatherings be face-to-face (Boyd and Leavitt, 2021). The church's website publicly argued that "having engaged in an immense amount of research," it feared COVID-19 was "being used to fundamentally alter society and strip us all of our civil liberties," adding, "By the time the so-called 'pandemic' is over, if it is ever permitted to be over, Albertans will be utterly reliant on government, instead of free, prosperous, and independent" (quoted in Climenhaga, 2021a). The JCCF came to the church's defense. Maxime Bernier also offered support.

Coates was arrested in February for violating public health orders and spent the next month in jail before being released. In the meantime, hundreds of the church's members continued to meet each week, without masks. In April 2021, Alberta health erected a metal fence around the church to prohibit services. Echoing Jesus, Coates said, "Any attempt to dictate to us the terms of worship is not in the government's jurisdiction, and I refuse to give the government what isn't theirs." Money poured into the church's offering plate.

Four days later, an unmasked crowd of 500 hundred marched to the church. Some sang hymns, read from the Bible, or prayed for the church to reopen; others carried Canadian flags or signs promoting conspiracy theories, warning of tyranny and communism. Racist insults were shouted at members of the Enoch Cree Nation, whose lands bordered the church property. The chief's car was vandalized. The fence was torn down. At least one protestor was arrested. The next day, April 10, Kenney pleaded with protesters to recognize "the sanctity of human life," that the virus was real, and that the province was experiencing its "biggest wave of infection ... to date" (quoted in Baig, 2021).

That same day, an angry mob of 750 people marched to the legislature grounds. The crowd was again dominated by people bearing tiki torches and Confederate flags, anti-vaxxers alleging the harm of vaccines, conspiracy theorists holding pictures of Bill Gates, and anti-government protesters. Mixed in, too, were placards denouncing Kenney, Shandro, and Dr. Hinshaw, to the repeated refrain of, "lock her up," a misogynistic riff on Hillary Clinton's 2015 presidential candidacy in the U.S. but also later used against Rachel Notley in 2019; now a regular threat directed at female politicians.

The next day, Kenney condemned the crowd's efforts at threats and intimidation, and defended Dr. Hinshaw, in particular:

> Albertans respect the freedoms of speech and protest. But breaking the law, trespassing, threats, and intimidation go too far. I condemn these actions and statements. It is increasingly clear that many involved in these protests are unhinged conspiracy theorists.

It was one of Kenney's better moments, but it was also a moment he helped create. He had long fostered anti-government populism, fed into notions of a federal conspiracy against Alberta, downplayed (at first) the virus' seriousness, and responded weakly to those not following health-care measures — all in hopes of keeping his fractious caucus happy.

They were far from happy. Days before the GraceLife riot, 17 UCP backbenchers, including the House Speaker Nathan Cooper, wrote a public letter criticizing their own government's health-care measures as going too far. Others posted their support for the letter on Facebook. Roughly half the letter-writers were former Wildrose MLAs, the others elected for the first time in 2019; all represented rural constituencies (Braid, 2021a). In short, they represented one of the two main partners in the UCP marriage over which Kenney had presided in 2017, but which now appeared increasingly headed for divorce.

In part, the dissension reflected polls showing much of the public was also not happy. An Angus Reid poll in late March showed 75 per cent of Albertans viewed Kenney as doing a "poor job" in handling the pandemic. The same poll showed a nearly equal split between those who believed health restrictions had not gone far enough and those believing they had gone too far — much of the latter making up the UCP's electoral base. Most worrying for the party, a series of polls showed the NDP would soundly defeat the UCP if an election was then held.

On May 1, the Alberta government announced a record 2,433 cases of COVID-19, a new record; in fact, per capita, the highest rates of infection in all of North America. On the same day, a defiant crowd of 3,000 held a rodeo on a private field just outside of Bowden, a small town heretofore best known for its minimum-security prison. A spokesperson for the event, promoted as the "No More Lockdown rodeo," said rural residents and local businesses had had enough of the province's endless restrictions. "They've had enough of the lies and they're ready to stand up [for] freedom" (quoted in Fedor, 2021).

At a regular COVID-19 briefing days later, Premier Kenney opined that Alberta had a "compliance problem," implying new restrictions wouldn't make any difference as people wouldn't follow them anyway; an argument he used several times during the pandemic. But the next day, he reversed himself, declaring a series of stronger health-care measures, including a return to online learning for grades K-12, and enhanced enforcement of those who defied the measures.

Kenney's messaging — vague, confusing, and sometimes contradictory — and lack of regulatory enforcement were a product of trying to keep his fractious coalition together. Early on in the pandemic, the broad Alberta public wanted him to take strong, decisive measures, but his party's libertarian-populist supporters wanted him to do the bare minimum.[6] In the end, Kenney satisfied neither the broad public nor his own base.

By the fall of 2021, the libertarian horse was well out of the barn and running free. It was soon joined by a stable of adherents from across the country.

The Return of the Truckers

Two years in, the pandemic had worn away many peoples' patience. Everyone wanted things to go back to normal. Typically stoic, most Canadians soldiered on, following the mandated restrictions. For some, however, anger and frustration replaced hope and endurance. The roiling anger found its vehicle — literally — on Jan. 23, 2022 when a few hundred people driving trucks and assorted other vehicles set out from Delta, B.C. headed for Ottawa. The "Freedom Convoy," as its participants called themselves, had one purported aim: to protest the federal government's ending of an exemption, effective Jan. 15, for truckers crossing the Canada-U.S. border. In effect, they would now be required to follow the same rules requiring everyone else to be fully vaccinated or else quarantine for 14 days upon entry; a requirement easily met by 90 per cent of Canadian truckers. Behind this simple story, however, was a more complicated and contrived one.

By the time the convoy reached Ottawa on Jan. 28, its numbers had grown, but its purpose had been rendered moot by the United States' introduction of a similar regulation. Like a puff of diesel smoke, the plight of unvaccinated truckers evaporated. Likewise, as governments signaled their intentions to end most of the health measures, the protests became increasingly moot. By then, however, the convoy's aims had been rhetorically re-purposed to serve a nebulous defense of freedom. Before it reached Ottawa, one convoy organizer declared its purpose to be defending the "rights" of people against the government's manipulation and oppression (Mason, 2022a). Another ringleader admonished protesters to stop talking about vaccines altogether and instead concentrate on the mantra of freedom; "freedom" was weaponized against opponents of the convoy's aims (Mason, 2022b). In the days and weeks following, the protesters appropriated the maple leaf flag — flying prominently amidst a sea of Make Canada Great Again hats and truck labels — as a symbol of their alleged superior nationalism compared with those who followed and supported health mandates.

Though it later acquired its own momentum, the Freedom Convoy was not an entirely spontaneous event. Its beginnings went back more than three years to the "Yellow Vest" protests (later, the United We Roll Convoy), which had driven to Ottawa in early 2019 in defense of the oil industry. Out of this protest was born Canada Unity, which brought together a host of right-wing groups, such as Hold Fast Canada and Action4Canada (Leedham, 2022b).[7] Canada Unity's founder is James Bauder, an admitted conspiracy theorist who has endorsed the QAnon movement and called COVID-19 "the biggest political scam in history," and contended vaccine mandates and passports were illegal under Canada's constitution, the Nuremberg Code, and a host of other international conventions (Leedham, 2022b).

Between December 2021 and mid-January 2022, Canada Unity supporters blockaded Ottawa-area media buildings, screamed "freedom" inside the entrance to the National Arts Centre, staged a protest inside a downtown Dollarama store, and drove around the gates of Rideau Hall shouting at Prime Minister Trudeau. All of this was a dry run for what followed.

It seems clear the Freedom Convoy of early 2022 was made up overwhelmingly of Conservative and People's Party of Canada voters — certainly not Liberal supporters — still smarting from an electoral defeat five months earlier. Not surprisingly, the convoy picked up verbal support along the way from several prominent Conservatives, notably MPs Andrew Scheer (former party leader), Candice Bergen, and Pierre Poilievre, though Erin O'Toole (then, the party's current leader) took a more cautious stance. Preston Manning (founder and former leader of the Reform Party) and Kenney also provided verbal support.[8] As Bergen was perhaps unwisely quoted, the Conservative party hoped the protests would create a political problem for the federal

Liberals; though Conservative support for the protesters was clearly meant also to stanch the bleed of conservative voters to Maxime Bernier's PPC, who — post-2021 election surveys showed — were particularly motivated by a perceived loss of freedom from vaccine mandates (Parkin, 2021).

The convoy arrived in Ottawa on Friday, Jan. 28. The ensemble included commercial big rig trucks and a larger array of pick-ups and personal cars; at its peak, perhaps 400 vehicles and maybe 2,000 people — men, women and children — on a regular basis, though the number on Parliament Hill on the first Saturday may have swelled to 18,000, and 8,000 on other days. The occupiers, no longer temporary protesters, took up residence on Wellington Street in front of Parliament (Ling, 2022). Most participants believed they were engaged in legitimate protest, protected by the Charter of Rights and Freedoms, though the protests featured a mixed bag of influences and beliefs. Several flags contained American historical references ("Don't tread on me") and the more odious Confederate flag. When later arrested, some protest leaders demanded they be read their Miranda Rights. Some signs displayed QAnon conspiracy theories; others, Bible quotes. Some occupiers said God had called them to join the convoy. Still other posters, bearing Trudeau's face, read, "Wanted for crimes against humanity," or more simply, "F*ck Trudeau." While most protesters engaged in peaceful, if highly inconvenient, protests, stories of threats to Ottawa citizens and of vandalism also grew.

The convoy's many participants had varied aims. Some simply wanted to vent anger, mainly at the federal (Liberal) government; many hoped (naively) to end vaccine mandates; a few saw the chance to make money; others were simply mischief makers; the online comments of some suggest they hoped the convoy's arrival in Ottawa would lead to a Canadian version of the Jan. 6 insurrection in Washington a year earlier. A memorandum of understanding (MoU) first posted by Canada Unity (2022) in December 2021 speaks to the group's undemocratic and delusional aims. Later reissued, with minor changes[9], and finally deleted from its website on Feb. 8, 2022, the MoU demanded the federal government resign and that all federal, provincial, and municipal mandates and measures dealing with the pandemic be immediately overturned.

Amidst growing public frustration with government and police inaction, Prime Minister Trudeau enacted the Emergencies Act on Feb. 14. The act gives the federal government broad powers to deal with a national emergency. The opposition Conservatives, their provincial counterparts, and civil liberties organizations immediately denounced the move as an over-reach. They were joined by 29 Christian Evangelical pastors belonging to Liberty Coalition Canada who implored the prime minister in an open letter, to "restore the constitutional freedoms of the people, respect the God-given rights of our citizenry and above all to humble yourself and take a knee before Christ the King lest you perish in the way" (all quotes in Leedham, 2022a).

By law, the Emergencies Act mandates that a commission of inquiry be convened within 60 days of the revocation or expiry of the declaration of an emergency (McLeod and Walsh, 2022). The inquiry was launched on October 13, 2022; its final report is scheduled for mid-February 2023. However, the public hearings held in the fall paint a picture of failure by government and police officials, at all levels, to coordinate their actions and an unwillingness of authorities to use the tools available to them to deal with a situation that was unprecedented, potentially volatile, and economically costly. On the protesters part, the evidence shows a group of individuals who might be best described as naïve, narcissistic, anti-social, publicity-seeking, and delusional; a few were openly racist and displayed a criminally violent intent.

Technically, the inquiry's final report may find the act's invocation was not constitutionally legal or justified. Politically, however, many Canadians will likely find it justified. It was largely a political response by the federal government under pressure to do *something*. The act's invocation did seem to push the police to finally take action and perhaps also to send a message to the protesters. Almost immediately, the border blockades ended. The Freedom Convoy's leaders were arrested and Ottawa's streets cleared.

Mass movements, whether protests or revolutions, are unpredictable. While the federal government had imposed a vaccine mandate for federally regulated workers and at the border, almost all COVID-19 restrictions fell under provincial jurisdiction. Except for British Columbia, the largest provinces imposing the measures were governed by conservative or conservative-leaning parties. But no province was so charged with mandate anger and calls for freedom as was Kenney's Alberta.

The Trucker Protests Come to Alberta

On Jan. 29, as celebrants gathered on Parliament Hill, a trucker blockade began at the Coutts, Alta. border crossing (Leavitt, 2022). Similar blockades soon spread to border crossings in Ontario, British Columbia, Manitoba, and Saskatchewan, while sporadic protests occurred also in other provinces and sites (Palmer, 2022). But the Coutts blockade was particularly important, economically and politically.

As columnist Don Braid (2022) pointed out, the protesters' demands to immediately end all COVID-19 restrictions were counter to the UCP's policy. Blockading big rigs backed-up traffic stretching several kilometres, causing vaccinated truckers to either sit with their goods or alter their travel route, an equally costly decision. Moreover, the border protest was both illegal and economically damaging. The government had at its disposal the Critical Infrastructure Bill against the Coutts blockaders, but did not use it.

The Coutts blockade starkly revealed Kenney and the UCP's dilemma. Conservatives provincially and federally saw the protestors as a useful club to be used against the Trudeau Liberals. But the UCP's rural rump, and many of its caucus members, also opposed their own government's health restrictions. Grant Hunter, the UCP MLA for Taber-Warner, joined the blockaders in support of their demands. Independent MLAs Drew Barnes and Todd Loewen added their support. But the majority of Albertans remained wary of ending all measures too quickly, and had no sympathy for the protesters. Additionally, the blockades were illegal; the government did not want to seem giving in to them. The government was caught on the horns of a political dilemma.

Kenney's government had a tool to use against the protesters. In spring 2020, it had passed Bill 1, the Critical Infrastructure Defense Act, aimed at protecting "essential infrastructure from damage or interference caused by blockades, protests or similar activities, which can cause significant public safety, social, economic and environmental consequences." It was clear the act was meant to deal with pipeline protestors. The fact the bill was not used to remove the Coutts blockades indicates more than a whiff of arbitrariness and, indeed, authoritarianism in the UCP's use of state power and its defense of freedom.

Kenney did finally condemn the Coutts protests and demanded the blockades come down. The protesters did not care. The number of big-rig trucks diminished, but were replaced by F-150 trucks, cars, and heavy farm equipment. Only after the Emergencies Measures Act was invoked did the blockade end. The RCMP arrested 13 individuals, four of them charged with serious weapons offences. (The trials are set for 2023.)

Whose Sovereignty?

In September 2021, amidst the ongoing pandemic and continued faltering of Alberta's economy, Rob Anderson, one of nine Wildrose MLAs who followed Smith in crossing the floor in late 2014 to join Jim Prentice's PCs, launched the Free Alberta strategy. He declared it "a strategic plan for Alberta to assert its sovereignty, offload the burden of Ottawa's tyrannical economic policies against this province, and secure self-determination for the people of Alberta within a reformed Canada" (Climenhaga, 2021b). The strategy proposed the Alberta legislature pass the "Alberta Sovereignty Act, granting the Alberta Legislature absolute discretion to refuse any provincial enforcement of federal legislation or judicial decisions that, in its view, interfere with provincial areas of jurisdiction or constitute an attack on the interests of Albertans" (story in Braid, 2021b). (Smith later picked up the Alberta Sovereignty Act proposal as a plank in her successful UCP leadership bid.) Two UCP MLAs, Angela Pitt and Jason Stephan, supported the Free Alberta strategy, as did Loewen and Barnes. The strategy is an extension of the Firewall demands of the early 1990s, regurgitated by the UCP's Fair Deal Panel. It calls for Alberta, among other things, to create its own police force and to take control over pensions, taxation, and immigration.

The strategy is based on arguments that Alberta needs to be freed from "foreign" oppression. This idea has a long history. It reemerges every time Alberta has an economic downturn (see Harrison and Acuna, Chapter 1), a time when the governing elite fear their power eroding. Convincing Albertans they are threatened by external forces strengthens the elite's grip. Fringe-driven anger aimed at the federal government is useful in achieving short-term political ends. But conservatives, whose first mandate is always law and order, also fear libertarian anarchy (Harrison, 1995, p. 125).

We must be clear. Neither the UCP, nor conservatives generally, are anti-government *per se*. They simply want more powers housed at the provincial level — removed from the Charter of Rights and Freedoms — where, one suspects, the government could pick and choose which policies are worthy of support and which freedoms are worthy of protection.[10] Right-libertarian defenses of private property require a state that can exercise coercion; hence, the Critical Infrastructure Defense Act.

The purpose of the Free Alberta strategy is not to free Albertans; it is to offer angry and alienated voters a pretend freedom from the federal level of government while binding them closer to the province's corporate state. In a word, conservative espousals of libertarian freedoms are a smokescreen for greater provincial power.

Consider the proposal that Alberta have its own police force or gain control over pensions, taxation and immigration. None of these enhance the actual freedom of individual Albertans; each would duplicate bureaucracy at a cost to Albertans — a curious result from a party constantly promoting itself as a custodian of people's money. Such proposals make sense, however, if viewed from the perspective of an entrenched provincial elite desperate to hold onto power through chimerical appeals to freedom.

But the freedom mantra also contains a poison pill. Having groomed their base to be angry freedom fighters, the UCP candidates during the race to replace Kenney as premier were unable to be anything but similarly angry defenders of Alberta's independence. The UCP's most extreme members demand nothing less than irrational homage to a slogan; the party's leadership, for reasons of self-preservation, are only too happy to oblige.

Conclusion

There is something quaintly naïve, even nostalgic, about libertarianism. It exudes a whiff of de Tocqueville's America, where communities are held together by voluntary associations; and perhaps ancient Greece which — ignoring such notable things as slavery and the exclusion of women from public life — some advocates view as a model society (Hanson, 1995). In the real world, however, libertarian policies can be disastrous, as discovered in 2008. Called before Congress to explain how the Great Recession had come about, Alan Greenspan, former Federal Reserve chairman and a devotee of Ayn Rand, acknowledged he had made a "mistake" believing that banks would do what was necessary to protect their shareholders and institutions. By way of a gross understatement, he said, there was "a flaw in the model ... that defines how the world works" (Associated Press, 2008). In effect, the self-regulating market — a major cornerstone of libertarian faith — had proved fragile in the face of such human frailties as greed, dishonesty, self-delusion, and stupidity.

Theoretical debates aside, few current conservatives use libertarianism as a genuine recipe; rather, as a slogan. As in the United States and the rest of Canada, Kenney's UCP used the rhetoric of libertarian populism before and after the 2019 election as a means of mobilizing supporters against imaginary enemies outside, and sometimes within, Alberta's gates. His aim was to keep voters within the conservative fold while maintaining and enlarging the power of Alberta's corporate elite.

But Kenney's dogmatic anti-government, anti-Ottawa, anti-Liberal stance and adherence to libertarian rhetoric proved problematic in dealing with the pandemic. His vacillating leadership in response to a collective crisis reopened fractures within his party while also lessening broad public support for his government. In his final days as premier, Kenney spoke often of the extremes of those seeking freedom in the absence of practicing social responsibility or respect for others. It was a lesson he had learned too late, hoisted on his own libertarian petard.

What is Alberta's political future? Kenney's successor as premier, Smith, is an even more ideological libertarian. She has proudly termed herself a libertarian populist whose "views on the role of government have been shaped by Friedrich Hayek, Adam Smith, John Locke, Ayn Rand and the US Constitution" (Smith and Finkel, 2018). In her acceptance speech as leader of the UCP on Oct. 6, Smith repeatedly emphasized her aim to make Alberta "prosperous and free" and to "double down" on a libertarian agenda that utilizes the freedom mantra as a political strategy for rallying angry and disgruntled voters. How many Albertans and Canadians will buy into empty sloganeering remains to be seen.

References

Adams, E. M. (2020, Dec. 12). Opinion: COVID restrictions aren't suspending Charter rights in Alberta. *Edmonton Journal.*

Archer, S. (2020, November). Failings of the Great Barrington Declaration's dangerous plan for COVID-19 natural herd immunity. *The Conversation.*

Associated Press. (2008, Oct. 23). *Greenspan admits "mistake" that helped crisis.* NBC News. https://www.nbcnews.com/id/wbna27335454

Baig, F. (2021, April 11). Expert says gathering outside Alberta church attended by many conspiracy theorists. *The Star.*

Bhattacharya, J., Gupta, S., & Kulldorff, M. (2020). *The Great Barrington Declaration.* Great Barrington, Massachusetts: American Institute for Economic Research.

Berlin, I. (1969). Two concepts of liberty. In I. Berlin, *Four Essays on Liberty*, 118-72. London: Oxford Press.

Boaz, D. (1997). *Libertarianism: A Primer*. New York: The Free Press.

Boyd, A., & Leavitt, C. (2021, April 14). How an Alberta church and pastor became flashpoints in the COVID-19 culture war. *The Star Calgary.*

Braid, D. (2021a, April 21). Explosive letter circulating within UCP calls for Kenney's resignation. *Calgary Herald.*

Braid, D. (2021b, Sept. 28). Group with MLA backing wants Alberta to flout federal laws, claim sovereignty. *Edmonton Journal.*

Braid, D. (2022, Jan. 31). The border crisis is a major challenge for Kenney and UCP. *Calgary Herald.*

Buchanan, J., & Tullock, G. (1962). *The Calculus of Consent: Logical Foundations of Constitutional Democracy*. Ann Arbor: University of Michigan Press.

Canada Unity. (2022). *Introduction to the Memorandum of Understanding*. Internet Archive, retrieved document. https://web.archive.org/web/20220122173201/https://canada-unity.com/wp-content/uploads/2022/01/Combined-MOU-Dec03.pdf

Climenhaga, D. (2021a, April 12). Hundreds of maskless demonstrators, apparently none from GraceLife congregation, protest COVID-defiant church's closing. *Alberta Politics Blog.*

Climenhaga, D. (2021b, Sept. 29). Volcano spews lava from Spanish island into Atlantic, huge bozo eruption reported in Alberta city north of Calgary. *Alberta Politics Blog.*

Cowan, M. (2016). *Fabian Libertarianism: 100 Years to Freedom*. Xibris.

Fedor, T. (2021, May 2). *Central Alberta rodeo goes on without a hitch but not everyone is on board.* CTV Calgary.

Friedman, M. (1962). *Capitalism and Freedom.* Chicago: University of Chicago Press.

Friedman, M., & Friedman, R. (1990). *Free to Choose: A Personal Statement.* New York: Harvest.

Gier, N. (n.d.). *The uneasy and contradictory alliance between libertarianism and Christianity.* https://www.webpages.uidaho.edu/ngier/libchristian.htm

Hanson, V. D. (1995). *The Other Greeks: The Family Farm and the Agrarian Roots of Western Civilization.* New York: The Free Press.

Harrison, T. (1995). Making the trains run on time: Corporatism in Alberta. In G. Laxer and T. Harrison (Eds.), *The Trojan Horse: Alberta and the Future of Canada,* 118-133. Montreal: Black Rose Books.

Hayek, F. A. (2007). *The Road to Serfdom: The Definitive Edition.* B. Caldwell (ed.). Chicago: University of Chicago Press.

Hinshaw, D. (2020). *Herd Immunity and the Great Barrington Declaration.* Government of Alberta. Oct. 28. https://www.alberta.ca/herd-immunity-and-the-great-barrington-declaration.aspx

Joannou, A. (2020, Nov. 24). Alberta's projected deficit drops nearly $3 billion, but still on track for record high. *Edmonton Journal.*

Keckler, C., & Rosell, M. J. (2015). The libertarian right and the religious right. *Perspectives on Political Science* 44(2), 92-99.

Kiley, J. (2014). *In Search of Libertarians.* Pew Research Center, Aug. 25.

Leavitt, K. (2022, Jan. 31). Tensions rise amid trucker blockade at Alberta border crossing. *The Star Calgary.*

Leedham, E. (2022a, Feb. 18). A network of far-right evangelical pastors helped occupy Ottawa and block US-Canada borders. *Press Progress.*

Leedham, E. (2022b, Feb. 5). Canada's "Freedom Convoy" is a front for a right-wing, anti-worker agenda. *Jacobin.*

Lemaire, F. (2022, Feb.). Bitcoin, even better than gold? *Le Monde Diplomatique,* 10-11.

Lewis, P., Boseley, S., & Duncan, P. (2019, May 10). Revealed: populists most likely to believe conspiracies. *Guardian Weekly,* 29.

Ling, J. (2022, Feb. 8). 5G and QAnon: how conspiracy theorists steered Canada's anti-vaccine trucker protest. *Guardian Weekly.*

Lipset, S. M. (1968). *Revolution and Counter-Revolution.* New York: Basic Books.

Long, R. (2008, Nov. 19). Keeping libertarian, keeping left. *Cato Unbound: A Journal of Debate.*

Mason, G. (2022a, Jan. 27). Trucker convoy has evolved into something far more dangerous. *Globe and Mail.*

Mason, G. (2022b, Feb. 8). How truck convoy supporters like Pierre Poilievre have weaponized "freedom." *Globe and Mail.*

McLeod, M., & Walsh, M. (2022, Nov. 4). Emergencies Act inquiry: What to know about the commission and what's happened so far. *Globe and Mail.*

Mont Pelerin Society. (n.d.). (year of the content). Retrieved Dec. 23, 2020

https://www.montpelerin.org/#:~:text=The%20Mont%20Pelerin%20Society%20is%20composed%20of%20persons,also%20its%20apparent%20decline%20in%20more%20recent%20times

Palmer, B. (2022, Feb. 17). Canada's alt-right "freedom' rage. *The Bullet.*

Parkin, A. (2021, Nov. 16). Who voted for the People's Party of Canada? Anti-vaxxers and those opposed to vaccine mandates. *The Conversation.*

Nozick, R. (1974). *Anarchy, State, and Utopia.* New York: Basic Books.

Raymer, E. (2021, July 13). Justice Centre head John Carpay steps down after hiring PI firm to surveil Manitoba judge. *Canadian Lawyer.*

Rothbard, M. (1985). *For a New Liberty: The Libertarian Manifesto.* Third edition. Macmillan.

Schansberg, D. E. (2002). Common ground between the philosophies of Christianity and libertarianism. *Journal of Markets and Morality* 5(2), Fall, 439-457.

Smith, D., & Finkel, A. (2018, Jan. 1). The rightful role of government. *Alberta Views.*

Stewart, D. K., & Archer, K. (2000). *Quasi-Democracy? Parties and Leadership Selection in Alberta.* Vancouver: University of British Columbia Press.

Walter, C. (2019). *Here is Premier-elect Jason Kenney's full victory speech.* DH News. https://dailyhive.com/calgary/premier-jason-kenney-victory-speech

Wesley, J. (2011). *Code Politics: Campaigns and Cultures on the Canadian Prairies.* Vancouver: University of British Columbia Press.

White, R. (2020, Nov. 23). *Majority of Albertans support stronger restrictions during pandemic: ThinkHQ poll.* CTV News Calgary.

World Health Organization. 2022. *WHO Coronavirus (COVID-19) Dashboard.* November 3. https://covid19.who.int

NOTES

1 The doctrine connects to Max Weber's famous thesis that certain tenets of early Protestantism provide the ingredient for capital accumulation.

2 Poilievre also enthusiastically supported crypto-currencies, favoured by libertarians because 1) they are outside state control and 2) they are (somewhat) anonymous — the same reasons money launderers and fraudsters also like crypto-currencies (Lemaire, 2022).

3 Federally, the Conservative party, Poilievre, has said that such things as the Canada Pension Plan and Employment Insurance are taxes.

4 Among other conspiracies, the same individuals were more likely to believe global warming is a hoax, the CIA invented AIDS, and the U.S. government was involved in the 9/11 attacks.

5 In one case, the centre hired a private investigator to surveil a Manitoba Court of Queen's Bench Chief Justice with the intent of catching him breaking public health rules. Carpay apologized to the chief justice and stepped down as director in July 2021 (Raymer, 2021), but returned as president the next month amidst the departure of half the centre's board members.

6 A poll conducted by ThinkHQ Public Affairs Inc. in November 2020 showed 51 per cent of Albertans believed restrictions did not go far enough, while only 13 per cent believed they had gone too far (White, 2020).

7 The former once claimed the Canadian Broadcasting Corporation's headquarters housed concentration camps, while the latter claimed the pandemic was part of a scheme by Bill Gates and a "New World (Economic) Order" to inject the population with 5G-enabled microchips.

8 Once entrenched in Ottawa, the protesters, now occupiers, picked up international support from right-wing libertarians such as Elon Musk, Donald Trump and Ron Paul, as well as Fox broadcasters Sean Hannity, Laura Ingraham and Tucker Carlson.

9 One notable change: The revised MoU did not mention cross-border truckers as had the December draft.

10 Of note here is the decision by the Ontario government of Doug Ford in November 2022 to invoke the Constitution's Not-Withstanding Clause to pre-emptively over-ride the bargaining rights of public sector education workers. This decision shows the growing willingness of right-wing governments to gut the Charter.

CHAPTER 6

"WE REJECT THE PREMISE OF YOUR QUESTION." THE MEDIA AND JASON KENNEY'S GOVERNMENT

David Climenhaga

"You cannot hope to bribe or twist, thank God! the British journalist.

But, seeing what the man will do unbribed, there's no occasion to."

— Humbert Wolfe, "Over the Fire" from *The Uncelestial City*, 1930

IF YOU WERE a news reporter with a tough question for the Alberta Government in 2021 or 2022 — years of pandemic and growing political uncertainty in Alberta, former premier Jason Kenney probably rejected the premise of your question.

It wasn't just Kenney.

Alberta's premier may have taken testy to a whole new level, but his entire United Conservative Party (UCP) cabinet and caucus were infected with a kind of disdain for professional media seldom seen in Canadian politics, and arguably never so openly expressed in Alberta political history.

This kind of open contempt for journalists is something new, even as other, more expected, pressures continued to add stresses to the relationship between media and politicians in Alberta — a relationship that, as in most places, traditionally could be fraught one moment and collegial the next.

If there was any comfort for the journalistic fraternity, though, it was that this reflected a more general hostility by the leaders of the UCP government toward anyone who dared to question what they were up to, often expressed openly on social media run by the UCP's aggressive "issues managers," "press secretaries," and other political staffers. This may have partly explained the provincial media's supine response to this treatment.

Asked during a news conference in December 2020 if he accepted responsibility for the way he had handled the second wave of COVID-19 up to then, Kenney began by telling reporter Sammy Hudes, then employed by the *Calgary Herald*, "that sounds more like an NDP speech than a media question, Sammy." He went on: "I reject the entire premise of your question" (Kenny-Hudes exchange, 2020).

The rest of Kenney's answer that day was chippy, tendentious and aimed at distracting from the number of deaths in Alberta at that point in the pandemic. Still, for a while, the premier managed to stay within the bounds of a traditional press conference answer — even back in the days when reporters could be found in the same room as the politicians they were questioning and the politicians could read the expressions on their interlocutors' faces.

Then Kenney couldn't resist making it personal one more time: "I know that you've just joined folks who are doing drive-by smears on Alberta," he jabbed, reminding Hudes of a long list of statistics that in the premier's interpretation showed Alberta was doing better responding to COVID at that time than other provinces.

This may not have been a very meaningful exchange in the long story of the relationship between frustrated politicians and unrelenting journalists. Nor was it in any way unique to Kenney's interactions with journalists. UCP elected officials and spokespeople were remarkably belligerent in their responses to almost anyone who challenged their decisions, be they opposition politicians, academics, union members or plain old citizens who took issue with a government policy. Nevertheless, the premier's one-sided exchange with Hudes starkly illuminates how things have changed between the media and the government in Alberta two decades into the 21st century. It expressed a mood that is something new, and unusual, in this province. In this chapter, I will consider how the traditional, comfortable relationship between the news media and Alberta premiers, their cabinets and caucuses came to change so strikingly under the Kenney government, and whether these changes are likely to persist under the new UCP leader.

The Not So Long-Ago "Old Days"

In recent years, we became — unfortunately — used to Donald Trump abusing, even threatening, reporters who asked questions he didn't like. Chippy responses by politicians to reporters who won't stop asking uncomfortable questions aren't exactly unheard of on Parliament Hill, either. Forty years ago, in Salmon Arm, B.C., then-prime minister Pierre Trudeau famously give the middle finger to a group of hecklers.

Still, Canadian premiers' relationships with journalists have traditionally been pretty respectful. After all, the press could cause problems for a politician who wouldn't play ball, and nothing encourages a media corps inclined to docility like the collegial relationship with powerful politicians typical of Canadian legislative press galleries, including Alberta's, where a few insider winks and nudges can go a long way.

But Kenney? Not so much. His sharp *ad hominem* attack on Hudes during a live broadcast became something of a watershed moment in the relationship between the UCP and the media. At least, it was a public watershed. It was emblematic of the premier's attitude toward media — soon reflected by his closest ministers, many members of caucus, and the legions of social media spokespeople employed by the government. Open disdain, that is, rather than the politely veiled variety that has been the norm in modern Canadian politics.

But to the UCP, *everything* is personal. And that includes its relationships with media. Indeed, foreign registration documents filed with the U.S, Department of Justice in May 2022 provided evidence of how deep the UCP government's hostility to media ran right from the start of its mandate. Reporting in *The Tyee*, a Vancouver-based online news site, New York-based investigative climate journalist Geoff Dembicki reported the Canadian Energy Centre, as the Alberta Energy War Room is known, hired Counterpoint Strategies Ltd. to provide strategic social media and digital advice on how to confront media. Counterpoint founder James Andrew McCarthy "renders services directly" to the War Room, the filings said.

The New York-based crisis management firm is known for its "counter-adversarial" fightback strategy that treats journalists as "predatory" enemies who are "hostile, aggressive and willing to disregard standards and collude with almost any antagonist," Dembicki reported, quoting Counterpoint's website. Counterpoint and McCarthy, wrote Dembicki, "gained notoriety from media watchers for pioneering the practice of using Google ads to draw negative attention to specific journalists who wrote stories that his corporate clients didn't like." This indicated a philosophy quite in sync with the War Room's claim on its website that media worked directly with environmental activists to "create self-sustaining controversy and inflict maximum damage to their subjects."

Common sense and conventional wisdom suggest treating reporters with good manners and simple decency is the smarter course. But then, that simple and well-understood formula for media relations dates to a day before social media, or, as some of us have come to think of it, anti-social media.

Social media has become the primary stream for political parties to communicate their messages directly to the public, bypassing traditional media. Reporters, on the front lines of traditional media work, were bound to get caught in the crossfire. This is nothing unique to Alberta, or even Canada.

Still, can you imagine recent Alberta premiers like Ralph Klein, Ed Stelmach, Dave Hancock, Jim Prentice or Rachel Notley — or for that matter Alison Redford, whose relationship with media was rockier — publicly accusing a reporter of being an ally of another party in the legislature? Sure, as in any relationship, things could get testy from time to time. Klein — a broadcast journalist before he was a politician — had an insider's understanding of how the media worked, and how media workers saw the world. That stood him in good stead in daily jousts with reporters. But neither Klein nor any of the other premiers who held office before Kenney since the 1930s would have thought such a belligerent approach was either appropriate or prudent.

In the fullness of time, we will find out whether this was a Kenneyesque aberration or a permanent change to the Alberta political scene. Consider Danielle Smith, the successful candidate to replace Kenney as leader of the UCP. The former Wildrose leader, who on Oct. 11, 2022, became the 19th premier of Alberta, was also a successful broadcast journalist. So she, of all candidates to lead the UCP, understood what was needed to maintain a successful relationship with political reporters. She of all candidates needed the care and attention of media to be successful in her renewed quest to lead the UCP and return to right-wing respectability from the shadow cast by her effort in 2014 to cross the floor with eight of her MLAs to the Progressive Conservative Party then led by Jim Prentice. She will need it again to hang on to her job as premier in the election expected in 2023.

Perhaps if she is more tolerant of the media than Kenney, it will send a signal things are returning to their historical patterns. Or, if not, that the world really has turned. The jury remains out on that question.

The standard, tried and true response when a politician rejected the premise of a reporter's question was for the politician to roll their eyes inwardly … and answer a different question. Indeed, not so long ago, seasoned politicians well understood the principle there's no need to answer the question you've just been asked if you'd rather answer another one that suits you better.

This requires a certain amount of skill, of course. But it's not rocket surgery, as Klein used to say.

No politician wants to be like Tyler Shandro, who — shortly after his appointment as health minister by Kenney — famously fluffed an attempt to stick to his talking points under aggressive questioning by a radio reporter (Shandro Press Conference, 2019). Nine times in three minutes he repeated the talking point that, "in due course," he'd get back to a group that wanted to ban conversion therapy.

Shandro's response was sufficiently embarrassing to draw international attention. As British media advisor Adam Fisher (2019) wrote in his client blog, the "consequence of this flawed approach was that it made it perfectly clear to both the media and the audience that he was stonewalling — a long way from the image of transparency he should have been striving for." This was also the moment a lot of Albertans began to realize Shandro might not be the sharpest tool in Kenney's kitchen cabinet cutlery drawer.

The traditional approach called for the politician to congratulate the reporter he was about to ignore for her perspicacious question before plowing on to answer the friendlier question he'd composed for himself. Everyone knew what was going on, but this was how the game was played. Usually, no one complained — at least publicly.

Kenney, obviously, understands this, and knows how to do it.

It's a simple skill, after all, and he is quite capable of rambling on at excruciating length. But something else has changed in Alberta and elsewhere in the world. It's an understanding, unfortunately possibly correct, that abusing media — and other citizens too — publicly scores points with elements of a party's base, groups who are then more likely to get out to vote.

This determination, it appears, is more common on the political right where, to quote *New York Times* economic columnist Paul Krugman (2021), "closed mindedness and ignorance have become core conservative values, and those who reject those values are the enemy…." Krugman argues a common thread links "this cross-disciplinary commitment to ignorance;" that, in every case where it is evidenced, the refusal "to acknowledge reality" serves "special interests." For example, climate change denial — apropos of Alberta — aids the fossil fuel industry. As a result, concludes Krugman, "right-wingers have gone all in on ignorance, so they were bound to come into conflict with every institution … that is trying to cultivate knowledge."

The media — as least as it was seen in the circle around Kenney — is one of those institutions.

Having seen Trump repeatedly abuse the media (usually excepting *Fox News*) at large public gatherings with apparently no negative consequences, this new approach was bound to infect Canadian politics. But the instinct to attack runs deep in Kenney's personality and in the political DNA of the UCP as a whole, which, contrary to the claims of his rival Brian Jean, Kenney has significantly shaped.

Still, there was during Kenney's tenure as premier a kind of clueless lack of self-awareness among many political operatives who surrounded him that goes to something deeper than the man himself. Perhaps its related to the rise of the lifelong politician — of which Kenney is a prime example; people, mostly men, who have spent their entire political lives working as students of politics, student politicians, political staffers, employees of political think tanks and lobby groups, and eventually, holders of actual elected office.

Kenney's career is a model of such a path — an attention-grabbing rabble rouser for right-wing causes before dropping out of university, a political assistant to a (Liberal) provincial minister in Saskatchewan, the founder of one anti-tax organization and mastermind of its merger with another to become a political force on the national stage — all before he was 30. He was elected as an MP at the age of 29, named Parliamentary Secretary to a highly ideological Conservative prime minister at 38, and elevated to cabinet by the same PM at 39, serving in various cabinet roles until the federal government he represented fell in 2015.

Kenney soon made his successful jump to provincial politics and was elected leader of the United Conservative Party in 2017. At the time, his move seemed to be part of a grand plan to eventually return to federal politics to occupy the highest office in the land. That scheme seems likely to have been derailed by his growing unpopularity as premier, though missteps did not hinder Smith's resurrection. Nevertheless, Kenney stands as a model of a new generation of such lifelong politicians, some of whom are part of the cadre of key political advisors that dominates the UCP today.

Not all were cut from the same cloth, of course, but through the pandemic, there was an exodus of older, experienced, more diplomatic political staffers from the UCP's employ.

A few, like Kenney chief of staff Jamie Huckabay, who took a mid-pandemic holiday in 2021, were pushed out because of errors. Others, such as political advisor Ariella Kimmell, left because of workplace conflicts. (Kimmel protested harassment of female co-workers in a cabinet that seemed at times to operate like a frat house, a matter now before the courts in a wrongful-dismissal suit.)

Other advisors to depart — probably because they could see the writing on the wall — included principal secretary Howard Anglin, Invest Alberta CEO David Knight Legg, communications director Katie Merrifield, chief of staff Larry Kaumeyer, and issues manager Matt Wolf, although no one would accuse the last person on that list of being too diplomatic.

In addition, a raft of experienced press secretaries and ministerial chiefs of staff exited as the Kenney government devolved into chaos, among them Blaise Boehmer, Lauren Armstrong, and Robin Henwood. By the time Kenney stepped down as premier in May 2022, probably more than half of the original occupants of important political advisory positions sworn in in 2019 had departed.

It seems no coincidence that, as these experienced hands streamed out the doors, the government's attitude toward the media kept getting worse and worse. At times, it seemed the government's remaining press secretaries were in a contest to see who could be most insulting and abusive to reporters on social media.

The boys in short pants were truly in charge!

Why Can't the Media Get No Respect ...

To blame social media for this increasingly nasty tone misses an important part of the larger story about the invention of social media and its impact. If journalists for respected media organizations *can't get no respect* anymore, it's also because media isn't what it was even two decades ago.

There was a day when the publishers of major Canadian metropolitan newspapers could boast, with some justification, that no local politician could get elected without their endorsement, or at least their acquiesce. But those days are gone. *Legacy* media — a term not much liked by newspapers and traditional broadcasters for what should be obvious reasons — is a pale shadow of what it once was. The invention of the Internet has undermined the newspaper industry's business model, stolen its core revenue sources, and sucked profit from its operations. The situation is not much different for traditional broadcasters.

For more than a century, it was hard *not* to make money in the newspaper business. But almost overnight, it seemed, that changed completely. Business began to turn away from print advertising — a medium that couldn't even produce the hyperlinked platforms advertisers wanted. Thanks to the Internet, advertisers had new and better alternatives.

The huge social media and data-vacuuming corporations like Google, Twitter and Meta (formerly, Facebook) that replaced traditional media did not deal with politicians in the same way. Instead of disdainfully providing a risky and often critical venue for their messages, they *offered* them a deal that only involved the sale of a small sliver of their souls. Where a politician once needed to approach powerful media corporations with cap in hand and a humble expression on their face for a meeting with the editorial board, that approach wasn't needed in the digital world. Now, for a modest fee, the means were at hand to bypass media completely if they wished. As a result, there is certainly less need to kowtow to them.

The techies who created the social media giants, moreover, offered politicians ways to direct their messages to targeted audiences — the people political organizations needed for a particular vote in a particular location; voters who could be motivated by fear, or bigotry; or a demographic most likely to be impacted by a policy decision. All you needed to do was cough up the dough for digital advertising and figure out how to exploit the algorithms used by social media giants to direct their messages to the desired audiences.

So if a politician had a nasty streak, why would he even try to be nice to reporters? There was probably more to be gained from many groups by siccing the wolves, as it were, on them. Faced with this new reality — incomprehensible to traditional media managers — what was to become of the great old media organizations of the past? Their answer, as the trusty business model imploded, especially in the print world, often carried more than a whiff of desperation.

In the spring of 2021, the *Toronto Star*, once one of the most influential newspapers in Canadian history, announced it wanted to open an *online casino!* (Climenhaga, 2021a). Someone called Torstar's chief corporate development officer announced in a news story that sounded remarkably like a press release:

> As an Ontario-based media business and trusted brand for more than 128 years, we believe Torstar will provide a unique and responsible gaming brand that creates new jobs, offers growth for the Ontario economy, and generates new tax revenue to help support important programs in our province.

One of the corporation's new owners was quoted as saying, "Doing this as part of Torstar will help support the growth and expansion of quality community-based journalism" (Climenhaga, 2021b).

Words fail. Surely it is obvious this isn't a good bet if survival is your objective!

Two months later, *Postmedia*, the largest newspaper chain in English Canada, announced it intended to start *a parcel delivery service* on the Prairies and in Ontario.

"At Postmedia we have a long, proud history of delivering to homes in communities across our country," Postmedia President and CEO Andrew MacLeod said in a press release. "Extending our offerings and trusted relationships in the communities we already serve, through this new partnership, aligns to our corporate strategy" (all quotes from *Postmedia*, 2021). One awaited breathlessly for news of pizza deliveries.

Months later, nothing more had been heard of those schemes, probably because all they were really suitable for was punchlines in a comedy sketch. Perhaps someone had figured out you couldn't roll up a legal document and toss it onto someone's doorstep.

Whatever they were thinking, in the same time frame, the same two competitors struck a deal to shut down 36 of the 41 smaller newspapers they'd swapped in 2019 when they cooked up a deal to divide up parts of Canada into largely competition-free zones. Torstar's free Star Metro newspapers in Edmonton and Calgary were shut down as part of that agreement. The closing in the fall of 2019 of the free Star dailies across Canada left Alberta's two largest cities with no newspapers that weren't published by Postmedia, by then owned by U.S. venture capital funds. The lost Torstar papers had their flaws, but they were better from a journalistic perspective than anything Postmedia published in Alberta.

Postmedia's cost-cutting strategy, meanwhile, seemed to be to jettison experienced reporters — a strategy that continues to be employed — and in many cases to replace them with no one at all. Columnists employed strictly to write opinions on the news of the day, seemed to have a safer berth. Opinion columnists like Rick Bell, Don Braid, Lorne Gunter and David Staples — some pretty good, some not so good — sailed serenely on, their sinecures apparently safe as long as they kept bloviating and didn't waste too much time and money on actual reporting. If they left — usually without being replaced — like the *Edmonton Journal*'s respected and thoughtful Graham Thomson or the *Calgary Herald*'s stridently conservative Licia Corbella, it was usually because they chose to retire or take a buyout.

When respected Edmonton sports reporter Terry Jones got a phone call from Toronto in June 2022 telling him his services were no longer required, he tweeted bitterly:

> My last scoop. At 1 p.m. today, after beginning my career at The Edmonton Journal in 1967 I received a phone call from Toronto informing me my position had been eliminated by Post Media. Thank you all so much for reading. Hardly the way I hoped it would end.

He concluded the tweet with "–30–," the traditional newspaper sign-off indicating the end of a story.

Meanwhile, Journal columnist and former sports writer David Staples, often reviled for his pro-UCP views in NDP-leaning Edmonton, remained secure in Postmedia's employ — a fact observed by many who responded to Jones's parting lament.

Like the proverbial frogs in that slowly heating pot of water, at least until the pandemic hit, journalists went along feeling that nothing much was changing; and, indeed, little did change from day to day. But every day, they were a little less important, their words conveyed a little less power, and they were treated with a little more contempt by the people they were paid — increasingly less in terms of purchasing power — to report on. It was inevitable that one day someone would come along and say out loud what everyone had quietly come to realize.

That someone just happened to be Jason Kenney.

The Pandemic Changes Everything – and Nothing

Then came the pandemic.

COVID-19, arriving in Year 2 of the Kenney government, changed everything, and nothing. When the virus arrived from the east — that is, Alberta's west — it not only created some entirely predictable political problems for the Kenney government, it also handed the UCP's strategic brain trust new opportunities to exploit.

The stresses imposed on public-sector health-care workers, surgical facilities, and intensive care units gave Kenney's UCP the opportunity to justify a rapid privatization of health care in the name of responding to the crisis. But the excuse of COVID-19's threat in public places could also be used as a mechanism for clamping down on media access to government officials, handing the UCP a powerful tool for media manipulation and control.

It wasn't just the Kenney government that took advantage of this, of course. To some degree, driven by the genuine need to "bend the curve" of infection downward, the term regularly used, all governments, and not just in Canada, used COVID-19 as an excuse to restrict media access. But the Kenney government — led by a man with a natural bent for secrecy and manipulation and a party in which a distrust of media already ran deep — embraced it with gusto.

It was soon obvious that news conferences using video conferencing software, moderated by a party employee with control over media questions, were an effective way to control and direct the narrative. The political staffer moderating the news conference now had the ability to move to another question by another reporter to cut off the first reporter who had asked a question the official providing the answers didn't like.

People who have not worked in media, who have not had to ask questions in the room at a news conference, don't realize how much influence the first question asked by a journalist at one of these events can have on the direction taken by the whole affair. This is why politicians of all parties have long preferred stage-managed news conferences, with a certain ritual decorum, to hallway "scrums" at which any reporter can hurl a hostile question captured by rolling media cameras. (It was just such a moment that launched Kenney in 1993 when he ambushed Klein, asking him about MLAs' pensions.)

Most of the time, if the organizer can get a news conference moving in the right direction, the reporters in attendance can be expected to adopt the tone set at the opening. Moreover, a friendly opening question — and in a properly staged newser, the organizers usually get to choose the first questioner — also gives the spokesperson the opportunity to run out the clock on reporters who are seen as likely to have less-friendly queries.

So here's a pro tip: If you're a journalist and find yourself in a traditional news conference organized by people who know what they're doing, always try to ask the first question. On days when there's no raging controversy with every reporter chomping at the bit to ask the same question anyway, there's usually a polite little pause right at the start of media questions when, if you have your wits about you, you can dive in and seize control of the narrative. Use it or lose it!

With COVID-19, of course, that was gone. Maybe lost *forever*. A political employee of the minister was acting as the moderator. Reporters were phoning in from remote locations with their questions. Like the host of a radio talk show, the moderators could pick who got to speak first and who didn't get to ask questions at all. Reporters were immediately limited to one question, a situation unusual in live news conferences.

Later, when the normally well-behaved members of the Alberta Legislature Press Gallery complained — doubtless politely — that was upped to one question and a follow-up. This was an excellent modification from the UCP's perspective, causing as it did no loss of control, but adding significant opportunities to run out the clock. Remember, no experienced reporter worth their salt will ever decline the opportunity in such circumstances to ask a second question. But independent reporters, bloggers and those that had built a reputation for asking the toughest questions like the CBC's Charles Rusnell, were soon having difficulty asking any questions at all.

Video-conferencing-only news conferences give the government almost total control. While the UCP government allowed anyone to sign up for its news releases — an improvement over the NDP's more traditional approach of vetting reporters' credentials first, which tended to exclude reporters from alternative media — the process thereafter was tightly controlled. The reporter would have to call in, provide a passcode emailed to them by the government, give their name, media outlet and phone number, then be assigned to a question queue. The process was not blind. Communications staff knew who had called in and was waiting. Reporters for well-known mainstream media organizations tended to get the nod. Independent journalists and those associated with small outlets — particularly those thought to be too liberal in their viewpoint or perhaps too strident in their questions — might or might not.

Duncan Kinney of Alberta's Progress Report was told in 2020 by government officials that he wouldn't be allowed to take part in a budget lockup because "your organization has been reviewed and determined to be an advocacy organization. As such, your request for media accreditation has been denied. The media embargo is for members of the media only." Four years earlier, mainstream media had participated in a huge brouhaha when a civil servant denied representatives of the far-right Rebel Media video site entry to a news briefing. Postmedia political columnist Lorne Gunter (2016) accused the NDP government of the day of "seeking to muzzle journalists." The negative news coverage in 2016 eventually caused the NDP government to cave, with Rachel Notley's communications director saying "it's clear we made a mistake" in deciding to keep the Rebel Media staffers out (Bellefontaine, 2016).

By contrast, Kinney's difficulties four years later resulted in little media backlash (story in Climenhaga, 2020). Kinney went to court alone. It worked, though. The next day, Court of Queen's Bench Justice Paul Belzil ordered the government to admit him to the lockup and even awarded Progress Alberta $2,000 in costs. The UCP government backed down and quietly paid up.

Since then, Kinney has been allowed on the line for government news conferences, but is rarely permitted to ask questions. He sat in on about 30 news conferences throughout the pandemic and was able to ask a question three times. However, representatives of right-wing publications like *Western Standard* (run by former UCP MLA Derek Fildebrandt) and *True North* seem to have no difficulty getting into the question queue.

What's more, says Kinney, there seems to be a pecking order: mainstream media, right-wing media, and everybody else, in that order. "There's an extra level of attention for the big boss," he observes. "I never got a sniff from Kenney."

Things have loosened somewhat since the government began easing COVID-19 restrictions. If a reporter shows up in person for the news conference after going through the government's fairly rigorous process to get media access to the legislature and the nearby building housing MLAs' offices — known confusingly as the Federal Building — it's harder to gate-keep them. As a result, reporters willing to persist asking a question tend eventually to get a response.

Of course, there are no statistics. Someone could file a freedom of information request, but what would be the point? This is a case when the government can honestly say there are no records, because the moderators were making their decisions about who spoke when, and who got to ask questions, by necessity *on the fly*.

During the pandemic's peak, if reporters managing to ask a tough question experienced a significant level of gaslighting and rhetoric, particularly when Kenney was at the podium, there was not much they could do about it. Without reporters in the room, the predictable tide of attacks on Prime Minister Justin Trudeau and assorted dog-whistles tuned to the shibboleths of the radical right just rolled on, unchallenged.

In the pandemic's early stage, the questions were mostly softballs, but as COVID-19 infections roared into their second year, even press gallery members — who enjoy privileges not available to other journalists and are inclined to behave themselves to hang onto them — showed signs of growing impatient with the level of manipulation.

This provides a good illustration of how COVID-era news conferences worked to the government's advantage. Again, the evidence is anecdotal, but as COVID-19 peaked, and reporters' questions grew more difficult, and in particular as previously UCP-friendly reporters started asking tougher questions, some of those reporters seemed to lose their privileged opening slots.

Postmedia's Rick Bell springs to mind. He was often the first in the lineup in the days when he and the premier seemed to have a collegial, even collaborative, relationship. That grew less so as Bell seemed to sour on the government's line. Still, prominence and a columnist's highly visible platform in mainstream media had its advantages. Bell was never relegated to those who didn't get to ask a question at all, and the premier's obvious impatience with his queries seldom took on the tone of the personal rebuke directed at Hudes.

Still, for a while, when government officials were subjected to aggressive investigative reporting by Rusnell and Jennie Russel, the prominent CBC reporters and some of their colleagues at the national broadcaster complained they were often left with their questions unasked. Rusnell and Russel, whose aggressive reporting had embarrassed Alberta governments for several years, were often not backed up by timid newsroom managers in Edmonton, and in December 2021 resigned from the CBC.

Their work has since appeared in *The Tyee*, a Vancouver-based online news organization, and occasionally with other Alberta broadcasters. But their profile has unquestionably declined, no doubt to the government's satisfaction.

It is impossible to measure how much of this change from respectful to adversarial questions, as the pandemic ground on, was tied to the decline in Kenney's personal popularity. Regardless, one thing is a certainty, and not just in Alberta: the pandemic will leave its mark on how all governments continue to manage their relationship with media.

Indeed, the pandemic provided an opportunity to institutionalize and exacerbate practices already emerging in the United States under the Trump administration: public rebukes of reporters for asking the "wrong" questions, uninterrupted gaslighting, decreased opportunities to question politicians and government officials in any context, the bypassing of mainstream media entirely through social media, and the cutting off of media access to politicians in the legislature by reducing media access to the legislature.

Kenney was not a premier like Klein, Hancock or Notley, who could all hold their own in public debate with journalists, and even enjoy it a little. And he is no Stelmach or Redford, who might not have enjoyed it, but were willing to talk with reporters in unscheduled scrums in the halls of the legislature because they felt it was an important democratic duty. Kenney doesn't like anyone who talks back and asks undiplomatic questions. The pandemic gave him something better than the earplugs he once handed out to his caucus when the NDP started talking in the legislative chamber.

That may change a bit now that Kenney is gone, but it's never going to go away completely. It's just become too convenient.

More than COVID: A matter of willful blindness

While COVID-19 provided legitimate cover for Alberta's UCP government to tighten control over media access, in hopes its narrative would become the only narrative, mainstream media also behaved in a fairly predictable way: willfully ignoring the government's rapid shift to the right and dragging the public definition of the centre along with it.

A certain amount of denial is understandable when we have no choice but to deal with unpalatable facts. Why would the human beings who work in media behave any differently? (History provides numerous examples of this behaviour.) So, a certain amount of willful blindness to the unpleasant trends unfolding before the journalists' eyes was probably inevitable — so much the better if it could be made to work to serve the short-term goals of mainstream journalism.

Anyway, as all good journalists know, it always makes for a better story if there's a certain tension between two sides of an argument, so why not go with the flow and portray the cut and thrust of public debate as a contest between two roughly equivalent arguments? When it comes to basic assumptions about society and government, there is a baked-in incentive for journalists to assume the two major political forces in any society are singing from the same hymn sheet. As we have seen dramatically in the United States in the aftermath of the Jan. 6, 2020 insurrection in Washington, this includes an unwillingness, perhaps even an inability, to perceive, or at least admit, a deeper reality about what is actually happening.

Inevitably, this leads to what Paul Krugman long ago termed, "the falsity of false equivalence." This can result in the tendency of journalists, who have been taught to seek balance in their reports, to report each side of a divisive issue as if they had equal weight. It can also take the form — which continually happens in Alberta — of downplaying the sins of a leader who lies all the time, while exaggerating the minor sins of a leader who behaves within the normal limits of civilized public discourse, as if to even things out for the sake of appearances.

Y'all know who I have in mind.

In Alberta we have a set of mainstream journalistic institutions that have consistently recaptioned the extreme, even outrageous, behaviour by the UCP and its leader as if the United Conservative Party was essentially the equivalent of the Progressive Conservatives they supplanted.

That is a kind of complicity.

Conclusion – A Government in Transition, its Media Relations Unchanged

Kenney's tendency to pick fights *with everyone*, combined with arrogance and a failure to be selective in the distribution of his angry retorts, got him in trouble with members of the UCP, giving them the means and the motivation to bring him down. Kenney's lack of respect for the UCP base was frequently raised in the days before the leadership review vote in spring 2022 that ended his premiership. He ultimately garnered 51.4 per cent support, a number too low even for a politician who had vowed that 50 per cent plus one would be good enough for him to carry on. Yet the message that should have been conveyed by that dismal result, ever-so-reluctantly announced by Kenney when giving the vote results on May 18, didn't seem to filter down to his media spokespeople as the procedure to find his replacement began to move into action.

In the first week of June 2022, fierce criticism about the state of Alberta's health care and emergency response systems erupted when an elderly woman who was mauled by three pit bulls in her Calgary neighbourhood died when the wait for an ambulance took half an hour. Health Minister Jason Copping's press secretary, Steve Buick, responded fiercely to the criticism, assailing the opposition, seeming to suggest the neighbour who called the ambulance was at fault and, naturally it would seem, striking out at the media for reporting on the tragedy and asking questions about why it took the ambulance so long to get there.

Responding to the NDP's health critic, David Shepherd, Buick's indignant tweets — "no error, no undue delay, just a 911 call that didn't initially indicate how serious the case was" — suggested nothing had changed with the folks in charge of communications for the lame-duck Kenney government (Buick-Shepherd Exchange, 2022).

The government's best response to the appalling death of Betty Ann Williams, a frail woman of 86, would have been to humbly accept the blame, to promise to get to the bottom of what happened, and to do better. Instead, the government's response soon became a powerful symbol of just how bad things had gotten. Indeed, the UCP government doubled down in its handling of the tragedy.

First, Buick told the CBC (Markus, 2022), in the words of the network's report, that his boss, Copping, "was 'relieved' to hear the AHS investigation confirmed there was no undue delay in the EMS response." Then he told the union representing most Alberta paramedics and EMTs, the Health Sciences Association of Alberta (HSAA), that Alberta Health Services "now confirms there was no error, no undue delay." To the HSAA's point, the union was "fed up with grieving the failures of this system," and its conclusion that "EMS should be there when you need it," Buick shot back: "This tweet and other comments have turned out to be grossly wrong," then adding sarcastically, "Over to you" (Buick-HSAA Exchange, 2022).

Calgary Herald political columnist Don Braid (Braid, 2022), no ill-mannered firebrand, tweeted that, "AHS and Copping say ambulance response was fine because dog attack was first treated as a police matter. They should immediately release transcripts or recordings of calls to 911 dispatch."

Buick's riposte again channelled Kenney's typical approach response to critical questions: "'Fine'? No one says it's fine, this is shockingly unfair. AHS said the initial 911 call came to EMS via police as non-life threatening. When EMS got info that it was life threatening, they were there in 9 mins. This is just too much politics, too much distortion. It's wrong."

Buick's tweets prompted real anger, and calls for a public inquiry into what went wrong.

The Kenney issues management machine appeared, however, unperturbed. Well, change will come … or it won't.

With Danielle Smith as UCP leader and premier of Alberta, the decision about how to manage relations with traditional media is now ultimately in the hands of an experienced print and television commentator.

Will this make a difference? Almost certainly.

As the United Conservative Party's leader and Alberta's top politician, Smith and her media-relations advisors are sure to want to put as much distance as possible as quickly as possible between their government and the former premier's three years of belligerent media mismanagement.

This is bound to involve a change in the tone of the UCP's issues management strategy for dealing with both social media and in traditional media. Perhaps we'll even get back to the days when journalists can ask the questions they wish, in the reasonable expectation they can get answers that are reasonably polite, even if they're not particularly informative. Still, we're likely struck with some of the advantages conferred on the governments by the response to COVID.

It's early days yet, but Smith's genial speaking style has already turned down the dial a little from the confrontational approach taken by Kenney, notwithstanding aggressive questioning by Press Gallery reporters who seem anxious not to let the new premier get away with some of the excesses of the old one.

Smith stumbled quickly, though, blurting out how she had never in her life witnessed a more persecuted group than Alberta's COVID vaccine refuseniks. Unsurprisingly, the response from the public and media to this was harsh. If it happens again, or keeps happening, the Smith government's relationship with media may sour.

At the same time, even with the recent return to in-person news conferences and the more congenial tone, the UCP government's two-question rule continues to be firmly applied. This is to be expected, and will continue as long as journalists are prepared to put up with it.

On the social media side, the most antagonistic attacks by Kenney's issues managers seem to have largely ceased. At the same time, Smith's social media staff have been aggressively blocking Twitter followers who criticize the new premier. They have been more careful about employing this response to critical comments by journalists, including commentators not affiliated with mainstream media, and other public figures.

Anecdotal accounts on social media by critics of the government suggest increased use of attack bots against critics of some of Smith's statements, but there is no evidence these originate with the government and there plenty of other suspects for such activities. Smith's most enthusiastic supporters, naturally, can be expected to be passionate in defence of their leader.

So, for now, it appears the uncivil relations pioneered by Kenney will be moderated somewhat by Smith and her government. How much remains to be seen. So far, at least, she has not rejected the premise of a single question!

References

Bellefontaine, M. (2016, Feb. 17). *Rachel Notley's NDP lifts ban on The Rebel, says it made a mistake.* CBC News. https://www.cbc.ca/news/canada/edmonton/notley-the-rebel-1.3451838

Braid. D. (2022, June 7). *Tweet.* https://twitter.com/DonBraid/status/1534368887898140672

Buick-HSAA Exchange. (2022, June 7). *Tweet.* https://twitter.com/SteveBuick2/status/1534300383098507264

Buick-Shepherd Exchange. (2022, June 7). *Tweet.* https://twitter.com/SteveBuick2/status/1534298640574951424

Climenhaga, D. (2020, June 11). It's not up to the premier's staff to decide who's a journalist — except when it is. *Alberta Politics Blog.* https://albertapolitics.ca/2020/06/its-not-up-to-the-premiers-staff-to-decide-whos-a-journalist-except-when-it-is/

Climenhaga, D. (2021a, March 2). Casino capitalism comes to newspapers and it sure looks like the dealin's done! *Alberta Politics Blog.* https://albertapolitics.ca/2020/06/its-not-up-to-the-premiers-staff-to-decide-whos-a-journalist-except-when-it-is/

Climenhaga, D. (2021b, May 21). State of the Media: Postmedia, looking for a future, post media as it were, wants to deliver your parcels! *Alberta Politics Blog.* https://albertapolitics.ca/2021/05/state-of-the-media-postmedia-looking-for-a-future-post-media-as-it-were-wants-to-deliver-your-parcels/

Fisher, A. (2019, June 3). *"Silly" interview highlights dangers of rigidly sticking to same response.* mediafirst. https://www.mediafirst.co.uk/blog/silly-interview-highlights-dangers-of-rigidly-sticking-to-same-response/

Gunter, L. (2016, Feb. 15). NDP seeking to muzzle opposing journalists. *Edmonton Sun.* https://edmontonsun.com/2016/02/15/ndp-seeking-to-muzzle-opposing-journalists

Kenney-Hudes Exchange. (2020, Dec. 8). CTV News. https://twitter.com/ctvnews/status/1336484648789139457?lang=en

Krugman, P. (2021, June 30). The Republican right goes all in on ignorance. *The Mercury News.* https://www.mercurynews.com/2021/06/30/krugman-the-right-goes-all-in-on-ignorance/

Markus, J. (2022, June 8). *"Are you sending an ambulance?" Neighbour says multiple calls made to 911 following dog attack.* CBC News. https://www.cbc.ca/news/canada/calgary/ahs-city-ems-dog-attack-1.6482275

Postmedia. (2021). Introducing postmedia parcel services. Media release. May. https://www.postmedia.com/2021/05/20/introducing—postmedia—parcel—services/

Shandro Press Conference. (2019, May 29). CBC News. https://twitter.com/CBCEdmonton/status/1133959158049300480

ECONOMY AND ENVIRONMENT

CHAPTER 7

EXTRACTION FIRST: THE ANTI-ENVIRONMENTAL POLICIES OF THE UCP GOVERNMENT

Laurie Adkin

"We have no other spare or replacement planet.

We have only this one, and we have to take action."

— Berta Cáceres, Lenca Indigenous activist, co-founder of the Council of Popular and Indigenous Organizations of Honduras (1971-2016)

IF THERE IS A PART of the Alberta economy the United Conservative Party (UCP) government helped during its time in office, it must be the sign printing business. Lawn signs mushroomed around the province declaring support for threatened public goods and services. Among these were signs defending provincial parks, watersheds, and the mountains from further enclosure by private interests and extractive development. The campaign hashtags #DefendABParks, #SaveOur Mountains, #WaterNotCoal, and #MountainsNotMines proliferated on social media alongside their longstanding and more universal predecessors, #ClimateEmergency and #JustTransition.

In this chapter, I examine key environmental conflicts that reveal the general orientations of the UCP government toward the profound ecological crises confronting us at local, national, and global levels. They include the battle over the UCP's attempt to privatize a portion of the provincial parks system, opening up some areas to commercial development; the UCP's rescission of the 1976 Coal Development Policy that had prohibited open-pit coal mining in the critical watershed areas of the eastern Rocky Mountain slopes; and the UCP's responses to pressures to regulate greenhouse gas emissions and other environmental harms created by oil and gas extraction.

The UCP's policy initiatives and approaches are shaped primarily by the ideological commitments of the party's leaders that are aligned with the interests of the civil society actors that in turn fund and mobilize support for the party. Among these actors — as we know from decades of social science research — are corporations in the extractive and agrobusiness sectors and the businesses that service or depend on these dominant sectors for their contracts and revenue. However, what the UCP government *has been able to achieve* with regard to environmental deregulation and privatisation is another story — one that, to be fully understood, requires a broader analysis of responses from a wide array of civil society actors. Thus, in this chapter, I describe not only what the UCP government tried to do, but also how its initiatives were in various ways obstructed by actors such as citizens' groups, environmental non-governmental organizations (ENGOs), and academics, using the resources available to them (e.g. the courts, public hearings, or campaigns to inform and mobilize public opinion). The UCP government was obliged, on occasion, to take public opinion into account, adjusting or even reversing some of its policies and its political discourse to avoid alienating large sections of the electorate.

Alberta's conservative governments have, since the mid-1980s, taken a neoliberal approach to regulation (Adkin, 2016), but the primary driver of environmental and energy policy has been the fusion of the state with petro-capitalist interests. This relationship can be traced to the Social Credit era, but took its current, unmediated form in the 1990s, as described by Kevin Taft in his chapter in this book (see Taft, Chapter 8). Thus, neoliberal tenets like "governments should not interfere in the economy by propping up failing industries or by public ownership" are tossed aside when what is at stake is ensuring the profitability of the corporations in the fossil fuel sector. After a brief New Democrat Party (NDP) interregnum, the UCP government resuscitated the approach to environmental regulation that characterized the Progressive Conservative Party (PCP) governments of Ralph Klein.

Neoliberalism and petro-politics do not fully capture the nature of environmental policy in Alberta, however, for this is also profoundly driven by extractive colonialism. In this regard, the "Alberta First" slogan of one of the contenders for the leadership of the UCP in 2022, Danielle Smith — like the decades of nativist discourse of previous Alberta politicians — performed two tricks. On the one hand, it mashed the complex, and sometimes conflicting interests of Albertans into one, simplified populist identity. For Smith, as for most Alberta politicians, this identity is characterized by dependence on fossil fuel extraction for a good quality of life, with no other kind of economy being imaginable. On the other hand, "Alberta First" renders invisible the colonial foundations of the province and its fossil fuel extraction-based economy. With every public interest (including ecological sustainability), as well as any move toward decolonization being subordinated to the interests of extractive capitalism, a more honest Conservative slogan would be "Extraction First!"

More challenging to explain than UCP policies is the hold fossil capital had over the centrist NDP during its time in office from May 2015-April 2019. There are significant differences between the NDP and UCP governments' environmental policies, but their political discourses concerning Alberta's dependence on oil and gas exports, their support for the oil and gas corporations, and their visions of economic diversification have been remarkably similar. While this chapter focuses on the environmental policies of the UCP, it speaks to the broader patterns of environmental regulation in a jurisdiction long characterized by extractive colonialism and petro-state politics.

#DefendABParks

> "Our landscapes and watersheds have been neglected; we expect too much of them and they are coming apart at the seams. It can't be a free-for-all anymore."
> —*excerpt from a letter from 37 biologists to Environment and Parks Minister Shannon Phillips, January 2019, supporting the NDP's plans to establish eight new parks covering 4,000 square kilometres of land in the central foothills (quoted in Weber, 2019a).*

In 2017-18 the NDP government established the Castle Provincial Park and Castle Wildland Provincial Park and announced a $20-million plan to improve recreation opportunities in these parks. In addition, over 100 parks received enhancements such as shelters, washrooms, trails, parking lots, or upgraded roads (Weber, 2019a). The NDP government also announced in November 2018 plans for a wildland provincial park, three provincial parks, four public recreation areas and two public land-use zones within a new multi-use region called Bighorn Country. This plan, which had the support of conservation associations, called for $40 million in infrastructure improvements. The wildland park designation was opposed, however, by some commercial interests as well as users of motorized recreation vehicles that would be prohibited from a wildland park (CBC News, 2018). Prior to the April 2019 provincial election, the UCP MLA for Rocky Mountain House, Jason Nixon, had called the Bighorn Wilderness Park proposal a "foreign-funded plot to wall off the back country to Albertans" (Nixon, quoted in Lewis, 2019). When he became the UCP's minister of environment and parks, Nixon shelved the Bighorn Country plan and returned decisions about the management of the area to the North Saskatchewan Regional Planning process.

Back in the 1990s, Klein's PCP government privatized provincial campgrounds as part of its debt and deficit slashing "revolution." Retired fish and wildlife biologist Lorne Fitch remembers "the net impact of that short-sighted decision was the erosion of services, widespread user dissatisfaction, declines in use, failure to maintain parks infrastructure and, a huge public rebuilding cost to bring facilities back to acceptable standards" (Fitch, 2020a). The UCP government of former premier Jason Kenney, emulating the Klein governments of the 1990s in so many ways, also turned to parks as a dispensable public expense. In conjunction with a 25 per cent reduction in parks funding over two years,[1] the UCP announced in February 2020 a plan not only to contract more parks maintenance to third parties, but to de-designate many parks altogether.

The Optimizing Alberta Parks plan announced Feb. 29, 2020[2] declared the government's intention to fully or partially close 20 provincial parks and seek third-party 'partners' to run another 164 of them. According to the Canadian Parks and Wilderness Society (CPAWS), these parks constituted 39 per cent of the provincial park system (CPAWS 2020a). The areas for which partners couldn't be found would lose park status and revert to general Crown land. This opened the possibilities that delisted parks (converted to public lands) would be sold to private owners or would be opened to activities such as coal mining, forestry, oil and gas extraction, or commercial facilities such as golf courses. That such motives might be the real drivers of the policy appeared more likely in light of the ministry's estimate that delisting would save the government only $5 million per year (Bell, 2020).

There was an immediate outcry from numerous civil society organizations involved in recreational uses of the parks, as well as from conservation organizations like CPAWS and the Alberta Wilderness Association (AWA). Within days, the ministry had removed the word "sale" from the website describing the Optimizing Alberta Parks plan and had begun to deny the sale of delisted land had ever been on the agenda (CPAWS, n.d.a).

A Leger poll commissioned by CPAWS in March 2020 found that 69 per cent of a representative sample of Albertans opposed or strongly opposed the removal of parks from the parks system; only six per cent "strongly supported" such an action (Leger, 2020). Of greater concern to the UCP government may have been Leger's finding that 68 per cent of *rural* Albertans opposed the delisting of parks.

In keeping with a pattern that was to become clear in the context of the coal policy debacle, discussed in the next section, the UCP cabinet chose not to consult the public about its plan to "right size" the parks system, although ministry staff forewarned that citizens might have concerns about "loss of/changes to sites and perceived loss of conservation focus," and that political support for the plan at the local level was "uncertain" (Alberta Environment and Parks [AEP], 2019, slide 22). Further, CPAWS discovered the Alberta Parks Consultation Framework had been altered just before the announcement of the Optimizing Alberta Parks plan by the removal of the requirement for public consultation prior to significant changes to the legal classification of a park, its uses, or other matters (CPAWS, n.d.b; Heelan Powell, 2020). The UCP government's failure to consult on changes to the parks system was particularly notable in light of the environment minister's previous accusations that his NDP predecessor had not adequately consulted Albertans about the creation of the Castle Provincial Park and the Bighorn Wildland Provincial Park (*Canadian Press*, 2019).

Documents obtained by CPAWS through a Freedom of Information and Privacy (FOIP) application revealed the government had indeed intended to sell up to 30 per cent of the delisted parks, and it had not made its preliminary selection based on data about parks usage or the financial costs and benefits of delisting the parks (CPAWS, 2020b). Thus, there was no evidence for its justification that the parks were "under-utilized." On the contrary, a slide presentation prepared for the minister described the purposes of the Right Sizing Alberta's Parks Project as: "financial sustainability, engag[ing] in partnerships, enabl[ing] economic opportunities, reduc[ing] stakeholder irritants" (AEP, 2019, slide 19). These goals were further facilitated by changes the UCP government made to the Public Lands Administration Regulation in November 2019. CPAWS' analysis of these changes concluded they would "make it easier for private partners to apply for and receive dispositions for commercial developments on public lands including hotels, lodges, shops, mini-golf, motocross tracks among others" (CPAWS, n.d.b). Such moves were consistent with the government's frequently declared intention to "open Alberta for business" by "cutting red tape" (regulations) inhibiting private investment, among other measures.

CPAWS, the AWA, and the Alberta Environmental Network (AEN) joined forces to inform the public and media of these threats to provincial parks. As the COVID-19 pandemic had reached Alberta by March 2020, and restrictions on public gatherings were taking effect, events took the form of online town halls and a virtual protest held in April 2020. Citizens were urged to contact their MLAs and the minister of environment and parks to demand a reversal of the "optimization" plan, with the ENGOs providing online websites to facilitate such messages. They reported that more than 20,000 letters were sent to Alberta politicians (Weber, 2020). As the opposition grew stronger, the government backtracked further, asserting there was nothing new about ministry partnerships with civil society groups and private contractors who play a role in maintaining the parks and providing other services. On May 4, the ministry removed from its website the list of sites to be delisted or closed (CPAWS, n.d.a).

However, on their reading of the initial government announcements and the documents obtained through the FOIP application, the conservation organizations were convinced the government's intentions were to delist parks, sell park facilities, or grant long-term leases for private management of the parks, thereby either removing them from the parks system or turning them over to private operators. In July, the NDP launched a campaign called Don't Go Breaking My Parks, with an online petition directed toward the UCP government. In August, CPAWS and the AEN launched the Defend Alberta Parks campaign, and signs with this statement began to appear on lawns across the province. A petition circulated by leadnow.ca obtained 57,736 signatures.[3]

In mid-September, the government announced it would spend $43 million to upgrade provincial parks and day-use facilities on public lands. The investment was framed as "an important part of Alberta's [Post-COVID-19] Recovery Plan" (Government of Alberta, 2020a). However, the minister continued to talk about delisting parks considered "remote" or underused (Smith, 2020).

In November 2020, the government announced Alberta's Crown Land Vision, a plan billed as "a common sense approach to Crown land management that finds the right balance between conservation, recreation and economic use" (Government of Alberta, 2020b, n.d.). Read with knowledge of the government's overall orientation, including its tendency to privilege resource development and private interests over ecological concerns and public goods, conservationists were concerned about where this "vision" might lead (Fitch, 2020b).

This time, the ministry set up a seven-week online consultation process to obtain public input on "how we can enhance trail experiences for a variety of users; supporting [sic] partnerships and funding opportunities; and how dollars can be reinvested into recreation as well as education and enforcement" (Government of Alberta, 2020c). In addition, it conducted "targeted discussions with key stakeholders and consultation with Indigenous people" (Government of Alberta, 2021, p. 1).[4] According to the ministry, 8,194 Albertans participated in the online survey. The results suggested that citizens with conservation concerns had taken the opportunity to push back against any ideas of delisting parks or increasing their use by off-highway vehicles (Riley, 2019b). A small majority of respondents agreed the government could charge user fees "based on the type of activity and intensity of use while ensuring considerations to [sic] individual's ability to pay," and to help finance the maintenance and policing of the parks (not for general revenue). Overall, respondents wanted public lands to be "held in trust" and "funded by public money."

The legislative outcomes of the Crown Land Vision, so far, include the Public Lands Amendment Act, which introduced fees for recreation on public lands, and the Trails Act, giving the environment minister wide discretion to designate trails and who has access to them. These fees will serve the UCP's goal of shifting funding for the parks to users, in keeping with its general approach to public goods. Conservationists criticized the Trails Act, fearing it would facilitate ecologically harmful uses of already compromised natural areas (*Canadian Press*, 2021).

Faced with the strong public reaction against proposals to lease out or de-list provincial parks, the UCP government announced in December 2020 that no parks would be de-listed and that "partners" for managing 170 parks had been found. It released a partial list of partners, but did not specify which were long-standing as opposed to new (Weber, 2021). On its website, Alberta Parks advertises opportunities for private interests to operate campgrounds, concessions, and tourism businesses, or deliver services in the parks.

#WaterNotCoal #MountainsNotMines

> "It seems [the Conservative government] would sell any part of our natural heritage if the price were right. Forests, rivers, mountains, and valleys: to them, none are too important to industrialize."
> — *Alberta Liberal Party election platform 1997*[5]

This statement taken from the Alberta Liberal Party election platform of 1997, speaking to the environmental policies of the Klein government, could easily have been written in 2020-21 during the battle against the opening of the eastern slopes of the Rocky Mountains to open-pit coal mining. However, even the Klein governments had not dared to overturn The Coal Development Policy adopted by the Lougheed conservatives in 1976 to protect certain areas of the eastern slopes from industrial development. On June 1, 2020, the Kenney government quietly rescinded the policy. This move came without prior public or Indigenous consultation and in response to requests from coal mining corporations for access to these formerly off-limits areas (Kalinowski, 2021; Nikiforuk, 2020a). When this action became known to the public, the ministers of energy and environment insisted the Coal Development Policy was "outdated," and "obsolete" because environmental policies had been "modernized" since the 1970s.

The 1976 Coal Development Policy had created four categories of land in relation to mining access. Category 1, where all coal development was forbidden, encompassed mostly the Rocky Mountains in an area stretching 700 kilometres from the U.S. border north to Kakwa Wildland Provincial Park. This prohibition recognized the need to protect watersheds and biodiversity, as well as recreation and tourism uses of the eastern slopes (Bankes, 2021). Category 2, where open-pit mines were off-limits, covered lands mostly to the east of Category 1 lands, but including mountains and foothills. A CBC journalist who followed the coal mining conflict described Category 2 as having three major "chunks" in the south, central, and northern Rockies, adding up to "a landmass the size of Jamaica" (Fletcher, 2021). Category 3 and 4 lands, where some mining would be permitted subject to environmental assessments, were mostly farther east, toward the plains (Fletcher, Anderson, and Omstead, 2020).

Coal mining companies — mostly Australian-based — have been keen to extract metallurgical coal for export from Category 2 lands, along with a section of Category 4 land located in the mountains. Companies holding leases in these areas include Benga Mining Ltd. (a Canadian subsidiary of Riversdale Resources[6]), Atrum Coal, Ram River Coal, Cabin Ridge, and Montem Resources. Four projects had received permits for exploratory activities prior to the UCP's rescinding of the Coal Development Policy; two others received exploratory approval afterward. As of May 2020, there were coal leases on Category 2 lands covering roughly 420,000 hectares, and hundreds of applications for leases.[7]

In August 2020 it was revealed that ministers in the UCP government had sent letters of support in October and December 2019 to at least one mining company, Valory Resources Inc., that was seeking to build a 16,000-hectare open-pit coal mine in a Category 2 zone near the Bighorn Wilderness area (Nikiforuk, 2020b). In his letter to Valory Resources, Minister of Environment Jason Nixon wrote that the government was seeking to attract investment through "efforts to streamline policy and processes to bring greater certainty and stability to our investment climate and make Alberta open for business." Atrum Coal reported that it had regular engagement with the Alberta government — as did the Coal Association of Canada — prior to the government's decision to rescind the Coal Development Policy (Nikiforuk, 2020b).

The first mine project to come up for environmental and social impact assessment following the revocation of the Coal Development Policy was Benga's proposed open-pit mine on Grassy Mountain. Although this site is classified in Category 4, it has important ecological values and became the fulcrum of a massive mobilisation against coal mining on the eastern slopes. A great deal was at stake in the decision about the Grassy Mountain Mine, because lined up behind it were the Montem Resources Mine at Tent Mountain and other coal mining projects.

Benga Mining had been conducting exploratory testing on the site and going through provincial approval processes for mining permits since 2013. The project included:

> the construction, operation, and reclamation of an open-pit metallurgical coal mine near the Crowsnest Pass . . . The production capacity of the Project would be a maximum of 4.5 million tonnes of metallurgical coal per year, over a mine-life of approximately 24 years. It would include surface coal mine pits and waste rock disposal areas, a coal handling and processing plant with associated

> infrastructure, water management structures, an overland conveyor system, a rail load-out facility, and other facilities. The surface mine area of the Project is approximately 2800 hectares (Government of Canada, 2020).

Impacts on fisheries and species at risk, as well as greenhouse gas emissions, are all matters governed by federal legislation. In 2018, the federal minister of the environment decided the Grassy Mountain project should be subject to review under the Canadian Environmental Assessment Act, and a joint provincial-federal review panel (JRP) was appointed to undertake the assessment and make recommendations (Impact Assessment Agency of Canada, 2018). The agreement to establish the JRP set out the composition and terms of reference for the joint review panel and a consultation process that would include a public hearing and the creation of a registry for the filing of comments from members of the public. These comments were to be reviewed and summarized as part of the final report. The JPR was also to "take into account any community knowledge and Aboriginal Traditional Knowledge" received during the assessment period (Impact Assessment Agency of Canada, 2018, pp. A1-A3).

The public hearing component of the process began on Oct. 27, 2020 and continued into December. Intervenors approved by the JRP included seven First Nations, the Metis Nation of Alberta (Region 3), the Canadian Parks and Wilderness Society (Southern Alberta Chapter), the Coalition of the Alberta Wilderness Association and the Grassy Mountain Group, the Livingstone Landowners Group, and the Municipality of Crowsnest Pass.

As we saw in the case of the proposal to delist provincial parks, civil society organizations went into action to inform the public about the government's rescission of the Coal Development Policy and the implications of the Grassy Mountain project and other proposed open-pit coal mine projects.

Scientists with expertise in areas such as the ecology of the Oldman River watershed, toxicology, and hydrology contributed briefs to the JRP, bringing to the fore the risks of coal mining for contamination of downstream waterways with selenium, extirpation of endangered fish species, and exposures to coal dust, among other harms to environmental and human health.[8] There were concerns about water allocations to the mine that would further stress fish habitat as well as downstream water supply for agriculture and ranching (Bankes and Bradley, 2020). Local landowners and tourism sector businesses were concerned about possible negative impacts of coal mining on their livelihoods. Such consequences had been well-documented from other coal mining sites in the Rocky Mountains and the United States. In addition, socioeconomic arguments in favour of the Grassy Mountain Mine (local employment, coal royalty revenue to the provincial government) were refuted by other analyses that incorporated considerations such as loss of livelihoods in agriculture, tourism, and recreation, the predicted environmental costs, the unpredictability of global demand, and lack of support for the claims Benga was making about future royalty payments (see, e.g., Urquhart, 2021). Not only was the credibility of the predicted economic benefits of the mine (400 jobs and $690 million in royalties over 23 years) called into question by testimony given to the JRP, but the necessity of this environmentally high-risk path of economic development was also challenged.

While Benga Mining had secured a benefits agreement with the band council of the Piikani First Nation and claimed it had met the requirements for Indigenous consultation set out by Alberta's Aboriginal Consultation Office, other members of the Blackfoot Confederacy protested that they had not been consulted by the authorities who made the agreement. An Indigenous group called Niitsitapi Water Protectors organized to stop the mine, creating social media platforms, a petition, and a letter-writing campaign to the federal minister of environment. The Alberta regional chief for the Assembly of First Nations stated that the Indigenous parties to treaties 6, 7, and 8 had not been consulted prior to the revocation of the Coal Development Policy. The Bearspaw, Kainai, Siksika, Ermineskin and Goodfish Lake First Nations requested a judicial review of the decision to rescind the 1976 Coal Policy.

Many people from the region worked hard to spread the word about the Joint Review Panel process and inform Alberta citizens and Indigenous communities about the ecological and social harms open-pit coal mines on the eastern slopes would cause. These watersheds fuel the river systems that provide drinking water to cities and towns throughout Alberta and Saskatchewan. Drawing on information provided by the conservation organizations, a broad citizens' coalition formed.

As with the provincial parks issue, mobilisation was constrained by the COVID-19 pandemic restrictions on public gatherings. However, in mid-December 2020 a handful of local organizers formed a Facebook group called Protect Alberta's Rockies and Headwaters. Within a month, this group constituted a "loose coalition" of more than 28,000 members, and as of April 2022 it had nearly 37,000 members.[9] A petition initiated by an area resident, calling on the federal minister of environment and parks to reject the Grassy Mountain Mine proposal, received no support from Conservative Party MPs but was sponsored by a Green Party MP, Elizabeth May; within three months, it obtained 27,720 signatures.[10] A Change.org petition calling on the provincial government to halt plans to open the eastern slopes to new coal mines was also initiated in December 2020, and gained 91,428 signatures by Feb. 9, 2021.[11] Meanwhile, between October 2020 and mid-January 2021, the Impact Assessment Agency of Canada's registry for public comments on this project received nearly 5,000 comments. A close analysis of the first 853 individual comments found that only 18 (two per cent) supported the mine project; these came from residents of Crowsnest Pass who felt the mine would bring needed employment and support local businesses.[12]

Both the scale and the breadth of this mobilization were unprecedented in Alberta. Country music artists from the southwest of the province participated in fundraiser events to support the legal costs of court actions against the government brought by local landowners and to cover other organizing costs. The opening of the eastern slopes to coal mines was debated in municipal councils and was raised as an issue of concern in the cities whose drinking water originates in these watersheds. Relationships were forged between Indigenous land and water protectors and settler landowners who discovered a shared desire to protect the foothills and mountains from further ecological degradation. Residents of a corner of the province that had traditionally been an electoral stronghold for the Conservative and Wildrose parties began to see environmental organizations and university academics as allies, and to question the motives of the UCP government. Some may have taken note of the refusal of the region's Conservative Party MPs to sponsor citizens' petitions to the federal government, and of the support they received, in contrast, from Green Party members of parliament.

A survey of 1,140 Albertans taken in February 2021 by ThinkHQ Public Affairs Inc. found that more than three-quarters of respondents were aware of the conflict over coal mining on the eastern slopes, and that 69 per cent were opposed to such development. The polling company reported that "a majority don't necessarily trust the current provincial government to find a reasonable balance between the economy and environment," and that "even a majority of UCP voters (56 per cent) say they disapprove of expanded mining in formerly protected areas of the Rockies . . . Almost 40 per cent of UCP voters didn't trust the government they supported in 2019 to do the right thing." The NDP opposition joined in, as well, starting its own petition in January 2021, and introducing a Private Member's Bill to ban new coal mining in the Rocky Mountains.[13]

If the UCP cabinet believed, in May 2020, that facilitating open-pit coal mining on formerly protected areas of the eastern slopes would meet with no opposition beyond that of the "usual suspects," the ENGOs and Indigenous environmentalists, they seriously miscalculated the attachment of their own electoral base to the lifestyles and landscapes of the southwest. The "economic development" argument advanced by the UCP was not persuasive in the coal context, in which the benefits of the mines would accrue predominantly to foreign-owned mining companies while the predictable costs would accrue to local residents and to the downstream populations reliant on the mountains for their drinking water.

Faced with a tsunami of opposition to the Grassy Mountain coal mine proposal and the revocation of the Coal Development Policy, the UCP government again backtracked in February 2021. Minister of Energy Sonya Savage held a press conference in which she admitted the government "didn't get this one right," and promised that no "mountain-top removal" would be permitted (Savage, 2021). The government would "pause" future coal development and coal lease sales in Category 2 lands pending public consultation on "a new, modern coal policy." However, Savage said the projects already in the exploratory phase would be allowed to continue, along with their significant damage to habitats and watersheds caused by road building, tree clearing, drilling, and blasting.

Professors of environmental and natural resource law at the University of Calgary observed that the vehicle for consultation already existed in the form of the land-use planning framework (Yewchuk, 2022). The UCP government chose instead to appoint a committee of five members to lead a public consultation on "strategic goals and desired objectives" for a "modern coal policy" whose mandate was restricted to matters within the jurisdiction of the energy minister (Minister of Energy, 2021). The committee began work in April and was asked to provide its report by the end of December 2021.

Before the Coal Policy Committee reported, there were other significant developments at the federal level. In response to a petition initiated by the Niitsitapi Water Protectors and sponsored by the federal NDP MP for Edmonton-Strathcona, Heather McPherson,[14] federal Minister of Environment and Climate Change Jonathan Wilkinson announced on June 16, 2021 that any coal project that could potentially release selenium into water bodies would henceforth come under federal environmental assessment. The following day, the JRP for the Grassy Mountain Coal Project issued its report and recommendations. The commissioners concluded that the project was likely to result in "significant adverse environmental effects on surface water quality, westslope cutthroat trout [an endangered species] and its habitat, whitebark pine, rough fescue grasslands, and vegetation species and community biodiversity [as well as] significant adverse effects on [the] physical and cultural heritage of some First Nations" (IAAC, 2021). The adverse impacts on surface water quality and westslope cutthroat trout alone outweighed any positive economic benefits of the project. Hence it was not in the public interest. Benga Mining Ltd.'s application was denied.

While tens of thousands of Albertans celebrated this decision, they also recognized that other mining proposals remained active. In keeping with his earlier commitment, however, the federal minister did designate the Tent Mountain Coal Mine Project for federal impact assessment later that month. Benga Mining sought judicial review of the JRP decision, but its application was rejected by the Alberta Court of Appeal. The company, along with the Piikani and Stoney Nakoda First Nations, then applied to the Supreme Court of Canada for leave to appeal the JRP's decision. This application, too, was denied.

In March 2022, Minister Savage finally released the reports of the Coal Policy Committee she had received at the end of December 2021. The summary of the engagement process revealed the mobilization of Albertans in defence of the watersheds and mountains had been sustained throughout the provincial consultation process (Coal Policy Committee, 2021a, 2021b). As for the recommendations, in addition to a list of more specific concerns highlighted in the public submissions, the committee reported that there exists a broad consensus that:

> [C]oal exploration and development should only be allowed on lands that conform to regional or subregional plans completed under the *Alberta Land Stewardship Act*. Such land use certainty should replace the existing coal categories for the purposes of land use decisions about where coal exploration and surface or underground development can and cannot occur. Regional plans and subregional plans, and associated implementation strategies should supersede the coal categories and be legally binding. In sum, a new coal policy for Alberta needs to include modernized land use guidance that is aligned with comprehensive, enforceable land use planning for the entire Eastern Slopes (Coal Policy Committee, 2021b, p. 6).

This citizen consensus mirrored the arguments advanced by ecologists and environmental law experts who had long held that decisions about industrial development should be taken in the context of holistic analyses of the ecological impacts of multiple land uses on a region-by-region basis. Whether or not any form of coal mining should take place in southern, central, or northern stretches of the eastern slopes of the Rockies should be decided within the framework of regional land-use planning and its consultation processes. "Consequently, regional and subregional plans for the Eastern Slopes must first be completed before any major coal project approvals are considered" (Coal Policy Committee, 2021b, p. 7).

The energy minister responded by extending the ministerial order (MO 002/2022) suspending applications for coal exploration or development throughout the eastern slopes, except on lands subject to "an advanced coal project or an active approval for a coal mine." In his analysis of this document, environmental law expert Drew Yewchuk (2022) observes that the bar for "advanced" is set very low, permitting projects that have not yet undergone any environmental assessment to proceed through the review process. Yewchuk also observes that the wording of the MO does not specify that it will remain in place until a new plan under the Alberta Land Stewardship Act (ALSA) exists for each area covered by the Coal Development Policy (as recommended by the Coal Policy Committee). The public consultation process for the ALSA plans had not yet been initiated as of April 2022. One of the obstacles to developing the land-use plans may be the UCP's cutting of the Land Use Secretariat's budget from $6.1 million in 2018-19 to $1.5 million in 2020-21.

The Protect Alberta Water and Rocky Mountains coalition has expressed little trust in the intentions or commitments of the UCP government, warning on its website that "Albertans need to remain vigilant and actively engaged in the land use planning and development frameworks, as it is clear that open pit coal mining may indeed be allowed under regional plans. The devastating effects of open pit coal mining will remain a threat until the Eastern Slopes are protected within these Land Use plans developed for each region."[15] The voting down of the NDP's private member's bill aiming to ban coal mining on the eastern slopes by members of the UCP caucus (including the minister of environment and parks) on March 22, 2022, did little to reassure coalition members that a UCP government will stop supporting the interests of the coal mining companies.

Despite the Benga Mine decision, it appears the battles over coal mining on the eastern slopes are not over. Atrum Coal, Cabin Ridge Coal, and Elan Coal Ltd. have all filed suits against the government of Alberta demanding billions of dollars in compensation for the losses they claim to have incurred as a result of the UCP's decision in March 2022 to declare an indefinite moratorium on coal exploration in Rockies (with the exception of the "advanced" projects that were already undergoing environmental review). As for the province's new premier, Smith, when asked in October 2022 whether she would continue the moratorium on coal mining, she said only that she would review "all the available options" (Smith quoted in Junker, 2022a).

#ClimateEmergency

> "Alberta will not accept production cuts in the insane climate plan released by the Liberal-NDP coalition."
> — *Jason Nixon, Alberta minister of environment and parks, April 2, 2022.*

While the UCP's drive to open more of Alberta's protected areas to industrial development pushed ENGOs and many citizens into permanent mobilization mode on multiple fronts, trench warfare was similarly being waged between oil and gas industry interests and the Indigenous environmental and climate justice movements. The UCP leadership and its allies in business organizations and media made the NDP government's Climate Leadership Plan the primary target of its election campaigning in 2018-19. The UCP promised to cancel the Alberta carbon tax on combustible transportation and heating fuels, fight federal carbon pricing measures, secure more pipeline capacity for oil and gas exports, cut corporate taxes, "streamline" environmental regulations, "cut red tape," and do anything necessary to attract more oil and gas investment to the province.

From the Klein era onward, the key components of Alberta's climate change plans have been emissions intensity (EI) reduction targets for large emitters, transfer of the revenue from the carbon levy on these emitters to a technology fund that would subsidize the emitters' R&D to decrease the carbon footprint of their production, and other streams of public investment in fossil-fuel-sector-related technologies such as carbon capture and storage (Adkin, 2019). In other words, the key pillar of Alberta's climate policy has always been a gamble on technologies to capture or reduce GHG emissions from the oil and gas sector (Adkin, 2017). Alberta politicians have shown little interest in other approaches advocated by environmentalists, such as progressively lowered sectoral caps on (absolute) emissions and public investment (direct or indirect) in the renewable energy sector, energy efficiency programs, and public transit.

Alberta's so-called climate change policy was in fact designed to permit the province's GHG emissions to grow indefinitely. And grow they did — from 224 Mt in 2000 to 284 Mt in 2015. Since then, they have dipped and risen from 268 Mt (2016) to 279 Mt (2019). In 2020, Alberta's GHG emissions fell by eight per cent, a decrease attributed mainly to the impacts of the COVID-19 pandemic on the transportation sector (ECCC 2022b). Emissions from oilsands mining and upgrading, which grew by 137 per cent between 2005 and 2019, declined by only 2.4 per cent in 2020 (ECCC 2022c). In Figure 7.1 (see next page) we see the trends in GHG emissions for the oilsands, the Canadian oil and gas sector as a whole, Alberta, and Canada.

The NDP government retained its predecessors' overall approach to the regulation of large emitters, but brought in some new initiatives in other areas. Ahead of a federal plan to do the same, the NDP created a "carbon levy" (tax) on combustible fuels (exempting electricity and fuels used in farm operations). Revenue from this levy was "rebated" to individuals or households based on their income, with the result that approximately 60 per cent of households received rebates (Maclean 2019). The NDP also established the Energy Efficiency Agency (EEA), a crown corporation funded by revenue from the carbon levy and mandated to help Albertans make their households and businesses more energy efficient. The NDP's Renewable Electricity Program (REP) aimed to increase the share of renewable energies in the Alberta market to 30 per cent by 2030 (Government of Alberta, 2022). This program had succeeded, by December 2018, in contracting for 1,360 MW at prices per kilowatt-hour that were attracting international attention for reaching new lows. The second auction for power grid share was open to bidders with a minimum of 25 per cent Indigenous equity ownership, thereby incentivizing renewable energy partnerships including First Nations. In addition, the NDP accelerated the phasing out of coal-fired electricity generation, a major source of provincial GHG emissions. The coal phase-out plan included $1.3 billion in monetary compensation to three corporations (financed from the carbon levy on large emitters, discussed below), a $40-million transition fund to support coal workers, and a $5-million Community Transition Fund (Jackson and Hussey, 2019). Overall, GHG emissions from electricity generation fell from 50 Mt in 2015 to 33 Mt in 2018 (McCarthy, 2021). Lastly, the NDP sought to reduce methane emissions from the oil and gas sector, in co-operation with the federal government's regulations in this regard, by 45 per cent by 2025.

One of the UCP government's first actions was to cancel the NDP's carbon levy on combustible fuels. (This led, predictably, to the imposition of the federal fuel charge in its place.) Once again, an environment minister joined the premier in campaigning against a federal climate policy. UCP Minister of Environment and Parks Jason Nixon called the federal carbon tax an "attack on Alberta's economy and jurisdiction" (Nixon, quoted in Brown, 2020). The UCP's challenge to the constitutionality of the federal Greenhouse Gas Pollution Pricing Act, supported by the governments of

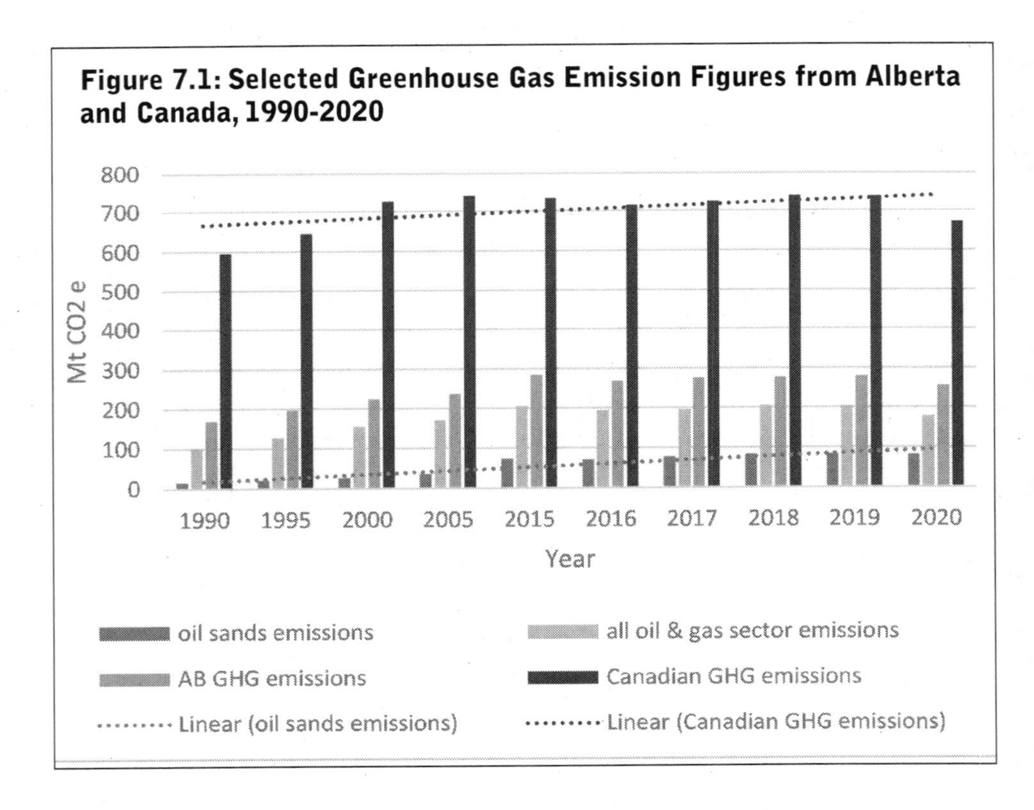
Figure 7.1: Selected Greenhouse Gas Emission Figures from Alberta and Canada, 1990-2020

Saskatchewan and Ontario, was rejected by the Supreme Court of Canada (SCC 2021). The EEA, the crown corporation created by the NDP government to roll out energy efficiency initiatives, was dissolved by the UCP, and the REP was discontinued. While not reversing the NDP's coal phase-out strategy, the UCP government did cancel the $200-million fund created late in the NDP government's term to support investment in small-scale, locally generated electricity projects, $50 million of which was earmarked for communities affected by the coal phase-out.

The UCP did not revoke the NDP's Oil Sands Emissions Limit Act, which had proposed a cap of 100 MT (not including various exempted emissions). The NDP government never issued the regulation required for the cap to take effect; nor did the UCP. Emissions from the oilsands were reported by the industry to total 70 MT in 2016 (ECCC, 2022b) — comfortably below the proposed "cap." However, a study by federal scientists published in 2019 estimated that GHG emissions from the oilsands were overall 30 per cent higher than the industry was reporting (Liggio et al., 2019). This adjustment would put oilsands emissions at 91 MT in 2016, 100 MT in 2017, and at levels exceeding 100 Mt ever since.

Understanding the UCP's changes to the carbon levy system on larger emitters requires an explanation of this system's prior design. The Specified Gas Emitters Regulation (SGER) and the Climate Change Emissions Management (CCEM) Fund were instituted in 2007 by then-premier Ed Stelmach's Progressive Conservative (PC) government. Under the SGER, facilities emitting more than 100,000 tonnes of CO_2 eq per year were required to reduce the emissions intensity (EI) of their output (i.e., the quantity of CO_2 eq emissions per unit of output) by an amount calculated on the basis of the facility's years of commercial operation and the corresponding targets set out in the regulation (Government of Alberta, 2007, Part 2). Regulated facilities include oilsands operations, gas producers, electricity generators, petrochemical and fertilizer plants, and cement factories. Newer facilities were given less stringent requirements than older ones. The most stringent requirement applied to a facility in it ninth or subsequent year of operation, which was to reduce its EI by 12 per cent over its baseline year[16] by 2015, 15 per cent by 2016, and 20 per cent by 2017. An oil producer, for example, would have to reduce the EI of its total output by the set amount, or pay a "carbon levy" for each tonne of CO_2 eq emitted above that target. The levy was initially set at $15 per tonne of CO_2 eq. Note that, under this system, emitters were not made to pay a tax on every tonne of CO_2 eq emitted, but only on the portion that exceeded the emissions intensity limit. Regulated facilities were given two other ways of complying with the regulation. They could purchase emissions offsets from sellers in Alberta or apply banked emission performance credits against a portion of their required EI reduction limit.

As critics predicted, the SGER/CCEMC system was ineffective in significantly reducing GHG emissions because the carbon price and the emission reduction requirements were too low. The carbon levy did not raise the cost of production sufficiently to incentivize investments in pollution abatement. Instead, facilities typically opted to pay the carbon levy into the CCEM fund. The CCEM Corporation then used this revenue to disburse grants to "innovation" projects that were supposed to reduce greenhouse gas emissions (Adkin, 2019).

When the NDP formed the government, it made some changes to the SGER/CCEMC system. The Carbon Competitiveness Incentive Regulation (CCIR) replaced the SGER in December 2017. The CCEMC was renamed Emissions Reduction Alberta (ERA), although little change was made to its mandate (Adkin, 2019). Instead of requiring facilities to reduce the emissions intensity of their output against a baseline year of their own operations, their required reductions were based on a sector-wide performance standard (or benchmark) set by the most efficient producers. The electricity sector was given a "good as best gas" benchmark, thereby pushing coal-fired electricity plants to either capture emissions or make the transition to gas. The oilsands in situ and mining operations were assigned a "top-quartile performance or better" standard. Net emissions of any facility that exceeded the "output-based allocation" for the year would incur a cost: the purchase of fund credits, emission offsets, or the application of emission performance credits. The NDP increased the carbon levy to $20/tonne in 2016 and $30/tonne from 2017 to 2019 and made other changes to incentivize investment in emission reductions (for details, see Saric, 2017). It also increased the number of gases that were regulated by the CCIR.

Although the CCIR system was thought to be more stringent in its requirements, overall, than its predecessor, research published in 2021on the CCIR's impact on 18 in situ oilsands facilities found that it had not been effective in reducing GHG emissions in this sector (Si et al., 2021). The emissions intensity of production averaged across the 18 operations decreased by four per cent in 2019, compared to the baseline. The authors emphasised, however, that *absolute* GHG emissions increased by 15 per cent and production increased by 10 per cent (Ibid., p. 8). Thus, a good part of this small sectoral decrease in emissions intensity was attributed to the increase in production and the maturity of the operations, rather than to the effect of the CCIR (Si et al., 2021, p. 9). Overall, emissions from in situ oilsands mining grew from 38 Mt in 2015 to 43 Mt in 2019 (ECCC, 2022a, p. 67). Emissions from the oilsands overall (mining, in situ, and upgrading combined) grew from 73 Mt in 2015 to 83 Mt in 2019, as reported in Canada's National Inventory report for 2022 (p. 67).[17]

Despite the apparent inadequacy of the CCIR as a means of reducing absolute emissions from large emitters, the UCP government sought to weaken the system further with the aim of reducing the compliance costs for the regulated emitters (Weber, 2019). After holding consultations only with industry stakeholders in the summer of 2019 (Riley, 2019), the UCP brought in its Technology Innovation and Emissions Reduction Implementation Act in late October, along with the Technology Innovation and Emissions Reduction (TIER) Regulation, replacing the CCIR effective Jan. 1, 2020. Essentially, the TIER reverted to the SGER method of establishing facility-specific performance benchmarks for EI, although electricity generators continued to be held to a "good as best gas" standard. For other facilities, the EI reduction target is set at 10 per cent below the facility's average EI performance as averaged over 2013-2015 (to be achieved in 2020) and by a further one per cent per year for every year after that, starting in 2021 (for details, see Olexiuk, Saric, Kennedy, and Wetter, 2019). To be considered equivalent to the federal output-based carbon pricing system (OBPS) for large emitters, the UCP was obliged to increase the price of a TIER fund credit to $40 in 2021 and $50 in 2022.[18]

How has the technological approach to GHG emission reductions worked so far? In its 2020-2021 annual report, the ERA (formerly the CCEMC) reported it distributed $821 million to 221 projects since 2010. ERA estimated these projects yielded 6.1 Mt in CO_2eq emission reductions as of 2020 and will reduce GHG emissions by a total of 42.3 Mt by 2030 compared to business as usual (ERA 2021, 11-12).[19] Journalist Jeff Gailus (2022) calculates that 42.3 Mt amounts to a mere 0.72 per cent of Alberta's cumulative GHG emissions from 2009-2020. The problem, in other words, is that these investments have not substantially and progressively reduced Alberta's annual GHG emissions. They have, however, displaced other, more effective approaches to decarbonization and provided political cover for the fossil fuel industry to continue extracting and exporting its products even in the face of repeated warnings from bodies like the IPCC that the combustion of fossil fuels is making Earth uninhabitable.

Moreover, not all the revenue from the UCP's redesigned TIER goes to ERA. The government may use a portion of TIER revenue for anything, and indeed, TIER revenue is being used to fund the UCP's Canadian Energy Centre (AKA Energy War Room). This corporation, created by the UCP government, is overseen by the minister of energy and serves as another propaganda arm of the oil and gas corporations.

When the federal government unveiled its Emissions Reduction Plan in March 2022, the response from Alberta politicians was predictable. While the leader of the NDP characterized the plan as "a fantasy," Minister Nixon called the plan "insane," and framed federal climate policy as an attack on the economy and sovereignty of Albertans. This political rhetoric ignores the enormous subsidies the federal Liberals have handed over to the fossil fuel industry for R&D in carbon capture, storage, and reuse and other technologies, in the forms (for example) of the Emissions Reduction

Fund ($750 million), the Carbon Capture Utilization and Storage Investment Tax Credit ($8.6 billion), and other federal financial support (Samson, Phillips, and Drummond, 2022). Instead, the UCP government has portrayed the decarbonization targets for the oil and gas sector as "catastrophic" for the Canadian economy and "a full-frontal attack on the 800,000 people who work in the energy sector" (Kenney, quoted in Drinkwater, 2022). Kenney and his caucus continue to represent Alberta's heavy crude as an "ethical" alternative to what, in 2022, he had begun to call "dictator oil" from "Putin's Russia and the OPEC dictatorships."

The ethical oil/petro-nationalist and Alberta nativist discourses are not unique to the UCP, having longer roots in Alberta's PC and Wildrose parties and the petro-turf organizations linked to the oil and gas industry (Adkin and Stares, 2016; Enoch, 2022; Gunster, Fleet, and Neubauer, 2021; Kinder, 2019). They are routinely used to mobilize the electoral base of the province's conservative parties, a section of which supports Alberta's separation from Canada. Danielle Smith, the new leader of the UCP selected in October 2022, continues the practice of fomenting Alberta "sovereigntist" sentiment among the party's base, characterizing the federal carbon tax as an outrageous imposition on average Albertans whose constitutionality her government will re-litigate. Like the federal Conservative leader, Pierre Poilievre, Smith consistently omits to mention that the federal government rebates the carbon tax revenue to low- and middle-income households. She also participated in misinformation campaigns started by a far-right publication that claimed the federal government was creating a corps of armed "climate cops" and that it intended to force farmers to stop using nitrogen fertilizers (Anderson, 2022; Quenneville, 2022; D. Smith, 2022a). Like her predecessors in the UCP, Smith believes that technological breakthroughs can "make continued fossil fuel use not only possible but preferable for fuelling the energy needs of this generation and the next" (quoted in Anderson, 2022). Thus, her premiership promises continuing conflict with the federal Liberal government — and with majority public opinion in Canada and Alberta — over the need to reduce Alberta's greenhouse gas emissions. Smith's rhetoric and tactics during the campaign for leadership of the UCP also indicate she will seek to shape public opinion about climate policy through a never-ending stream of misinformation in collaboration with populist far-right media outlets.

The Oilsands Region: National Sacrifice Zone

> "Our lives and territories have been eviscerated by this industry."
> — *Eriel Tchekwie Deranger, member of the Athabasca Chipewyan First Nation and executive director of Indigenous Climate Action, Feb. 11, 2020.*[20]

In 2005, Alberta journalist Andrew Nikiforuk described the oilsands region as a national sacrifice zone. By this, he meant that there was a national-provincial political consensus to permit the region to be environmentally devastated for the extraction of mineral wealth, with the promise of eventual "reclamation." Sacrificed to these ends were landscapes, ecosystems, and the well-being of the non-human species and human communities depending upon them. These impacts have been well documented over many years.

The environmental science, regulatory, and monitoring capacity of the ministry of environment has been notoriously weak since the 1980s. Many of the functions the ministry ought to have carried out in defending the public interest have been "delegated" to multi-stakeholder bodies with the aim of reducing government staff numbers and direct expenses. These bodies were typically dominated by representatives from the extractive industries and the captured provincial ministries. Examples include the delegation in 1997 of responsibilities under the Wildlife Act to the Alberta Conservation Association, the Regional Aquatics Monitoring Program created in 1997 to monitor aquatic pollution in the oilsands, and the Cumulative Environmental Management Association created in 2000 to monitor and manage the cumulative impacts of oilsands development on land, water, air, and biodiversity. The Alberta Energy Regulator was established in 2013 to consolidate and streamline the regulatory functions previously performed by the Energy Resources Conservation Board and the ministry of environment (related to the energy sector). While the AER reports to the ministers of energy and environment, its operations are funded by levies on firms in the energy sector.

Numerous reports, as well as testimonies or articles by environmental scientists and Indigenous knowledge keepers, have characterized Alberta's environmental regulatory system as fragmented and ineffective, and the ministry responsible as lacking both the resources needed to do its job and independence from industry "stakeholders" (Carter, 2016, 2020; Dubé, Dunlop, Davidson, Beausoleil, Hazewinkel, and Wyatt, 2021; Quinn, Alexander, Kennett, Stelfox, Tyler, Vlavianos, Passelac-Ross, Duke, and Purves-Smith, 2016; Riley, 2022). The research shows that Indigenous concerns have not been taken seriously; Indigenous knowledge has not been incorporated in the design and execution of monitoring. There have been problems with integrating the data from multiple studies and making it available to the public and to policy makers (AEMP, 2011; Cronmiller and Noble, 2019). While oilsands extraction expanded rapidly after 2000, environmental monitoring, regulation, and enforcement were treated as inconveniences, with profound consequences. Little progress has been made on regional land-use plans that are supposed to build on cumulative environmental effects data. Dubé et al. (2021, p. 324) describe the land-use framework for the Lower Athabasca Region as being "years behind in its reporting" and say it has not been "effectively integrated with environmental monitoring programs in the region."

Prior to 2015, NDP politicians appeared to share environmentalists' view that the AER's mandate to oversee environmental approval processes, monitoring, and enforcement conflicted with its mandate to promote fossil fuel extraction. However, once elected, the NDP government decided not to transfer responsibilities for monitoring pollution from the energy sector back to the ministry of environment (Carter, 2020, p. 41). The UCP criticized the AER for taking too long to approve oil and gas projects (Stephenson, 2019b). The heads of the Explorers and Producers Association of Canada and the Canadian Association of Petroleum Producers amplified the demand that the environmental regulatory framework for fossil fuel projects be made more "efficient" and "competitive" (CBC News, 2019). Thus, in September 2019 the UCP government fired the board of directors and initiated a review of the AER's functioning. The reconstituted board is chaired by a former oil and gas industry executive, while a close associate of Kenney was appointed VP for Technical Science & External Innovation (Alberta Energy Regulator, 2022).

Remediation of Abandoned or Orphaned Oil and Gas Wells

Under the existing regulatory regime, progress on the remediation of abandoned or orphaned wells and of the oilsands mining sites has been glacial, as described in the chapter on energy policy (see Taft, Chapter 8). Alberta's Mine Financial Security Program (MFSP) is supposed to protect taxpayers from being stuck with the costs of cleaning up mining sites by collecting security deposits from the mining companies. Yet, the MFSP is designed — like all other aspects of regulation in Alberta — to protect the bottom line of the corporations rather than the economic and environmental future of citizens (Auditor General of Alberta, 2015; and 2021, pp. 29-34). The $913 million in securities collected as of September 2021 from oilsands producers (Yewchuk, 2021) are wholly inadequate to the task, amounting to only three per cent of the low-ball estimate of the $30 billion remediation cost for the oilsands mines and 0.7 per cent of the higher estimate of $130 billion (Riley, 2021). The auditor general, policy analysts at the Pembina Institute and other ENGOs, Indigenous leaders, and environmental law experts have all called on Alberta governments to require the payment of full securities and to establish timelines for progressive reclamation of the mine sites (Riley, 2021).

In 2020, however, the UCP took a step in the opposite direction, altering the MFSP rules so the mining companies could avoid making security deposits that would otherwise have been required under the existing asset-to-liability formula (Yewchuk, 2021). This action — like the decision to suspend some environmental monitoring in the oilsands in 2020, ostensibly because of the COVID-19 pandemic — was taken without any consultation with affected Indigenous communities.

How likely is it that in the remaining years of profitability for the oilsands companies they will set aside $30 billion (let alone $130 billion) to cover the environmental liabilities they have created unless they are required to do so? Alberta governments evade this question by covering their eyes and ears when the IPCC or International Energy Agency issues a new report. As environmental law expert Drew Yewchuk (2021) points out:

> The MFSP includes an assumption that oilsands mines will operate until they run out of reserves — that is why the MFSP does not require security to be in place until the mine has less than 15 years of reserves remaining (through the Operating Life Deposit system). The assumption that the oilsands mines will operate until they run out of reserves is clearly no longer a safe one. The world is going to take some action to control climate change . . . The oilsands will be operating for some time yet, but few or none will be able to use up their reserves, which in many cases was forecasted to occur in the latter half of this century. Oilsands mines are more likely to close when international carbon pricing and other measures make their operation uneconomical, and the MFSP does not take that into account.

Another environmental law expert, Martin Olsynzksi, fears taxpayers will almost inevitably be stuck with the bill for cleaning up the oilsands. "There's an incredible level of frustration watching this unfold. It's like watching a slow-motion train wreck" (quoted in Riley, 2021).

We should remember that these are the same corporations that have demanded enormous public subsidies for the development of carbon capture and sequestration infrastructure to capture GHG emissions from upgraders and refineries, as well as for other technologies to reduce energy consumption and water use in unconventional oil extraction. The industry's goal is to reduce its environmental compliance costs, as we see with its currently favoured approach to getting rid of the tailings lakes in the oilsands region.

Removal of Toxic Tailings Lakes

For decades environmentalists have called for regulatory authorities to set medium-term timelines for tailings treatment or containment (Lothian, 2020). Instead, the tailings "ponds" were permitted to grow to their current volume of about 1.4 trillion litres of processed water, contaminated with naphthenic acids, polycyclic aromatic hydrocarbons, benzene, cyanide, toluene, heavy metals (including arsenic), salt, and other substances. It has been known for years that the ponds have been leaching into groundwater and "leaking" into the adjacent Athabasca River, yet the federal Fisheries Act was not enforced (Riley, 2020).

It was reported in 2021 that the federal ministry of environment and climate change was developing regulations to permit the discharge into the Athabasca River of partially treated tailings liquids, based on research funded by the Natural Sciences and Engineering Research Council and the Canadian Oil Sands Innovation Alliance (COSIA) involving dozens of university researchers (COSIA, 2021). A scientist leading the tailings treatment research suggested it is technically possible to purify the fluids to the standard of drinkable water, but the industry does not find this solution "economically feasible" (Mohamed Gamal El-Din, quoted in Bakx, 2021). In other words, the corporations don't want to pay for it. In August 2022, following strong opposition from Indigenous and environmental organizations to the plan to discharge the pond tailings into the river, the (new) federal Minister of Environment Steven Guilbeault said other options were being considered (such as injecting the water from the tailings underground) (Weber, 2022).

Unsurprisingly, the response from Indigenous communities to the proposal to discharge treated tailings water into the Athabasca River was less than enthusiastic. Jesse Cardinal, executive director of Keepers of the Water and a member of the Kikino Metis Settlement, argues that the industry and government should "do what's right, not what's fast and easy" (quoted in Bakx, 2021). Fort McKay First Nation has expressed many concerns about the AER's directive for fluid tailings management (Alberta Energy Regulator, 2017) and about the discharge/dilute solution for the mine effluent (Bori Arrobo, quoted in Bakx, 2021). The chief of Smith Landing First Nation, Gerry Cheezie, called the plan environmental racism (quoted in Zingel, 2021). But they are told this is the only alternative to the continued expansion of the tailings ponds, with the attendant risk that violent weather events could cause dam breaches and result in catastrophic contamination of the river. The assumptions underlying such regulatory decisions are: (a) the oilsands mines will be permitted not only to continue production but to expand in size, and; (b) the oil and gas corporations will not be required to pay the full costs of environmental remediation. Instead, the health and livelihoods of communities in the oilsands region will be put at risk, and the environmental costs of oil and gas extraction will be shunted onto future generations.

Overall, the economic benefits to Albertans of oilsands extraction have been hugely exaggerated. The environmental liabilities (cleanup of abandoned or orphaned wells and pipelines, remediation of in situ and surface mine sites, tailings ponds removal) for the oil and gas sector alone were estimated in 2018 to be as high as $260 billion

(De Souza, Jarvis, McIntosh, and Bruser, 2018). By comparison, the total of all the non-renewable resource revenue received by the Government of Alberta from 1970/71 to 2020/21 is $252 billion (Government of Alberta, 2022b). Obviously, the $252-billion figure does not capture all the economic benefits oil and gas production have provided — most importantly, in employment income. Nor does the $260 million figure include an estimate of costs related to the impacts of climate change on Alberta, or the costs of oilsands pollution and its draw-down on the volume of the Athabasca River for wildlife and human populations living in the region and downstream. Any way we look at it, the balance sheet is sobering and raises important questions about the value and opportunity costs of this extractive model of development so vociferously promoted by the UCP government.

Conclusion

During its term in office, the UCP government sought to restore the *status quo ante* in environmental policy in Alberta by reversing the NDP government's plan to create new protected areas on the eastern slopes of the mountains as well as the new GHG emission reduction measures that had been implemented by the NDP. The UCP went further than any previous conservative government in its attempts to open the eastern slopes of the Rocky Mountains to extractive and commercial development. In the resulting struggles to protect parks and watersheds, many citizens were introduced to the framing of what was at stake as a conflict between private interests (mining companies and investors) and public goods (a healthy environment, clean water, free access to natural spaces protected for the common benefit). They were politicized in a new way by the realization that the government many of them had helped to elect represented not their interests, but those of the extractive industries. We have yet to see whether these experiences will have long-lasting effects on the political loyalties of citizens who have long constituted a secure voting base for the province's conservative parties.

Rural landowners expressed their profound attachment to the landscapes of the southwest, viewed as being integral to their identity as Albertans. There was some recognition among them that Indigenous communities share this attachment, and that they, too, have cultural identities rooted in these lands and waters. Some participants in these campaigns recognized that the Indigenous communities inhabiting the oilsands region are every bit as attached to their lands and waters — whose destruction they have suffered for decades — as the southern ranchers are to the foothills. As Indigenous leaders have explained repeatedly, in many venues, their cultures cannot thrive on poisoned land and rivers. Can the defenders of the Oldman, Livingstone, and Red Deer rivers also be rallied to defend the Peace and the Athabasca rivers and delta? There is, as yet, no #WaterNotOil, #SaveTheAthabasca, or

#StopTheTMX hashtag trending in Alberta. How far will the nascent solidarities of the coalition seeking to protect the Rocky Mountains and their watersheds stretch?

The petro-nativist response of the UCP government to the federal Emissions Reduction Plan (ERP), released March 29, 2022, has been as virulent as that of the Klein government's war against the ratification of the Kyoto Protocol 20 years ago. But Alberta is not the same place in 2022 as it was in 2002. Progressive civil society is better organized and has access to alternative media as well as the research produced by numerous non-governmental organizations and think tanks. In each of the case studies, we see the important role played by ENGOs and Indigenous environmental activists in communicating ecological knowledge to media and to citizens' groups. First Nations and Metis Settlements have also deployed their constitutional and treaty rights to challenge the extractive model of development. Young people who have learned about the causes and consequences of global warming in school have organized climate strikes and marches — even bringing Greta Thunberg to Alberta in October 2019 for a tour that included one of the largest marches ever to happen in the province with about 10,000 in attendance. Climate justice groups have formed in cities and post-secondary education institutions and are working in solidarity with Indigenous communities that are demanding the restoration of treaty lands and sovereignty over these lands.

There is some evidence that young people no longer see the oil and gas industry as a desirable or secure option for life-long employment (e.g., Seskus, 2021). The workforce in this sector has been shrinking because of automation, the maturity of the extraction projects, and the low prices for Alberta crude that prevailed from 2014-2021. With declining global demand for fossil fuels (starting with those most costly to produce and most carbon intensive, like Alberta's heavy crude), young people see the writing on the wall and are looking for careers in sectors that provide solutions to social and environment problems. In other words, Albertans may not be where petro-politicians think they are. For a growing number of Albertans, #GreenTransition and #JustTransition campaigns speak to the desires for income security, a good life, and a healthy environment for the generations to come.

Push-back from Albertan civil society actors prevented the UCP government from implementing some of its objectives with regard to the privatization of nature and the deregulation of coal mining. On the climate change front, however, Alberta slipped further backward under UCP rule. In cancelling the carbon tax and the Energy Efficiency Agency, the UCP reversed NDP policies that promised to help Albertan households and farmers reduce their energy costs and that had a moderately redistributive effect (Winter, Dolter, and Fellows, 2021).[21]

By offering loan guarantees for Indigenous corporations to invest in resource extraction projects through its Alberta Indigenous Opportunities Corporation, the UCP sought to drive a wedge between Indigenous leaders concerned to find sources of revenue for their bands, on the one hand, and Indigenous climate justice activists and their allies who are resisting such projects, on the other hand (Atleo, Crowe, Krawchenko, and Shaw, 2022; Iamsees, 2020; Narine, 2020). Meanwhile, the UCP's Critical Infrastructure Defence Act aimed to repress social movement opposition to the building of new fossil fuel infrastructure that will further lock-in the province's reliance on oil and gas revenues and significantly increase GHG emissions (Plotnikoff, 2020). Instead of investing in economic diversification fuelled by renewable energies, the UCP government committed billions of dollars to the Keystone pipeline extension project (see Chapter 8) and other forms of subsidies to the oil and gas sector (CCS, tax credits, royalty credits, corporate tax cuts) (Adkin, 2019; Adkin and Cabral, 2020). The GHG emissions from the oilsands continued to rise during the UCP's term in office, and the government has testily rejected proposals to cap these emissions, as we see in the quoted statement by the province's environment minister, Jason Nixon. Smith has characterized the proposed cap on the oil and gas sector's emissions as an "unprovoked assault on Albertans" (D. Smith, 2022b) that she would attempt to nullify by passing an Alberta Sovereignty Act (Junker, 2022b).

In addition, environmental regulation of local pollution from the oil and gas sector has been weakened under the UCP in the name of "cutting red tape," and the huge problem of environmental liabilities created by oil and gas extraction has been shunted down the road. Indigenous communities continue to bear the brunt of the toxic environment and biodiversity loss created by this extractivist, colonial regime. Albertans remain on the hook for billions in environmental costs, while opportunities to advance a just transition to a greener economy have been foregone at a critical juncture in the global climate crisis.

The outcome of the next election in Alberta will be determined, in part, by the success of the province's civil society in highlighting ecological crises and realistic solutions to these salient issues. While these are not priority concerns for the UCP's core electorate, research suggests most Albertans do care about preserving a livable planet for future generations (Pembina Institute, 2021). And we know from the fight over coal mining that both rural and urban Albertans care deeply about protecting landscapes, watersheds, and clean water. As Alberta is increasingly affected by climate destabilization and the global shift to decarbonization, a party seeking to govern will need to produce a credible blueprint for an ecologically sustainable path of development.

References

Adkin, L. E. (2016). Alberta's neoliberal environment. In L. E. Adkin (Ed.), *First World Petro-politics: The Political Ecology and Governance of Alberta* (pp. 78-113). Toronto: University of Toronto Press.

Adkin, L. E. (2017). Crossroads in Alberta: Climate capitalism or ecological democracy? *Socialist Studies* 12(1). Retrieved from https://socialiststudies.com/index.php/sss/article/view/27191/20045

Adkin, L. E. (2019). Technology innovation as a response to climate change: The case of the Climate Change Emissions Management Corporation of Alberta. *Review of Policy Research* 36(5). https://onlinelibrary.wiley.com/doi/full/10.1111/ropr.12357

Adkin, L. E. and Cabral, L. (2020). *Knowledge for an Ecologically Sustainable Future? Innovation Policy and Alberta Universities*. Edmonton, AB: Parkland Institute. June 24. https://www.corporatemapping.ca/knowledge-for-an-ecologically-sustainable-future/

Adkin, L. E. and Stares, B. J. (2016). Turning up the heat: Hegemonic politics in a first world petro-state. In L. E. Adkin (Ed.), *First World Petro-politics: The Political Ecology and Governance of Alberta* (pp. 190-240). Toronto: University of Toronto Press.

Alberta Energy Regulator (AER). (2017). *Directive 085: Fluid tailings management for oil sands mining projects: Stakeholder feedback and AER responses*. October. Retrieved from https://static.aer.ca/prd/documents/directives/Directive085-EngagementSummary.pdf

Alberta Energy Regulator (AER). (2022). *Organizational Structure*. Retrieved from https://www.aer.ca/providing-information/about-the-aer/who-we-are/organizational-structure

Alberta Environment and Parks (AEP). (2019). *Rightsizing Alberta parks: Confidential advice to minister*. Dec. 18. Document obtained by CPAWS through a Freedom of Information and Privacy Act application to the ministry of environment and parks. Retrieved from https://cpawsnab.org/wp-content/uploads/2020/08/Parks_FOIP_compressed.pdf

Alberta Environmental Monitoring Panel (AEMP). (2011). *A world-class environmental monitoring, evaluation and reporting system for Alberta*. The report of the Alberta Environmental Monitoring Panel. Edmonton, AB: Government of Alberta.

Anderson, D. (2022, Oct. 7). What a Danielle Smith government could mean for the environment in Alberta. *The Narwhal*. https://thenarwhal.ca/ucp-danielle-smith-alberta-environment/

Auditor General of Alberta. (2015). *Environment and Parks and the Alberta Energy*

Regulator—Systems to ensure sufficient financial security for land disturbances from mining. Report of the Auditor General of Alberta, July. Edmonton, AB: Government of Alberta. Retrieved from https://www.oag.ab.ca/wp-content/uploads/2020/05/EP_PA_July2015_AER_Systems_Ensure_Fin_Security_Land_Disturb.pdf

Atleo, C., Crowe, T., Krawchenko, T., & Shaw, K. (2022). Indigenous ambivalence? It's not about the pipeline . . .: Indigenous responses to fossil fuel export projects in Western Canada, pp. 157-176. In H. Boudet and S. Hazboun, Eds. *Public Responses to Fossil Fuel Export*. Amsterdam: Elsevier.

Auditor General of Alberta. (2021). *Report of the Auditor General*. Edmonton, AB: Government of Alberta. June. Retrieved from https://www.oag.ab.ca/wp-content/uploads/2021/06/oag-june-2021-report.pdf

Bakx, K. (2021, Dec. 6). *Banned for decades, releasing oilsands tailings water is now on the horizon*. CBC News. https://www.cbc.ca/news/business/bakx-oilsands-tailings-release-mining-effluent-regulations-1.6271537

Bankes, N. (2021, Feb. 8). Coal law and policy in Alberta, Part One: The Coal Policy and its legal status. *The University of Calgary Faculty of Law Blog*.

Bankes, N., & Bradley, C. (2020, Dec. 4). Water for coal developments: where will it come from? *The University of Calgary Faculty of Law Blog*.

Bell, D. (2020, March 3). *20 Alberta parks to be fully or partially closed, while dozens opened up for "partnerships."* CBC News. https://www.cbc.ca/news/canada/calgary/alberta-park-funding-slashed-1.5484095

Brown, C. (2020, Dec. 12). *Nixon: Carbon tax increases an "attack on Alberta's economy and jurisdiction."* Everything Grand Prairie. https://everythinggp.com/2020/12/12/nixon-carbon-tax-increases-an-attack-on-albertas-economy-and-jurisdiction/

Canadian Oil Sands Innovation Alliance (COSIA). (2021). *2020 Water Mining Research Report*. June. https://cosia.ca/sites/default/files/attachments/Park%20-%20COSIA%20-%20Mining%20Report%20-%20June%208.pdf

Canadian Parks and Wilderness Society (CPAWS). (n.d.a). *Changes to Alberta's provincial parks under the "optimizing Alberta parks" plan: A timeline*. Retrieved April 9, 2022, from https://cpawsnab.org/defend-alberta-parks-timeline/

Canadian Parks and Wilderness Society (CPAWS). (n.d.b). *Fact Check: 13 truths and a lie about Alberta parks changes*. Retrieved April 8, 2022, from https://cpaws-southernalberta.org/fact-check-13-truths-and-a-lie-about-alberta-parks-changes/

Canadian Parks and Wilderness Society (CPAWS). (2020a). *Strong majority of Albertans oppose government changes to parks*. March 19. Retrieved from https://cpawsnab.org/strong-majority-of-albertans-oppose-government-changes-to-parks/

Canadian Parks and Wilderness Society (CPAWS). (2020b). *Internal government documents reveal concerning new details about the plan for Alberta's parks: land sales, no comprehensive analysis of costs, and exclusion of the public from the decision*. July 23. Retrieved from https://cpawsnab.org/concerning-new-details-about-the-plan-for-albertas-parks/

Canadian Press. (2019, May 7). *Bighorn parks proposal will not go ahead, says Alberta's environment minister*. CBC News. https://www.cbc.ca/ news/canada/edmonton/alberta-bighorn-parks-proposal-cancelled-environment-minister-1.5126389

Canadian Press. (2021, Nov. 14). *Scientists say new Alberta trails act threatens already-stressed environment*. CBC News. https://www.cbc.ca/news/canada/calgary/alberta-trails-act-jason-nixon-1.6248609

Carter, A. V. (2016). The petro-politics of environmental regulation in the tar sands. In L. E. Adkin (Ed.), *First World Petro-politics: The Political Ecology and Governance of Alberta* (pp. 152-189). Toronto: University of Toronto Press.

Carter, A. V. (2020). *Fossilized: Environmental Policy in Canada's Petro-provinces*. Vancouver: University of British Columbia Press.

CBC News. (2018, Sept. 14). *Park designation proposal triggers turf war in Bighorn backcountry*. https://www.cbc.ca/news/canada/edmonton/bighorn-backcountry-alberta-wildland-provincial-park-1.4818669.

CBC News. (2019, April 29). *Former Pembina Institute head quits AER before Kenney could fire him, alleging "smear campaign."* https://www.cbc.ca/news/canada/calgary/ed-whittingham-aer-board-pembina-resignation-1.5115373

Coal Policy Committee. (2021a). *Engaging Albertans about Coal*. December, n.d. https://open.alberta.ca/dataset/78cfffec-e5dc-4474-8617-72b1ca2f4ab2/resource/604fd294-49ba-4942-88b4-5a8fd4d1d191/download/energy-coal-policy-committee-engaging-albertans-2021-12.pdf

Coal Policy Committee. (2021b). *Final Report: Recommendations for the Management of Coal Resources in Alberta*. December, n.d. https://open.alberta.ca/dataset/cabeccc3-3937-408a-9eb5-f49af85a7b3f/resource/75d241f9-5567-4a86-91e7-3ed285e42f18/download/energy-coal-policy-committee-final-report-2021-12.pdf

Cronmiller, J. G., & Noble, B. F. (2019). The discontinuity of environmental effects monitoring in the Lower Athabasca region of Alberta, Canada: Institutional challenges to long-term monitoring and cumulative effects management. *Environmental Reviews* 26(2), 169-180.

Cryderman, K. (2002, Sept. 21). Alberta's response: We don't believe you. *The Edmonton Journal*.

De Souza, M., Jarvis, C., McIntosh, E., & Bruser, D. (2018, Nov. 1). Alberta regulator privately estimates oilpatch's financial liabilities are hundreds of billions more than what it told the public. *Canada's National Observer*. https://www.nationalobserver.com/2018/11/01/news/alberta-regulator-privately-estimates-oilpatchs-financial-liabilities-are-hundreds

Drinkwater, R. (2002, April 4). *Alberta premier calls Ottawa's greenhouse gas targets "nuts"; pledges to fight them*. CBC News. https://www.cbc.ca/news/canada/edmonton/climate-change-plan-alta-1.6407436

Dubé, M. G., Dunlop, J. M., Davidson, C., Beausoleil, D. L., Hazewinkel, R. R. O., & Wyatt, F. (2021). History, overview, and governance of environmental monitoring in the oil sands region of Alberta, Canada. *Integrated Environmental Assessment and Management 18*, pp. 319-332. Retrieved from https://setac.onlinelibrary.wiley.com/doi/full/10.1002/ieam.4490

Dykstra, M. (2015, Nov. 24). Carbon tax revenue to be recycled back into the economy immediately, says Alberta Premier Notley. *Edmonton Sun*. https://edmontonsun.com/2015/11/24/carbon-tax-revenue-to-be-recycled-back-into-the-economy-immediately-says-alberta-premier-notley

Emissions Reduction Alberta. (2021). *Annual Report 2020-2021*. N.d. Retrieved from https://www.eralberta.ca/wp-content/uploads/2022/01/ERA_AnnualReport_Jan19.pdf

Enoch, S. (2022, Oct. 14). The oil industry's Frankenstein. *Briarpatch*. https://briarpatchmagazine.com/articles/view/the-oil-industrys-frankenstein

Environment and Climate Change Canada (ECCC). (2022a). *National Inventory Report 1990-2020: Greenhouse Gas Sources and Sinks in Canada*. April. Gatineau, QC: ECCC. www.canada.ca/en/environment-climate-change/services/climate-change/greenhouse-gas-emissions/inventory.html

Environment and Climate Change Canada (ECCC). (2022b). *Greenhouse gas emissions from the oil and gas sector*. Retrieved from https://www.canada.ca/en/environment-climate-change/services/environmental-indicators/greenhouse-gas-emissions.html

Fitch, L. (2020a, April 1). The (not so) great Alberta parks clearance sale. *Lethbridge Herald*. https://lethbridgeherald.com/commentary/opinions/2020/04/01/the-not-so-great-alberta-parks-clearance-sale/

Fitch, L. (2020b). Common sense: Really? *Wildlands Advocate* 28(3). September. Retrieved from https://albertawilderness.ca/wp-content/uploads/2021/01/Common_Sense_Really.pdf

Fletcher, R. (2021, Jan. 21). *Answers to questions about Alberta's coal policy that, at this point, you're too afraid to ask*. CBC News. https://www.cbc.ca/news/canada/calgary/alberta-coal-policy-faq-frequently-asked-questions-1.5880659

Fletcher, R., Anderson, D., & Omstead, J. (2020, July 7). *Bringing coal back.* CBC News. https://newsinteractives.cbc.ca/longform/bringing-coal-back

Gailus, J. (2022, March 1). Hot air: TIER is Alberta's world-class carbon reduction sham. *Alberta Views.* https://albertaviews.ca/hot-air-2/

Government of Alberta. (2007). *Specified Gas Emitters Regulation, Alta Reg 139/2007.* Retrieved from https://www.canlii.org/en/ab/laws/regu/alta-reg-139-2007/latest/alta-reg-139-2007.html

Government of Alberta. (2020a). *Creating jobs and improving Alberta's parks.* Media release, Sept 15. Retrieved from https://www.alberta.ca/release.cfm?xID=73233BA0BCDBD-01EC-4F01-47805F5AAB3C8AB6

Government of Alberta. (2020b). *Alberta crown land vision.* November n.d. Retrieved April 10, 2021 from https://www.alberta.ca/alberta-crown-land-vision.aspx

Government of Alberta. (2020c). *Seeking input on outdoor recreation and trails.* Nov. 26. Retrieved from https://www.alberta.ca/release.cfm?xID=7577942010381-960E-7E4A-0E863614A7662974

Government of Alberta. (2021). *Sustainable outdoor recreation: What we heard report.* April.

Government of Alberta. (2022a). *Renewable electricity program.* Retrieved from https://www.alberta.ca/renewable-electricity-program.aspx

Government of Alberta. (2022b). *Historical royalty revenue.* Retrieved from https://open.alberta.ca/opendata/historical-royalty-revenue#summary

Government of Canada. (2020). *Public notice: Grassy Mountain coal project ¾ Notice of hearing.* June 29. Retrieved from https://iaac-aeic.gc.ca/050/evaluations/document/136429

Gunster, S., Fleet, D., & Neubauer, R. (2021). Challenging petro-nationalism: Another Canada is possible? *Journal of Canadian Studies* 55(1) (Winter), 57-87. doi: 10.3138/jcs.2019-0033

Heelan Powell, B. (2020). *Optimizing Alberta parks without public consultation.* Alberta Environmental Law Centre [blog], Aug. 10. Retrieved from https://elc.ab.ca/optimizing-alberta-parks-without-public-consultation-the-government-has-announced-removal-of-175-sites-from-the-parks-system/

Iamsees, C. (2020). *The Alberta Indigenous Opportunities Corporation: A critical analysis.* Yellowhead Institute, Feb. 13. Retrieved from https://yellowheadinstitute.org/2020/02/13/alberta-indigenous-opportunities-corporation-analysis/

Impact Assessment Agency of Canada (IAAC). (2018). *Agreement to establish a Joint Review Panel for the Grassy Mountain Coal Project between the Minister of the Environment, Canada, and the Alberta Energy Regulator, Alberta.* Signed May 23,

2018 by Minister Catherine McKenna and July 9, 2018 by Jim Ellis, CEO of the AER. Retrieved from https://iaac-aeic.gc.ca/050/documents/p80101/124445E.pdf

Impact Assessment Agency of Canada (IAAC). (2021). *Joint Review Panel for Grassy Mountain Coal Project concludes its review*. Media release, June 17. Retrieved from https://iaac-aeic.gc.ca/050/evaluations/document/139410?&culture=en-CA

Jackson, E. & Hussey, I. (2019). *Alberta's Coal Phase-out: A Just Transition?* Edmonton, AB: Parkland Institute, Nov. 20. https://www.parklandinstitute.ca/albertas_coal_phaseout

Junker, A. (2022a, Oct. 22). UCP candidates weigh in on coal mining. *Calgary Herald*. https://calgaryherald.com/news/politics/ucp-leadership-candidates-weigh-in-on-coal-mining/wcm/d3532ff7-ac02-4130-b5fe-3d0ffc74dd7c#Echobox=1664665503

Junker, A. (2022b, Sept. 6). Danielle Smith releases overview of proposed Alberta sovereignty act. *Edmonton Journal*. https://edmontonjournal.com/news/politics/danielle-smith-releases-overview-of-proposed-alberta-sovereignty-act

Kalinowski, T. (2021, March 9). Records show excessive lobbying to rescind 1976 Coal Policy. *Medicine Hat News*. https://medicinehatnews.com/news/local-news/2021/03/09/records-show-excessive-lobbying-to-rescind-1976-coal-policy/

Kinder, J. (2019). *Liquid ethics, fluid politics. The cultural and material politics of petroturfing*. Doctoral diss. Department of English and Film Studies, University of Alberta. Retrieved from https://era.library.ualberta.ca/items/8ced22b2-2ea5-4780-b11d-e24b5fded742

Leger (Polling). (2020). *Alberta omnibus survey*, March 18. Retrieved from https://cpawsnab.org/wp-content/uploads/2020/03/CPAWS-OMNI.pdf

Lewis, J. (2019, Jan. 7). On the Rockies' edge, frictions form over Alberta's plan for new provincial park. *The Globe and Mail*. https://www.theglobeandmail.com/canada/alberta/article-on-the-rockies-edge-frictions-form-over-albertas-plan-for-new/

Liggio, J., Li, S-M., Staebler, R. M., Hayden, K., Darlington, A., Mittermeier, R. L., O'Brien, J., McLaren, R., Wolde, M., Worthy, D., & Vogel, F. (2019). Measured Canadian oil sands CO_2 emissions are higher than estimates made using internationally recommended methods. *Nature Communications* 10, article no. 1863. https://doi.org/10.1038/s41467-019-09714-9

Lothian, N. (2020). *Government action holding oilsands operators to account on tailings management long overdue*. Media release, Sept. 4. Drayton Valley, AB: Pembina Institute. Retrieved from https://www.pembina.org/media-release/government-action-holding-oilsands-operators-account-tailings-management-long-overdue

Lowey, M. (2003). Alberta snubs federal strategy on Kyoto. *Business Edge* 3(33), Sept. 18.

Maclean, Rachel. (2019, April 8). *Alberta's carbon tax brought in billions. See where it went.* CBC News, April 8 https://www.cbc.ca/news/canada/calgary/carbon-tax-alberta-election-climate-leadership-plan-revenue-generated-1.5050438

McCarthy, S. (2021). *Net-zero report card: How future-friendly are Canadian provinces?* Corporate Knights, April 19. Retrieved from https://www.corporateknights.com/climate-and-carbon/net-zero-report-card-how-future-friendly-are-canadian-provinces/

Minister of Energy, Government of Alberta. (2021). *Coal policy committee: Terms of reference.* March 29. Retrieved from https://www.alberta.ca/assets/documents/coal-policy-committee-terms-of-reference.pdf

Narine, S. (2020, Jan. 27). *Come together for TMX ownership bid, and funding will flow, Alberta Premier tells First Nations consortiums.* Windspeaker.com. https://www.windspeaker.com/news/windspeaker-news/come-together-tmx-ownership-bid-and-funding-will-flow-alberta-premier-tells

Nikiforuk, A. (2005, Sept. 28). If Ralph's a friend, who needs enemies? *The Globe and Mail.*

Nikiforuk, A. (2020a, Aug. 3). Alberta coal grab: What is the sound of one group lobbying? *The Tyee.* https://thetyee.ca/Analysis/2020/08/03/Alberta-Coal-Grab-Sound-One-Group-Lobbying/

Nikiforuk, A. (2020b, Aug. 12). Alberta's environment minister cheered on coal mining in new areas before restrictions were dropped. *The Tyee.* https://thetyee.ca/News/2020/08/12/Alberta-Environment-Minister-Cheered-Coal-Mining/

Nixon, J. (2022, April 2). Alberta won't accept a climate plan that destroys its economy. *Edmonton Journal.* https://edmontonjournal.com/opinion/columnists/jason-nixon-alberta-will-not-accept-a-climate-plan-that-destroys-its-economy

Olexiuk, P., Saric, D., Kennedy, J., Wetter, C. (2019). *The more things change, the more they stay the same: Alberta revamps carbon pricing regime for large emitters.* Osler, Hoskin & Harcourt LLP, Nov. 26. Retrieved from https://www.osler.com/en/resources/regulations/2019/the-more-things-change-the-more-they-stay-the-same-alberta-revamps-carbon-pricing-regime-for-large

Pembina Institute. (2021). Strong majority of Albertans support accelerated climate action. Feb. 1. Retrieved from https://www.pembina.org/blog/strong-majority-albertans-support-accelerated-climate-action

Plotnikoff, E. (2020). *Activists or active threats? How the state securitization of critical infrastructure impacts environmental and Indigenous activists in Canada and the United States*. MA diss., Dept. of Political Science, University of British Columbia. http://hdl.handle.net/2429/75730

Pulido-Guzman, A. (2022, March 23). Road open to further coal mining after bill blocked, says Notley. *Lethbridge Herald*. https://lethbridgeherald.com/news/lethbridge-news/2022/03/23/road-open-to-further-coal-mining-after-bill-blocked-says-notley/

Quenneville, G. (2022, Sept. 1). *Environment Canada says online reports of "climate police" are false*. CBC News. https://www.cbc.ca/news/politics/climate-police-misinformation-1.6569812

Quinn, M., Alexander, S. M., Kennett, S. A., Stelfox, B., Tyler, M-E., Vlavianos, N., Passelac-Ross, M., Duke, D., & Purves-Smith, N. (2016). The ecological and political landscapes of Alberta's hydrocarbon economy. In L. E. Adkin (Ed.), *First World Petro-politics: The Political Ecology and Governance of Alberta* (pp. 114-151). Toronto: University of Toronto Press.

Riley, S. J. (2019a, Jan. 23). "It can't be a free-for-all anymore": The battle for Bighorn Country. *The Narwhal*. https://thenarwhal.ca/it-cant-be-a-free-for-all-anymore-the-battle-for-bighorn-country/

Riley, S. J. (2019b, July 10). Alberta government only invites industry to consultation on new emission regulations. *The Narwhal*. https://thenarwhal.ca/alberta-government-only-invites-industry-to-consultation-on-new-emissions-regulations/

Riley, S. J. (2020, Dec. 14). It's official: Alberta's oilsands tailings ponds are leaking. Now what? *The Narwhal*. https://thenarwhal.ca/tailings-ponds-leaking-alberta-oilsands/

Riley, S. J. (2021, May 18). Alberta "undermining" system meant to ensure oilsands companies pay for cleanup, critics say. *The Narwhal*. https://thenarwhal.ca/alberta-oilsands-security-deposit-changes-critics/

Riley, S. J. (2022, March 10). Stonewalled: Alberta ignored warnings about oil and gas cleanup, ex-government scientist says. *The Narwhal*. https://thenarwhal.ca/alberta-oil-gas-wells-reclamation-scientist/

Saric, D. (2017). *Carbon competitiveness incentive regulation replaces and adds rigour to Alberta's existing industrial carbon emissions regulation*. Dec. 22. Osler, Hoskin & Harcourt LLP. Retrieved from https://www.osler.com/en/resources/regulations/2017/carbon-competitiveness-incentive-regulation-replac

Samson, R., Phillips, P., & Drummond, D. (2022). *Cutting to the chase on fossil fuel subsidies*. Report from the Canadian Climate Institute, Feb. 9. https://climatechoices.ca/wp-content/uploads/2022/02/Fossil-Fuels-Main-Report-English-FINAL-1.pdf

Savage, S. (2021, Feb. 8). Press conference. https://www.cbc.ca/news/canada/calgary/alberta-coal-policy-changes-press-conference-1.5905484

Seskus, T. (2021, July 8). *University of Calgary hits pause on bachelor's program in oil and gas engineering*. CBC News. https://www.cbc.ca/news/canada/calgary/university-of-calgary-engineering-1.6092648

Si, M., Bai, L., & Du, K. (2021). Fuel consumption analysis and cap and trade system evaluation for Canadian in situ oil sands extraction. *Renewable and Sustainable Energy Reviews* 146, May 10. Retrieved from https://www.sciencedirect.com/science/article/abs/pii/S1364032121004342?via%3Dihub

Smith, A. (2020, Sept. 15). UCP moves forward with plans to "delist" Alberta parks as government commits $43 million to improve existing sites. *Calgary Herald*. https://calgaryherald.com/news/politics/ucp-moves-forward-with-plans-to-delist-alberta-parks-as-government-commits-43-million-to-improve-existing-sites

Smith, D. (2022a). Twitter post, July 23. @ABDanielleSmith.

Smith, D. (2022b). Twitter post, July 20. @ABDanielleSmith.

Stephenson, A. (2019a, July 4). UCP has killed off Renewable Energy Plan, but wind companies see a bright future. *Calgary Herald*. https://calgaryherald.com/business/energy/renewable-energy-plan-officially-dead-but-ab-wind-companies-still-see-a-bright-future

Stephenson, A. (2019b, Sept. 6). UCP cans AER board, launches promised review of regulator's mandate. *Calgary Herald*. https://calgaryherald.com/business/local-business/ucp-cans-aer-board-launches-promised-review-of-regulators-mandate

Supreme Court of Canada (SCC). (2021). *References and decision re Greenhouse Gas Pollution Pricing Act. 2021 SCC 11*, March 25. Retrieved from https://decisions.scc-csc.ca/scc-csc/scc-csc/en/item/18781/index.do

ThinkHQ Public Affairs, Inc. (2021). *Sizable majority opposes expanded coal mining in the eastern Rockies*. Media release, Feb. 8. Retrieved from http://thinkhq.ca/wp-content/uploads/2021/02/ThinkHQ-Media-Release-February-8-2021.pdf

Urquhart, I. (2021). *Royalties: False Promise? Submission to the Impact Assessment Agency of Canada regarding the Grassy Mountain Coal Project*. Reference #1334. Jan. 8.

Weber, B. (2019a, Jan. 2). *Biologists pen letter over Alberta MLA's "misinformation" on conservation plans*. CBC News. https://www.cbc.ca/news/canada/edmonton/biologists-letter-bighorn-conservation-1.4964002

Weber, B. (2019b, Oct. 29). *UCP government tables climate plan for industry; retains key parts of old legislation*. Global News. https://globalnews.ca/news/6099073/alberta-greenhouse-gas-emissions-environment-ucp/

Weber, B. (2020, Dec. 24). UCP says it won't close any provincial parks. *The Globe and Mail*, A6.

Weber, B. (2021, Feb. 22). Critics ask why no details on government plan for parks management. *Pique Newsmagazine*. https://www.piquenewsmagazine.com/national-news/critics-ask-why-no-details-on-government-plan-for-parks-management-3446128

Weber, B. (2022, Aug. 17). *Releasing oilsands tailings into river only one solution being considered: Guilbeault.* CTV News. https://www.ctvnews.ca/climate-and-environment/releasing-oilsands-tailings-into-river-only-one-solution-being-considered-guilbeault-1.6031066

Winter, J., Dolter, B., & Fellows, G. K. (2021). *Carbon pricing costs for households and the progressivity of revenue recycling options in Canada*. Clean Economy Working Paper Series, Smart Prosperity Institute. June. Retrieved from https://institute.smartprosperity.ca/sites/default/files/Winter_Dolter_Fellows__working_version_June2021.pdf

Yewchuk, D. (2021, Oct. 19). Another year gone under the mine financial security program. *University of Calgary Faculty of Law Blog*. https://ablawg.ca/2021/10/19/ another-year-gone-under-the-mine-financial-security-program/

Yewchuk, D. (2022, March 15). Coal law and policy part eight: the results of the Coal Consultation and the return to the Alberta Land Stewardship Act. *University of Calgary Faculty of Law Blog*. https://ablawg.ca/2022/03/15/coal-law-and-policy-part-eight-the-results-of-the-coal-consultation-and-the-return-to-the-alberta-land-stewardship-act/

Zingel, A. (2021, Dec. 15). *Some northern residents vow to oppose federal regulations to release treated oilsands tailings water*. CBC News. https://www.cbc.ca/news/canada/north/oilsands-regulations-tailings-nwt-1.6285940

NOTES

1 Actual expenditures on parks fell from $89 million in 2018-19 to $66.6 million in 2020-21.

2 Details of the UCP's plan were described in numerous media and ENGO reports from March 2020 onward. The original government document, however, has been removed from the government website, along with news releases referring to it. A portion of the plan, showing the parks scheduled for closure, partial closure, or "partnerships," may be found here: https://web.archive.org/web/20200307163751/https://albertaparks.ca/media/6496183/parks-impacted-list.pdf.

3 The petition was initiated in March 2020; the closing date is unknown, although the webpage is dated 2021. See https://www.leadnow.ca/dont-let-kenney-privatize-nature/.

4 There is no reporting of the outcomes of the Indigenous consultations in the two-page summary.

5 The digitalized copy of this document is available at https://www.poltext.org/sites/poltext.org/files/plateformesV2/Alberta/AB_PL_1997_LIB.pdf (accessed April 8, 2022).

6 Riversdale is in turn a subsidiary of Hancock Prospecting, owned by Gina Rinehart, the Australian billionaire.

7 A map of coal leases on Category 2 land is available in Nigel Bankes (2021). The 420,000 hectares figure is from Fletcher (2021). A lease alone is not a permit to explore for, or to extract coal; these need to be obtained separately.

8 All the hearing transcripts and briefs may be obtained from the website of the Impact Assessment Agency of Canada, https://iaac-aeic.gc.ca/050/evaluations/proj/80101.

9 Protect Alberta Water and Rocky Mountains, https://protectalbertawater.ca/; Protect Alberta's Rockies and Headwaters, https://www.facebook.com/groups/albertansagainstcoal; Niitsitapi Water Protectors, https://www.facebook.com/NiitsitapiWaterProtectors/.

10 Parliamentary petition e-2912, https://petitions.ourcommons.ca/en/Petition/Details?Petition=e-2912.

11 Petition url: https://www.change.org/p/alberta-government-stop-open-pit-mining-from-happening-around-the-canadian-rockies. The author tracked the progress of the Facebook group and the petitions until the end of February 2021. As of April 12, 2022, the Change.org petition had 96,497 signatures.

12 The author read 1,261 of the comments that had been submitted as of Jan. 12, 2021, 853 of which were from individuals. The registry was closed for comments on Jan. 16, 2021.

13 Bill 214, Eastern Slopes Protection Act, was sponsored by MLA Rachel Notley and introduced in the legislature in April 2021. Retrieved from https://docs.assembly.ab.ca/LADDAR_files/docs/bills/bill/legislature_30/session_2/20200225_bill-214.pdf. For the fate of this bill, see Pulido-Guzman (2022).

14 Parliamentary petition e-3178, tabled March 22, 2021, with the minster's response (May 5, 2021), https://petitions.ourcommons.ca/en/Petition/Details?Petition=e-3178.

15 This statement is found at the bottom of the Updates section of the webpage: https://protectalbertawater.ca/.

16 For facilities in operation before Jan. 1, 2000, the emissions-intensity baseline was set as the average emissions intensity over the years 2003-2005. For facilities beginning operation after that date, the emissions-intensity baseline was determined by the EI of its production in its third year of operation. See Part 4 of the Specified Gas Emitters Regulation (Alberta Regulation 139/2007).

17 Note: these figures are likely under-estimated, as per Liggio et al. (2019).

18 Under the federal Greenhouse Gas Pollution Pricing Act, both the fuel charge and the OBPS will be applied in a province unless a provincial system for pricing GHG emissions has been deemed equivalent.

19 The Climate Change Emissions Management Corporation was created only in 2009, and the first grants were approved in 2010.

20 Eriel Tchekwie Deranger, interview with the APTN News Face to Face program, Feb. 11, 2020, https://www.youtube.com/watch?v=YfjMPa4SHkY.

21 Interestingly, when the carbon tax was brought in by the NDP government, the interim leader of the Progressive Conservative Party of Alberta at the time, Ric McIver, called the tax an "NDP wealth redistribution program" (Dykstra, 2015).

CHAPTER 8

THE FUTURE IS PAST: A POLITICAL HISTORY OF THE UCP ENERGY POLICY

Kevin Taft

"One does not establish a dictatorship in order to safeguard a revolution; one makes the revolution in order to establish the dictatorship."

— George Orwell, *1984*

THE PETROLEUM INDUSTRY held a special place in the United Conservative Party's (UCP) platform in Alberta's 2019 general election. The industry had been treated generously by the sitting NDP government, which had backed down on royalty increases, committed $3.6 billion to ship crude oil by rail, and feuded vigorously with provinces that threatened the industry's interests, even imposing a boycott of B.C. wine. The federal Liberal government paid $4.5 billion to buy the TransMountain pipeline and would commit a further $12.6 billion to expand its capacity to carry Alberta oil and bitumen to tidewater. (In 2022, the federal Liberals would also provide a $10-billion loan guarantee to help complete the project.)

Even so, the UCP platform made it seem Alberta's oil and gas sector was the victim of an unending stream of provincial, national, and international grievances. It claimed Alberta's petroleum industry had been "gutted" by the "NDP's ideological policies and red tape, and their 'social license' alliance with Justin Trudeau." Even bigger enemies lurked abroad: the "number one reason" for Alberta's economic stagnation and decline was, according to the UCP, "...the success of the foreign funded campaign to landlock our energy, a consequence of the Trudeau-Notley alliance." This alliance had scrapped pipelines, banned tankers, and regulated oil industry projects "out of existence." On top of that was the cash grab of the carbon tax, mentioned 31 times in the platform.

The platform cast the UCP as the industry's champions. A UCP government would immediately repeal the NDP's "job-killing carbon tax" and challenge Ottawa's carbon tax in court. It would spend $20 million a year on a new energy "war room" to "share the truth about Alberta's resource sector" and would commit $2.5 million to launch a public inquiry "...into the foreign sources of funds behind the anti-Alberta energy campaign." It would "pursue every possible pipeline to get our oil and gas to market," while cancelling the NDP government's "irresponsible" oil-by-rail scheme. The board of directors of the Alberta Energy Regulator would be replaced and timelines for approving projects reduced "by at least 50 per cent." Corporate taxes would be cut; a special minister appointed to slice through red tape; service rigs would be reclassified as "off-road vehicles, such as farm equipment;" laws would be passed to freeze royalty rates "in perpetuity;" processes for dealing with abandoned oil and gas wells would be streamlined to reduce reclamation costs to industry; new investment would be attracted; and thousands of jobs created.

Things did not go as planned for the UCP government. The energy war room, named the Canadian Energy Centre, stumbled through a series of embarrassments, was condemned as a waste of money by then-premier Jason Kenney's former employer the Canadian Taxpayer's Federation, and had its budget chopped sharply (Fedor, 2021). The Public Inquiry into Anti-Alberta Energy Campaigns was dogged by a sole-source contract controversy, went $1 million over budget, needed four deadline extensions, and found no wrongdoing. The court challenge of Ottawa's carbon tax failed. Corporate tax rates were cut and royalty rates kept low, but the industry shed 18,000 jobs from 2018 to early 2020, even before the impact of COVID was felt (Alberta Department of Energy, 2020, p. 28). In a spectacularly bad investment, the UCP government committed $1.5 billion of taxpayer money and $6 billion more in loan guarantees to Calgary-based TC Energy to build the Keystone XL pipeline (Government of Alberta website). Ten months later, when U.S. President Biden fulfilled a campaign promise and cancelled the pipeline's permit, the Alberta government lost $1.3 billion.

In short, the UCP's dreams met up with reality, and reality won. Reality, it turned out, was catching up with all of Alberta.

Alberta is an astonishingly fortunate place. Its farmland and forests are abundant, its cities inviting, its population young and well-educated, and its citizens generally expect to live in peace and prosperity. These benefits are supercharged by the province's immense petroleum resources that, under the constitution of Canada, are owned by the Alberta government. With a population one-third smaller than greater Toronto, Alberta owns the third largest oil reserves on the planet, more than all of Russia or the U.S. If it were a country, in 2020 it would be the world's seventh largest oil producer, and climbing. It is also one of the world's largest exporters of natural gas. Yet it has piled up huge public debt and has declining credit ratings; has high unemployment and real estate vacancy rates; and is at serious risk for an estimated $260 billion in unfunded environmental liabilities from the petroleum industry.

Perhaps the pre-eminent role model for the UCP is Ralph Klein, Alberta's premier from 1992 to 2006. In 2016, UCP leader Jason Kenney heaped adulation on "the late, great Ralph Klein," calling him "the most remarkable leader" for building "the Alberta Advantage" through deficit-busting, tax cutting, privatizing, and government downsizing. As Jason Markusoff (2016) wrote for *Maclean's*, in the policies and legacy of Ralph Klein, "Kenney has found a ...convenient template to neatly fill."

The irony is that the origins of the profound problems facing Alberta lay largely with the much vaunted "Klein revolution" of the 1990s, which gave the petroleum industry a decades-long lock on Alberta politics, allowing it to strip untold public resource wealth from the people of Alberta while simultaneously shouldering them with staggering environmental liabilities. The Klein revolution did something more profound than bring a new government to Alberta; it secured in place a new public-private structure of political power and a new regime of elected officials and corporate executives. There weren't just different people in the positions of premier, cabinet, and the legislature, there was a substantially different relationship between Alberta's government and its corporate leaders, who were heavily concentrated in the petroleum industry. This new "petroleum regime," which had been brooding for years, brought new norms and rules to government decision-making. It has remained in place continuously since 1993, through eight premiers, seven general elections, and the governments of three different parties, but it has been weakening since at least the 2012 general election. The ambition of the UCP government was to counter these weaknesses and secure the regime in place for years to come.

This chapter examines three closely entwined policy areas where Alberta government failures originating with the Klein revolution and continued by the UCP are now so immense they jeopardize the province's future and make it virtually impossible for the UCP to fulfill its agenda: climate change, the environmental liabilities of the oil industry, and royalties.

Prelude to a Revolution

Because the template for the UCP was the early Klein government, it's important to know the conditions that gave rise to that government. The 1980s brought political and economic whirlwinds to Alberta. Iran's Islamic Revolution drove world oil prices to highs that peaked in 1981 at $125US/barrel in 2021 dollars. Alberta boomed, while global inflation climbed to worrying heights, exceeding 12 per cent in Canada in 1981. Central banks raised interest rates to unprecedented levels, deliberately tipping their economies into recession to bring down inflation; mortgage rates in Canada exceeded 18 per cent in 1981 and 1982. As economies slowed, unemployment rose, passing 11 per cent in Canada in 1983.

The governments of Margaret Thatcher in Britain and Ronald Reagan in the U.S. responded to the economic turmoil they'd helped create with harsh policies that shrank governments, reduced taxes, weakened unions, de-regulated industries, promoted globalization, and privatized services. These policies were justified by economic theories celebrating the roles of markets, private businesses, and the singular pursuit of profit, while denigrating the value of government. As Reagan said in his first inaugural address, "In this present crisis, government is not the solution to our problem; government is the problem."

By the early 1990s these policies formed an internationally entrenched political, legal, and economic framework known variously as Thatcherism, the Washington consensus, and market fundamentalism. It prevailed as orthodoxy until its deep flaws were exposed by the 2008 global financial crisis, the growing catastrophe of climate change, and the COVID-19 pandemic. In its early years, market fundamentalism provided fertile ground for the Klein revolution; by the time of the UCP government, it was thin and barren.

The turmoil of the 1980s hit Alberta hard. In 1986, Don Getty's first year as premier, the price of oil collapsed, throwing Alberta into deep and prolonged recession. From 1987 to 1992, corporate profits in Alberta fell 36 per cent (Statistics Canada, 2015). Job losses and high interest rates caused widespread mortgage failures, and as the provincial government borrowed money to meet its obligations, it faced surging interest payments of its own. Provincial spending on public services was cut to well below national averages (Taft, 1997). To create private sector jobs the Getty government made a series of business investments, many of which were embarrassing and expensive failures. In the 1989 general election, the Progressive Conservatives (PC) managed to win 59 of 83 seats, but Getty lost his own seat and had to run in a by-election. Controversies kept plaguing the government, while opposition parties gained strength. By the time Klein won the PC leadership and replaced Getty as premier in December, 1992, Alberta was primed for change.

Revolution

Klein described his political strategy as finding a parade and getting in front of it. The parade created by market fundamentalists was so big that Klein and his main rival in the 1993 general election, Alberta Liberal leader Laurence Decore, were vying to lead it, one promising brutal cuts and the other massive cuts. When the votes were counted, Klein won. The parade followed fast on his heels (see Acuña, Chapter 2).

Of all the upheavals the Klein revolution brought, one is perhaps more enduring and costly than any other: the infiltration by the petroleum industry into the government of Alberta. "The lines between government and private business are breaking down," wrote *Edmonton Journal* columnist Mark Lisac (1995, p. 111) a few weeks after the election. Within months, the speed of the government's decision-making, combined with its bullying, polarization, and distortions, had Lisac wondering if he "...were watching the creation of a sugar-coated fascism" (p.142). Others had the same concern. Eventually, Lisac ruled against that concern, in part because "there was no measure of belief in violence as a political tool" (p. 158).

There was no question, though, that the Klein team opened the doors of government to corporate executives and welcomed them in, giving them direct influence on government decision-making. Lisac gave pages of examples in his book *The Klein Revolution*. Business leaders were invited onto committees reviewing the annual plans of government departments, including Mel Gray, president of Resman Oil and Gas, who reviewed the plans for energy and environmental protection. The government's financial review commission was led by Marshall Williams, a retired chairman of TransAlta, and he and two other TransAlta executives, Ken McCready and Harry Schaefer, held key roles in reviewing post-secondary education plans and drafting the province's economic strategy. The CEO and president of Syncrude, Eric Newell, reviewed the business plans of the public works and treasury departments, and advised government on longer-term strategies. As we'll see, Newell's impact on royalties would be of historic importance. Norm Wagner, chair of Alberta Natural Gas Company, managed policy discussions for the premier, had key roles on the government's "round tables" on both health and the budget, and became chair of the government's audit committee. Art Smith, chair of the petroleum division of SNC-Lavalin, became co-chair of the recently formed Alberta Economic Development Authority. Gwyn Morgan, a senior vice-president with Alberta Energy Company, helped review government business plans; he would become CEO of EnCana and then chair of scandal-plagued SNC Lavalin Group. Sherrold Moore, a vice-president of Amoco Canada Petroleum, held various positions, including on the government's round table on environment and economy. Jack Donald, chair of independent refiner and gasoline retailer Parkland Industries, chaired the tax reform commission. Dee Parkinson, a vice-president with Suncor, took on several roles for the government, including as a member of the tax reform commission. In short, members of Alberta's tightly knit corporate community, well-known to one another, centred in downtown Calgary and concentrated in the energy industry, became a *de facto* steering committee of the provincial government, the vanguard of the new petroleum regime. As Lisac wrote, Klein "clearly ran government in partnership with business — government as a joint venture" (p. 152).

No one at the time could know exactly where the new regime would lead, but its logic was inexorable. The notions that government and industry should be separate, and that arm's length norms should be respected, were left behind. It was made clear to civil servants and anyone else watching that in the world of market fundamentalism, corporate executives held sway and the way to build public prosperity was to serve private interests. The concentration of influence and power in such a narrowly based elite fused the values and interests of the petroleum industry onto the operations of the government. The individual and corporate membership in this elite changed as years passed, with the influence of oil sands players rising, but once the circuits of access and influence were established, they became permanent.

Above all these positions was the minister of energy, who sat on the most powerful cabinet committees; steered legislation and regulations crucial to the petroleum industry; was responsible for both the department of energy and the industry regulator; and oversaw the royalty regime. Historically, this position had been occupied by people independent of the industry. From the 1914 discovery of petroleum at Turner Valley, all the way through to 1989, only two people from the oil industry served as the minister responsible for the industry's governance: Charles Ross, from 1935 to 1937; and Don Getty, from 1975 to 1979. This pattern reversed in 1989, when then-premier Getty appointed Rick Orman to the position; Orman came from and returned to the oil industry (and would eventually play important roles in the UCP). Klein followed Getty's example and posted oil industry veteran Patricia Nelson to the job from 1992 to 1997.

In total, eight ministers of energy or provincial treasurers from 1989 to 2022 came from the energy industry, or joined it after politics, or both.[1] The most notable exception to this pattern was Margaret McCuaig-Boyd, an educator and college vice-president who served as minister of energy for four years under the NDP. The UCP replaced McCuaig-Boyd with industry insider Sonya Savage. A lawyer, Savage had worked nine years at pipeline giant Enbridge before becoming a senior executive at the Canadian Energy Pipeline Association. As energy minister, Savage helped negotiate the end of the NDP's crude-by-rail plan, which cost the government $2.1 billion in losses (Anderson, 2020), and the $7.5-billion cash-and-loan package for the Keystone XL pipeline, which cost $1.3 billion when it collapsed.

The pattern of industry insiders in key public positions extended to the various iterations of the industry regulator, especially the Alberta Energy Regulator (AER). Even so, the industry never stopped wanting more. A July 2017 report by the powerful Canadian Association of Petroleum Producers (CAPP) recommended the establishment of a joint government-industry steering committee comprised of the premier's office, three powerful cabinet ministers, the AER, and senior industry representatives, to provide "a whole-of-government approach" to protect investment and minimize industry costs. Whether this committee was established, likely informally to provide deniability, is not a matter of public record. "The whole of government in service to the oil industry" could be the mission statement for the regime running Alberta.

These incursions into government were in addition to the industry's ongoing political lobbying, influencing, and donations. As well, groups within the industry undertook direct political organizing. A group calling itself Protect the Patch, for example, got behind the Wildrose Party after the 2008 election, occupied key positions on the party executive, raised large sums of money, shaped party policies, and was tied to 13 candidates for the 2012 election (Taft, 2017, pp. 172-173). The political structure of the Wildrose Party became crucial to the rise of the UCP.

The Klein revolution coincided with dramatic technological improvements in oilsands extraction, changes in federal taxes that benefited the oil industry, and climbing prices for oil and natural gas. These created a prolonged economic boom in Alberta but set the stage for long-term decline that put the UCP in an impossible bind.

The Climate Crisis

The climate crisis has been damaging Alberta for years. Warming temperatures have contributed to pine beetle infestations; worsening wildfires, including those at Slave Lake in 2011 and Fort McMurray in 2016; Calgary's 2013 flood and 2020 hailstorm; and the 2021 heat dome that reduced crop yields and melted glaciers so rapidly the North Saskatchewan River changed colour from the run off. The likelihood and severity of events like these will increase as greenhouse gas emissions continue. The climate crisis is also a defining economic issue for Alberta, because the province's huge petroleum sector is exposed to the risk of obsolescence as the world strives to reduce the use of petroleum.

Rather than adapting to these challenges, the UCP commenced what Daub and others call "new denialism," avoiding outright denials of global warming while continuing to enact policies that make it worse (Daub, Blue, Rajewicz, and Yunker, 2021). The 2019 UCP platform never mentioned global warming and used the term "climate change" only twice. It stated the UCP was "committed to responsible energy development" including "action to mitigate greenhouse emissions and reduce their contribution to climate change." The only policy it listed was an explicit step backward: replace the 2017 NDP large emitters program by "restoring an updated version" of a 2007 program, including lowering the compliance price charged to large CO_2 emitters. This would, said the platform, "reassure investors."

Through the 1960s, 1970s, and 1980s, research conducted by governments, universities, research institutes, and the petroleum industry confirmed fears that rising CO_2 levels were driving up global temperatures, and in 1989 James Hansen, one of NASA's top scientists, led the news when he testified to the U.S. Congress that fossil fuel and other emissions were almost certainly causing the atmosphere to warm dangerously. The risks of global warming were headline news before the Klein revolution was launched.

In 1990, Alberta's ministers of energy and environment, Orman and Klein respectively, announced the Clean Air Strategy for Alberta. Spurred by this strategy, the department of energy produced "A Discussion Paper On The Potential for Reducing CO_2 Emissions in Alberta," confirming the science of global warming and presenting a preliminary plan to reduce Alberta's emissions. The Alberta government seemed onside with climate science.

It did not last. The Klein revolution, guided by Klein's coterie of oil executives, reversed the government's position on global warming. The discussion paper was quickly buried and the office that prepared it closed. Climate science was ringing a death knell for the petroleum industry and the new regime in Alberta was refusing to listen. Three decades later the UCP, true to the same regime, was still refusing to listen.

By 2001, the Klein revolution had done its intended work: Alberta's spending on public programs was relatively low; a flat tax benefiting the wealthy was in place; royalty and tax regimes were rewritten to the advantage of industry; and opposition of all kinds was marginalized. While oil prices were flat, natural gas prices were high, filling the government's treasury and boosting the economy. The seamless merging of government with the oil industry was perfectly symbolized in 1996, when a group of four government MLAs and three cabinet ministers formed their own oil company, nicknamed Tory oil, operating in the full knowledge of the premier and cabinet.

The climate change issue, however, was a problem. The federal government signed the Kyoto Protocol, committing Canada to reducing emissions, a direct threat to the petroleum industry. Polling showed more than 70 per cent of Albertans supported the Kyoto Protocol (Chase, 2002).

As we'll see in a later section, the Alberta government had recently made royalty changes that were opening floodgates of capital, and managing public opinion for this explosive expansion was crucial. To maintain its legitimacy, Alberta's petroleum regime had to subvert public confidence in climate science. It took to the task enthusiastically.

The main component of the government's efforts was a $1.5-million campaign of television, radio, and newspaper advertisements discrediting the Kyoto Protocol, launched in September 2002 by (ironically) then-environment minister Lorne Taylor (CBC News, 2002). It was a large campaign for the time and it worked. Within months, support for Kyoto had collapsed to 27 per cent of Albertans, while 60 per cent were opposed (Ipsos Reid, 2002).

Alberta's campaign to shift the public against Kyoto was a turning point in Alberta's recent history. It was what sociologist Linsey McGoey (2019) would call a campaign of strategic ignorance. The growth of the industry depended on the ignorance of the public, but the public were knowledgeable, so replacing the public's valid knowledge with misinformation became a strategic priority.

Once the regime turned a majority of Albertans against the science of global warming, it kept them in ignorance with a steady stream of reinforcing messages, from Klein's joke about dinosaur farts in 2002, through the many pronouncements of the Harper conservatives, to the combative statements of the UCP, backed up constantly by the industry. For many Albertans, climate change skepticism seemed to become an issue of identity, a way to belong to the community of Albertans and differentiate from other Canadians. It was also a way to show support for an industry that had made many of them prosperous, and it tapped into the politics of grievance taking hold in the province. Alberta politicians were quick to reinforce this identity for their own agendas: the PCs against Ottawa and Quebec; the NDP against British Columbia; and the UCP against a long list of governments and organizations.

The centrepiece of UCP climate change policy is the Technology Innovation and Emissions Reduction (TIER) program. In 2007, the PC government had begun charging the province's largest emitters $15/tonne for emissions above 100,000 tonnes annually, and the NDP raised the charge to $30/tonne. These programs were replaced by TIER, which the UCP said "protects [emitters] from the full cost of complying" with federal regulations.[2] Under pressure from federal climate change plans, TIER increased the charge for emissions to $40/tonne, but allowed facilities to choose "less stringent" benchmarks and enabled the government to grant exemptions to corporations on the basis of "economic hardship," which it did six times for oilsands projects by 2020.[3]

TIER and previous programs raised vast amounts of money (in 2020 over $500 million). Before TIER, these funds had to be spent on projects that reduced emissions or helped Alberta adapt to climate change.[4] TIER broadened the use of these funds. They could be channelled back to oil companies to help them increase production, and in 2021 TIER paid $20 million for the UCP's energy war room. TIER had become part of the UCP's service to the oil industry, as former environment minister Jason Nixon made clear on Sept. 23, 2020: "At the end of the day, it is industry money that they have contributed to that fund. Our commitment to the industry was that we would use it to be able to help advance the industry."[5]

Regardless of the UCP's rhetoric, the physical reality of the climate crisis continues. Advances in green energy technologies and shifting politics beyond the borders of Alberta and Canada mean Alberta is often politically isolated. The UCP's reluctance to take substantial action on global warming sets up the province to be unnecessarily vulnerable.

Unfunded Environmental Liabilities

In 2018, I was invited to speak at the Petroleum Club in Calgary about my book, *Oil's Deep State*. I was surprised: my book criticized the oil industry, and this was the club that once revoked premier Lougheed's membership when he raised royalty rates (Simpson). The invitation was from the Petroleum History Society and was arranged by Nick Taylor, a self-styled "renegade oilman" who liked to stir up things.

By chance, on the day I spoke, the *Edmonton Journal* and *Calgary Herald* carried full-page ads paid by undisclosed supporters of the oil industry, with giant red STOP signs and bold red and black headlines: "Tell University of Alberta No honorary degree for David Suzuki." Suzuki already had 28 honorary degrees, including one from the University of Calgary, but that didn't matter now: he had become an outspoken critic of the oil industry. The ad was part of a vitriolic campaign intended to force a major university into submission, and when the university refused to submit, one estimate was it lost $200 million in funding.[6*]

My book's cover carried an endorsement by Suzuki, and as I put copies for sale at tables deep in the Petroleum Club, I enjoyed the irony, even as some patrons grumbled. My talk was politely received, though sales barely covered expenses. (No fee was offered.)

As the small crowd dispersed, one person mentioned a presentation two months prior by a vice-president of the AER named Robert Wadsworth, who said Alberta's petroleum industry was sitting on $260 billion in unfunded clean-up liabilities. "You mean $260 *million*," I said. "No. I mean $260 *billion*," the person confirmed, and told me I could find the details on the society's website, maintained by a retired oil executive. I exchanged a glance with Regan Boychuk, an independent energy policy researcher who had listened to my presentation. This was a giant issue that hadn't been made public.

At home, I went straight to the website. There was no trace of the presentation slides. I inquired and was told someone had "neglected" to write up the notes from that meeting, effectively keeping the $260-billion liability off the public record. To his credit, Boychuk pursued the matter, and with a team of journalists including Mike De Souza of the *National Observer* and others, eventually obtained the presentation through freedom of information requests and made it public (De Souza, Jarvis, McIntosh, and Bruser, 2018).

When the story of the $260-billion liability broke, the governing NDP, opposition UCP, AER, and industry all deflected the issue. It was "hypothetical" and "a worst-case scenario," they claimed, which the media dutifully reported. The presentation said something quite different: "The liability displayed is likely less than the actual cost," and "The number is expected to grow as more data becomes available." In the legislature, David Swann, the solitary member of the Alberta Liberals, moved for an emergency debate on the issue, but his request was denied after both the NDP government and UCP opposition spoke against his motion, and the UCP did not raise the matter in question period (Alberta Hansard, Nov. 5 and 6, 2018).

The campaign against Suzuki's degree, the missing information from the website, and the co-ordinated spin from politicians, regulators, and industry, revealed the fervour with which the industry was protected. A network of decentralized but like-minded cells sustained the revolution begun in Alberta decades earlier.

Wadsworth's $260-billion presentation needed to be taken seriously, because the AER was responsible for regulating the industry's environmental liabilities and had comprehensive and detailed information. The presentation exposed the scale of a problem that had built up for decades, and which the UCP government would soon inherit. Here are some of the presentation's main points, dated February, 2018:

- The AER's internal estimate for cleaning up Alberta's oilsands mines, including tailings ponds, was $130 billion. This was a "rough estimate" and was "expected to grow." The presentation compared this to the $27.8 billion reported to the AER by industry for the same liability. The presentation said the security collected from industry to cover these costs was $1.4 billion.
- The AER's internal estimate for cleaning up conventional oil and gas wells and sites, and in situ oil sands operations, was $100 billion. The liability reported to the AER by industry for this was $30.1 billion. The actual security collected from industry to cover this was $0.2 billion.
- The AER's internal cost estimate for remediating the province's pipelines was $30 billion. The liability estimated by industry was zero, and no security was collected.
- These problems were getting larger because the number of inactive and marginal oil and gas wells in Alberta was rising and far exceeded the number of commercially productive wells; a growing number of companies did not have the capacity to meet these liabilities; there was "insufficient collection of security to meet closure commitments;" and there were no timelines for abandonment.
- Alberta did not measure up to New Mexico or Texas in terms of security deposits or timelines for abandonment.

In sum, as of 2018 there was an estimated $260 billion cost to clean up the oil industry, likely more; there was $1.6 billion collected from the industry to cover it, less than one per cent of the liability; and a rising number of companies were unable to meet their obligations. These are not paper liabilities that will disappear with a write-off; these are physical liabilities with serious risk to public safety, farmland, forests, water, and air. These liabilities represent *almost five times the cost of the Alberta government's total spending in 2018 and are growing by an estimated $8 billion to $11 billion every year* (Ascah, 2021).

The unfunded liability reveals a failure at the heart of the regime's business model and puts the UCP in a bind between two of its largest political bases, the oil industry and rural Albertans. Markusoff's report for CBC in May, 2022, noted, "The top 12 ridings for UCP memberships are all outside the main cities." As the scale of the failure becomes apparent, trust in the industry, government and regulator declines, a decline that seems likely to hit the UCP. "In Alberta today we have a broken social contract," was how the director of a landowner group described the situation in 2020. "The reason we got into this situation is there's no effective timeline requiring these companies to reclaim these wells. There are rules on the books but they're not enforced" (Glen, 2020).

Requirements to reclaim well sites were formalized under the provincial industry regulator, the Energy Resources Conservation Board (ERCB), in the 1970s, the same period when reclamation was first required for oil sands operations. But the ERCB struggled to balance the public's concerns for the environment with an aggressive industry. After retiring, ERCB chairman Vern Millard reflected that regulatory decision-making created confrontations between industry and the public, "with the public almost always a loser" (Jaremko, 2013, p.106).

Since then, there have been attempts to hold the industry to account for its environmental liabilities, such as the ERCB's 1989 plan titled Recommendations to Limit the Public Risk from Corporate Insolvencies Involving Inactive Wells, but every attempt has been defeated by industry and its political allies.

By 1999, the number of inactive wells passed 34,000, many of them inactive for more than 10 years (Robinson, 2014, p.4). Exacerbating the situation, by 2003 most well sites were rarely if ever visited by inspectors, the regulator instead relying on an industry honour system (Way and Simpson-Marran, 2019; Riley, 2018). By 2005, there were almost 45,000 inactive wells, more than 10,000 of which had been inactive longer than 25 years. By 2012, there were 65,000 inactive wells. As Barry Robinson (2014, p. 10) wrote, the regulator's inactive well programs were not about enforcement, they were "amnesty programs." High world oil prices meant the years 2002-2014 were perhaps the most profitable in the history of Alberta's petroleum industry; if the industry wasn't going to pay reclamation costs in these years, it likely never would.

In 2014 the ERCB was replaced by the AER. The AER was based on the Responsible Energy Development Act of 2012, which was the culmination of years of work on "regulatory enhancement" by the industry and the department of energy (Alberta Hansard, 2012). The AER was a benchmark of success for the petroleum regime.

While the ERCB had been required under legislation to consider if energy projects were in the public interest, the AER had no such obligations (Low, 2011). Rather, its purpose as declared in the act was "to provide for the efficient, safe, orderly and environmentally responsible development of energy resources in Alberta." The AER seemed less a regulator and more a facilitator. Its budget was fully paid by the industry, granting the industry immense sway over its leadership, and its first chairman was a former executive at EnCana and former president of CAPP. Under the new legislation, the government transferred to the AER all department of environment staff working on issues relating to the energy industry, gutting the government's own capacity to monitor and police the industry. The AER looked like institutionalized conflict of interest.

In the legislature, the Wildrose voted against the bill to create the AER, not because it went too far, but in part because it did not "streamline" regulatory processes enough (Alberta Hansard, debate on The Responsible Energy Development Act, November-December 2012), a precursor to the 2019 UCP platform promise to "cut red tape."

By 2018, the number of inactive wells had soared to 90,000, in addition to which there were 107,000 marginally productive wells. When the UCP formed government in 2019, the situation was breaking into crisis: landowners were suing oil companies for failing their lease obligations; untended wells were leaking onto farms, forests, and water;[7*] oil companies were suing each other for concealing well liabilities; oil companies were declaring bankruptcy; bankruptcy trustees were in court battles with the Orphan Well Association; and rural municipalities were up in arms over unpaid taxes. The UCP's political base was fracturing into a mosaic of skirmishing pieces, a dynamic that no doubt contributed to the UCP's leadership struggles.

The UCP tried to paper over some of these cracks, but its first loyalty always seemed to be with the oil and gas industry. In October 2020, after tense negotiations involving the oil industry, which wanted rural property tax assessments sharply reduced, and rural municipalities, which relied on these assessments to pay their costs, the UCP announced a three-year property tax exemption for new wells and pipelines, and a 35 per cent cut in tax assessments for a range of oil and gas wells.[8] While the industry regarded these breaks as "a positive step," the municipalities were more cautious. Their caution was justified. On March 8, 2022, the Rural Municipalities of Alberta reported that oil and gas companies owed $253 million in unpaid taxes, up 47 per cent in two years, "even as Alberta's economy improves, oil prices skyrocket, and new oil well drills increase substantially."[9]

The UCP government faced even bigger problems with Alberta's oilsands mines, among the biggest open-pit mines in the world. When these mines separate the bitumen from the grains of sand it coats, they create acutely toxic liquid waste called tailings that can kill birds, fish, and mammals within hours of contact. These tailings are stored in ponds behind some of the world's largest earthen dams. From the early days of oil sands commercialization, both industry and government struggled to cope with tailings and invested heavily in the search for solutions. None have been found. The total area covered by tailings ponds and structures exceeds 250 square kilometres and is growing (Commission for Environmental Cooperation, 2020, p.25). The mines and ponds are close to a major waterway, the Athabasca River — some are on its banks — and have been leaching contaminants into it for many years (Kelly, Schindler, Hodson, Short, Radmanovich, and Nelson, 2010).

Oilsands mine operators are legally responsible for reclaiming these mines and tailings, and the AER is to make sure they do so. Signs the AER is failing at this have been growing for decades. Alberta's auditor general raised concerns in 1998, 2000, 2004, and 2009, but the concerns went unresolved. In 2015, the auditor general published a more detailed and worrying report, and in 2021 raised the alarm louder. When the report was made public and tabled in the legislature, the official opposition NDP were quiet on the issue. In response to media inquiries, the UCP government said it was undertaking consultations, but provided little detail (Weber, 2021).

Oilsands companies do not pay into security funds if a project's assets are at least three times the size of its liabilities, and if the project has more than 15 years of operating life remaining. (The resource is so vast that many projects could operate 50 years or more.) This approach enables companies to extract most of the value of the resource while deferring clean-up costs to the end of the project's life. If a company were to fail before then, or if demand for oil dropped, the cost of clean-up could easily fall on the public.

In 2021, the auditor general described ways companies "inappropriately" extend a mine's estimated life and defer security payments (pg. 32; 34). For example, companies combine the value of non-mining in situ operations with their mines to increase the calculated value of the mines, or combine a new mine operation with an old one to delay security payments for the old mine.

As well, oil companies estimate the value of their mines' remaining reserves by assuming the price of oil will steadily rise for the life of the mines, sometimes into the 2060s. For example, Suncor's 2020 financial statements value the reserves at its Fort Hills mine assuming the price of Western Canada Select oil will rise two per cent a year "over the life of the project" to the year 2061 (Suncor). This is where the industry's efforts at framing the climate crisis are particularly crucial. If the industry can convince enough Albertans the climate crisis is a distant worry or even a hoax, then oilsands operators can more easily claim the price of oil will rise indefinitely. This allows them to inflate the future values of their mines and avoid paying security funds. If, on the other hand, the climate crisis is real and demand for oil begins to decline in the later 2030s (Mercure, Salas, Vercoulen, Semieniuk, Lam, Pollitt, Holden, Vakilifard, Chewpreecha, Edwards, and Vinuales, 2021), the mines may already be nearing the last 15 years of their viable operations, requiring companies to start paying their reclamation costs much sooner than they want.

The industry has quietly anticipated the risk that the climate crisis will cause a permanent collapse in oil prices, and with virtually no one in the public being aware, the Government of Alberta has willingly taken it on. "If an abrupt financial and operational decline were to occur in the oil sands sector," warned the July 2015 auditor general's report, "it would likely be difficult for an oil sands mine operator to provide this security...It is important to recognize that the department has accepted the risk of not protecting against a broad based and rapid structural decline in the oil sands sector..." (pg. 27-28). A broad-based and rapid structural decline in the oilsands industry is exactly what an effective response to the climate crisis will bring, and taxpayers will be on the hook for the industry's mess. The UCP completely ignores this risk.

Royalties

In most of the world, including Alberta, the state owns mineral resources in the ground. If a private corporation wants to develop those resources it pays the state a royalty to buy them. In a healthy democracy, there is an inherent tension between the state as owner and seller of the resource, and the corporation as buyer, the former wanting a high long-term price and the latter wanting to pay the least amount possible.

The UCP 2019 platform made an unequivocal promise to conventional oil and gas producers: "A United Conservative government will guarantee in law that the royalty regime in place when a well is permitted will remain in place for that project in perpetuity." To the much larger oil sands industry, the platform made clear there would be no change to the oilsands royalty regime.

Ottawa transferred ownership of natural resources to Alberta in 1930, and the United Farmers' government under premier John Brownlee immediately levied a five per cent royalty on the industry and began efforts to improve regulation. (Finch, 2008; auditor general of Alberta 2006-2007, p.97). The industry mounted intense campaigns against these moves, including billboards and court challenges (Breen, 1993), but governments stood their ground. In 1935, the new Social Credit government of William Aberhart doubled the royalty rate to 10 per cent, and in 1938 established the Petroleum and Natural Gas Conservation Board, distant forerunner of the AER (Jaremko, 2013, p.160).

In 1941, the Socreds under Ernest Manning proposed raising the maximum royalty again, to 12.5 per cent. "Alberta oilmen went on strike," wrote historian David Finch (2008), "and refused to pay the higher rates," claiming the increase was illegal and submitting payments based on the old rates. After a debate in the legislature in which an independent MLA led the cause of the oil industry, the legislature voted to impose the higher rates but promised to review them every 10 years. When the first 10-year review was conducted in 1951, Alberta's oil production was soaring. Once again, the industry campaigned aggressively against higher royalty rates, and once again the Socred government stood its ground and raised the maximum to 15 per cent. When rates came up for review in 1961, industry again lobbied hard against their increase, and again the Socreds raised them, this time to a maximum of 16.66 per cent.

By the next 10-year review, the Progressive Conservatives under Peter Lougheed were in government. After famously bitter disputes with the industry, royalties were further raised, and from 1974 to 1985 the Lougheed government collected an average of 34 per cent of the value of total petroleum production through royalties and sales of crown leases (Anielski, 2001).

The Lougheed government emphasized growth in oilsands production to offset long-term decline in conventional production. The two major oilsands producers through the 1970s and 1980s were the open-pit mines of Suncor and Syncrude. These projects were so big and complex the government required them to go through their own application processes and negotiated individual royalty systems for each, which evolved with changing conditions. After oil prices fell in 1986, royalties were lowered, but the objective of the system was still to collect from 21 per cent to as much as 40 per cent of the selling price (Alberta Department of Energy, 1987). Then in October 1992, energy minister Orman reduced royalties further in "the most sweeping changes since 1974" (Alberta Department of Energy, 1995).

There would be no return to the royalty levels of the 1970s, not even when oil and natural gas prices soared again, as gas did from 2000 to 2009, and oil did from 2004 to 2014, and in 2022. Instead, royalty rates went into long-term decline, falling below the level of the Manning Socreds, then below the Aberhart Socreds, to some of the lowest levels since Alberta gained ownership of the resource in 1930, which is where the UCP promises to keep them.

Economist Mark Anielski has analyzed the share of total industry sales the Alberta government collects through royalties and bonus sales from all petroleum products.[10] In the last year of Social Credit, 1970, the government collected 20.4 per cent of the value of total production. The highest portion was collected under the Lougheed PCs in 1978: 48.9 per cent. From 1998 to 2011, the average portion collected by government was 16.9 per cent. From 2012 to 2020, the average dropped to 7.2 per cent.

It is sobering to compare royalties with environmental liabilities, which increased an average of $9.7 billion per year from 2012 to 2018 (Ascah, 2021, p.13). Without some combination of higher royalties and an immediate resolution to industry's unfunded environmental liabilities, the Government of Alberta may well be going deeper into debt with each barrel of oil produced.

The acquiescence of the UCP government to the petroleum industry has its roots in the period 1993-1997, the first years of the petroleum regime and the time Nelson served as energy minister. Nelson (Patricia Black before 1998) had worked 15 years at Suncor and other oil companies before winning a seat in the legislature. Premier Klein appointed her his first energy minister in December 1992.

Syncrude, Suncor, and smaller oilsands operations were producing 350,000 barrels of synthetic crude a day and were nicely profitable but chafing under the agreements they'd signed with the Lougheed and Getty governments. The leading industry association in those days was the Alberta Chamber of Resources, and sensing political opportunity, it ramped up its National Task Force on Oil Sands Strategies, led by Syncrude CEO Eric Newell. The task force worked relentlessly, and in 1995 produced its plan for oilsands development (Alberta Chamber of Resources, 1995; Glenbow Museum, n.d.; Urquhart, 2018).

Energy minister Nelson, eager to help, converted the industry's plan into public policy. Working closely with industry leaders such as Newell, she standardized approval processes for new projects; convinced the federal Liberal government to allow bitumen producers much more generous tax write-offs for capital investments; and laid the groundwork for the Generic Oil Sands Royalty Regime, enacted in 1997 by Steve West, Nelson's successor as minister of energy. This was a one-size-fits-all approach to oilsands projects, bringing an end to negotiating each project on its own merits. The new royalty regime collected only one per cent of the gross revenues of oilsands producers until the investors had recovered the full capital cost of their investment, at which point the royalty climbed to 25 per cent of net revenues. In effect, the government paid for the oil company's investment by foregoing almost all royalties until the capital cost of an oilsands operation was fully recovered. It was a stark contrast to the original Syncrude deal, in which the Alberta government got a 50 per cent royalty on net profits from year one (Hustak, 1979, p.178).

Newell was delighted. "(P)art of the task force work was to identify the key barriers," he said. "...it was amazing how easy it was, relatively easy it was to figure out how to knock down those barriers or modify them..." (Glenbow Museum, n.d.).

There was a reason it was easy: Nelson effectively turned the leadership of her portfolio over to the industry. "They were very good, the industry," she told an interviewer from the Glenbow Museum in 2011.

> I created a kitchen cabinet that I relied on heavily. I had people from...every aspect of the industry...I would bring in and they would be part of this kitchen cabinet, and we would meet every Saturday morning. And they would sometimes just beat me up fiercely, and at other times I could go to them and say, "What do you think of this?" and they'd either say, "Perfect!" or "Oh, my word. You better have a look at this, this and this." ...all of these people were fundamentally key to helping with the change that we went through.

The Klein government had become an instrument of the oil industry, a role the UCP intended to continue.

Alberta was selling its oilsands at bargain prices and investors rushed from around the world to cash in, overheating the economy. In 2005, TD Economics reported that per capita GDP in the Calgary-Edmonton corridor was a "gigantic" 47 per cent above the Canadian average.[11] In 2006, the municipal council of Fort McMurray/Wood Buffalo, where most of the oilsands were located, formally requested the EUB to slow down approvals of new oilsands projects because the region was overwhelmed. The EUB, operating under legislation brought in by Nelson, said it did not have the mandate to do that (Taft, 2017, p. 162). The pace of expansion accelerated.

The following year, 2007, concern that the royalty regime had tilted too far in favour of industry tore into political debate. Klein had stepped down as premier and candidates to replace him declared the royalty regime favoured industry over Albertans. The new premier, Ed Stelmach, appointed a royalty review panel that came to a blunt conclusion: "Our review revealed that Albertans do not receive their fair share from energy development and they have not, in fact, been receiving their fair share for quite some time... Albertans own the resources. The onus is on government to re-balance the royalty and tax system to ensure a fair share is collected both currently and as circumstances change" (Government of Alberta, 2007, p.4).

The report of the royalty review panel was independently reinforced by the auditor general's *2006-2007 Annual Report*, which included a 42-page review of the royalty system. The royalty system was collecting less than was originally intended, wrote the auditor general, and internal royalty reviews had "design flaws" (p.111). It was left to the reader to ponder who had designed those flaws into the system, and why. The audit found internal department of energy estimates that said it could collect "$1 billion or more per year without stifling industry profitability" (p.92), but was not doing so. The report pointed the reason for this toward "those responsible for the royalty regimes" (p.91), presumably the minister of energy and his or her colleagues.

The petroleum regime, complacent from a decade of easy success, went ballistic. (Full disclosure: as leader of the official opposition through this period I was witness to the industry's outrage in private meetings, phone calls, and public events.) The members of the royalty review panel were subject to verbal attacks on their professional and personal qualifications, as was the auditor general. As the 2008 general election approached the three major political parties — PC, Liberal, and NDP — called for royalty increases, while the industry mounted counter campaigns and drastically curtailed donations to both governing and opposition parties (Taft 2017, Chapter 13). The Stelmach PCs won the election handily, but in the face of intense resistance premier Stelmach was unable to enact significant royalty increases and, despite leading his party to a majority, was pushed out before the end of his first term as premier.[12]

If the 2002 public relations campaign that converted many Albertans into climate change skeptics was a turning point in public opinion, the intense controversy over royalties was a turning point for political parties. After the 2008 election, a segment of the oil industry worked to spur the rise of the Wildrose party; the PCs were splintered by intense internal feuds; and the Alberta Liberals fell into rapid decline. The extraordinary twists of the 2015 election produced an NDP majority that caught everyone, including the NDP, by surprise. Backed heavily by smaller oil and gas companies (Laxer, 2021, p. 36), Kenney resigned from federal politics to lead a process of merging the Progressive Conservative and Wildrose parties into the UCP. By 2019, the PC party of Lougheed and Klein did not exist, and Alberta had become a two-party province governed by the UCP.

But the political stability of the past remained entirely elusive. In May 2022, Kenney reluctantly announced he would step down as UCP leader and premier of Alberta after getting just 51 per cent approval from party members in a leadership review. The results were made public by UCP chief returning officer Orman, an enduring member of Alberta's petroleum elite who glided between politics and industry and had served as energy minister under PC premier Don Getty.

Many things were blamed for Kenney's failure, including his COVID policies, his hyper-partisan abrasiveness, and corruption in his 2017 leadership bid, but something larger was at play. Kenney was Alberta's sixth premier in 16 years. They came from three different parties and only Rachel Notley served a full term. The contradictions created by the Klein revolution mired these premiers and their political parties in fractures caused by austerity among abundance; deficits among profits; concealed liabilities among empty regulations; conspiracy theories and grievance politics, and the UCP's energy policy did little but increase the political pain. The prolonged turmoil was a sign that Alberta's petroleum regime was in a long-term crisis, a crisis that was not soon to end.

Prelude to the Next Revolution

Alberta's petroleum regime, firmly established in government during the Klein revolution and currently manifest in the UCP, rests on three public policy legs: denying the threat of climate change (in deed, if not in word); side-stepping responsibility for environmental liabilities; and paying royalties at rates below fair value. Like the three legs of a milking stool, these policy legs provide a stable base. Should one leg fail, however, the whole structure is destabilized. The role of the UCP government is to keep the three legs intact, as its 2019 platform made clear. Ironically, it is a role made impossible by policies of the very governments the UCP wants to emulate.

Doing nothing is not an option for the UCP. The industry-friendly policies of the petroleum regime have allowed many weak companies to operate for decades. Detailed analysis by the Alberta Liabilities Disclosure Project indicates that 49 per cent of Alberta's oil and gas well licensees are, according to the AER's own measures, effectively insolvent, unable to pay the costs of cleaning up their own wells and facilities (Boychuk, Anielski, Snow Jr., and Stelfox, 2021, p.4). While these companies milk the system, the number of inactive and orphan wells increases every year. And that's only one of several problems. The environmental liabilities of the oilsands grow daily. Provincial government finances are mired in debt only temporarily disguised by oil price spikes such as in 2022. As the climate crisis intensifies, the future of the industry dims and flickers. Fractures beset the UCP's political base. For the UCP, to stand still is to crumble.

But to do something meaningful seems politically impossible for the UCP. If the UCP requires Alberta's oil and gas companies to more realistically account for the risks of climate change; or to clean up their almost 200,000 inactive and marginal wells; or to address the auditor general's concerns about unfunded environmental liabilities of oil sands operations; or to return royalty levels even to those of the Socreds, it will tip many of these operations into bankruptcy and cause a cascade of crises.

There are always options, but they require standing up to the petroleum industry; an old habit Alberta governments have lost. The provincial government could, for example, strike a panel to examine and report publicly on the full extent of environmental liabilities of Alberta's petroleum industry, including wells, oilsands, pipelines, and all associated sites and facilities. The panel could be modelled on the 2007 Royalty Review Panel, consisting of independent experts, perhaps mostly from outside the province. The panel would make recommendations for action and would be required to share its entire report with the public. This will not be done in the foreseeable future.

There is probably no lasting way forward for the UCP — or any other party — that does not entail a complete overhaul of the Alberta government's approach to the petroleum industry, including an overthrow of the petroleum regime. The turmoil of the 1980s laid the groundwork for the Klein revolution, and the Klein revolution led to the UCP. The turmoil of the 2020s is laying the groundwork for the next revolution, for which the UCP seems utterly unprepared.

References

Alberta Chamber of Resources. (1995). *The oil sands: A new energy vision for Canada*. Report of the National Oil Sands Task Force.

Alberta Department of Energy. (1987). *Oil and gas fiscal regimes comparison of British Columbia, Alberta and Saskatchewan*, July.

Alberta Department of Energy. (1990). *A discussion paper on the potential for reducing CO2 emissions in Alberta, 1988-2005, executive summary*. Energy Efficiency Branch, Alberta Department of Energy, September 1990. Alberta Legislature Sessional Paper 547/2002, 2 Session, 25 Legislature.

Alberta Department of Energy. (1995). *Oil and gas regimes of the western Canadian provinces and territories*, May.

Alberta Department of Energy. (2020). *Annual report 2019-2020*.

Alberta Federation of Labour. (2016). *Royalty policy is the biggest decision any Alberta government has to make*. January. Retrieved from https://d3n8a8pro7vhmx.cloudfront.net/afl/pages/2770/attachments/original/1454352012/AFL_Advice_on_AB_energy_and_royalty_policy_26Jan201r2FINAL.pdf?1454352012

Alberta Hansard. (2012, Oct. 31). Legislative Assembly of Alberta.

Anderson, D. (2020, Aug. 28). *Plenty of finger pointing, little transparency, in province's $2.1 billion crude-by-rail losses*. CBC News. https://www.cbc.ca/news/canada/calgary/crude-by-rail-losses-blame-alberta-1.5704350

Anielski, M. (2021). *Alberta oil and gas rent capture 1962-2021*. Updated Nov. 20, 2021. Edmonton: Anielski Management Inc.

Ascah, R. (2021). *Alberta's public debt: Entering the third crisis*. The School of Public Policy Publications, SPP Pre-Publication Series. University of Calgary. July. https://www.policyschool.ca/wp-content/uploads/2021/07/AF22_AB-Public-Debt_Ascah.pdf

Auditor General of Alberta. (1998-1999). *Annual report*.

Auditor General of Alberta. (2006-2007). *Annual report*, Vol 1.

Auditor General of Alberta. (July 2015). *Report of the Auditor General of Alberta*.

Auditor General of Alberta. (June 2021.) *Report of the Auditor General of Alberta*.

Breen, D. H. (1993). *Alberta's petroleum industry and the conservation board*. Edmonton: University of Alberta Press.

Boychuk, R. (2020). *Alberta's idyllic moment: 1991-92*. Unpublished paper.

Boychuk, R., Anielski, M., Snow Jr., J., & Stelfox, B. (2021). *The big cleanup*. The Alberta Liability Disclosure Project, Calgary.

Burleton, D. (2003). *The Calgary-Edmonton corridor: Take action now to ensure tiger's roar doesn't fade.* TD Economics Special Report. TD Bank Financial Group. April 22.

Canadian Association of Petroleum Producers. (2017). *A competitive policy and regulatory framework for Alberta's upstream oil and natural gas industry*, July.

CBC News. (2002, Sept. 18). *Alberta launches campaign against Kyoto*. http://www.cbc.ca/news/canada/alberta-launches-campaign-againstkyoto-1.349305

Chase, S. (2002, May 31). Albertans support Kyoto accord, poll says. *The Globe and Mail*. Retrieved from http://www.theglobeandmail.com/news/national/albertans-support-kyoto-accord-pollsays/article4135778/

Commission for Environmental Cooperation. (2020.) *Alberta tailings ponds II. Factual record regarding submission SEM-17-001*. Montreal. http://www3.cec.org/islandora/fr/item/11861-alberta-tailings-ponds-ii-factual-record-north-american-environmental-law-and-en.pdf

Daub, S., Blue, G., Rajewicz, L., & Yunker, Z. (2021). Episodes in the New Climate Denialism. In Wm. Carroll (ed.), *Regime of Obstruction: How Corporate Power Blocks Energy Democracy*, pp. 225-248. Edmonton: AU Press.

De Souza, C. J., McIntosh, E., & Bruser, D. (2018, Nov. 1). Alberta regulator privately estimates oilpatch's financial liabilities are hundreds of billions more than what it told the public. *National Observer*. https://www.nationalobserver.com/2018/11/01/news/alberta-regulator-privately-estimates-oilpatchs-financial-liabilities-are-hundreds

Energy Resources Conservation Board. (1989). *Recommendations to Limit the Public Risk from Corporate Insolvencies Involving Inactive Wells*, December.

Fedor, T. (2021, March 17). *Is Alberta's $30M energy war room living up to its intended purpose?* CTV News Calgary. https://calgary.ctvnews.ca/is-alberta-s-30m-energy-war-room-living-up-to-its-intended-purpose-1.5351477

Finch, D. (2008, March 1.) The great royalty debate. *Alberta Views Magazine*. https://albertaviews.ca/great-royalty-debate/

Glen, B. (2020, Feb. 27.) Tensions on the rise in Alberta's oil country. *The Western Producer*. https://www.producer.com/news/tensions-on-the-rise-in-albertas-oil-country/

Glenbow Museum, Calgary. (n.d.). *Oil sands oral history project*. Interviews with Patricia Nelson; Eric Newell; Anne McLellan; Paul Precht.

Government of Alberta. (2007). *Our fair share.* Report of the Alberta Royalty Review Panel.

Government of Alberta. (2020-21). *Final results year-end report.* https://open.alberta.ca/dataset/9c81a5a7-cdf1-49ad-a923-d1ecb42944e4/resource/732c465a-196e-488f-8b79-c774197dedf9/download/2020-21-final-results-year-end-report.pdf

Government of Alberta. *Pipeline project — Keystone XL.* Retrieved Oct. 15, 2021, from https://www.alberta.ca/keystone-xl-pipeline-project.aspx

Hustak, A. (1979). *Peter Lougheed: A biography.* Toronto: McClelland and Stewart.

Ipsos Reid. (2002). Canadian's stance on the Kyoto Accord. *The Public Policy Landscape Vol 17, no. 4.* November-December 2002. http://www.ipsos.ca/common/dl/pdf/tr/publicpolicylandscape1102.pdf.

Jaremko, G. (2013). Energy resources conservation board. *Steward: 75 Years of Alberta Energy Regulation.* Retrieved from https://static.aer.ca/prd/documents/about-us/Steward_Ebook.pdf.

Kelly, E. N., Schindler, D. W., Hodson, P. V., Short, J. W., Radmanovich, R., & Nelson, C. C. (2010, Sept. 14.) *Oil sands development contributes elements toxic at low concentrations to the Athabasca River and its tributaries.* Proceedings of the National Academy of Sciences of the United States of America. https://www.pnas.org/content/107/37/16178.short

Laxer, G. (2021, December). *Posing as Canadian: How big foreign oil captures Canadian energy and climate policy.* The Council of Canadians, Canadian Centre for Policy Alternatives-British Columbia, and Canadian Centre for Policy Alternatives-Saskatchewan. https://canadians.org/sites/default/files/publications/Posing%20as%20Canadian%20-%20Gordon%20Laxer.pdf

Lisac, M. (1995). *The Klein revolution.* Edmonton: NewWest Press.

Low, C. (2011). *The "public interest" in section 3 of Alberta's energy resources conservation act: Where do we stand and where do we go from here?* Canadian Institute of Resources Law, University of Calgary, September. Retrieved from https://prism.ucalgary.ca/bitstream/handle/1880/48757/PublicInterestOP36w.pdf;jsessionid=59A8C157F31A56D1412459E68E35E43A?sequence=1

Markusoff, J. (2016, July 16). Jason Kenney vies to become Ralph Klein, reincarnate. *MacLean's.*

Markusoff, J. (2022, May 11). *This is where the Albertans deciding Jason Kenney's future live (think small).* CBC News. https://www.cbc.ca/news/canada/calgary/jason-kenney-united-conservative-party-leadership-review-ridings-1.6448382

McGoey, L. (2019). *The Unknowers: How strategic ignorance rules the world.* London: Zed Books.

McMillan, M. (2019, April). Deficit free by 2023. *Alberta Views.* https://albertaviews.ca/deficit-free-2023/

Mercure, JF., Salas, P., Vercoulen, G., Semieniuk A., Lam, H., Pollitt, P. B., Holden, Vakilifard, N., Chewpreecha, U., Edwards, N.R., & Vinuales, J.E. (2021, Nov. 4). Reframing incentives for climate policy action. *Nature Energy.* https://doi.org/10.1038/s41560-021-00934-2

Nichol, J. R. (1991). *Orphan wells: Who is responsible – for how long and at what cost?* Energy Resources Conservation Board.

Orwell, G. (2017/1949). *1984.* Penguin Canada.

Reagan, R. (1981). *First inaugural address.* Retrieved from https://avalon.law.yale.edu/20th_century/reagan1.asp

Riley, S. J. (2018, Dec. 6). Many of Alberta's reclaimed wells aren't actually reclaimed: government presentation. *The Narwhal.* https://thenarwhal.ca/many-of-albertas-reclaimed-wells-arent-actually-reclaimed-government-presentation/

Robinson, B. (2014, Nov.). The inactive well compliance program: Alberta's latest attempt to bring the inactive well problem under control. *Ecojustice.* https://ecojustice.ca/wp-content/uploads/2014/12/IWCP-Paper-FINAL-20-Nov-2014.pdf.

Rural Municipalities of Alberta. (2021). *Rural municipalities continue to struggle as unpaid tax amounts owed by oil and gas companies increase.* Retrieved Oct. 20, 2021, from https://rmalberta.com/news/rural-municipalities-continue-to-struggle-as-unpaid-tax-amounts-owed-by-oil-and-gas-companies-increase/

Simpson, J. (2006, July 7). Call a halt, Albertans. *The Globe and Mail.* https://www.theglobeandmail.com/news/politics/call-a-halt-albertans/article730540/

Statistics Canada. (2015). *Gross domestic product, income-based, Alberta.* Retrieved from https://www150.statcan.gc.ca/n1/pub/13-018-x/2011001/t/tab0148-eng.htm

Suncor. (2020) *Annual report.* Note 16 to financial statements, p. 115.

Taft, K. (1997). *Shredding the public interest.* Edmonton: University of Alberta Press.

Taft, K., McMillan, M., & Jahangir, J. (2012). *Follow the money.* Calgary: Detselig/Brush Publishing.

Taft, K. (2017). *Oil's deep state.* Toronto: Lorimer.

Timoney, K. (2021). *Hidden scourge: Exposing the truth about fossil fuel industry spills.* Montreal: McGill-Queens University Press.

United Conservative Party. (2019). *Alberta strong and free. United conservatives platform.* Retrieved Oct. 14, 2021, from https://static.unitedconservative.ca/2020/07/Alberta-Strong-and-Free-Platform-1.pdf

Urquhart, I. (2018). *Costly fix.* Toronto: University of Toronto Press.

Way, N., & Simpson-Marran, M. (2019). *Landowners' primer: What you need to know about unreclaimed oil and gas wells.* Pembina Institute, Nov. 21. Retrieved from https://www.pembina.org/pub/landowners-primer-what-you-need-know-about-unreclaimed-oil-and-gas-wells

Weber, B. (2021, June 10). Auditor scolds Alberta over mine cleanup fund. *Canadian Press.* https://www.cbc.ca/news/canada/edmonton/alberta-auditor-mines-cleanup-oilsands-1.6061474

NOTES

1 The ministers were Rick Orman, Patricia Black/Nelson, Jim Dinning, Murray Smith, Shirley McClellan, Greg Melchin, Mel Knight, and Sonya Savage. After politics, McClellan went to the board of PennWest Energy, which in 2013 alone paid her $150,612. Melchin went to Baytex Energy, which in 2013 paid him $154,892, and in which he held $1,571,254 in "director equity ownership" in 2014. See corporate filings for the two companies.

2 TIER Regulation Fact Sheet, July 2020, Government of Alberta. https://www.alberta.ca/assets/documents/ep-fact-sheet-tier-regulation.pdf

3 Jeff Gailus. Hot Air: TIER is Alberta's world-class carbon reduction sham. *Alberta Views*, March 1, 2022. Online at https://albertaviews.ca/hot-air-2/

4 Carbon Competitiveness Incentive Regulation Fact Sheet, April 2018, Government of Alberta. https://www.alberta.ca/assets/documents/cci-fact-sheet.pdf

5 Ashley Joannou. Alberta empties TIER fund in an attempt to create jobs, energy efficiency. *Edmonton Journal*, Sept. 23, 2020. https://edmontonjournal.com/news/politics/kenney-promises-tech-money

6 I saw emails from the campaign that were disturbingly hostile. The estimate of the loss to the University of Alberta is from a private conversation with a well-placed source.

7 In an extraordinary effort, biologist Kevin Timoney has compiled records of 30,329 crude oil spills and 26,699 saline spills within Alberta from 1975 to 2018, an average of about 25 per week (Timoney, 2021, pp. 24 and 261).

8 Working together to boost Alberta's recovery. Oct. 19, 2020. Government of Alberta. https://www.alberta.ca/release.cfm?xID=7450191EACDD1-D715-3078-59DF6480E106FC9E

9 As the industry booms, rural municipalities continue to face mounting unpaid property tax bills from oil and gas companies. March 8, 2022 media release. Rural Municipalities of Alberta. https://rmalberta.com/news/as-the-industry-booms-rural-municipalities-continue-to-face-mounting-unpaid-property-tax-bills-from-oil-and-gas-companies/

10 Thanks to Mark Anielski, president of Anielski Management Inc., for updating his data for use in this paper. The data is drawn from publicly available royalty and bonus sale data published by the Government of Alberta, and industry data published in the CAPP Statistical Handbook. Anielski's database has been published in various forms; see for example *Royalty policy is the biggest decision any Alberta government has to make.* January 2016. Alberta Federation of Labour.

11 *An Update on the Economy of the Calgary-Edmonton Corridor.* TD Economics Topic Paper, Oct. 3, 2005. TD Bank Financial Group.

12 One indication of the campaign within the PCs against Stelmach was an infamous series of emails attacking Stelmach, sent in September 2009 by Hal Walker, a prominent PC Party organizer. Walker deliberately showed the vast list of people to whom he sent the emails, including ministers in Stelmach's cabinet, reporters, PC Party organizers, and oil industry leaders. Emails in author's possession.

CHAPTER 9

ALBERTA'S BRAIN DRAIN REDUX: THE MIGRATION OF ALBERTA'S YOUTH UNDER THE UCP

Richard E. Mueller

"Alberta is back in a big way, but one of the biggest challenges to sustaining that amazing growth is having enough people who are filling the jobs that are being created."

— Then-premier Jason Kenney

FLANKED BY FIVE large posters extolling the virtues of moving to Alberta, and reminiscent of the pithy motivational sayings that once populated offices throughout the country, the Alberta premier made the above statement at a news conference on Aug. 15, 2022, launching the Alberta is Calling campaign aimed at attracting workers from Toronto and Vancouver (Derworiz and Graveland, 2022). The irony is the province is losing some of the exact types of workers it is trying to attract as once again an Alberta government is counting on talent arriving from outside the province, rather than retaining existing and nurturing new talent. Public sector jobs are cut, wages in that sector fall in real terms, and the outright contempt the government displays toward appropriately funding these workers and public services in general are leading many to leave the province, or at least contemplate doing so. This all looks eerily familiar to those who lived through a similar situation in the 1990s, the time when Ralph Klein led the province.

The "brain drain" of the mid-1990s resulted in thousands of Albertans — mainly public sector workers — leaving the province, many destined for the United States. This was largely the result of government cutbacks at provincial and federal levels of government following years of government budget deficits. Some argue it took years for Alberta to regain the talented physicians, nurses, teachers, and other public servants who departed. Following this gutting of the public sector, the economic boom of the next 20 years filled government coffers, spending on public services increased, and many Canadians again began moving to Alberta. The Alberta Advantage of low provincial taxes was oft-touted as the reason for this in-migration.

Fast forward to the present, and the United Conservative Party (UCP) is also attacking the public sector under the guise of deficit and debt reduction. And, unlike in the 1990s, other provinces are already actively recruiting Alberta's displaced public sector workers. Given that interprovincial migration is much more common than international migration, and the fact the U.S. border is no longer as open to temporary or permanent immigration as it was in the 1990s, Alberta's loss of talent is likely to benefit other provinces. While the UCP has only been in power since April 2019, and recent migration data are largely unavailable, anecdotal evidence or limited to certain professions suggests this migration is happening, especially among Alberta's young people. Furthermore, if we view migration as a lengthy process, which begins with dissatisfaction with the province of origin, then the reports of marginalization, especially among young Albertans, suggest it is only a matter of time until the official statistics reflect this migration. The survey evidence suggests, as do other anecdotes, that many young people in Alberta no longer share the province's "values" and for this reason are considering relocating. Much of this has to do with the dissatisfaction with the perceived direction the UCP has taken the province and the lack of optimism regarding its future.

This chapter will address what we know about the state of migration to and from the province where relevant data are available, and speculate as to what may transpire in the future, given what is known both theoretically and empirically about the key factors and motivators behind migration. The next section discusses some of the previous theoretical and empirical work related to this topic, in the context of viewing migration as a multi-stage process. The third section presents some existing, albeit sparse, data quantifying the flows of people in an out of the province. The fourth section uses contemporary news and social media sources to add some texture to the data presented in the previous section. Talking about "Alberta's image problem" and how it affects migration is the topic of section five. The concluding section argues that if Alberta's youth migration outflow is to be turned around, Alberta's perceived image (both internally and externally) must be changed to better reflect reality and not stereotypes. There must also be more attention paid to changing the culture of post-secondary education in Alberta to better reflect the post-energy economy in Alberta (see Spooner, Chapter 14).

Migration as a Process

Migration happens for several reasons, from the economic, to the political, to the personal. But what all migration processes have in common is the view that one or all these factors are better in the destination province than the province of origin. In essence, there are push and pull factors involved in the migration decision and these are multi-faceted.

Hiller's (2009, p. 133) thesis is that migrants are people who feel some form of marginalization in their communities of origin who seek to resolve it by migrating to a new community. To paraphrase Hiller, the first step in this marginalization process consists of dissatisfaction; the second step consists of the cognitive interpretation the potential migrant gives to their social context in the community of origin, or *cognitive marginalization*, which is correlated with *distancing* whereby individuals begin to have both an intellectual evaluation and an emotional separation of the origin (Hiller, 2009, p. 154). This leads the potential migrant to the edge at a time when migration may or may not occur. The point is the process takes time, from the initial feelings of dissatisfaction to the actual physical migration.

The interviews conducted by Hiller (2009, p. 414), revealed that "moving for work" simplifies a complex process because it only acknowledges economic factors. A wide range of reasons for moving came to the surface, reasons the simple response "moving for work" masked, as only about one-quarter of migrants said they had moved primarily for reasons of employment. It was also a sense of dissatisfaction that emerged regarding how the migrant viewed his/her social position within the community of origin.

For most people in Hiller's survey (2009, pp. 394-5), a negative political and economic atmosphere appears to be the strongest causal factor in out-migration:

> There are grounds to conclude that relocation was, in large measure, prompted by the perception of politico-economic conditions in which a variety of other factors were also at work . . . it is clear that, for many, a perception of more compatible thinking with people in Alberta was not a critical factor in the choice to move there. Interestingly though, one-third of the respondents did indicate that such ideological similarities were important to them.

Discussing the in-migration of people from other provinces in the 2000s, Hiller (2009, p. 218) writes, "Ideology/politics was a major factor for many migrants because all provinces from Ontario westward had recently elected governments on the left. Alberta, in contrast, always elected governments on the right, and migrants often talked about fleeing left-wing governments."

Further, migrants to Alberta were self-selected; those who were most motivated. Not only did Alberta attract a particular type of person, but it retained them as well (e.g., those who succeeded and embraced Alberta's apparent values) (Hiller, 2009, p. 428).

It is obvious Alberta may be attracting people who share what is (at least perceived as) the current political culture in the province, while others are leaving or planning to leave because of it. Recently, Hirsch (2021) echoed the same sentiment saying Alberta has attracted in-migration over the years because of, not despite, its perceived socially conservative and libertarian values. Of course, this argument is also symmetric, with those who do not share the perceived ethos of the province more apt to either leave Alberta or not migrate to the province.

That individuals vote with their feet was outlined more than 60 years ago by Tiebout (1956), who argued individuals migrate to a community that offers the public services and the taxation they like best. In other words, while narrowly focusing on economic factors, individuals choose a location based on these services and taxes. While the current Alberta government is focused on the taxation part of this, it is clear from what follows that the services are also of importance to Albertans and do figure into their decisions to stay or to migrate.

Rheault (2019) also discusses the importance of policy in determining migration. He conducted a survey of 1,523 English-speaking Canadians to test three central hypotheses about the migration process: (1) if satisfaction with provincial taxation and services affects the decision to move; (2) whether people who support centralization are more likely to consider moving across provinces; and (3) the classical hypothesis of migration theory, which posits that labour market incentives drive mobility. He finds satisfaction with provincial services (such as education and health care) reduces the incentive to migrate, and this factor appears much more robust than satisfaction with tax rates and perceptions of job market opportunities. Canadians who support a stronger federal government are more likely to move across provinces, and these people tend to be young and more educated. Finally, the economic variables related to perceived better relative job satisfaction and wages do not have any statistically significant effect on the intention to migrate once the other factors are controlled. This result, if it as applicable to Albertans as it is to all English-speaking Canadians in the sample, does not bode well for Alberta youth as public services important to young people are being cut. It is the factors not as important to them — jobs and taxes — the UCP government has promoted.

Day and Winer (2012) point to the complexities in studying the effects of policy shocks on interprovincial migration. While large policy shocks, such as the election of the PQ in Quebec in 1976 or the closing of the cod fishery off the coast of Newfoundland in 1992, the incremental ones make the relationship between migration and policy more difficult to study. Whether the election of the UCP government in April 2019 and the introduction of new policies is considered large or incremental is left for the reader to decide, but the point is these policy shocks are important determinants in the migration decision, even if they are not easy to quantify.

In what follows, I follow the logic that migration is a process (Hiller, 2009), where policy shocks such as those implemented by the Kenney government may be difficult to quantify (Day and Winer, 2012), but whose impact is nonetheless real. While available data are used to determine the extent of any interprovincial migration to and from Alberta, these data are inherently backward-looking. Given that migration is a process and not a single decision, other relevant information, such as surveys and social media posts, will also be used to give the reader a clearer vision of what future migration from Alberta may look like. It is these current individuals' negative perceptions about the state of Alberta that are likely to be "leading indicators" of future migration to and from the province, migration which will ultimately be reflected in the interprovincial migration statistics.

The next section presents data on the historical patterns of migration with special emphasis on young people coming to and leaving the province.

Quantitative Evidence of Albertans Leaving the Province

Figure 9.1 (below) presents data on net migration as a percentage of the population within various age groups. Numbers greater than zero show more in-migration than out-migration to Alberta in the given year. The pattern in these data is obviously

Figure 9.1: Net interprovincial migration as a percentage of population
Alberta, 1971-72 to 2021-22, various age groups

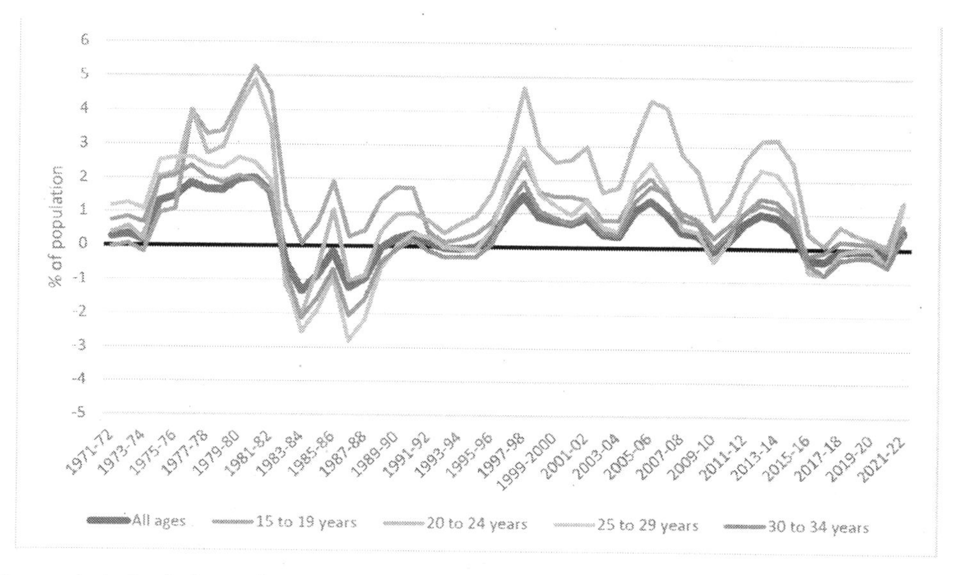

Source: Author's calculations from Statistics Canada Tables 10-17-0005-01 and 10-17-0015-01. Percentages are obtained by using the migration figures for each year and dividing by the population figure at the beginning of the year. For example, 1989-90 is the net migration between July 1, 1989 and June 30, 1990, divided by the corresponding population as of July 1, 1989.

related to the state of the Alberta economy, with larger in-migration waves in the mid- to late-1970s, the late-1980s, and then again from the late-1990s through to 2014-15, the year when oil prices declined rapidly. Since this time, with a few exceptions, negative net migration has been the norm. The exception is the most recent year when there has been an increase in net migration. This not does imply, however, that out-migration has ceased, rather that in-migration has exceeded it. A more nuanced look at these data shows that out migration continues.

It is clear from this figure that net migration tends to follow the same pattern, regardless of the age group, but the variability between groups differs. If we look at all ages, the variability is relatively small, reflecting the trend that older people have put down roots in their current province of residence and are therefore less likely to migrate between provinces.

Lane, Laverty, and Finch (2022) also show the province has seen the loss of many young people in recent years. Similar to the Hiller (2009) argument that migration is a process, the authors write that previous research shows youth consider moving at three main life stages (or "moments of truth"). The first is high school graduation, the second is post-secondary graduation, and the third is a moment in time when individuals choose to advance their career or settle down as young professionals. They argue these are related roughly to the three sub-cohorts in the 15-19, 20-24 and 25-29 age groups. The third moment of truth can occur multiple times beyond the age of 30.

While the low or negative net migration in the early-1980s, and the international financial crisis in the late-2000s, can largely be blamed on the economy (and low energy prices, in particular), there are interesting parallels between the most recent out-migration, and the decline in net migration during the Klein years. Klein was elected in December 1992 and talked about the Alberta deficit being too high. In the budget of March 1993, Klein promised to eliminate the deficit in four years without raising taxes. This was to be done by reducing expenditures by 20 per cent or more. This budget was brought to the voters in the June 1993 election, which Klein's Progressive Conservative Party proceeded to win. By 1995, Alberta was spending much less per capita on programs than the six other provinces with balanced budgets, and the province was projected to be spending between nine and 18 per cent less when the budget was scheduled to be balanced in 1996-97. "These are remarkable differences, especially when recognizing the much greater fiscal capacity of Alberta and the fact that all of these six provinces, except British Columbia, are among the 'have-not' provinces . . ." (McMillan and Warrack, 1995, p. 149). Of course, increasing natural resource revenues allowed the Klein government to balance the budget ahead of schedule (similar to the UCP's most recent 2022-23 budget).

While deep budget cuts have become synonymous with the Klein and Kenney governments, spending on public programs in Alberta peaked in the mid-1980s and spending levels were below the national average for Canadian provinces by the time then-premier Don Getty resigned. Yet Klein told a different story when he became premier in December 1992: a story of runaway public spending and unsustainable deficits. "In other words, the severe cuts of the Klein government *began* on budgets that were already relatively low" (Taft, 1997, p. 23). The government cuts did not begin in earnest until after the general election of June 1993, with more than $800 million cut from spending in the first year. Public sector job losses were large, and a five per cent pay cut was imposed on MLAs and public sector workers including civil servants, teachers, nurses, and university staff. The government's plan for "reinvestment" was presented in the fall of 1996, but did little to turn things around. Funding did increase in 1996-97 but did not keep pace with inflation and population growth. Similar tactics are being used by the Kenney government: first taking large amounts of funding out of the system before partially restoring funding in key areas, but not restoring funding in real terms.

The Kenney government's plan to balance the Alberta budget largely follows the recommendation of the MacKinnon report in 2019, which also favoured deep cuts to public expenditures to tame to deficit and debt, with no talk about how to enhance revenues. And, like the Klein government in the mid-1990s, it was natural resource revenues that came to the rescue and balanced the provincial budget in fiscal year 2022-23, and are expected to reach record heights in that same fiscal year (Markusoff, 2022).

What is interesting is the current period appears to be the worst in terms of negative net migration since the 1980s, a time when the energy market was in trouble and exploration and drilling activity was severely curtailed. The parallels between the mid-1990s and the past four to five years are obvious, especially with the net loss of young people.

It is worthwhile to look at how Alberta compares to other western provinces in terms of net migration, especially amongst youth migrants. For our comparators, we use British Columbia, Saskatchewan and Manitoba. Figure 9.2 (below) shows the historical patterns of net migration from each of the four provinces. It is immediately apparent there is much more variability in net migration to Alberta over this period. Not surprisingly, and as noted above, this net migration tends to be positive when the price of oil is high, and negative during periods when the price is low (Lane et al., 2022). In general, both Alberta and British Columbia have had positive net migration over this period, while Saskatchewan and Manitoba are more likely to be on the negative side. Of note is the most recent year for which data are available (2021-2022) which shows a year-over-year decline all provinces with the exception of Alberta.

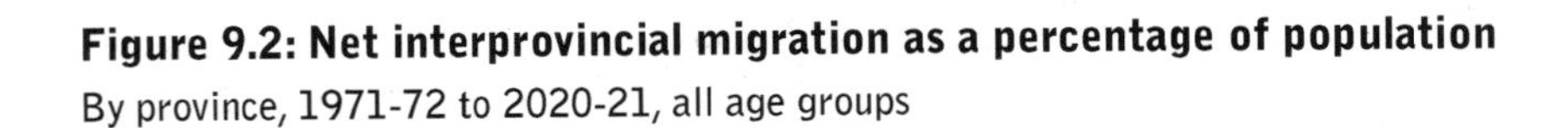

Figure 9.2: Net interprovincial migration as a percentage of population
By province, 1971-72 to 2020-21, all age groups

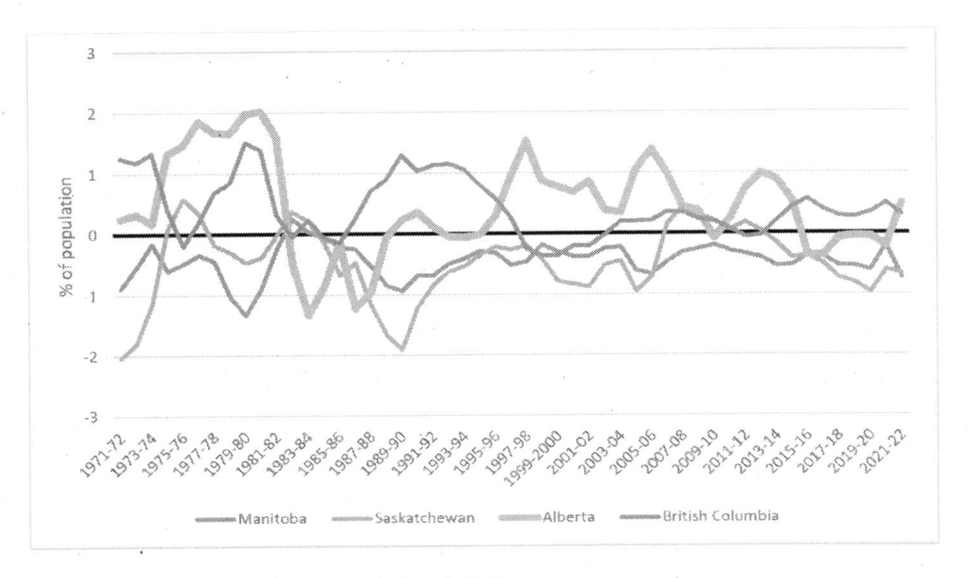

Source: Author's calculations from Statistics Canada Tables 10-17-0005-01 and 10-17-0015-01.

The net migration figures tend to camouflage the in- and out-migration patterns. Figure 9.3 (below) shows in-migration among young people was at historical lows as a percentage of the corresponding Alberta population until the most recent year (2021-22), when again young people started migrating to Alberta. For comparison purposes, the same trends for the other three western provinces (not shown here) have seen increases in in-migration over the most recent year in most cases after being flat for the previous few years.

In terms of out-migration, Figure 9.4 (below) shows there was an increase for all age groups in the most recent year. In the other three western provinces (not shown here), the rates were either flat or slightly positive in certain age groups. The exception is Saskatchewan, where all age groups of out-migrants are trending downwards over the most recent year.

There are two important points to make about these migration data. The first (not shown here due to space limitations) is that of the 21,660 net migrants from other provinces to Alberta in 2021-22, some two-thirds (14,143) is due to more Ontarians moving to Alberta in that year than moved in the opposite direction.[1] Also, over this period has been the drastic increase in migration from British Columbia which has almost equalled the out-migration of Albertans to that province, so that net migration

Figure 9.3: In-migrants as a percentage of the population
Alberta, 1971-72 to 2021-22, various age groups

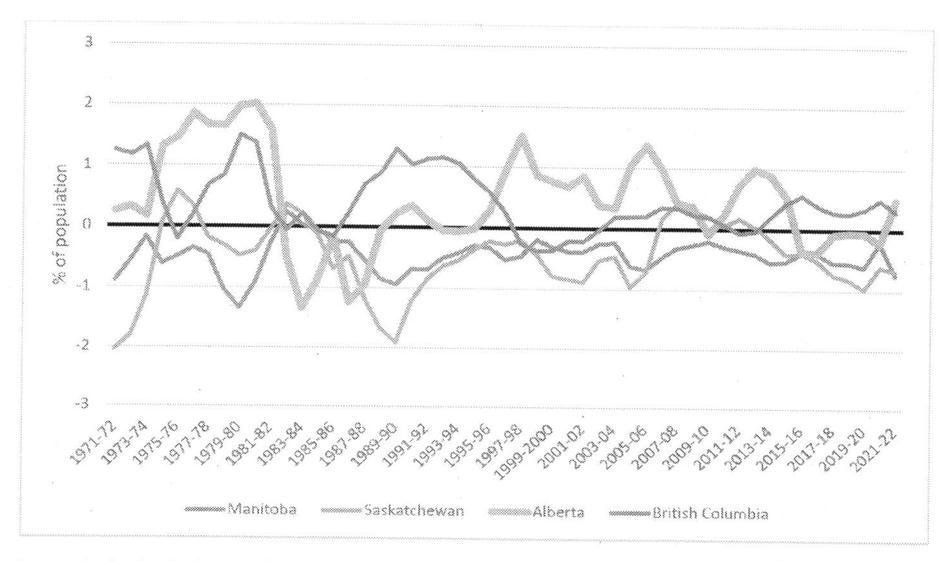

Source: Author's calculations from Statistics Canada Tables 10-17-0005-01 and 10-17-0015-01.

Figure 9.4: Out-migrants as a percentage of the population
Alberta 1971-72 to 2021-22, various age groups

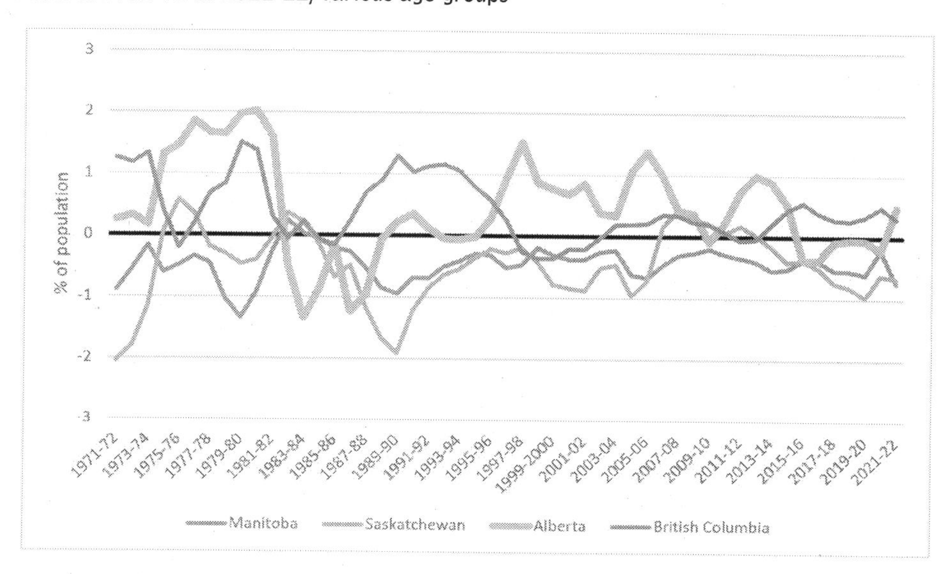

Source: Author's calculations from Statistics Canada Tables 10-17-0005-01 and 10-17-0015-01.

from Alberta was only 1,175 individuals in 2021-22, much lower than the average net loss of 8,755 in the previous seven years from 2014-15 (when oil prices fell dramatically) through 2020-21. Together, it is the anomalous migration from these two provinces — at least by recent historical standards — that has resulted in the most recent uptick in net migration to Alberta. In fact, if the net migration figures to Alberta from British Columbia and Ontario were at the same level as in 2020-21, the overall net migration shown in Figure 9.2 above would be negative — not positive — in the most recent year. Second, the most recent year has revealed that *both* in-migration and out-migration have increased for most of the young age groups and in the four Western provinces. This is suggestive of a churning of the young population and that there may be some common factor driving these increased migrations, perhaps the pent-up demand for migration following the relatively low levels of interprovincial migration during the pandemic. Whether the migration patterns in the most recent data will continue or they are outliers will be ascertained as newer data are released.

That young people have been leaving Alberta is further reflected in the aging of Albertans vis-à-vis the other provinces. While Alberta once had the lowest median age in the country — a place it held since 1971 — by July 1, 2021, Statistics Canada estimates showed the median age in Alberta was 37.9 years, slightly older than Manitoba at 37.8 years, "driven by large numbers of young adults leaving the province to seek opportunities elsewhere" (Alberta Treasury Board and Finance 2021, p. 4). Commenting on this report, Varcoe (2021) said ". . . more young adults are leaving — or not moving — to Alberta, amplifying concerns that a brain drain has begun." In the same piece, David Finch at Mount Royal University is quoted, "Decisions are being made that are draining the talent out of the province or, on the flip side, not attracting talent to the province." Janet Lane of the Canada West Foundation said, in the same article, "We've got a bit of an early warning here that there could be a longer-term problem." While the final two quarters of 2021 show positive interprovincial migration to Alberta, the data are not detailed enough to determine how much of this may be accounted for by young people.[2]

Young, educated people tend to be concentrated in urban areas, and the evidence here points to an aging population, relative to the national average, in Alberta's two major urban areas. Between 2014 and 2020 (the year when energy prices fell, through to the year of the most recent available data), the median age of Calgary and Edmonton census metropolitan areas (CMAs) increased by 1.6 years and 1.1 years, respectively. This compares to the Canadian average of 0.4 years. Furthermore, each of the four age groups under consideration here either decreased in absolute numbers, or increased at a lower rate, than the overall increase in population in both cities.[3]

Young Albertans feel they are shouldering a disproportionate share of Alberta's fiscal burdens with a lower level of support from the province. This disenchantment and perceived lack of economic opportunities, coupled with the reality that young people tend to be educated and geographically mobile, means many have left the province, or are at least considering doing so. In the words of one focus group participant: "I can't feel like I love a place that doesn't love me back" (Williams, 2021).

Lane et al. (2022) argue the large declines in net migration, evidenced above, are concerning, but it is more worrisome when looking at the 30-34 cohort, since this age group tends to settle down in their province of residence and is less likely to return to Alberta. Further, the remedy for negative youth net migration in the past has been to wait for another oil boom. This may not be realistic in future as oil and gas production have become much less labour intensive, and the recent increases in oil prices will not generate the positive net migration of youth they have in the past. Hussey (2020) agrees, noting that the 2000-2018 growth phase of the oilsands industry has transitioned to its mature production phase, where capital spending and related employment are not expected to return to the highs of the early-2010s. Adding new technology and modernization trends into the mix will further dampen increases in employment.

Given the lag in official statistics, and the model projecting migration as a process taking some time to be realized, it is not surprising that (as of writing) there have not been higher rates of negative net migration from Alberta. No one expected a mass outmigration of Albertans following the UCP's election in April 2019. Rather, this was the beginning of the process of marginalization in the community (Alberta), which Hiller (2009, p. 145) says is the first step in the migration process and ". . . occurs in the community of origin when a person is increasingly dissatisfied with some aspect of his/her status, role, or ties in the community." The next section presents evidence this dissatisfaction is occurring in Alberta and has been for some time, even if official statistics will likely take some time to manifest this unhappiness with the province amongst certain groups.

What We Know About Who is Leaving the Province and Why

The above data shows people are leaving the province, especially young people, but it says little about who exactly is exiting. In the absence of appropriate data, we must turn to media outlets and various professional associations who keep statistics on their members and have administered membership surveys on their professional intentions.

The Business Council of Alberta (2021) says Alberta is increasingly losing in the competition for attracting young families from other provinces, noting those in their

20s and 30s drive the majority of migration to and from the provinces, and this has historically contributed to the relatively young provincial population. But this has changed over the past few years, as young people are now more likely to leave and fewer are choosing to move to Alberta.

While it is obvious younger people are more likely to relocate compared to older individuals, age is not the only factor that differs between migrants and non-migrants. Higher income professionals almost certainly face wider labour markets, which are more commonly associated with movement across provincial boundaries (Finnie 2004). In other words, it is the highly educated and the well compensated who are more likely to leave the province.

This demographic can be seen in certain public sector occupations the Kenney government targeted, often using the rhetoric of the report by the Blue Ribbon Panel on Alberta's Finances (2019), commonly referred to as the MacKinnon Report. That report uses questionable data to argue Alberta's public sector wages are out of line with those in the public sectors of the three largest provinces. A more recent and rigorous analysis of public sector wage differentials shows no systemic difference between Alberta's public sector when compared to her three comparator provinces (British Columbia, Ontario and Quebec), nor when comparing public sector to private sector wages within each province (Mueller, 2021). Still, the assault on public sector workers continues, and the repercussions of such actions are beginning to manifest as labour unions are voicing the concerns of their members regarding the longer-term consequences of these government actions.

Arguably, physicians have been the group receiving the most attention in Alberta. This is largely due to their collective response to how the government has treated them, as well as current reports the shortage of family physicians in the province will get worse should there be an exodus of physicians. This group is worthy of attention, as they are the types of educated Albertans who are highly mobile, and statistics on this group are often more up-to-date than the previous cited Statistics Canada data.[4] In essence, physicians are the canary in the coal mine.

About 15 months after the UCP entered office in April 2019, the Alberta Medical Association (AMA) released results of a survey showing 87 per cent of physicians would be making changes to their medical practices as a result of the funding framework imposed on them by then-health minister Tyler Shandro (AMA, 2020). Of this group, 49 per cent (or 42 per cent of all Alberta physicians) had made plans, or were considering, looking for work in another province. Many others in the survey said they would respond to the funding changes by withdrawing services, cutting back on hours, and laying off staff. Responding to this survey, then-AMA president Paul Boucher (2020) wrote:

> Make no mistake, this is having an impact on our physician workforce. I have had a number of personal communications from members who are closing their practices and leaving as a direct result of our relationship with government. Others have told me of new recruits to their practice that have decided to go elsewhere. I have heard of students and residents making active choices to train outside of the province. Every one of these accounts represents a loss to a clinic, a community, a program and to Albertans. It is also the loss of a colleague or a bright young mind that might otherwise have contributed to our health care system.

Though sparse, evidence to date does seem to show an increase in the number of physicians leaving Alberta. The College of Physicians and Surgeons of Alberta (CPSA, 2022a) reported 140 doctors left the province in 2021, up from 87 in 2020 and 54 in 2019. The same table also shows the number of new registrants has been on a downward trend over the past number of years while the number of physicians removed from the registry has been increasing. The most recent quarterly data (CPSA, 2022c) shows 48 physicians left the province in the first three quarters of 2022, down slightly from 50 in the same period in 2021 (CPSA, 2022b). The net increase in the number of registrants (which includes all new registrants less those whose left Alberta, retired, etc.) was only 45 in 2021, down from 161 in the previous years, and between 262 and 328 in the years 2017-19. The number of registrants has also decreased by 188 in the first quarter of 2022 compared to a loss of 254 in 2021. The number of registrants has increased in the second and third quarters of 2022, reversing some of the earlier declines in 2020 and 2021.[5] Whether these numbers will continue to improve remains to be seen, but there is little doubt that this has contributed to fears that Alberta will suffer a physician shortage — a shortage exacerbated by the backlog of patients seeking medical attention for issues foregone during the pandemic (Henschel, 2022).

What are the motives for leaving the province and/or the profession? Gunter (2022) has heard from dozens of Alberta physicians, and identifies two factors that contributed most to the doubling in the number moving away or quitting in 2021 compared to 2020: the pandemic and Tyler Shandro. Similarly, independent Alberta journalist Kim Siever has an active list of physicians closing their practices, and many cite the government as being their primary reason for leaving the profession and/or the province.[6]

The loss of physicians is likely to be concentrated among the youngest members of the profession, since young professionals tend to be the most mobile, but we will await the release of more recent statistics. Also confounding the interpretation of the limited evidence is the COVID pandemic, which has postponed many physicians' plans to move, retire, or reduce services. For example, private talks with an Alberta physician indicated she could not morally shut down her practice during the pandemic, but that many would likely do so following the pandemic. She also mentioned talk of Calgary having a hard time attracting physicians, something unprecedented.

This real and potential future of medical professions is not limited to physicians, but also nurses and others involved in health care. Responding to a May 2022 announcement by Premier Kenney that the province would provide more hospital beds without the necessary nurses or others to staff them, United Nurses of Alberta (UNA) president Heather Smith pointed out many nurses are exhausted and disillusioned from sacrificing so much during the pandemic. She went on to say many are seriously contemplating retiring, leaving the profession for other work, or departing for other jurisdictions with better working conditions or pay. Smith noted further that, among other changes, Alberta must stop acting as if nurses are "enemies" (UNA, 2022). Indeed, other provinces, suffering nursing shortages and other similar pandemic issues as Alberta, are offering attractive signing bonuses and other perks to attract nurses (Young, 2021). The proposed wage cuts to nurses in 2021 simply exacerbated the situation, with nurses continuing to entertain leaving the province, despite the recent contract settlement. UNA's Smith said Alberta nurses are not prepared to deal with the disrespect they are feeling in the province, and are prepared to practice in other locations. Premier Kenney, taking a seemingly narrow view of working conditions, said, "I wouldn't agree that people would move from Alberta to receive lower pay in other provinces and to pay higher taxes. That wouldn't add up" (Lee, 2021).

As with Alberta's physicians and nurses, the province's teachers too have expressed their desire to leave the profession and the province. For example, a survey of about 1,300 members by the Alberta Teachers' Association (ATA) in November 2021 revealed 37.6 per cent of respondents said they would not be teaching in Alberta in the next school year, while 16.4 per cent will be retiring, with the remaining 21.2 per cent leaving the province and/or the profession altogether. According to ATA president Jason Schilling, much of this is due to the perceived bungling by the UCP with respect to lack of substantive government-imposed regulations during the global pandemic, as well as the controversial proposed changes to the province-wide school curriculum (Karstens-Smith, 2022).

Perhaps most worrisome is this displeasure with the province is not limited to those in the labour force, but also threatens the ability of the province to retain its own young people before they have started their careers. A March 2021 survey, for example, from the University of Alberta Students' Union (UASU, 2021), showed 61 per cent of student respondents had at least a 50/50 chance of leaving the province following graduation.

Among those who responding they would probably or definitely leave, 71 per cent said it was due to their strong feelings and opinions about the provincial government. Among this same group of probable or definite leavers, almost one-half of those who intended to leave the province are doing so to pursue further studies, perhaps the result of the perceived or real impact of cuts to the post-secondary education sector.

Rowan Ley, president of the University of Alberta Students' Union (UASU) in late-2021 said:

> There's a narrative out there that people who leave Alberta are privileged folks who are moving to Vancouver for lifestyle reasons. Essentially, we know from our data this is actually not true. The students who plan to leave are the ones that don't feel welcome in Alberta or don't feel they can make it in Alberta (Williams, 2021).

Rachel Zimmerman, then-vice president external for the Students' Association of Mount Royal University (SAMRU), echoed this sentiment in a public town hall in May 2021. She said students are leaving the province — she is one of the few staying in Alberta. Young Albertans were brought up to believe Alberta is the land of opportunity but are less optimistic today. Furthermore, the students who are leaving are not saying good things about the province.[7]

As the Alberta energy industry is unlikely to return to its former importance as a source of employment in the Alberta economy, education to pivot the labour force into new jobs is essential. Yet education too — especially post-secondary education (PSE) — has been subject to large government expenditure reductions. This has led to the elimination of many jobs in the sector, with the likelihood this will negatively impact the quality of education and research in the province. This dynamic, coupled with the coming demographic bulge of students of students in the 18-25-year-old group who normally attend PSE, along with increasing student discontent and unprecedented tuition hikes, has led some (e.g., Acuña, 2021) to warn of an impending exodus of post-secondary students from the province, many of whom are unlikely to return. This has been compounded by policy changes, such as the elimination of the student tax credit (akin to an increase in fees), which, according to student groups and education advocates, "makes the province's postsecondary schools less attractive to students, who are already leaving Alberta in droves" (Pike, 2021). In short, Alberta is not only losing current labour to other jurisdictions, but also potential replacement labour for when the Alberta economy turns around.

In the news conference to introduce the party's post-secondary education plan, New Democratic Party (NDP) leader Rachel Notley said:

> We have always been a magnet for highly educated, young people to move to Alberta. That's what has driven our demographic advantage over the rest of the country. Now, not only are they not coming here, but our young folks are leaving (Cowley, 2022).

These trends are at odds with University of Alberta president Bill Flanagan (2002), who has the goal of increasing enrolment at the university by over 25 per cent over the next five years. He notes that over 6,000 students per year leave the province to pursue post-secondary studies, and rhetorically asks, "Once gone, how many of these talented students will return to Alberta upon graduating?" He then goes on to laud the modest $48 million his institution will receive from the provincial budget to expand enrolment in high-demand programs. Whether these funds are sufficient to stem the potential outflow of students caused by the draconian cutbacks to U of A since the UCP rose to power remains to be seen.

What is factual is more undergraduate students have been leaving the province than coming. In his daily blog, Alex Usher (2021) wrote the net intake of undergraduate students to Alberta was almost -6,000 in 2018-19, meaning Alberta was a net exporter of students, whereas 15 years before this academic year, Alberta was a net importer. In a subsequent blog (Usher, 2022), he tries to uncover the reasons for this. He notes this cannot be a UCP-related phenomenon since the data do not cover the UCP-governing period, but neither can it be blamed on the NDP government since, the student deficit started before the Notley government. Nor is it related to energy prices or to any decrease in funding. Finally, it does not seem to be related to a shortage in the number of available spaces overall in Alberta universities. Usher speculates the actual reason for the decline has been the growing Alberta population's greater difficulty getting into the most competitive programs at the universities of Alberta and Calgary; the exception perhaps being a non-trivial number of students able to afford attending equally prestigious out-of-province universities.

Will this shortage of spaces in competitive programs be reversed, and thus stem the net outflow of students? The UCP government is putting more money into targeted programs. That is the good news. The bad news is the government is trying to pick the winners by targeting programs such as information technology, finance, energy, health, and aviation, in addition to providing more money for apprenticeship training and work-integrated learning in its 2022-23 budget (Adkin, 2022). Also, the injection of new funds has not begun to compensate for the slashing of the Campus Alberta Grant, which has decimated programs not targeted in the budget. This speaks to the UCP's very narrow focus on what it considers important. But, if the Usher hypothesis is correct, and the UCP have targeted programs correctly, more Alberta students will be staying home for post-secondary and (ideally) after graduation. However, those whose thoughts of leaving are being driven by feelings of marginalization or education goals that cannot be accommodated will continue to leave, despite the government's efforts.

Since the UCP government was elected in April 2019, there has been plenty of mainstream and social media coverage about the reasons individuals are leaving, or contemplating leaving, Alberta. Many respondents outlined their displeasure with the current government, its policies, and the general rise in the party's right-wing ideology as factors leading to their disenchantment with the province.

For example, Ford (2021) reflects this general pessimism regarding the future of the province: "I left Alberta in the summer of 2019. Jason Kenney, leading a triumphant United Conservative Party, had taken the reins of power just three months earlier. It wasn't a government I was keen to live under." He asserts it is increasingly clear he is not alone, proceeds to document the decisions of others to leave the province, and ends the article by asserting, "There are only so many times you can be told you are unwanted, before you finally believe it."

In late 2021, a series of Reddit posts resulted from a request by a Radio-Canada reporter in Edmonton soliciting opinions from Albertans who had recently moved, or were considering moving, to British Columbia.[8] Many of the people commented with reasons pointing at better economic opportunities in British Columbia, but many others suggested the political climate was responsible. Often the two reasons co-existed, such as with cutbacks to public sector employment. Here is a small sampling of the (verbatim) responses:

> One person gave many reasons for going to BC, his last point was: "And, yep . . . the remaining 50% . . . Kenney politics. I'm not even going to open that garbage bag in this thread."
> "Moved to BC in '95... the reasons... Ralph Klein and snow. Alberta politics have moved even further to the right and the snow/winter still exists." And a response to this said: "Same. Except substitute Jason Kenney for Ralph Klein."
> "We moved mid-way through 2019 after Jason Kenney won the election . . . "
> "I got offered a full-time teaching job here in Vancouver whereas in Alberta, specifically, Calgary, jobs and salaries were being cut left right and centre due to the political turmoil Jason Kenney brought along."
> "I feel like the economy in Alberta is unstable, and Kenney makes that worse. Sometimes I think about moving back, but I know Kenney could pull the rug out from under me at any time."
> "I moved to BC in 2019 from Calgary because Kenney killed my public sector industry. There were no jobs! But I've had plenty of opportunity in BC."

> " . . . not all of us support the actions of the AB government or majority. It's kind of why you see a lot fleeing."
> " . . . after the way UCP handled the pandemic, I can't wait to get out of here. I just graduated from university last year so I know there will be more opportunity in BC, but I'll definitely miss the sunshine."
> "I'd love to be closer to my family, all of whom are still in AB, but won't move back as long as Albertans continue to support the UCP. It's utter insanity that a majority of people in AB refuse to change their politics but expect different results for the province."
> "Especially UCP politics."
> "Alberta is becoming a conservative dump without a long term future think 20 to 40 years out."
> "Alberta has fallen apart politically the past few years, a lot of doctors have left because they do not have contracts."
> "The ocean and to get away from the idiot conservatives was why I did it. Not in that order."
> "I've lived in Calgary my whole life and I'm in the final year of a professional degree program at U of C . . . The lack of progressive politics or vision for the future of Alberta is making me feel less inclined to set down roots here after I graduate . . . I would rather pay a premium for a good quality of life and a city that is more in line with my values than stay in Alberta."
> "Alberta is a dumpster fire right now, and not in a 'this is temporary' situation. The decisions the current Gov Has made during their term will impact Alberta for decades."

These perceptions of Alberta are important, since many individuals have stated their disenchantment with the current Alberta government and its policies and, as we have argued, it is this feeling of marginalization in the province that is the first step in the migration process. While current Alberta residents may feel the current political climate does not reflect their values, some are willing to stay. Chand (2022), for example, expressed her displeasure in the Kenney government but decided to stay, writing:

> It's not that Alberta has been good or hospitable to me. And it's not like I haven't noticed how much this province has declined under the Jason Kenney-led United Conservative Party, whether in terms of the cratering economy or in feeling safe from the hostile alt-right. But I'd rather pour my energy into making Alberta a better place.

While some of those living in Alberta may be committed to the province in the long term, and willing to weather the current political storm, it must be remembered that migration works both ways, and those who might otherwise migrate to Alberta are seemingly not willing to do so, as shown in the preliminary migration numbers and comments above. Thus, we are losing young people on two fronts: those who are leaving, and those who are not coming to the province, at least in the numbers historically seen. Much of this is undoubtedly due to the negative perception of Alberta throughout the rest of Canada.

Alberta Has an Image Problem

The perceptions of Alberta inside and outside the province seem to differ widely, a gap likely causing some Canadians to rethink moving to Alberta. While the lure of high-paying and stable energy sector jobs may have been enough historically to have some overlook any (perceived) negative aspects of life in Alberta, those days are over. As this sector matures, the earning premium may no longer be sufficient to attract those from outside the province. In short, Alberta has an image problem, and has had for some time. This may be exacerbating the negative net migration outlined above, especially among the younger age groups, many of whom feel alienated by conservative politics and policies.

Hiller (2009, p. 424) writes that: "Outside the province, particularly in the largest metropolitan areas in Canada, the image of Alberta as an unsophisticated hinterland lingers." This is interesting since, when measuring the attitudes of Albertans toward a wide range of public policy issues, they are not strikingly different from those of people in other regions. "However, the *perception* that Albertans are ideologically distinct in a political sense, or that they possess a unique political culture, is a view that is widely shared . . . national media may have promoted a discourse of differences through constant repetition" (Hiller 2009, p. 426).

University of Alberta political scientist Jared Wesley studies what it means to be Albertan and how Albertans are perceived both inside and outside of the province. He notes:

> There's just this idea that to be Albertan means you have to be a hyper-masculine male. We see this in government advertising. We see it in not just in our focus groups, but on TV and everywhere else. . . . To the extent that that's your view of what it means to be Albertan, maybe you don't feel like you belong (quoted in Williams, 2021).

Thus, not identifying with the Average Joe Albertan, as Wesley calls the stereotypical Albertan, is one less reason to stay in Alberta. This, coupled with a stagnant economy and budget cuts — many of which are target young Albertans — makes Alberta less appealing.

In related work, focus groups in Alberta, Toronto, and Vancouver in mid-2021 by Lane et al. (2022) found young people in these locations perceive Alberta as not offering enough career choices (e.g., outside the energy sector), and lack vibrancy as well as inclusion and diversity. They say that while youth do migrate for economic reasons, quality of life is important for them and they "work to live," not live to work. Arguably, these negative perceptions have intensified over the UCP government's reign.

A recent poll in November 2021 from Maru Public Opinion (2021), in collaboration with CBC and Janet Brown Opinion Research, titled Brand Alberta pointed to this disconnect (Dawson, 2021). Based on this work, Brown said Alberta must think about how its reputation and brand are playing in the rest of Canada, while noting Alberta, "is a far more diverse and interesting society than people in the rest of Canada give it credit for, sometimes the stereotypes weigh stronger than the reality."

Speaking specifically of young Calgarians, Finch, Lukey, Althouse, Schaufele, Swiston, and Lane (2021, p. 22) argue many in the 18-29 age group are questioning their future in the city for a variety of social and economic reasons. They write:

> A potential consequence is accelerating outward migration of this population group to cities they perceive have more diverse employment opportunities and more alignment with their progressive values associated with embracing diversity and acting on the environment and climate.

Conclusion

The above analysis has relied largely on qualitative evidence, partially because of the lack of current quantitative data, and partially because of the limitations in these data even if they were available. Here we concur with Finch, et al. (2021, p. 21) in their case study of Calgary:

> Today, the data required to track young adult talent mobility is at best fragmented and not publicly available, or at worst, non-existent. It is essential that both Calgary and Alberta refine their data to track leading indicators of young adult talent mobility. Without access to leading indicators, we will be both reactive and fall victim to anecdotes.

Following Hiller (2009), who argued migration is a process that begins with disenchantment with one's current residence, gauging the attitudes of Albertans, especially young Albertans in this current case, is likely an important bellwether in determining the future out-migration from the province, as well as in-migration from other provinces. It is unlikely the out-migration of Albertans under the UCP has ceased. Arguably, the exodus of Albertans would have been higher had it not been for the global pandemic. As restrictions continue to ease and individuals move further into the migration process, if clearly feeling marginalized in Alberta, we expect the official migration tallies to be more representative of what is currently happening.

Alberta has historically attracted people to the province — and induced others to stay — based on the strength of its economy, high wages, and low unemployment rates. Much of this was driven by the booming energy sector, which offered lucrative employment to tens of thousands of Albertans and others and enticed people to come to or stay in the province, sometimes despite their displeasure with Alberta's provincial politics. Given that the energy sector is unlikely to achieve its former glory — the results of the development of alternative energy sources and a post-labour-intensive construction phase in the oil sands — the many other amenities Alberta has to offer will be what attract and retain young people.

The changing Alberta economy also means advanced education needs to be given a higher degree of priority if Alberta is to retain its young people and attract other young people from outside of the province. The energy sector, which acted as a magnet for this demographic for years, can no longer be relied upon to do so. High quality post-secondary education, coupled with accessibility and affordability, will produce the talented workforce of the future, according to ATB chief economist Todd Hirsch (2021). Unfortunately, accessibility and affordability have been moving in the opposite direction, says Hirsch, and the whole education system, from early childhood through post-secondary, could use a rethink. More funding is the starting point; however, this does not appear to be forthcoming, at least under UCP governments. Rather, the recent campaign to attract skilled workers from Vancouver and Toronto is simply a new chapter in the old playbook of importing talent from outside the province, rather than nurturing and developing it at home. The difference this time is that other jurisdictions — arguably those with political climates more congruent with the political learnings of the very people Alberta is trying to attract — are also in need of this same talent.

Alberta also has a cultural problem with respect to post-secondary attendance. Until recently, young Albertans could find lucrative and stable employment in the energy sector with minimal formal education. This perception too must catch up with reality. The NDP (2022), in its post-secondary education platform, reiterates the well-known fact that Alberta has the lowest post-secondary education participation compared to the three largest provinces and the Canadian average. It argues a well-educated and skilled workforce is more important than the corporate tax rate when companies look to move or expand their operations. Similarly, it says research has found that among the most important considerations for investment attraction are a skilled workforce, a quality education system, and liveability. The document also questions the UCP's cutbacks to education given that enrolments are projected to increase dramatically over the next few years. Young people in Alberta may be forced to look elsewhere for opportunities and potentially leave the province for good.

Alberta has an image problem, and it needs to change. In a phrase: the province has a branding problem (Hirsch, 2021, p. 35). It is portrayed — fairly or unfairly — "as nothing more than dirty, fossil-fuel spewing industries with social-conservative minded people." In addition to education, fostering inclusion and diversity is important, according to Hirsch, since bright, young, talented people can live anywhere in the world they please. Alberta must compete for this talent.

Yet, it appears the UCP government is caught in a time warp, thinking if only the past energy economy was restored, everything else would be OK. The UCP's 2019 election platform alludes to this desire, with its commitment to support post-secondary education and promote experiential and vocational learning. What has transpired since is a much narrower view of education, as mainly the acquisition of skills directly related to current labour market needs. Lee Easton, president of Mount Royal University's Faculty Association, called the government's post-secondary education plan, Alberta 2030, nothing more than 1990s thinking; or, as the University of Regina's Marc Spooner quipped, "Alberta 2030: preparing people for the jobs of yesterday!" David Stewart, president of the University of Calgary Faculty Association, noted that Alberta 2030 says we need more talent and must keep and attract the best students from around the world, yet the government has cut funding to post-secondary, forcing universities and colleges to increase student fees, while also increasing industry funding.[9] Perhaps ironically, the same 2019 UCP platform says (p. 62): "The NDP have failed these young people. We don't want them leaving Alberta. United Conservatives want to help prepare our young women and men to go to work and give them hope for a great future – right here in Alberta."

Yet, that is exactly what is happening. Young people are leaving the province, and the UCP's plans for post-secondary are out of touch with today's reality, especially among these young people who "work to live" and not "live to work." As Hiller (2009, p. 149) wrote "If there is no dissatisfaction, and no sense of marginalization, there is little incentive to migrate." It is the causes of what is clearly a sense of marginalization among Alberta's young people that must be addressed if the province is to continue to thrive in the post-energy economy, and the province has plenty of existing and potential assets to make this happen.

Changing Alberta's reputation, internally and externally, and changing the culture of post-secondary participation so more young people want to attend the province's world-class institutions, and these institutions have the capacity to accept them, are essential for Alberta to attract and retain young people to a province that is much more diverse and interesting than the outdated stereotypes suggest.

References

Acuña, R. (2021). Alberta's future needs a strong post-secondary system. *ATA News* 55(11), 3. https://mydigimag.rrd.com/publication/?m=61458&i=702769&p=2&ver=html5).

Adkin. L. (2022). Retelling the Story of the UCP Government's Budget and its Meaning for Post-Secondary Education in Alberta. Parkland Institute blog. March 15. https://www.parklandinstitute.ca/retelling_the_story_of_the_ucp_governments_budget

Alberta Medical Association (AMA). (2020). Looming physician exodus from Alberta caused by failed provincial funding framework. Press Release, July 10. https://www.albertadoctors.org/news/news-archives/july-10-2020-news-release

Alberta Treasury Board and Finance. (2021). Annual Population Report: Alberta 2020-21. https://open.alberta.ca/dataset/1050cf0a-8c1d-4875-9800-b7d2f3199e41/resource/608764dd-e247-4fc4-a99d-cbc59fc78a0c/download/2020-21-population-report.pdf

Blue Ribbon Panel on Alberta's Finances (J. MacKinnon). (2019). Report and recommendations. https://open.alberta.ca/publications/report-and-recommendations-blue-ribbon-panel-on-alberta-s-finances

Boucher, P.E. (2020). Understanding physician supply numbers. Alberta Medical Association President's Letter (October). https://www.albertadoctors.org/services/media-publications/presidents-letter/pl-archive/understanding-physician-supply-numbers

Business Council of Alberta. (2021). *Alberta's economy: An overview.* https://www.businesscouncilab.com/wp-content/uploads/2022/03/Albertas-Economy-Economic-Overview-FULL-REPORT-UPDATED.pdf

Chand, A. (2022, Jan. 6). Why I'm not leaving Alberta. *The Tyee*. https://thetyee.ca/Opinion/2022/01/06/Why-Not-Leaving-Alberta/?utm_source=facebook&utm_medium=social&utm_content=010622-f&utm_campaign%E2%80%A6

College of Physicians and Surgeons of Alberta (CPSA). (2022a). *Changes in physician workforce: 2017-2021.* https://cpsa.ca/wp-content/uploads/2022/01/Changes-in-physician-workforce-2021-2017.pdf

College of Physicians and Surgeons of Alberta (CPSA). (2022b). *Physician resources in Alberta*, quarterly update: Oct 01, 2021 to Dec 31 2021. https://cpsa.ca/wp-content/uploads/2022/01/Q4-2021-Quarterly-Report.pdf

College of Physicians and Surgeons of Alberta (CPSA). (2022c). *Physician resources in Alberta*, quarterly update: July 01, 2022 to Sept 30 2021. https://cpsa.ca/wp-content/uploads/2022/10/Q3-2022-Quarterly-Report.pdf

Cowley, P. (2022, Feb. 16). Alberta risks brain drain if post-secondary education underfunded, says NDP leader. *Red Deer Advocate*. https://www.reddeeradvocate.com/news/alberta-risks-brain-drain-if-post-secondary-education-underfunded-says-ndp-leader/

Dawson, T. (2021, Dec. 5). When it comes to living in Alberta, half the Canadians outside province say 'no thanks,' poll finds. *National Post*. https://nationalpost.com/news/canada/when-it-comes-to-living-in-alberta-half-of-canadians-outside-province-arent-keen-to-find-out-poll-finds

Day, K.M., & Winer, S.L. (2012). *Interregional Migration and Public Policy in Canada*. Montreal and Kingston: McGill-Queen's University Press.

Derworiz, C., & Graveland B. (2022, Aug. 15). *Kenney kicks of campaign to attract skilled workers to Alberta*. CBC News. https://www.cbc.ca/news/canada/calgary/jason-kenney-skilled-worker-campaign-1.6551658.

Finch, D., Lukey, K., Althouse, C., Schaufele, A., Swiston, A., & Lane, J. (2021). *Why Calgary? Competing for young mobile talent*. Mount Royal University, Institute for Community Prosperity. https://static1.squarespace.com/static/5b8748eb372b96145e73af65/t/603315c0f35fa24e7d8bb87e/1613960645484/Why+Calgary.+Starting+a+Discussion.pdf

Finnie, R. (2004). Who moves? A logit model analysis of interprovincial migration in Canada. *Applied Economics* 36(16), 1759-79.

Flanagan, B. (2022, May 14). Post-secondary investment helps keep young people in Alberta. *Edmonton Journal*. https://edmontonjournal.com/opinion/columnists/opinion-post-secondary-investment-helps-keep-young-people-in-alberta

Ford, T. (2021, Jan. 22). Meet the Albertans ditching the Jason Kenney government. *The Tyee*. https://thetyee.ca/Analysis/2021/01/22/Meet-Albertans-Ditching-Jason-Kenney-Government/

Gunter, L. (2022, March 25). Exodus of Alberta doctors can be blamed on the pandemic and Tyler Shandro. *Edmonton Sun*. https://edmontonsun.com/opinion/columnists/gunter-exodus-of-alberta-doctors-can-be-blamed-on-the-pandemic-and-tyler-shandro

Henschel, M. (2022, May 29). *Alberta Medical Association warns of physician shortage*. Global News. https://globalnews.ca/video/8879511/alberta-medical-association-warns-of-physician-shortage

Hiller, H. H. (2009). *Second Promised Land: Migration to Alberta and the Transformation of Canadian Society*. Montreal and Kingston: McGill-Queen's University Press.

Hirsch, T. (2021). The state of the Alberta economy and the path forward. In K.J. McKenzie, & R.L. Mansell (Eds.). *Alberta's economic and fiscal future* (pp. 25-38). Calgary: School of Public Policy, University of Calgary. Retrieved May 16, 2022, from https://www.policyschool.ca/wp-content/uploads/2021/11/Albertas-Economic-and-Fiscal-Future.pdf

Hussey, I. (2020). *The Future of Alberta's Oil Sands Industry: More Production, Less Capital, Fewer Jobs*. Edmonton, AB: Parkland Institute. https://d3n8a8pro7vhmx.cloudfront.net/parklandinstitute/pages/1785/attachments/original/1583615491/futureofalbertasoilsands.pdf?1583615491

Karstens-Smith, B. (2022, Jan. 26). *Alberta doctors, teachers, express interest in leaving the province*. Global News Edmonton. https://globalnews.ca/video/8541469/alberta-doctors-teachers-express-interest-in-leaving-the-province

Lane, J., Laverty, S., & Finch, David. (2022). *Work to Live: Alberta Youth Mobility*. Calgary: Canada West Foundation. https://cwf.ca/wp-content/uploads/2022/03/CWF_WorktoLive_Report_MAR2022-1.pdf

Lee, J. (2021, July 26). *Burned out and demoralized: Some Alberta nurses look to leave amid province's bid to cut pay*. CBC News. (https://www.cbc.ca/news/canada/calgary/alberta-nurses-1.6114721

Markusoff, J. (2022, June 1). *The next Alberta premier's big decision: What to do with a tsunami of cash*. CBC News. https://www.cbc.ca/news/canada/calgary/next-alberta-premier-decision-royalty-revenue-1.6472980

Maru Public Opinion. (2021). Brand Alberta 2021. Retrieved May 16, 2022, from https://www.marugroup.net/public-opinion-polls/canada/brand-alberta-2021

McMillan, M.L., & Warrack, A.A. (1995). One-track (thinking) towards deficit reduction. In T. Harrison & G. Laxer, (Eds.). *The Trojan Horse: Alberta and the Future of Canada* (pp. 134-62). Montreal/New York, London: Black Rose Books.

Mueller, R.E. (2021). Public and private sector wages: How does Alberta compare to the "big 3" provinces? In K.J. McKenzie, & R.L. Mansell (Eds.). *Alberta's economic and fiscal future* (pp. 265-90). Calgary: School of Public Policy, University of Calgary. https://www.policyschool.ca/wp-content/uploads/2021/11/Albertas-Economic-and-Fiscal-Future.pdf

New Democratic Party (NDP). (2022). *Strengthening post-secondary for a resilient future*. Retrieved February 17, 2022, from https://www.albertasfuture.ca/albertas-future/albertas-future-campaigns/post/strengthening-post-secondary-for-a-resilient-future

Pike, H. (2021, Feb. 6). *Advocacy group says Alberta's tuition tax credit cut isn't a student issue — it's a provincial one.* CBC News. https://www.cbc.ca/news/canada/calgary/alberta-tax-credit-rowan-ley-taylor-hides-university-1.5904157

Rheault, L. (2019). Why are Canadians reluctant to leave their province? *American Review of Canadian Studies* 49(3), 428-48.

Taft, K. (1997). *Shredding the Public Interest: Ralph Klein and 25 Years of One-Party Government.* Edmonton: University of Alberta Press and Parkland Institute.

United Conservative Party (UCP). (2019). *Alberta strong and free: Getting Alberta back to work.* UCP Election Platform. Retrieved December 14, 2021, from https://albertastrongandfree.ca/policy/

United Nurses of Alberta (UNA). (2022). *Kenney's hospital announcement lacks what's needed to solve Alberta's health care crisis: UNA President.* Retrieved June 3, 2022, from https://www.una.ca/1350/kenneys-hospital-announcement-lacked-whats-needed-to-solve-albertas-health-care-crisis-una-president

University of Alberta Students' Union (UASU). (2021). *Survey report: Your life after UAlberta.* Retrieved February 17, 2022, from https://www.su.ualberta.ca/media/uploads/1143/Survey%20Report_%20Your%20Life%20After%20UAlberta.pdf

Usher, A. (2021). *Interprovincial student mobility.* Retrieved May 25, 2022, from https://higheredstrategy.com/inter-provincial-student-mobility/.

Usher, A. (2022). *The Alberta exodus.* Retrieved May 25, 2022, from https://higheredstrategy.com/the-alberta-exodus/#comments

Varcoe, C. (2021, Oct. 2). No longer the youngest – Alberta's 'brain drain' sparks concern. *Calgary Herald.* https://calgaryherald.com/opinion/columnists/varcoe-no-longer-the-youngest-albertas-brain-drain-sparks-concern

Williams, E. (2021, Nov. 19). Alberta's identity crisis: What does it mean to be a young educated professional in Alberta? *The Gateway.* https://thegatewayonline.ca/2021/11/albertas-identity-crisis-what-does-it-mean-to-be-a-young-educated-professional-in-alberta/

Young, L. (2021, June 16). *Facing COVID-19 staffing crunch, some Ontario hospitals offer cash bonuses to new nurses.* Global News. https://globalnews.ca/news/7954398/nurse-shortage-signing-bonus/

NOTES

1 All figures in this paragraph calculated by the author using data from from Statistics Canada Table 17-10-0022-01.

2 See Statistics Canada Table 17-10-0020-01 for the latest quarterly statistics on interprovincial migration.

3 These are author's calculations from Statistics Canada Table 17-10-0135-01.

4 For example, the College of Physicians and Surgeons of Alberta (CPSA) publishes timely quarterly data on the number of physicians registered in the province, including the additions and removals from the registry and the reasons for the changes (e.g., leaving or returning to the province, retirement, etc.).

5 The number of registrants include those who hold a licence to practice medicine and may differ from the number of physicians actively practicing in the province (CPSA, 2022b and 2022c). Still, assuming the proportion of non-practicing registrants is relative stable over time, these numbers indicate the trends suggested in the text. The first quarter tends to have the highest number of physicians who do not renew their licences but have ceased to practice medicine in Alberta before this time, simply delaying notifying the CPSA until their licences are to be renewed in the first quarter of each year.

6 https://albertaworker.ca/2020/04/21/a-list-of-alberta-communities-losing-doctor-care/

7 "Post-secondary Education and the Alberta Brain Drain Town Hall," May 27, 2021, an online forum hosted by the Mount Royal University Faculty Association.

8 https://www.reddit.com/r/britishcolumbia/comments/qu3d9d/albertans_moving_to_bc_why/?utm_term=38255179837&utm_medium=comment_embed&utm_source=embed&utm_name=&utm_content=header

9 "Post-secondary Education and the Alberta Brain Drain Town Hall," May 27, 2021, an online forum hosted by the Mount Royal University Faculty Association.

CHAPTER 10

TURNING THE SCREWS, TURNING BACK THE CLOCK: THE UCP AND LABOUR

Jason Foster, Susan Cake and Bob Barnetson

"[T]he Kenney government has effectively turned Alberta into Canada's conservative policy laboratory and in so doing is expanding the country's other centre-right governments' sense of what is possible."
— Sean Speer, *National Post*, May 2, 2022

IN APRIL 2020, one of the worst COVID outbreaks on record began at the unionized Cargill meatpacking plant in High River, Alberta. Union efforts to address the hazard posed by COVID were thwarted by employer intransigence and collusion with politicians and civil servants. Provincial occupational health and safety officers inspected the plant on April 14 by video call, declaring the plant they were unwilling to enter to be as safe as reasonably practicable. On April 18, Agriculture Minister Devin Dreeshan, Chief Medical Officer of Health Deena Hinshaw, and other government officials attended a teleconference employee meeting called by Cargill. Here, government officials declared the plant safe and urged workers to continue to go to work, despite knowing 200 Cargill workers tested positive for COVID. The plant was eventually closed for two weeks, but only after hundreds of workers engaged in direct action by invoking their right to refuse dangerous work under the *Occupational Health and Safety Act*. More than 1,000 workers would eventually test positive and two would die during this outbreak. An RCMP investigation and a class-action lawsuit are underway (Foster, Cake and Barnetson, 2022). Outbreaks at meatpacking plants (including several more at Cargill) occurred throughout the pandemic.

The government's failure to protect meatpacking workers' health is a stark example of how Alberta's United Conservative Party (UCP) has consistently prioritized employer profitability over workers' health and lives. This chapter explores the UCP's approaches to labour policy and labour relations during its first term. Overall, the UCP's key promise to get Albertans back to work via tax cuts was overwhelmed by COVID-19 disruptions and a Canada-wide economic recovery in 2022. The UCP has been more successful at rolling back worker rights to cheapen labour for employers.

The UCP has also engaged in a protracted attack on unionized workers' rights as well as their wages and working conditions, including introducing U.S.-style picketing and union finance laws. Despite the UCP's efforts, the pandemic has limited the UCP's success at driving wage rollbacks into public-sector collective agreements.

Background

Alberta has historically had the most regressive regime of worker rights in Canada, including the lowest rate of unionization in Canada, one of the highest rates of workplace injury, and the lowest minimum wage. These outcomes reflect decades of conservative governments prioritizing the interests of business (Finkel, 2012). This policy orientation has had an uneven impact on workers. Many men earned high wages in Alberta's periodically booming oil and gas industry. Women typically fared worse economically, and Alberta has the second highest gender pay gap in Canada (McIntosh and Graff-McRae, 2020).

The election of a New Democratic (ND) government in 2015 resulted in changes to Alberta's labour and employment laws that brought Alberta in line with other provinces. Farm workers received basic employment rights for the first time. Workers in all sectors could unionize simply by showing membership card support, unions could access first-contract arbitration, and public-sector workers were granted the right to strike (Barnetson, 2017). Alberta's Occupational Health and Safety (OHS) legislation was updated for the first time since 1976 and joint health and safety committees were made mandatory in medium and large workplaces. Alberta's workers' compensation system was overhauled to address the culture of claims denial as well as impose an obligation on employers to re-employ injured workers upon recovery.

The UCP rolled back many of these changes following its 2019 election and, in places, has gone much further, once again making Alberta the most worker-unfriendly jurisdiction in Canada. The central plank of the UCP's election platform was getting Albertans back to work. The three main strategies the UCP promised would create jobs were:

- Lowering taxes: A corporate tax reduction of one-third (from 12 per cent to eight per cent) would create at least 55,000 new full-time jobs. How this tax cut was to be funded was not explicitly addressed but, given the proportion of government expenditures related to civil service salaries, it would require significantly reducing public-sector wages, jobs, or both.
- Lowering labour costs: Under the guise of red-tape reduction for employers, the UCP promised to roll back many employment rights, including lowering the minimum wage for youth, eliminating many farmworker rights, and changing the rules around overtime premiums.

Reducing workers' power: Under the guise of increasing workplace democracy, the UCP promised to eliminate card-check certification and replacement-worker bans and impose new conditions on union dues collection and spending (UCP, 2019).

The UCP's election platform was consistent with Premier Jason Kenney's long history in federal politics of favouring the interests of employers over those of workers. As a federal minister, Kenney oversaw a radical expansion of the temporary foreign worker (TFW) program, which flooded Alberta's labour market and allowed employers to dampen worker demands for increased wages and better working conditions while, at the same time, exposing TFWs to wage theft and other forms of exploitation. The Harper government also intervened to the benefit of the employer in work stoppages at CN Rail, Canada Post, Air Canada and CP Rail, legislated a wage freeze for federal employees in 2009 (including overriding previously negotiated increases), and made union certification votes mandatory in 2014. Just before being defeated in 2015, the Harper government also enacted onerous new financial disclosure requirements on unions that were subsequently repealed by the Trudeau government (Barnetson, 2019). This approach foreshadowed much of Kenney's agenda as premier.

Job Creation Efforts

Job creation featured heavily in the UCP's 2019 election campaign. This is not surprising given Alberta had been through five years of economic turmoil, including stubbornly high unemployment rates. Whether the UCP delivered on that promise is a difficult question to answer. Overall, it appears the effects of the COVID-19 pandemic, a Canada-wide post-pandemic rebound, and a surge in oil prices in 2022 explain more of the overall change in employment than the UCP tax cuts (see Figure 10.1 below).

Unemployment went up in the first part of the UCP's term. The province lost 89,000 full-time jobs between July 2019 and February 2020. COVID-19 accelerated these losses, with unemployment spiking to 15.9 per cent by June 2020. Employment levels reached their lowest level in April of 2020 (right after the COVID-19 pandemic first hit) with 386,400 (-16.8 per cent) fewer Albertans employed than when the UCP was elected. Employment began to pick up slowly in the spring of 2021, only to see those gains wiped out when the third wave of COVID enveloped the province following Kenney's "best summer ever." By the end of 2021, there were still 57,000 fewer employed Albertans than when the UCP were elected in May 2019. Through this period, unemployment in Alberta was approximately one per cent higher than the national average and highest among the four largest provinces (Statistics Canada, 2022a).

Things started to turn around in the spring of 2022 as the Canadian economy

Figure 10.1. Employment (000s) and selected events
January 2019 to December 2022

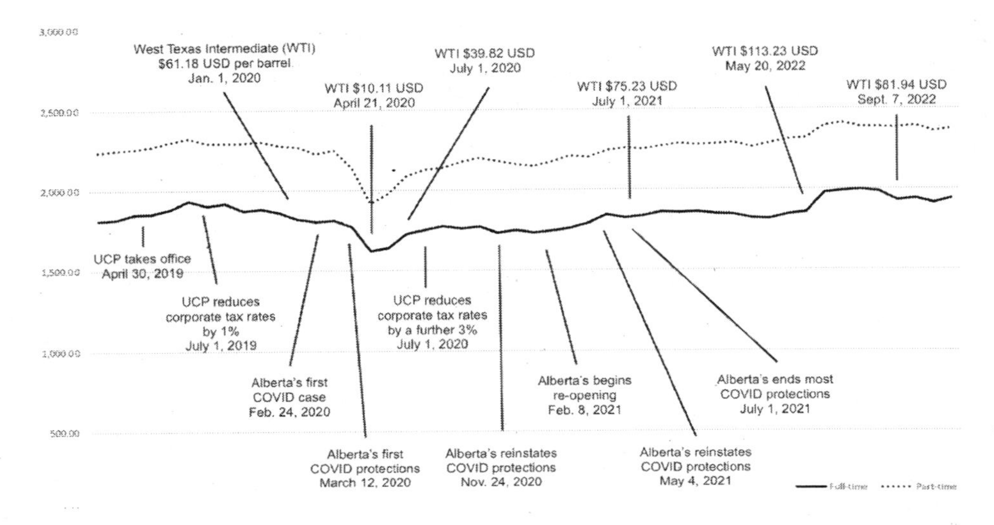

Source: Statistics Canada (2022a)

rebounded from the effects of the pandemic and the price of oil spiked to above $100 a barrel for the first time since 2014. By December 2022, employment was up 70,900 jobs since May 2019, including 56,600 new full-time jobs. It should be noted these gains were not uniform across sectors of the economy. As of May 2022, health care and social services was the sector with the largest net gain of jobs since 2019 (Statistics Canada, 2022b). Most of these jobs are in the public sector and the increase is likely due to resources put into the health-care system to combat the pandemic. Other areas of job growth included utilities, finance, and professional and scientific sectors. It is noteworthy that employment in oil and gas decreased during this time. Between May 2021 and May 2022, the largest areas of job growth were mostly in industries badly hit by COVID-19, such as accommodation and food services and arts and entertainment, which were mostly regaining jobs lost due to the pandemic. There was also significant growth in oil and gas and construction jobs, both hallmarks of past surges in oil prices.

The signature promise of the UCP platform (2019, p. 12) was to create "at least 55,000 new full-time jobs" via a corporate tax cut. At the time of writing, the province seems slowly moving towards this goal. It is, however, unclear how much of the gain is attributable to the corporate tax cut. The first cut of one per cent in 2019 was followed in subsequent months by a net loss in employment, even before the pandemic hit. In June 2020, Alberta accelerated its tax cuts and made substantial infrastructure investments to "create tens of thousands of jobs right now" (Alberta, 2020, p. 6). As shown in Figure 10.1, that effort did little to alter short-term employment trends. Employment only began to rise as the pandemic ebbed and the price of oil shot up due to international market factors. It is estimated this corporate tax cut removed up to $4.7 billion from government revenues in the short term.

The UCP has been selectively resistant to federal wage-subsidy programs designed to support workers during the pandemic. For example, Alberta was eligible for up to $348 million in federal aid for low-income workers performing essential work. To access this money, Alberta needed to provide $115 million in matching funds, which it failed to do (AFL, 2020). The province was eventually embarrassed into applying for this funding when it was revealed the UCP, as a political party, applied as an employer for a different federal wage subsidy (the Canada Emergency Wage Subsidy) to maintain its own party operations (Rieger, 2020). It is unclear why Alberta was reluctant to access federal funding for low-income workers performing essential work. One explanation is that Kenney may have viewed federal wage subsidies as undermining his goals of ending equalization payments and reducing public-sector wages. In the spring of 2021, Alberta opened its first intake for the Critical Worker Benefit that allowed employers to apply for up to $1,200 per low-wage worker to recognize their work providing care and critical services during the pandemic. Approximately $355 million was distributed to public- and private-sector employers. A second intake (concluding in September 2021) saw a further $98.8 million being distributed, mostly to private-sector employers (Alberta, n.d.). It is unclear if some of Alberta's job losses could have been prevented had the government moved faster to take advantage of the federal support.

Lowering Labour Costs

Complying with statutory employment rights typically entails a cost for employers. The UCP (2019, p. 19) platform promised that "measures like red tape reductions, will command the attention of job creators across Canada and beyond" and would result in job creation. The UCP was quick to enact legislative and policy changes that lowered labour costs for employers. "Our government ran on a promise to get Albertans, especially young people, back to work. … With Bill 2 and the youth minimum wage, we are restoring fairness and balance to the workplace and getting 'Help Wanted' signs back in the windows of Alberta businesses," said Kenney, when announcing the first of many changes intended to reduce costs to employers, often by shifting them to workers (Alberta, 2019a, p.1).

Upon election, the UCP immediately fulfilled an important election promise to rural supporters by rolling back farmworker rights. The 2019 *Farm Freedom and Safety*

Act eliminated industry-specific safety rules implemented by the NDs after months of consultations with stakeholders. Framed as addressing the realities of working on small farms, this change also ensured 80 per cent of farmworkers were not subject to employment standards and, for those that were, excluded them from rules around hours of work, rest periods, overtime pay, and child labour. Farm workers were also no longer allowed to join or form a union. According to Dreeshen, "Farmers would tell us there's no real place for a union on the farm" (Stephenson, 2019). The opinions of farmworkers were, apparently, of little importance to Dreeshan. Farmers could also opt out of workers' compensation if they purchase private injury insurance, a situation that historically resulted in under-insurance.

The UCP took a more nuanced approach to the minimum wage. The New Democratic government increased Alberta's minimum wage to $15 per hour over four years. Although this 47 per cent increase still meant workers were living below the poverty line in major cities, it did meaningfully improve the lives of low-wage workers who were primarily young women in permanent jobs, many of whom were parents (Alberta, 2018a). Despite growing consensus among economists that minimum-wage increases do not result in job losses, Kenney told the Legislature exactly the opposite in 2018:

> Mr. Kenney: ...What do you think a 50 per cent increase in the minimum wage results in? Well, according to the Bank of Canada 60,000 job losses across the country. According to the C.D. Howe Institute 25,000 job losses in Alberta. Think about how—oh, my goodness—when New Democrats get on their moral high horse and pretend they have a monopoly on compassion, and then because union bosses tell them to, they bring in a policy that, according to the think tanks will kill 25,000 jobs for immigrants and youth. Where is the compassion for those who lost their jobs, Mr. Speaker? There is none. There's no regard (Alberta, 2018b, p. 433).

Upon taking power, the UCP decided to freeze the minimum wage, despite calls by some employer groups to lower it. The freeze allowed inflation to erode the purchasing power of these workers. Alberta also reduced the minimum wage for students under 18 to $13 per hour in the expectation it would increase youth employment (Mertz, 2019). According to Kenney, "This is still a very generous wage — $13 an hour is a lot more than $0" (Keller, 2019). There is, however, no evidence this reduction created new jobs for youth. The UCP also struck a panel to assess the evidence in favour of re-implementing a lower minimum wage for servers (something the NDs had eliminated). The panel reported in 2020 but no action was taken. In 2022, the UCP announced the report would not be released because labour-market conditions had changed too much since the report was written.

The UCP also made changes to overtime (OT) rules that make it possible for employers to evade overtime premiums almost entirely. Workers in Alberta receive 1.5 times their regular pay if they work more than eight hours in a day or more than 44 hours in a week (whichever calculation is better for the worker). Historically, Alberta allowed employers and workers to voluntarily enter overtime averaging agreements. Averaging agreements allow an employer to average hours worked over a period of time when calculating whether the worker has met the weekly OT threshold. This allows workers and employers to agree to compressed work weeks (e.g., four 10-hour days instead of five eight-hour days) without triggering the OT premiums. It also provides flexibility for weather-dependent work. Any overtime still owed at the end of the averaging period is paid at OT premium rate.

In 2020, the UCP introduced changes that allowed employers to impose averaging arrangements with two weeks of notice, increase the averaging period from 12 weeks to 52 weeks, and eliminate the two-year limit on such arrangements. These changes provide employers opportunities to evade paying OT premiums. For example, the weekly overtime threshold is 44 hours. If a worker works a 60-hour week (say, six 10s), they would normally be eligible for 16 hours of pay at overtime rates. Under an overtime averaging arrangement, those 16 hours could be averaged (i.e., spread across) up to 52 weeks (roughly 20 minutes per week). This would spread the OT far enough not to trigger weekly OT payments. A moderately skilled payroll clerk could arrange for a worker to work up to 208 OT hours a year (i.e., more than five extra weeks) without ever receiving any OT premiums. The changes effectively guarantee that very few, if any, non-union Alberta workers will ever receive overtime pay unless the employer agrees to pay it as an act of altruism or a job perk. Further, when a worker is entitled to be paid OT under an averaging arrangement, that pay may be delayed until the end of the averaging period (now as long as 52 weeks). Labour Minister Jason Copping nonsensically framed these changes as "expanding choice for workers" (Copping, 2020) when, in fact, it grants the choice of whether or not to pay overtime solely to employers.

This 2020 change compounds 2019 changes made by the UCP to how banked overtime is paid out. Banking means workers can save their OT and have it paid out (at premium rates) when there is no work (such as during slow times in seasonal industries). Under the UCP's 2019 changes, a worker who enters an OT banking arrangement (which is notionally voluntary, but practically determined by the employer) and wishes to take banked time as time off with pay (instead of being paid out), does so at straight time. Together, these changes allow employers to essentially avoid paying overtime premiums to non-unionized workers (roughly 80 per cent of the labour force). In 2018, there were $3.3 billion in overtime premiums paid in Alberta, so this small change could transfer hundreds of millions of dollars from workers to employers each year. Particularly hard hit will be workers in seasonal industries, such as upstream oil and gas and construction.

Making Workplaces Less Safe

By the end of 2021, the UCP had made significant changes to the *Occupational Health and Safety Act*. According to Labour Minister Jason Copping,

> This is about improving safety for Alberta workers and making workplaces safer. We are also restoring balance and fairness to the workers' compensation system to meet the needs of workers and job creators now and into the future. Workers continue to have rights and protections while job creators face less red tape (Alberta, 2020).

A who's who of Alberta business leaders lined up to endorse these changes. Despite Copping's promises, the actual changes implemented by the UCP made workplaces less safe while also shifting the cost of workplace injuries from employers to workers (see next section). Of particular concern were the UCP's changes to work refusals and the weakening of joint health and safety committees (JHSCs).

JHSCs were introduced to Alberta in 2018 by the New Democrats. Alberta was the last Canadian province to require JHSCs in medium and large workplaces. Joint committees are the key mechanism by which workers participate in ensuring workplaces are safe. The UCP removed the rights of joint committees to investigate worker concerns and complaints, identify hazards during quarterly inspections, develop and evaluate mitigation strategies, and participate in investigations of serious injuries and incidents. Under the UCP, joint committees can only receive worker complaints and concerns, review inspection results, and make non-binding recommendations. The UCP changes mean there is no longer any requirement for regular workplace inspections. Employers have also been given much more control over the operation of JHSCs (Barnetson, Foster, Cake, and Matsunaga-Turnbull, 2021).

These changes shift the role of joint committees on worksites from active participants to passive observers, thereby undermining the principle of shared responsibility built into the Internal Responsibility System. While weakening joint committees may seem like a technical detail, they are actually a key component of Canada's health and safety system. They are supposed to be the mechanism where workers can exert influence to make workplaces safer by exercising their right to participate. While they often do not work as well as intended, their centrality to the system means the UCP changes undermine workplace safety in general.

The UCP also weakened workers' rights to refuse unsafe work without retaliation from their employers. The right to refuse is one of the only ways (and often the last way) a worker can avoid injury on the job and thus is one of workers' most powerful rights in the safety system. Prior to the UCP's changes, a worker could refuse work if the worker believed, on reasonable grounds, there was a dangerous condition at the worksite or the work constituted a danger to the worker's health or safety or to the health or safety of another worker or another person. While this provision has existed for decades, the NDP strengthened it by expanding the scope of refusals. The UCP reversed those changes and, arguably, made the right even weaker than before the NDP. The UCP replaced the term "dangerous condition" with "undue hazard." An undue hazard is one that poses a serious and immediate threat to the health and safety of a person. This means refusals can only occur when the danger is "serious" (an undefined term) and "immediate." This change suggests workers may not be able to refuse work where the potential injury is not serious or where the injury will occur over time (e.g., most occupational diseases).

The UCP also allowed employers to assign the refused work to another worker without informing the second worker about the previous refusal. Employers are also no longer explicitly required to pay workers while they are refusing unsafe work. And, while employers are still barred from disciplining workers for refusing unsafe work, the UCP's changes now allow employers to subject workers to discriminatory treatment (e.g., assigning refusers all night shifts). Overall, these changes weaken workers' ability to refuse unsafe work (Barnetson et al., 2021). Practically speaking, had they been in place in March of 2020, the UCP's changes would have made it difficult for workers at Cargill to enact the work refusals that stopped the COVID outbreak in their workplace (Foster et al., 2022).

In addition to legislative changes, government enforcement of injury-prevention laws has changed during the UCP's term. Alberta has never enforced safety laws particularly well, even under the NDP. However, the UCP has backed away from this responsibility even more. The number of workplace inspections were roughly stable under the UCP but, the number of orders to remedy violations written to employers fell by 49.6 per cent under the UCP, while the number of tickets written by inspectors dropped by 93.4 per cent. Charges laid against employers and fines levied also declined. Not surprisingly, the rate of lost-time claims (which is a proxy for serious injuries) rose by 16 per cent between under the UCP (Barnetson, 2022). Alberta continues to have one of the highest rates of occupational fatalities in Canada (Tucker and Keefe, 2022).

Cheapening Workplace Injuries

The UCP also enacted significant changes to the workers' compensation system, justifying it as protecting the viability of the WCB. "At a time when Alberta's economy and job creators are struggling the most, employers have told us that some of these changes have led to rising costs, additional red tape, and has put the system's future state of sustainability in doubt," said Labour Minister Jason Copping (Bellefontaine, 2020). No actual evidence that the WCB's accident fund was unsustainable or that cutting worker benefits was the best policy solution was advanced by the UCP to substantiate Copping's claims. In fact, the WCB continues to be one of the financially healthiest WCBs in the country (AWCBC, 2022).

The changes made by the UCP to the workers' compensation system transferred half a billion dollars in workplace injury costs from employers (by lowering their premiums) to injured workers (via reduced benefits). These changes were projected to save employers $111.2 million in 2021 and an additional $393 million in future years (Alberta, 2020b). Projected savings included approximately $240 million transferred from injured workers to employers by de-indexing wage-loss payments to injured workers from the cost of living. This projection is a clear indication that the WCB is expected to allow the purchasing power of injured workers' benefits to decline via inflation.

The UCP also removed the requirement that an employer continue providing health benefits for an injured worker. This change means that injured workers and their enrolled family members no longer have access to health benefits they would have enjoyed prior to the workplace injury happening. The WCB only covers medical costs stemming from the workplace injury, so this change shifts health-care costs onto the worker and their family which may, in fact, prolong and worsen a workplace injury.

The UCP also reintroduced a maximum insurable earnings cap that works in conjunction with a 90 per cent replacement of pre-injury earnings. These changes mean higher-waged workers in Alberta shoulder a greater financial burden when they are injured, as less of their income is replaced by the WCB system. This change is projected to transfer $33 million from injured workers to employers. Since workers who receive wage-replacement benefits are generally unable to work, they have no opportunity to replace this lost income.

Another important change the UCP implemented was eliminating the requirement for employers to re-employ injured workers once they are ready and able to return to work. This means if an employer decides to not reinstate a worker because of their injury, the worker must file a complaint with the Alberta Human Rights Commission to seek remedy. Human Rights complaints take, on average, two years to conclude. During this time, injured workers would be forced to seek other employment (which is often a difficult because of the lingering effects of an injury) and there is no guarantee the worker would be reinstated at the end of a human-rights process.

Appeals in the WCB system were also made more difficult by the UCP eliminating the Fair Practice Office introduced by the NDP in 2018 to help workers and employers navigate the complex claims process (a $1.8 million savings). The UCP also shortened the timeline for any parties to appeal WCB decisions from two years to one year. In addition to shortening the timeline for appeal, the WCB now has the power to suspend or even terminate wage-loss replacement for workers going through an appeal, which would severely harm workers utilizing this right.

One last change the UCP made to the workers' compensation system was narrowing presumptive coverage for psychological injuries. Presumptive coverage means that a claim for a particular type of injury is automatically accepted if a worker is employed in a specified occupation. Certain kinds of claims receive presumptive coverage because of the difficulty workers can have establishing injury causation (i.e., that the injury arose from and during the course of work). The New Democrats extended presumptive coverage to psychological injuries to all workers in 2018, reflecting the difficulty workers have in proving causation. The UCP narrowed presumptive status for psychological injuries to first responders, which are defined in legislation to mean firefighters, police, paramedics, peace officers, correctional officers, and emergency dispatchers. Workers in other occupations no longer have presumptive coverage for psychological injuries. They can still file a claim for such an injury but face a difficult battle having their claim accepted (which is why these injuries were given presumptive coverage in the first place). The elimination of presumptive coverage is projected to transfer $230 million from workers to employers between 2020 and 2023.

The timing of limiting presumptive coverage for psychological injuries also coincided with a dramatic increase in psychological injuries for health-care workers and others dealing with the COVID pandemic. Nurses and other health-care workers are not included as first responders in the legislation and therefore are not given presumptive coverage for their psychological injuries. Anecdotally, many health-care workers who do experience psychological injuries often turn to other negotiated benefits, such as long-term disability, rather than going through the WCB system. Workers who lack such benefits have largely been left to deal with their psychological injuries on their own.

Reducing Workers' Labour Market Power

Employers use their economic power and legal rights to direct work and workers, thereby advancing employers' interests (generally speaking, to make a profit). Workers attempt to have their interests, such as fair wages and safe work, accommodated by exerting collective power in the form of a union. Unions do not equalize power in the workplace, but they increase the ability of workers to negotiate better working conditions. Unions' leverage stems from their ability to withdraw (or threaten to withdraw) workers' labour (i.e., strike) and thereby inflict economic harm on the employer. Not surprisingly, the rules regulating the formation and operation of unions is a site of political contention.

Restoring Alberta's place as the least union-friendly jurisdiction in Canada under the guise of "restoring workplace democracy" was a UCP platform promise (UCP, 2019, p. 21). The UCP was quick to repeal modest NDP changes to the Alberta *Labour Relations Code*, including replacing card-check certification with a mandatory certification vote, eliminating mandatory timelines for certification votes, removing automatic certification as a potential remedy for violations of the code, and watering down first-contract arbitration provisions. The UCP also re-instituted a provision previously struck down by the Labour Relations Board (LRB) as unconstitutional. This provision allows an employer and a union to sign a new agreement early to avoid an "open period" when workers have the right to choose a different union. Avoiding open periods is a strategy used by unions such as the Christian Labour Association of Canada (CLAC) to avoid the threat of workers replacing it with a union more willing to challenge the employer. Taking away workers' periodic opportunity to choose to be represented by a more effective union seems at odds with the UCP's promise of greater workplace democracy.

The goal of these changes is to make it harder for workers to join a union. For example, requiring a vote (when an overwhelming majority of workers have already signed union cards) and then removing timelines for certification votes gives employers an opportunity to interfere in workers' decisions about unionization. Research has shown the delays in holding the vote led to fewer successful applications (Campolieti, Hebdon, and Dachis, 2014; Johnson, 2004), largely due to employer interference with the election through communication with members (Bentham, 2002; Martinello and Yates, 2004; Riddell, 2001). How allowing employers more opportunity to interfere in worker certification votes makes workers' decisions about joining or not joining a union more democratic was never explained by the UCP.

It is difficult to measure the impact of these changes in the short term, as many factors go into the success or failure of a union organizing drive. Alberta's unionization rate remained steady during the UCP's first term, and actually saw a small COVID-related bump in 2020 and 2021 due in part to the fact that unionized workers were less likely to lose their jobs during the pandemic. Still, Alberta's union density remains the lowest in Canada. The most compelling evidence relates to certification application. As set out in Figure 10.2, the NDP's 2017 changes led to modest increases

Figure 10.2. Certification application in Alberta, 2015/16 to 2021/22

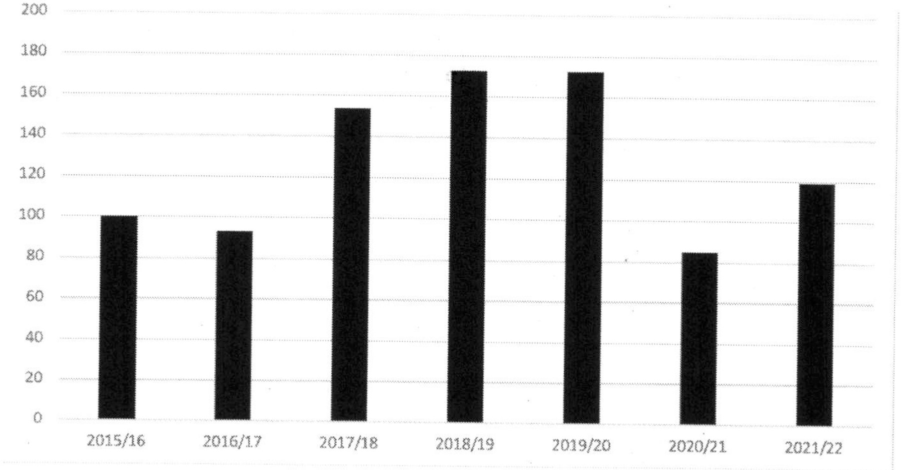

Source: Barnetson (2022)

in the number of certification applications. The UCP's 2019 changes were followed by a marked decline of in the number of applications. Longitudinal data on success rates is not available, but there is some evidence that UCP changes have reduced them.

Caution is required in interpreting these changes as the impacts of COVID-19 on union organizing are unknown. The effects of the UCP changes are likely to be seen in the longer term, as suppression of workers' democratic wishes compounds over time. Lower unionization is correlated with lower average wages, higher levels of inequality, and increased work precarity, which cheapens the cost of labour for employers (Banerjee, Poydock, McNicholas, Mangundayao, and Sait, 2021).

The UCP also introduced significant new rules regulating internal union operations and picketing that represent an Americanization of Alberta's labour relations system. First, the UCP implemented a dues opt-in requirement for all unionized workers. Unions were required to identify the percentage of dues spent on core activities (e.g., negotiations, handling grievances and educating members) and the share spent on "political activities and other causes." Activities in this latter category include donations to charities and community groups, funding for advocacy campaigns and, according to the regulations, any activity that "does not directly benefit dues payers in the workplace." Union members are only required to pay the portion of dues devoted to core activities. Unions must get each member to opt-in for non-core activities each year (a time-intensive undertaking). This dues opt-in was foreshadowed at the federal level by the Harper government in 2015 but exists nowhere else in Canada. It is similar to measures in many U.S. states that restrict payment of union dues to undermine unions' ability to represent their members, to organize new workers, and to participate in public debate and politics (Foster, 2021).

The second U.S.-style provision is a limitation on picket-line activity during a strike or lockout. Picket lines during a labour dispute serve multiple purposes, including creating awareness, providing opportunity for workers to communicate with the public about the dispute, and to create a physical presence to make it more difficult for the employer to engage in business operations. Stopping or delaying traffic in and out of the workplace is one of the most effective functions of picket lines because it causes economic consequences for the employer. The UCP has prohibited strikers obstructing or impeding a person who wishes to cross a picket line and, in effect, made it impossible for striking workers to impose economic consequences. This is consistent with the tighter picketing restrictions seen in U.S. jurisdictions.

Finally, the UCP further narrowed picketing by requiring workers to receive prior approval from the labour board for picketing anywhere other than their place of work. The practice of picketing other locations, such as the employers' customers/clients, associated businesses, or other public spaces, is called secondary picketing. Restricting the kinds of activity that can take place at a secondary location may violate the *Charter of Rights and Freedoms* (Buchanan, 2020; Foster, 2021). Essentially, the UCP has made legal picketing ineffective and effective picketing illegal. While the labour movement promised fierce resistance to picketing changes, no legal challenge with any legs had emerged at the time of writing. The significant and immediate consequences of breaking the law coupled with the protracted and uncertain outcome of legally challenging the resulting penalties have contributed to workers avoiding strikes or complying with the law during the strikes.

The UCP leaned in heavily to the anti-union measures of Bill 32. Premier Kenney bragged that "Bill 32 fulfills a big promise we made: no longer will union workers be forced to fund political campaigns of union bosses!" (Kenney, 2020), conveniently forgetting that the provisions also affect union donations to charities, community organizations, and disaster relief agencies. Union leaders were quick to criticize the bill, threatening Charter challenges and defiance of its anti-union provisions. United Food and Commercial Workers Local 401 president Tom Hesse offered an alternative rationale for the bill: "It's not intended to protect democratic principles. It's intended to be disruptive, to foment dissent, and to encumber unions so they can't operate" (French, 2021). The labour movement's response to the new rules remains unclear at the time of writing.

Public Sector Labour Relations

The UCP's platform promised to "balance the budget by 2022/23 without compromising core services" (UCP, 2019, p. 104). This required the government to find approximately $6.7 billion in savings. Given that approximately 55 per cent of Alberta's operating budget is allocated to salaries and benefits, a balanced budget under these parameters necessitated job cuts and/or wage reductions. The only discussion of this in the UCP platform were oblique references to moving Alberta "closer to the provincial average in program spending per capita" (p. 104) and saving "$200 million by letting the private sector deliver laundry services to [Alberta Health Services]" (p. 105).

In August of 2019, the UCP-appointed Blue Ribbon Panel on Alberta's Finances issued a series of recommendations that included establishing legislative bargaining mandates for government and its agencies, boards, and commissions (ABCs) that would bring public-sector compensation, including that of doctors, in line with public-sector compensation in comparable provinces (Alberta, 2019b). If the resulting negotiations precipitated a strike, the panel recommended back-to-work legislation. The panel made extensive use of interprovincial provincial salary comparisons but ignored the higher wages Alberta workers earn in many sectors compared to other provinces and that public-sector employers must compete with other employers in Alberta for workers. It is clear the panel's mandate was to set the stage for ongoing conflict with public-sector workers and their bargaining agents for the rest of the UCP's term.

An early signal of the UCP's orientation regarding public-sector workers came weeks after their election with the passage of the *Public Sector Wage Arbitration Deferral Act* (commonly referred to as Bill 9). The previous NDP government had negotiated collective agreements with many public-sector unions that included a "wage re-opener." A wage-reopener entails negotiating some salary increases during the life of the contract. If an agreement on the wage-reopener could not be reached, the matter was to be resolved through arbitration by June 30, 2019. Bill 9 unilaterally amended these collective agreements by changing the deadline to Dec. 15, 2019. The government's rationale was to provide time for the Blue Ribbon Panel to complete its work. Bill 9 was a clear indication the UCP would not let long-standing legal conventions interfere with its efforts to impose public-sector austerity. Unilaterally altering these contracts angered public-sector unions that launched, and subsequently lost, a court challenge regarding its legality. The eventual result of the arbitrations was a mixture of small salary increases and wage freezes.

Further evidence that the UCP was determined to achieve wage restraint on public-sector workers came in the fall of 2019. The *Public Sector Employers Act* allowed the government to impose a bargaining mandate on all public-sector employers (except, interestingly, religious-based post-secondary institutions). Employers were required to keep this mandate a secret from the unions with which they would bargain. These secret mandates sent a clear message the provincial government, and not individual employers, would determine the terms of any settlement proposed by an employer.

In the same omnibus bill (Bill 21), the UCP also gave itself the legal authority to unilaterally cancel the province-wide agreement with physicians, which it did in February 2020 (see Acuña, Chapter 3). The new agreement imposed by the government contained significant rollbacks in physician fees and other rights. The UCP's rationale, in line with other cutbacks and without evidence, was that physician compensation needed to be brought into alignment with other provinces. Response from physicians was one of fury. Dozens of doctors left the province and others retired earlier than planned (Siever, 2020). It is noteworthy this battle with the doctors took place during the first phases of the COVID pandemic, adding additional strain to Alberta's health-care system. Backlash from doctors and the public resulted in the government partially reversing its fee cuts. After a year of tense bargaining, Alberta Medical Association members narrowly rejected a tentative settlement for a new province-wide agreement in March 2021. A new agreement, including a below-inflation pay raise, was ratified in September of 2021.

The UCP turned to privatization to help achieve its targeted goals of balancing the provincial budget. In the health-care system, these recommendations were developed further in a report by Ernst and Young for Alberta Health. This report recommended embedding "just-in-time" labour practices into collective agreements, including health-care staffing and changing skill mix and staffing levels. The report also recommended privatizing practically all non-clinical staff, including food services, laundry, protective services, transcription and translation as well as patient transportation. In the fall of 2020, the UCP announced their plan to lay off 11,000 health-care support staff, which included cleaning, food, laundry, and protective services at Alberta Health Services. In July 2022, all community lab services began to be provided by DynaLIFE Medical Labs (Alberta Health Services, 2021).

In addition to the job losses in the health-care sector, the UCP cut funding to Alberta school boards, resulting in the loss of 25,000 education workers and education assistant positions (announced on a weekend over Twitter). During this announcement, Education Minister Adrianna LaGrange stressed that this was a temporary measure in response to COVID school closures and funding was to be restored once in-person classes resumed (Omstead, 2020). In Budget 2021, LaGrange stated 2,000 of those jobs would never return (Boothby, 2021). The UCP government also made drastic cuts to the post-secondary system (see Spooner, Chapter 13), disproportionately singling out the University of Alberta for a 33 per cent cut between 2019 and 2022. As a result, more than 1,100 jobs (mostly support staff) were eliminated while tuition increased by more than 20 per cent (Adkin, Carroll, Chen, Lang, and Shakespear, 2022).

The contracts for all major public-sector unions in core government services, education, health care and post-secondary education expired in 2020. Declining per-capita program funding and secret bargaining mandates resulted in initial employer proposals for three or four per cent wage rollbacks followed by several years of wage freezes, and a series of other sector-specific cuts. Union responses to the UCP's agenda for public-sector workers were rhetorically aggressive. Union leaders decried the government's actions in strong language and unions began preparing and mobilizing members. Many unions, especially in health care, talked openly of going on strike. In October 2020, thousands of AUPE health-care workers engaged in a one-day wildcat strike to express opposition to the government's plans to privatize 11,000 health-care jobs.

As contracts began to settle in late 2021 and 2022, public-sector workers managed to achieve modest wage increases, backloaded to later in the agreements, ranging between 2.75 per cent and 4.25 per cent over a four-year period. While better than initially expected, these settlements continued to fall far behind inflation and contributed to the government's goal of reducing public-sector wages toward the national average. Seen through the lens of high inflation in 2022, these settlements seem particularly meagre for public-sector workers struggling to meet the rising cost of living. Not all segments of the public sector avoided labour disputes. Post-secondary employers, through a combination of fiscal pressure due to cutbacks and the government's secret mandate, took particularly aggressive positions with their unions during collective bargaining. In early 2022, the first post-secondary strike in Alberta history took place at Concordia University of Edmonton, a private university with religious origins and in receipt of significant public funding. (Until 2018, post-secondary workers in Alberta were prohibited from striking.) A five-week strike followed at the University of Lethbridge that eventually ended with an agreement that followed the pattern set by an earlier agreement between the government and the Alberta Union of Provincial Employees. The strike also provided clear evidence that the government, not the board of governors, was calling the shots at the bargaining table.

Much of the shift in the government's approach between 2019 and 2022 can be attributed to the impacts of the COVID pandemic. The pandemic revealed the importance and scarcity of health-care workers. While the UCP continued to do less than other provinces to support health-care workers during the pandemic, including refusing to provide bonus pay to nurses, circumstances required them to back away from a full confrontation with public-sector workers. An improving fiscal situation due to rising oil prices also took away much of the rationale for drastic austerity.

Conclusion

It is likely in the 2023 election that the UCP will claim its policies led to the thousands of new jobs created in the latter part of their mandate, attributing them to corporate tax cuts and "cutting red tape." It is unclear if such a claim is true. Alberta's recent job growth is more likely attributable to post-COVID recovery (evident across Canada) and an unexpected spike in the price of oil. What is clearer is the UCP approach to labour has tilted both the playing field in favour of employers and set up the UCP to be referee in the game.

The UCP was successful in rolling back worker rights, including most of the improvements made under the former NDP government. Particularly notable were the UCP's efforts to freeze the minimum wage, eliminate overtime costs, reduce employers' workers' compensation premiums, and make it harder to join a union and negotiate a contract. Restrictions imposed on the internal operation of unions by the UCP enters new territory in Canadian labour relations and signal an Americanization of the system. Restricting unions' access to dues for political purposes strikes at democratic participation in Alberta. Regardless of whether someone supports or opposes unions as an institution, it is widely recognized they are a legitimate voice within political debate. Unions bring a distinct worker perspective to policy debates to counter the voices of corporations and their associations. The dues restriction measures are a direct attack on union members' rights to participate in political debate, and this weakens democracy. As a result, the measures indirectly affect all Albertans.

COVID had a major impact on the UCP's relationship with public-sector workers. The original UCP agenda was to engage in confrontation with public-sector unions to extract wage concessions and reduce per-capita program costs in the province. However, nearing the end of the term, little of the expected conflict had occurred. This is due mostly to the government shifting away from its hard wage-rollbacks agenda. COVID constrained the government's ability, both politically and operationally, to impose changes through conflict or legislated settlements. In particular, COVID contributed to a shift in political capital from the UCP to health-care workers. The government simply could not afford a strike during a time when the health-care system was under significant strain. Political dissention within the UCP may have also played a factor in the government's inability to follow through on its wage-concession agenda. The government has, however, continued to impose austerity through declining per-capita funding throughout the public-sector.

Overall, the UCP met its goal of significantly cheapening labour for employers. It has failed to achieve its goal of reducing public-sector compensation, although public-sector settlements continue to lag inflation. And on job creation, its most important priority, the UCP appears to have simply lucked out by being in power when the economy rebounded.

References

Adkin, L., Carroll, W., Chen, D., Lang, M., & Shakespear, M. (2022). Higher education: Corporate or public? Edmonton: Parkland Institute, May. https://assets.nationbuilder.com/parklandinstitute/pages/1974/attachments/original/1661762422/Higher_Education_report.pdf?1661762422 Accessed January 30, 2023.

Alberta. (n.d.). *Critical worker benefit funding: $465 million.* Infographic. https://www.alberta.ca/assets/documents/lbr-critical-worker-benefit-funding-infographic.pdf Accessed March 23, 2022.

Alberta. (2018a). *Alberta minimum wage profile, April 2017 to March 2018.* Edmonton.

Alberta. (2018b). *Hansard.* (Edmonton, April 5, 2018), 433.

Alberta. (2019a). *Alberta is open for business.* Press release. https://www.alberta.ca/release.cfm?xID=63946CE7066C0-D851-482C-A7C514615F7D31E6 Feb. 16, 2022.

Alberta. (2019b). *Report and Recommendations: Blue Ribbon Panel on Alberta's Finances.* Edmonton.

Alberta. (2020a). *Alberta's Recovery Plan.* Edmonton.

Alberta. (2020b). *Changes to Workers' Compensation Laws: Information for Albertans.* Edmonton. November.

Alberta. (2020c). *Improving safety for workers and cutting red tape.* Edmonton. Nov. 5. https://www.alberta.ca/release.cfm?xID=7564406FFA335-D369-B069-3A078427EBEEE562 Accessed June 3, 2022.

Alberta Federation of Labour. (2020). *Alberta frontline workers are losing out on more than $400 million in "hero pay" because the UCP refuses to hold up its end on a cost-shared wage program.* Press release, Nov. 27. Alberta Federation of Labour.

Alberta Health Services. (2021). *Negotiations to proceed with preferred proponent for community lab services.* Edmonton.

AWCBC. (2022). *Detailed key statistical measures report.* Ottawa: Association of Workers Compensation Boards of Canada. https://awcbc.org/en/statistics/ksm-annual-report/ accessed June 24, 2022.

Banerjee, A., Poydock, M., McNicholas, C., Mangundayao, I., & Sait, A. (2021). *Unions Are Not Only Good for Workers, They're Good for Communities and for Democracy*. Economic Policy Institute.

Barnetson, B. (2017). *"Not a Cutting-Edge, Lead-the-Country Reform": An Overview of the Changes Proposed in Bill 17*. Parkland Institute. May 29. https://www.parklandinstitute.ca/not_a_cutting_edge_lead_the_country_reform Accessed Sept. 28, 2021.

Barnetson, B. (2020). Kenney's war on workers: Contracts broken, wages cut and unions undermined. *Alberta Views* 23(1): 27-31.

Barnetson, B. (2022). *Alberta's 2021/22 Annual Report*. https://albertalabour.blogspot.com/2022/08/alberta-labours-202122-annual-report.html Accessed August 2, 2022.

Barnetson, B, Foster, J., Cake, S, & Matsunaga-Turnbull, J. (2021). *Why is Alberta Making Workplaces Less Safe: Alberta Undermines Internal Responsibility System with New Rules*. https://lawofwork.ca/albertaohschanges/ Accessed Feb. 17, 2022.

Bellefontaine, M. (2020, Nov. 5). *New bill would rollback NDP changes to Alberta workplace safety rules*. CBC News. https://www.cbc.ca/news/canada/edmonton/alberta-workplace-safety-1.5791510 Accessed June 3, 2022.

Bentham, K. J. (2002). Employer resistance to union certification: A study of eight Canadian jurisdictions. *Relations Industrielles / Industrial Relations* 57(1), 159–187.

Boothby, L. (2021, Feb. 26). Alberta budget 2021: Province says it won't "penalize" school boards for lower enrolment, but nearly 2,000 jobs lost to pandemic won't return. *Edmonton Journal*. https://edmontonjournal.com/news/politics/alberta-budget-2021-province-will-not-penalize-school-boards-amid-covid-19-enrolment-drop Accessed March 24, 2022.

Buchanan, D. (2020, July 27). Restricting a union's political activities: The constitutionality of Alberta Bill 32. *Canadian Law of Work Forum*. https://lawofwork.ca/bill32-charter/

Campolieti, M., Hebdon, R., & Dachis, B. (2014). The impact of collective bargaining legislation on strike activity and wage settlements. *Industrial Relations: A Journal of Economy and Society* 53(3), 394–429. https://doi.org/10.1111/irel.12063

Copping, J. (2020, July 16). Restoring balance in Alberta's workplaces. *Calgary Herald*. https://calgaryherald.com/opinion/columnists/restoring-balance-in-albertas-workplaces Accessed June 3, 2022.

Ernst and Young. (2019). Alberta Health Services Performance Review.

Finkel, A. (ed). *Working People in Alberta: A History*. Edmonton: Athabasca University Press.

Foster, J. (2021). *Tipping the Balance: Bill 32, the Charter and the Americanization of Alberta's Labour Relations System*. Parkland Institute. July 15.

Foster, J., Cake, S., & Barnetson, B. (2021). Profits First, Safety Second: Canada's Occupational Health and Safety System at Fifty. *Labour/Le Travail*. In review.

French, J. (2021, Sept. 13). *Alberta union leaders consider ignoring new legal restrictions on finances, picketing*. CBC News. https://www.cbc.ca/news/canada/edmonton/alberta-union-leaders-consider-ignoring-new-legal-restrictions-on-finances-picketing-1.6172181

Hall, A., Forrest, A., Sears, A., & Carlan, N. (2006). Making a difference: Knowledge activism and worker representation in joint OHS committees. *Relations industrielles/Industrial Relations* 61(3), 408-436.

Johnson, S. (2004). The impact of mandatory votes on the Canada-U.S. union density gap: A Note. *Industrial Relations* 43(2), 356–363.

Keller, J. (2019, May 27). Alberta slashes wages for teen students. *Globe and Mail*. https://www.theglobeandmail.com/canada/alberta/article-alberta-slashes-minimum-wage-for-teen-students/ Accessed June 3, 2022.

Kenney, J. (2020, August 1). Tweet by @jkenney. https://twitter.com/jkenney/status/1289354658205532163

Martinello, F. F., & Yates, C. (2004). Union and employer tactics in Ontario organising campaigns. In *Advances in Industrial & Labor Relations* 13, 157–190. Emerald.

McIntosh, A., & Graff McRae, R. (2020). *A Basic Income for Alberta*. Parkland Institute. Jan. 21. https://www.parklandinstitute.ca/a_basic_income_for_alberta Accessed Sept. 28, 2021.

Mertz, E. (2019). *Lower minimum wage for Alberta youth takes affect Wednesday*. Global News. https://globalnews.ca/news/5434502/lower-minimum-wage-alberta-youth/ Accessed Sept. 28, 2021.

Omstead, J. (2020). *Alberta education cut expected to lay off thousands during pandemic*. CBC News. https://www.cbc.ca/news/canada/edmonton/funding-reduction-alberta-k-12-covid-1.5513803?cmp=rss&fbclid=IwAR30xdACFAtfMTCIrn-QkrCPuA6VobGoovFa1Es6AoVxoLJYc7BtmtuB58A Accessed Nov. 24, 2021.

Riddell, C. (2001). Union suppression and certification success. *Canadian Journal of Economics* 34(2), 396–410.

Rieger, S. (2020). *Alberta UCP Criticized by Taxpayer Group as one of Few Provincial Parties to Apply for Federal Wage Subsidy*. https://www.cbc.ca/news/canada/calgary/alberta-ucp-wage-subsidy-1.5587053 Accessed Sept. 28, 2021.

Siever, K. (2020). A list of Alberta communities losing doctor care. https://kimsiever.ca/2020/04/21/a-list-of-alberta-communities-losing-doctor-care/ Accessed Feb. 17, 2022.

Statistics Canada. (2022a). *Employment and Unemployment Rate, Monthly, Unadjusted for Seasonality*. Table 14-10-0374-01. Accessed October 11, 2022.

Statistics Canada. (2022b). *Employment and Average Weekly Earnings (Including Overtime) for all Employees by Province and Territory Monthly, Seasonality Adjusted*. Table 14-10-0223-01. Accessed October 11 2022.

Stephenson, A. (2019, November 20). UCP unveil Bill 6 changes: New rules create exemptions for small farms. *Calgary Herald*. https://calgaryherald.com/business/local-business/ucp-unveils-bill-6-changes-new-rules-create-exemptions-for-small-farms Accessed June 2, 2022.

Tucker, S., & Keefe, A. (2022). *2022 Report on Work Fatality and Injury Rates in Canada*. Regina: University of Regina.

UCP. (2019). *Alberta Strong and Free: Getting Albertans Back to Work*. United Conservative Party.

CHAPTER 11

ALBERTA'S JOB CREATION TAX CREDIT: A HIDDEN GIFT TO OILSANDS PRODUCERS

Robert Ascah

"Economics is not an exact science. It's a combination of an art and elements of science. And that's almost the first and last lesson to be learned about economics: that in my judgment, we are not converging toward exactitude, but we're improving our data bases and our ways of reasoning about them."

— Paul Samuelson

THE UNITED CONSERVATIVE PARTY (UCP) released their policy platform Alberta Strong & Free during the 2019 provincial general election (UCP, 2019). The platform articulated five top commitments, the first being a job creation plan that included a "job creation tax cut" to reduce the tax on "job creators." The proposed tax cut was directed at restoring Alberta's tax advantage the UCP argued was eroded when the New Democrat government increased corporate taxes, causing lower overall corporate tax revenue. Citing two prominent University of Calgary economists, Jack Mintz and Bev Dahlby, the UCP's platform promised the corporate tax cut would result in significant gains in both nominal and real gross domestic product (GDP), as well as additional government revenues and gains in employment (UCP, pp.12, and 19-20). Shortly after its election victory, the UCP government introduced Bill 3, the *Job Creation Tax Cut (Alberta Corporate Income Tax Amendment) Act*, to take effect July 1, 2019.

This chapter examines the political arguments for and against cuts to corporate income tax (CIT). I briefly touch on the theory supporting the cuts and report on conflicting departmental advice on the tax cut. Next, I examine the actual experience of job creation, investment attraction and GDP growth over the past three years. First, I outline the views of economic development officers concerning the efficacy of the tax cut in luring investment to Alberta. Second, I offer a brief assessment of the case of the 2017 tax cuts in the United States. Third, I present preliminary evidence on how much revenue has been foregone by the provincial government, finding a serious underestimate of the cost of the tax cuts. Much of the empirical analysis presented is from financial statements of the Big Four oilsands producers.[1] Fourth, I draw upon current economic data on investment jobs and wages since 2019 and conclude there is little evidence supporting the purported benefits of tax cuts to CIT in Alberta. The chapter concludes by identifying the principal beneficiaries of these cuts.

The Political Debate

Alberta's Hansard records the two sides of the legislative debate held between the incoming UCP government and the outgoing NDP government. Introducing Bill 3, Travis Toews, Treasury Board president and Finance minister, claimed the province could no longer boast of having the most competitive business environment in Canada. The Alberta Job Creation Tax Cut (AJCTC), he declared, was designed to remedy this problem by helping attract investment and stimulate economic activity "at a time that is sorely needed" (Alberta Hansard, May 28, 2019, p. 111; June 12, 2019, p. 770).

At the bill's second reading, Toews explained the tax cuts would be staged in over three years: July 1, 2019 (12 to 11 per cent), Jan. 1, 2020 (11 to 10 per cent), Jan. 1, 2021 (10 to nine per cent), and Jan. 1, 2022 (nine to eight per cent). Toews characterized these changes as "bold action" that would reverse the flow "southward" of investment funds (Alberta Hansard, May 29, 2019, p. 236). Central to the rationale for the tax cuts was the assumption that tax structure "is a big part of the business environment." Toews cited research by Dahlby and Mintz that indicated tax cuts would create additional jobs and investment (Alberta Hansard June 5, 2019, pp. 473-74; June 11, 2019, p. 692; June 12, 2019, p. 760). Like former premier Ralph Klein, who believed in the virtues of free markets, a level-playing field, red tape reduction and low taxes, Toews dismissed NDP arguments that specific incentives were more efficient and effective ways to attract investment and jobs. Toews relied on testimonials from business-friendly organizations, such as the Canadian Taxpayers Association and the Alberta Enterprise Group, to justify the importance of the tax cut (Alberta Hansard, June 12, 2019, p. 760).

Many of the government speakers attacked the previous NDP government for chasing investment out of the province. Minister of Transportation Ric McIver cited the sale of assets by "Total, Murphy Oil, and a myriad of other oil companies" taking eighty billion dollars of investment out of the province (Alberta Hansard, June 5, 2019, p. 465). This claim erroneously suggested the assets sold by these companies had suddenly vanished from the province when in fact these assets were purchased by large Canadian producers[2] who would operate these assets and continue to employ Albertans.

Jason Stephen, (MLA, Red Deer South) emphasized mobility of capital as a principal reason for the tax cuts. He reprised the research of Dahlby, claiming "the job creation tax cut will generate a $12.7-billion increase in nominal GDP, a six-and-a-half per cent increase in per capita real GDP, and $1.2 billion in additional government revenue by 2023." The work of Mintz was cited showing an increase of 55,000 jobs consistent with the platform promise (Alberta Hansard, June 4, 2019, p. 352). Grant Hunter from Taber Warner ridiculed "democratic socialism" and how the previous NDP administration removed Alberta's competitive advantage by the 20 per cent tax hike in 2016 (Alberta Hansard, June 4, 2019, p. 353; June 11, 2019, p. 692).

New Democrats view public finances differently than the UCP. According to the UCP government, spending was too high for what the public was receiving in service quality. Public-sector workers were too highly paid when compared with their provincial counterparts (MacKinnon, 2019, pp. 44-50). In contrast, the NDP viewed public services and public spending as investments leading to longer-term gains in well-being and economic growth (Alberta Hansard, June 4, 2019, p. 343).

The NDP's criticisms boiled down to four arguments with variations. First of all, the AJCTC won't work because it's been tried elsewhere unsuccessfully. The experience of Kansas in the mid-2010s and Donald Trump's 2017 tax cuts were cited as failures (Alberta Hansard, June 4, 2019, pp. 344-345).

A second argument was that a tax cut would seriously weaken the province's fiscal position. Variations on this theme included the impact on the creditworthiness of Alberta and pressures leading to greater spending cuts necessary to address the revenue loss (Alberta Hansard, June 4, 2019, 341-2, p. 351; June 5, 2019, p. 459). Specifically, the NDP questioned the wisdom of giving up $4.5 billion in revenue over four years in light of the need to protect citizens who rely on key programs like health and education (Alberta Hansard, June 5, 2019, p. 263). The negative economic impact of proposed public spending cuts and layoffs cancelling out the benefit of the proposed CIT cuts was mentioned in reference to a report by The Conference Board (Alberta Hansard, June 4, 2019, p. 345; June 5, 2019, p. 468). To address the fiscal concerns, the NDP proposed an amendment to slow the pace of tax cuts to evaluate the effectiveness while reducing revenue loss. This amendment was defeated (Alberta Hansard, June 5, 2019, pp. 475-558).

A third theme was CIT rate cuts were unnecessary because Alberta had the lowest taxes in the country. (Figure 11.1, below, sets out Alberta's massive tax advantage as of February 2022.) Variations on this theme were the notion of an unnecessary "race to the bottom" (Alberta Hansard, June 5, 2019, p. 471) and a "massive giveaway to friends and donors of the government bench" (Alberta Hansard, June 5, 2019, p. 481).

The NDP held that Alberta was already "very competitive" in its tax regime (Alberta Hansard, June 4, 2019, p. 350; June 5, 2019, p. 467; June 12, 2019, p. 761). Alberta's corporate tax rate of 12 per cent in 2019 was the same rate as British Columbia and

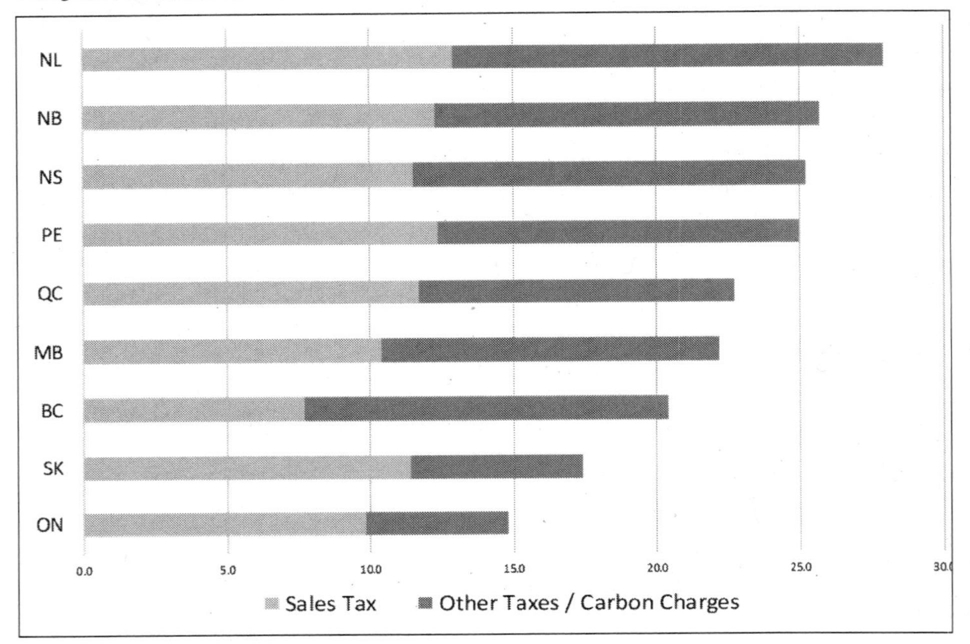

Source: Alberta Treasury Board and Finance, *Fiscal Plan 2022-25*, p. 173.

Saskatchewan and within one per cent of Ontario and Quebec. A related argument was that other policy instruments, besides tax cuts, would also foster investment in the province (Alberta Hansard, June 4, 2019, p. 348).

A fourth theme dealt with distributional effects of the CIT rate cuts. Besides accusations of rewarding friends of the UCP — "this massive giveaway" — (Alberta Hansard, June 4, 2019, p. 347; June 5, 2019, p. 468, June 11, 2019, p. 696), the NDP caucus argued the cuts would not assist lower-income families. Unlike large corporations, lower-income Albertans were more likely to spend the tax savings in the local economy. The NDP identified the beneficiaries of the cuts as the shareholders of corporations, noting this policy change would exacerbate existing inequality (Alberta Hansard, June 11, 2019, p. 681; June 12, 2019, p. 765).

The political debate concerning the impact of the corporate tax cut was renewed in June 2020 when, in response to the COVID-19 pandemic and based on private advice from the Alberta Economic Recovery Advisory Council chaired by Mintz[3], the government accelerated the implementation of the tax reduction from 11 per cent to eight per cent effective July 1, 2020. As Toews stated in introducing Bill 35, the *Tax*

Statutes (Creating Jobs and Driving Innovation) Amendment Act, 2020, the goal was "to further position Alberta's economy for diversification, recovery, and growth." He went on to repeat that a competitive business environment will attract new business to Alberta and incentivize businesses to invest and create jobs (Alberta Hansard, October 20, 2020, p. 2616). The bill was debated extensively and made effective for the 2020 tax year.

The next section briefly summarizes the economics case for the AJCTC and reports on conflicting departmental advice to successive governments about the case for tax cuts.

The Case for Tax Cuts

The political marketing for corporate tax cuts is the promise of investment, jobs and economic growth. The economic theory rests on the intuition of the Laffer curve that as marginal tax rates rise, the incentive to invest, save or work diminishes. The literature emphasizes the importance of mobility of capital relative to labour, suggesting labour too will benefit from tax cuts. The supportive literature almost entirely ignores the government as a winner or loser. Importantly, conventional economics has accepted the claim that private-sector investment is always superior to public-sector spending, dismissing the possibility that public-sector infrastructure spending or education spending can be as efficient as private-sector investment. In Alberta's situation, the relative immobility of oil and gas capital is a central factor in determining the effectiveness of tax cuts. Immobility of capital refers to oilsands mines and production facilities that cannot be practically moved or removed from the site of the resource. A healthy skepticism is crucial to assessing the fulfillment of the promises of a tax cut.

Departmental Advice

The role played by Alberta's public service in the debate is also of interest. After the provincial election, a freedom of information request asked the government for all studies or documents that referenced studies on the effects of the corporate tax cut. An April 2, 2019 briefing by Alberta Treasury Board and Finance came days after the UCP released their 2019 election platform. The department observed the Dahlby-Ferede, McKenzie-Ferede, and Mintz studies use Canadian data for estimates of Alberta tax responsiveness or specific tax rates, making the findings "unreliable in the Alberta context." The document pointed out that Alberta's economy is heavily weighted to the oil and gas industry such that estimates of tax rate change responsiveness from many other jurisdictions "may be unreliable in the Alberta context." The backgrounder noted, "Oil and gas taxation relative to key competing jurisdictions does not appear to be a major barrier to investment in Alberta" (Alberta Treasury Board and Finance, 2019, pp. 1-2).

With a change in government, budget documents referenced these studies as *justification* for the policy change (Alberta Treasury Board and Finance, *Fiscal Plan 2019-2023*, pp. 143-144). This behaviour highlights the political sensitivities of giving advice when providing policy analysis on highly visible political issues. Policy advice in such a context is not value neutral.

What do Economic Development Agencies Think?

This section reports findings of four interviews conducted with regional economic development officers whose job is to attract new investment and jobs to their regions. The four economic development organizations canvassed were Edmonton Global, the Industrial Heartland (Fort Saskatchewan), the Wood Buffalo Economic Development Corporation, and the City of Lethbridge Economic Development.

According to the qualitative survey, the AJCTC was not identified as one of the top three factors considered by companies investing in Alberta. Key factors depended on the nature of the regional economy and the type of corporations attracted. The top three factors were: access to markets and feedstock, including public infrastructure to support this; a skilled, trained workforce; and financial returns for the project, which are affected by government incentives, business fees and all taxes. Most respondents preferred specific industry incentives over a generalized tax cut.

Edmonton is considered globally competitive in six sectors: agrifood; health and life sciences; energy/hydrogen; manufacturing; artificial intelligence and tech; and transportation and logistics. For the Industrial Heartland, its focus is heavy industry, while Wood Buffalo is focused on resources and companies servicing the energy sector. In Lethbridge, the economic development authority focuses on light industrial manufacturing, including food processing.

Top factors attracting investment identified by Edmonton Global were human capital, innovation ecosystems, and a cluster of other factors specific to the companies, including market access, utilities, real estate and taxes in general (e.g., business taxes, CIT, property taxes, sales taxes[4]). For the Industrial Heartland, top factors were availability and cost of feedstock, capital costs to build the facility, and access to markets. For Wood Buffalo, top factors were a positive return on investment, the ease of doing business (regulatory burden), and regional infrastructure to provide market access. In Lethbridge, staff pointed to United States survey data listing the local work force, transportation infrastructure, availability of incentives, and access to markets as the top drivers for investment attraction. Further down the list at number eight were state and local tax regimes (2021 - Site Selectors most Important Location Criteria). Although the survey is American, given that Canadian provinces compete for investment against U.S. states, factors other than low taxes seem to drive more investment decisions than economic theory might suggest.

These agencies had not seen a significant pick-up of inquiries after the AJCTC was announced. The emergence of COVID-19 would certainly be a factor in this low responsiveness. However, the fact the tax cut was not cited as a top factor in luring investment validates the lower efficacy of the widely publicized AJCTC and the importance of factors other than tax reductions.

Reference was made to announcements of large investments such as Dow Chemical in the Industrial Heartland and the Air Products' hydrogen development facility in southeast Edmonton. However, announced "plans" are subject to engineering studies, board approvals and regulatory approvals. Announcements are not the same as workers clearing work sites and paying income taxes.[5] Figure 11.1, above, taken from the most recent provincial budget, shows how generous and competitive Alberta's overall tax regime is for individuals and corporations.

How many jobs have been created since the AJCTC was announced? The quality of data is spotty, but Edmonton Global identified seven investments over $20 million, including Air Products, creating 2,500 construction jobs and 506 permanent jobs. In Lethbridge, two investments totalling $350 million have been made. Economic Development Lethbridge estimated an uptick of 222 jobs in 2019, falling to 77 in 2020 and rising to 143 in 2021 (2022, p. 2). For projects that are under study or being planned, many of the jobs, primarily in engineering, are not necessarily performed in Alberta.

It was difficult for respondents to answer whether these investments would have occurred without the AJCTC. Respondents noted a tax cut didn't hurt and added positively to the profitability of existing operations and new capital investment. Another noted the AJCTC was a help but probably not enough of a sweetener to incent companies to move into the province. For example, it was pointed out that over the past two years, the Government of Saskatchewan, with a 12 per cent CIT rate, has successfully attracted three new canola crushing facilities (Cargill, Viterra and Richardson International), all large facilities costing $350-$500 million to build.[6] Even without government assistance, Saskatchewan's CIT rate did not seem to discourage new investment in Saskatchewan.

Economic development officers identified specific incentives or grant programs as more critical to attracting "greenfield" investments. Specifically, the Alberta Petrochemicals Incentive Program (Government of Alberta, 2019), announced in October 2019 to replace the NDP's 2016 Petrochemicals Diversification Program,[7] is a program that pays a grant equal to 12 per cent of eligible capital expenditures on the project. Another specific program cited was the interactive digital media tax credit, introduced by the NDP in 2018 and cancelled by the UCP in October 2019. In the face of strong opposition. the UCP introduced the Innovation Employment Grant program (Interactive Arts Alberta, 2019).[8] However, the province still does not have a digital media tax credit, which puts Alberta at a severe disadvantage compared to other provinces. This is a notable weakness for Alberta because intangible capital (intellectual property) is growing at faster rates than tangible capital.

The corporations these agencies are targeting are typically large international or interprovincial companies that have choices where they locate new production facilities. For example, in the area of food processing, where Lethbridge is pursuing, typical investors are western Canadian and some international, like Cargill. For the Industrial Heartland, its heavy industry focus means mostly out-of-province corporations like foreign petrochemical or chemicals producers.

Clearly, taxes alone are not sufficient to bring investors to Alberta. Location, access to markets and availability of talent is higher on the list. Still, taxes do matter and if Alberta CIT taxes had been considerably higher than competitor provinces, investors might balk at coming into the province. However, Alberta's overall tax advantage is significant at individual and corporate levels, as provincial budgets reveal. Alberta remained competitive at 12 per cent, so the AJCTC was not necessary to attract new businesses to Alberta.

United States Experience

Much of the hype for tax cuts originates in the United States. In 2017, federal legislation slashed corporate income tax rates from 35 per cent to 21 per cent. The corporate tax cut was justified by the desire to patriate trillions of dollars held offshore by American multinationals, with the ultimate goal to encourage these companies to invest in the United States and hire U.S. workers. The title of *Tax Cuts and Jobs Act (TCJA)* bears similarity to the Alberta Job Creation Tax Cut rhetorically linking the concept of jobs and tax cuts.

Evidence from a 2018 survey by the National Association of Business Economics indicated that 81 per cent of 116 firms surveyed had *not* changed investment or hiring plans because of Trump's tax cuts (Alberta Hansard, June 4, 2019, p. 345; June 11, 2019, pp. 690, 698). NDP MLA Shannon Phillips cited a JP Morgan report (Cembalest, 2017) that found roughly one-half of the tax cuts were going toward share buybacks rather than investments (Alberta Hansard, June 4, 2019, pp. 345-6; June 5, 2019, p. 487; June 11, 2019, p. 699).[9]

A July 2019 Congressional Research Service report found that for 2018, it could not be demonstrated how much of 2018's GDP growth was due to the tax cut. The report noted only a modest and less certain growth effect in the long run. The report added the long-run effects on investment would likely be offset because a growing government debt borrows resources that could otherwise be used by the private sector for investment (Congressional Research Bureau, 2019, 2-4). A 2021 Brookings Institution study found the act clearly reduced revenue and it was difficult to disaggregate the effect on GDP from the data. Growth in business formation, employment and median wages slowed after enactment. International profit shifting fell only slightly, and the boost in repatriated profits primarily led to increased share repurchases rather than new investment (Gale and Haldeman, 2021, p. 2).

Evidence

Data from Alberta Finance and Treasury Board provided to Statistics Canada (Statistics Canada, 2023) shows the breakdown of major sectors paying CIT to Alberta. In 2021-22, the largest contributor to Alberta's CIT take is from the finance and insurance sector (23 per cent), manufacturing, which includes refining (11 per cent), followed by oil and gas extraction (11 per cent), and then construction (7 per cent). Most of the taxpayers in the finance sector are large Canadian banks and insurers with head offices and shareholders outside Alberta and shareholders mostly outside Alberta. Construction and real estate firms are capable of shifting capital, but the site of their actual work, employment and sales is specific to a taxable jurisdiction. These companies tend to be local with interprovincial and international operations, like Ellis-Don and PCL Construction.

The UCP platform estimated foregone revenue of the AJCTC at $3.4 billion. Budget 2019 estimated a loss of $4.7 billion, with a net fiscal cost of $2.4 billion after accounting for "favourable impacts" (Alberta Treasury Board and Finance, 2019, p. 144). My arithmetical estimate of projected lost revenue from July 1, 2019 to March 31, 2023 assumes average corporate revenue of $4 billion a year roughly equivalent to the average of CIT receipts in the 2015-2020 period, a period of low oil prices. This is a period of low oil prices.The $4 billion number assumes all corporate taxes are paid at the general tax rate. Since about 10 per cent of CIT revenue is derived from small businesses paying the two per cent small business rate, only 90 per cent of the $4 billion could be considered revenue loss, or $3.6 billion (Alberta Treasury Board and Finance, 2019, p. 142). This number is close to the calculation of the UCP of a revenue loss of $3.4 billion before the acceleration was legislated. These numbers, however, would vary considerably given the sensitivity of CIT revenue to increases or decreases of corporate accounting profits and especially the sensitivity to oil and gas price movements, discussed later.

Before proceeding, it is important to know how tax law and accounting impacts who are the chief beneficiaries of this tax cut. Provisions in *the Income Tax Act* allow for accelerated depreciation to induce capital investment. Provisions permit current taxable income to be deferred to a later date. The difference between the tax and accounting treatments creates what are known as "timing differences." These differences create deferred tax assets or liabilities. When tax rates are changed, companies with large, deferred tax liabilities adjust their financial statements to record the benefit of paying deferred taxes at the lower rate in the future.

The beneficial effects of the AJCTC on the financial position of the Big Four oil and gas and oilsands producers: Cenovus; Canadian Natural Resources Limited (CNRL); Imperial Oil; and Suncor are material. By understanding the magnitude of the tax cuts' effects on Alberta's biggest energy producers, I can estimate from the bottom up the range of revenue loss based on other available public sources.

Table 11.1 shows the favourable impact of the corporate tax changes announced by the UCP government on the financial accounts of these four oilsands producers at the end of 2021.

Table 11.1: Favourable Tax Adjustment ($ Millions)

Company	Tax Adjustment
Cenovus	572
CNRL	1,618
Imperial Oil	795
Suncor Inc.	1,222
Total	**4,207**

Sources: Cenovus, Annual Reports 2019, 92, 2020, 96, Audited financial statements 2021, 41. CNRL annual report 2019, 82. Imperial Oil, Annual Report pursuant to section 13 or 15(d) of The Securities Exchange Act of 1934 (10-K report), 2021, p. 88. Suncor Annual Reports, 2020, p. 109; 2021, p. 105.

To estimate what portion of Alberta tax is paid by these producers, the portion of Alberta production by these four companies was estimated. Together, they produce approximately 2.6 million barrels a day in their oilsands operations alone, or equivalent to 66 per cent of the total daily production in 2021, as reported by ATB Financial.[10] With the Big Four producing at least two-thirds of Alberta's bitumen production, all things being equal, they would pay about two-thirds of the taxes paid by the whole oil and gas industry to the Government of Alberta. With $4.2 billion representing at least two-thirds of CIT tax adjustment benefit, the oil industry's total benefit could be as high as $6.3 billion.

I next confirm how much of Alberta's corporate income tax is paid by the oil and gas industry. Using the most up to date information from provided by Alberta Finance to Statistics Canada, on average over the last twenty years, the oil and gas extraction sector accounted for 12 per cent of Alberta CIT revenues. Finance and insurance paid 14.7 per cent of CIT revenue over that 20-year period, while manufacturing contributed 15.1 per cent of Alberta CIT.

Why is the oil and gas industry paying so little tax to the Alberta government given its predominance in gross domestic product? This result is due primarily to the huge oilsands capital investments which are deductible from taxable income. However, given that some of these producers' operations consist of refining operations, some of the tax payments would fall under also the manufacturing component. Between 2015 and 2021, the Big Four oilsands producers (Cenovus, CNRL, Imperial Oil, and Suncor) paid a total of $4-billion in taxes of which I attribute $1.6-billion payable to the Alberta treasury. This equates to about $240 million per year taxes payable to Alberta's corporate (Audited Financial Statements, 2015-2021 of Cenovus, CNRL, Imperial Oil, Suncor). The Big Four paid about six per cent of Alberta CIT over the past six fiscal years, suggesting the Big Four pay about half the industry's taxes. As noted, some the CIT paid would be attributed to manufacturing (refining), which would take the Big Four's share of the oil and gas sector's tax payment down significantly. The Big Four's percentage of the manufacturing segment accounted for by refining operations would be significant because of the high value produced by refining gasoline, diesel and aviation fuel. I assume the refining sector pays about one-half the tax paid by manufacturers,. To give a full accounting of the revenue loss over the four years, I add the $4.2 billion booked by the Big Four to the estimated one-time tax benefits from the other one-third of oil and gas taxpayers ($2.1 billion) for a total of $6.3 billion.[11] The $2.1-billion estimate could be an overestimate. Financial statements of these other producers and pipelines reveal smaller amounts of favourable adjustments from the AJCTC in spite of large deferred tax liabilities. The biggest beneficiaries are Enbridge and TC Energy, who have net deferred tax liabilities of $16.9 billion between them.

To complete the picture, we assume the remaining industries do not have any deferred liabilities.[12] Adding the $3.6-billion estimate for 2019-2022 produces a total loss of up to $10 billion, double the estimate of Alberta Treasury Board and Finance and nearly triple the estimate from the UCP's platform. To gauge the significance of the amount, $10 billion represents a little over one year of Alberta's education budget. This estimate does not take into account "favourable impacts" from increased investment, jobs and ultimately revenue purported in Budget 2019. The UCP and Budget 2019 appear to exclude the accounting adjustments, whose timing incidence may be years into the future.

In 2021, corporate profits grew by 63 per cent compared to 2020 (Alberta Treasury Board and Finance, 2022, pp. 15 and 25-26). In May 2022, CNRL reported a quarterly profit of $3.1 billion, with taxes payable (current and deferred) of $976 million (CNRL, 2022, p. 2). Assuming 30 per cent of taxes were payable to the Alberta government, the loss to Alberta taxpayers from the one-third reduction in taxes is approximately $100 million in one quarter for one company. With the tax base growing to an estimated $45 billion or an increase from about $20 billion in 2020, instead of $4 billion in revenue with an eight per cent rate, a 12 per cent rate would have raised an additional $1.4 billion or roughly equivalent to the budget of the Justice and Solicitor General's department in 2022-23.

The greatest irony is that the tax cuts will certainly generate higher CIT revenue from oilsands producers, given lower forecast levels of investment triggering the payment of deferred tax liabilities. In other words, some of the higher revenue is an artifact of the existing preferential tax laws and not the tax cuts policy. Given the Finance department's opinion that oil and gas taxation levels were competitive, it is astonishing that roughly one-half of the gain would fall to four major oilsands producers. To confer such a significant tax benefit in the context of large and rising environmental liabilities is contrary to sound public policy and an affront to ordinary Alberta taxpayers.

Investment

A key promise of the AJCTC was the growth of investment spurred by this policy measure. The October 2019 budget estimated the CIT reductions "will increase investment by about $4 billion per year by 2022-23 and real GDP growth by 0.3 to 0.4 percentage points annually between 2020 and 2023" (Alberta Treasury Board and Finance, October 2019, p. 143).[13] Has this promise shown up in the actual investment numbers? This question is fraught with numerous difficulties, principally because of the impact of the COVID-19 pandemic. A further difficulty is the available data only goes to 2022, and 2021 and 2022 data could be revised significantly.

Figure 11.2 (below) offers a 11-year glimpse of non-residential capital investment in Alberta. The numbers exclude the housing sector. The figure shows capital investment peaked in 2014 at the height of the oilsands construction boom then suffered step declines in 2015 and 2016. Capital investment stabilized at a much lower level from 2016 to 2019 and then fell again in 2020 with the onset of COVID. The most recent estimate for 2022 show that at $56 billion, investment must recover significantly just to reach the lower levels achieved after the 2015 oil price crash. This figure provides a sense of magnitude with respect to the hoped-for $4 billion in additional investment. Even if achieved it will take an additional three years of this growth to reach the levels

Figure 11.2: Alberta Capital and repair expenditures, non residential
2021-2022 (Current $millions)

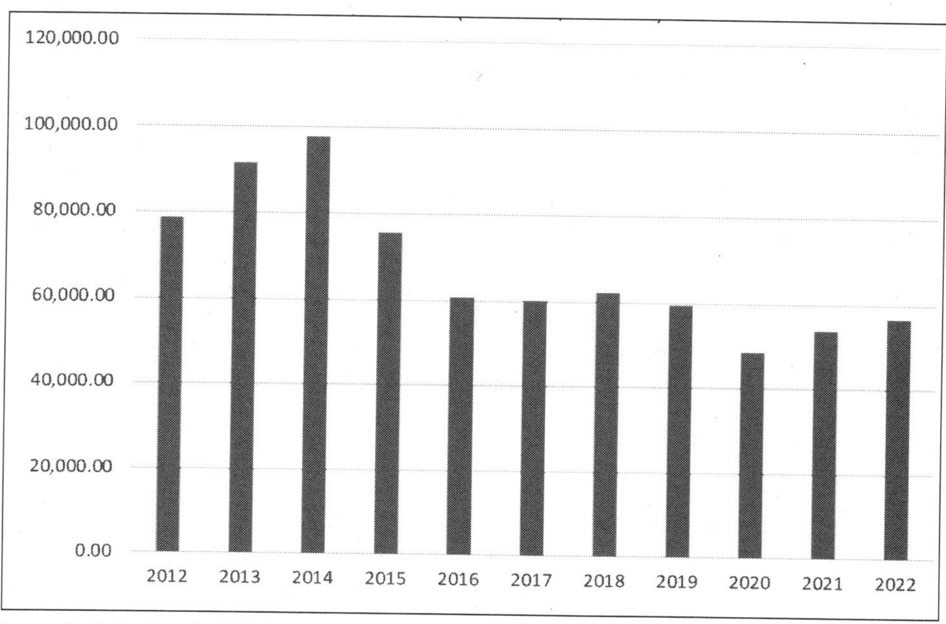

Source: Statistics Canada, Table: 34-10-0035-01 (formerly CANSIM 029-0045).

seen in 2015. Adjusting for inflation, these purported investment gains would do little to regain Alberta's reputation for high levels of per capita investment. Without the prospect of oilsands expansion or a major discovery of a new oil reservoir, the practicalities of this promise coming true are suspect.

Employment and Wages

In Budget 2019, the 2017 research paper by Professors McKenzie and Ferede was cited in support of the wisdom of the CIT cuts. McKenzie and Ferede's research suggests that for every $1 *increase* in corporate taxes, aggregate wages will *fall* by 95 cents (Alberta Treasury Board and Finance, October 2019, p. 143).

The most reliable information about employment and wages comes from Statistics Canada's Survey of Employment, Payrolls and Hours (SEPH). Figure 11.3 shows total employment of non-Indigenous Albertans from 2010 to December 2021. The chart illustrates employment in Alberta peaked in September 2014 at nearly 2.1 million. The trough, caused primarily by the oil crash, was reached in November 2016, with total job losses of 130,000. By the time of the provincial election in April 2019, almost half

Figure 11.3: Alberta = Employment Monthly
Seasonally Adjusted 2010-April 2022 (Persons)

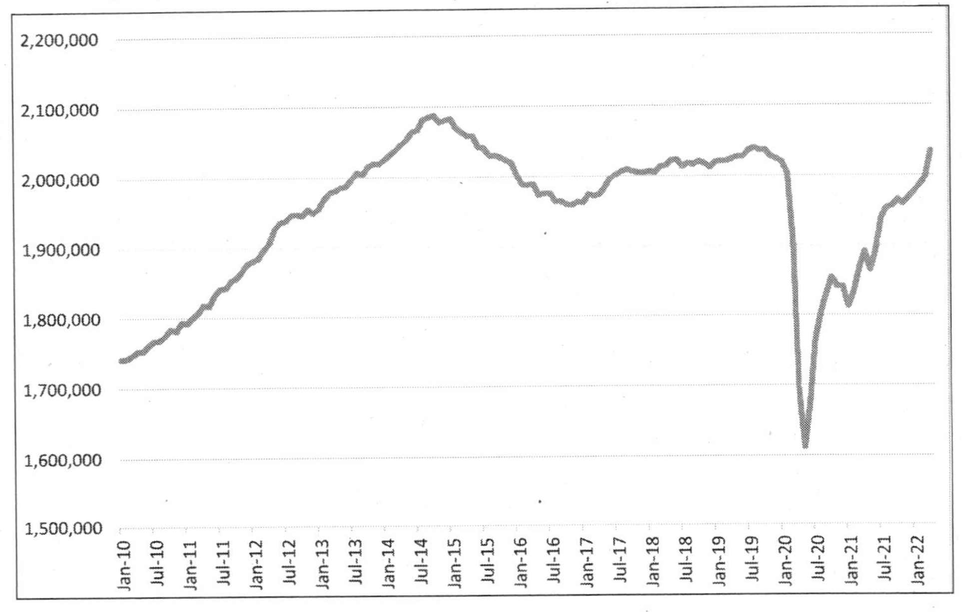

Source: Statistics Canada, Table: 14-10-0223-01 (formerly CANSIM 281-0063). Employment and average weekly earnings (including overtime) for all employees by province and territory, monthly, seasonally adjusted.

the lost ground had been made up. With the onset of COVID, employment plummeted temporarily by nearly 400.000 jobs. Since May 2020, the economy has recovered to 2.064 million jobs (November 2022) still below the previous nadir under the NDP. Again, it is difficult to judge the efficacy of the AJCTC in improving the jobs picture.

How did wages respond to CIT rate cuts? Alberta average weekly wages peaked in February 2015 then fell for another 26 months, reflecting problems in the oil and gas sector. Since that time, during which the minium wage was gradually raised to $15 an hour, average weekly wages have risen unevenly in Alberta (see Figure 11.4, below). However, Alberta's lead in average wages relative to the rest of the country has deteriorated from 121 per cent of the national average in April 2015 to 109 per cent of the national wage in December 2021. This is a significant relative drop over six years. Since July 2019, when the AJCTC came into force, the average wage in Alberta has increased by 3.8 per cent (Canada 4.8 per cent), with the Alberta CPI increasing by 10 per cent (Canada 9.9 per cent) during that period (Statistics Canada, Consumer Price Index, monthly, not seasonally adjusted Table: 18-10-0004-01). It is therefore doubtful the UCP promise of increased employment or wages has yet accomplished the goals

Figure 11.4: Average Weekly Earnings

Alberta: 2010-April 2022

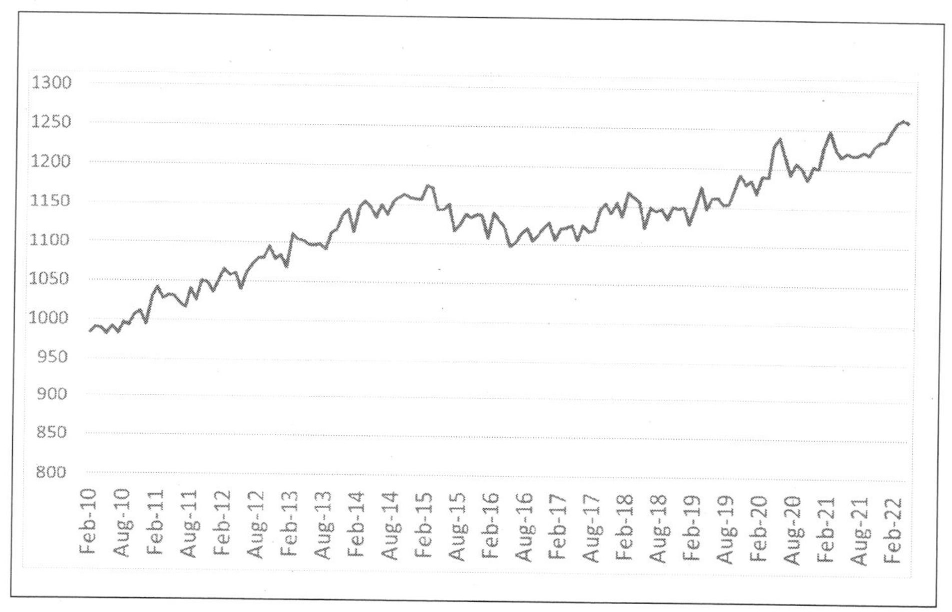

Source: Statistics Canada. Table: 14-10-0203-01 (formerly CANSIM 281-0026) Table 10-10-0158-01 Employment and average weekly earnings (including overtime) for all employees by province and territory, monthly, seasonally adjusted.

promised in its 2019 campaign literature. However, rising oil prices, which are outside the control of provincial politicians, has certainly contributed to a rebound in Alberta's GDP growth and a miraculous turnaround in the province's finances.

Another measure of the health of the labour market is the annual reported hours worked. Figure 11.5 (see below) shows the fall in hours worked caused by the recessionary conditions facing Alberta workers during 2015 and 2016. This was followed by a modest recovery in 2017 and 2018 succeeded by a dramatic plummet in 2020 due to the COVID pandemic. This data underscores the difficulty in carrying out any final determination of the AJCTC's impact when factors completely outside the control of governments, like the pandemic or rapid oil price movevements, intervene.

Conclusion

The 2019 UCP election victory signalled a radical shift in fiscal and tax policy for Alberta. The AJCTC formed a critical piece of the campaign and it followed in the tracks of policies promoted in the United States under President Donald Trump. The policy carried the seal of approval of some economists from the University of Calgary's

Figure 11.5: Alberta Total Hours Worked-all industries
2010-2021 (thousands of hours)

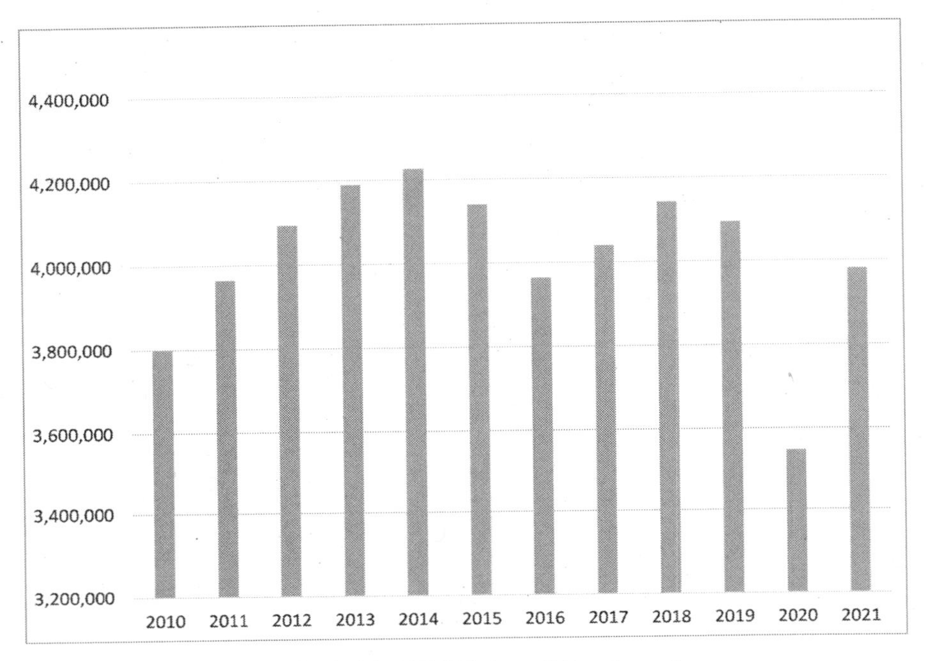

Source: Statistics Canada, Hours worked, by NAICS Industry Table: 36-10-0489-02.

School of Public Policy. The principal objective of the policy was to attract investment and create jobs. The political debate featured one side promoting its presumed growth features, while the NDP labelled the cuts as an unnecessary giveaway.

Estimating foregone revenue using both a top-down and bottom-up approach determined lost revenue could be up to two or three times the amount forecast by the UCP and Alberta Treasury and Finance. This difference is primarily due to including the $4.2-billion gain garnered by oilsands producers when restating their deferred tax liabilities. This phenomenon appears not to have been included in other estimates. The approximate loss of up to $10 billion excludes higher revenues due to increased oil and gas sector profitability of about $1 billion annually which the UCP government claims is due to the tax cuts..

The beneficiaries of the AJCTC were not small businesses.[14] The main beneficiaries were the largest oilsands corporation with an immediate benefit of over $4.2 billion. Financial institutions, most of whom are based in central Canada, manufacturers, and construction firms were also beneficiaries. These corporations are mostly interprovincial or international in operations and whose shareholder base is national and international.

Current statistical evidence indicates the cuts did not have a material impact on investments, job creation or wages. The qualitative survey of economic development agencies provides support that the AJCTC did little to attract investment. Indeed, even major investment announcements made in the fall of 2021 were silent about the merits of the AJCTC.

The COVID-19 pandemic makes it virtually impossible to discern the impact of the rate cuts. Methodological problems inherent in disaggregating the effects of the rate cuts on corporate investment and hiring behaviour suggest any attempt to confirm such projections is virtually impossible. Politics, including interest group and academic advocacy, rather than economics, was the driving force in its creation. Finally, as stated earlier, to confer such a significant tax benefit in the context of large and rising environmental liabilities is contrary to sound public policy and an affront to ordinary Alberta taxpayers.

References

Alberta Treasury Board and Finance (n.d.). *Response to FOIPP request 2019-G-1190-applicant copy*.

Ascah, R. (Bob) (2022). *Energy company shareholders getting paid back after years of famine*. http://abpolecon.ca/2022/03/08/energy-company-shareholders-getting-paid-back-after-years-of-famine/ March 8. Accessed April 2, 2022.

Cembalest, M. (2017). *Tax Cuts and Jobs Act of 2017*. J.P. Morgan Asset & Wealth Management. Dec. 19. https://www.google.ca/url?sa=t&rct=j&q=&esrc=s&source=web&cd=&ved=2ahUKEwiaoceXpsv2AhWXuKQKHQZIC3EQFnoECBEQAQ&url=https%3A%2F%2Fprivatebank.jpmorgan.com%2Fcontent%2Fdam%2Fjpm-wm-aem%2Fglobal%2Fpb%2Fen%2Finsights%2Feye-on-the-market%2Ftax-cuts-and-jobs-act-of-2017.pdf&usg=AOvVawoCpF2XsXBUHzJ4-efwHt2b Accessed March 16, 2022.

City of Lethbridge Economic Development (2021). *End of Year Report 2021*.

Congressional Budget Office (2019). *How the 2017 Tax Act has Affected CBO's GDP and Budget Projections Since January 2017*. Washington, Feb. 28. https://www.cbo.gov/publication/54994 Accessed March 17, 2022.

Ferede, E., & Dahlby, B. (2019). The effect of corporate income tax on the economic growth rate of the Canadian provinces. University of Calgary School of Public Policy. *SPP Technical Paper* 12(29) and accompanying commentary, September.

Gale, W. G., & Haldeman, C. (2021). *The Tax Cuts and Jobs Act: Searching for Supply-Side Effects*. Washington. Economic Studies at Brookings, July. chromextension://efaidnbmnnnibpcajpcglclefindmkaj/viewer.html?pdfurl=https%3A%2F%2Fwww.brookings.edu%2Fwp-content%2Fuploads%2F2021%2F07%2F20210628_TPC_GaleHaldeman_TCJASupplySideEffectsReport_FINAL.pdf&clen=605358&chunk=true Accessed April 3, 2022.

Government of Alberta. (2016). *Petrochemicals Diversification Program.* https://www.alberta.ca/petrochemicals-diversification-program.aspx Accessed Feb. 23, 2022.

Government of Alberta (2019). *Alberta Petrochemicals Incentive Program.* https://www.alberta.ca/alberta-petrochemicals-incentive-program.aspx Accessed Sep., 2022.

Government of Alberta. Treasury Board and Finance. (2019). *2019-23 Fiscal Plan,* October 24.

Government of Alberta. Open Government. Treasury Board and Finance. (2022). *Net Corporate Income Tax Revenue, Alberta.* https://open.alberta.ca/opendata/net-corporate-income-tax-revenue. Accessed March 1, 2022.

Interactive Arts Alberta. (2019). Loss of interactive digital media tax credit will cost Alberta jobs. *News Release*. Oct. 28. https://interactiveartsalberta.org/2019/10/28/loss-of-interactive-digital-media-tax-credit-will-cost-albertan-jobs/ Accessed Feb. 14, 2022.

Laxer, G. (2021). *Posing as Canadian: How Big Foreign Oil Captures Canadian Energy and Climate Policy*. Council of Canadians, CCPA Saskatchewan Office and CCPA BC Office, December. chrome-extension://efaidnbmnnnibpcajpcglclefindmkaj/viewer.html?pdfurl=https%3A%2F%2Fcanadians.org%2Fsites%2Fdefault%2Ffiles%2Fpublications%2FPosing%2520as%2520Canadian%2520-%2520Gordon%2520Laxer.pdf&clen=1340892&chunk=true Accessed 28 March 2022.

McKenzie, K. J., & Ferede, E. (2017). Who pays the corporate tax? Insights from the literature and evidence for Canadian provinces. Calgary: University of Calgary, The School of Public Policy, *SPP Research Paper* 10(6) (April).

Mintz, J. M. (2022, April 29). Time for a tax revolt. *Financial Post.* https://financialpost.com/opinion/jack-m-mintz-time-for-a-tax-revolt Accessed May 8, 2022.

O'Kane, J. (2022, March 29). Amazon has big plans in Alberta, but benefits of data centres challenged. *The Globe and Mail.* https://www.theglobeandmail.com/business/article-amazon-data-centre-alberta-kenney/ Accessed March 28, 2022.

Statistics Canada 2023. Net corporate income tax revenue, Alberta. https://open.canada.ca/data/en/dataset/ca44381a-5392-48cf-a659-9f2e2eea7ae8

Tableau Public Combined Tax rates (Corporate) — North American Jurisdiction (2021). Posted by Edmonton Global. https://public.tableau.com/app/profile/

jeff6244/viz/TaxRatesCorporate-ByNorthAmericanJurisdiction2021/taxbyjur-filter Accessed Feb. 14, 2022.

United Conservative Party. (2019). *Alberta Strong & Free*, https://www.unitedconservative.ca/2019-platform/ Accessed Feb. 10, 2022.

Annual Reports and Audited Financial Statements

Birchcliff Energy. (2019, 2021).

Cenovus Energy Inc. (2015-2021).

Canadian Natural Resources Limited. (2015-2021).

Enbridge Inc. (2019, 2021).

Imperial Oil Limited 2015-2021

MEG Energy Corp. (2019, 2021).

Suncor Inc. (2015-2021).

TC Energy Corporation (2019, 2021).

Tourmaline Oil Corp. (2019, 2021).

Whitecap Resources Inc. (2019, 2021).

Interviews

Alberta's Industrial Heartland Association, Mark Plamandon, Chief Executive Officer. February 17, 2022.

ChooseLethbridge. Trevor Lewington, Chief Executive Officer. February 28, 2022.

Edmonton Global. Malcolm Bruce, Chief Executive Officer and Jeff Bell, Director, Research and Business Intelligence. February 14, 2022.

Fort McMurray-Wood Buffalo Economic Development. And Tourism. Kevin Weidlich. President and CEO. February 23, 2022.

NOTES

1 Cenovus Energy Inc., Canadian Natural Resources Limited, Imperial Oil, and Suncor Inc. market capitalization of these four companies on July 22, 2022 was approximately $210 billion.

2 See Laxer (2021) which questions how Canadian the energy industry is dominated by foreign owners.

3 Other members of the council included former prime minister Stephen Harper and a roster of corporate CEOs. https://www.alberta.ca/economic-recovery-council.aspx

4 Alberta has no payroll or sales taxes.

5 A more definitive announcement was by Amazon Web Services (AWS) in Calgary. AWS promises to invest $4 billion in the new AWS region, which adds almost $5 billion to Canada's GDP and more than 950 jobs by 2037, according to the Daily Hive. Another perspective however suggests that the investment and the number of jobs created is much less than advertised. See Josh O'Kane (2022).

6 Cargill. April 22, 2021. "Cargill unveils plans for new canola processing facility in Regina, Saskatchewan." https://www.cargill.com/2021/cargill-unveils-new-canola-processing-facility-in-regina Accessed March 3, 2022. Richardson International. March 22, 2021. "Richardson Yorkton Crush Plant to double annual crush Capacity to 2.2 million metric tonnes." https://www.richardson.ca/richardson-yorkton-crush-plant-to-double-annual-crush-capacity-to-2-2-million-metric-tonnes/ Accessed March 3, 2022. Viterra. April 26, 2021. "Viterra announces intent to build world class canola crush facility." https://www.viterra.com/Media/News/Viterra-announces-intent-to-build-world-class-canola-crush-facility Accessed March 3, 2022.

7 This program was mentioned by Joe Ceci in the debate on third reading of Bill 3 (Alberta Hansard, 12 June 2019, 761).

8 https://www.alberta.ca/innovation-employment-grant.aspx Accessed March 23, 2022.

9 In the JP Morgan report on page 25, the research of Jack Mintz and Michael Bazel is referenced. The chart on page 28 from the Morgan report was derived from a survey conducted by BoA Merrill Lynch Corporate Risk Management in July 2017. https://www.cnbc.com/2017/07/13/companies-have-big-plans-foroverseas-cash—if-tax-reform-ever-happens.html (Cembalest, 2017). This survey included 302 respondents and showed that paying down debt was the first choice of 65 per cent of companies and share repurchases were favoured by about 45 per cent of companies. Capital investment (34 per cent) and mergers and acquisitions (43 per cent) were followed by dividends at nearly 30 per cent The overwhelming response from the corporate tax changes was to reward shareholders in one form or another rather than invest.

10 https://www.atb.com/company/insights/the-owl/alberta-oil-production-to-october-2021/ Cenovus, News release, Feb. 8, 2022 "Cenovus announces 2021 fourth-quarter and full-year results CNRL, Imperial, and Suncor https://www.oilsandsmagazine.com/news/2022/2/3/imperial-sets-new-production-records-kearl-despite-cold-december and https://www.oilsandsmagazine.com/news/2022/2/10/cenovus-second-largest-producer-canada-2021-overtaking-suncor-energy

11 The deferred tax liabilities of four large senior producers — Birchcliff Energy, Whitecap Resources, MEG Energy, and Tourmaline — and two of the largest pipeline companies — TC Energy and Enbridge — totalled $27 billion versus $8.5 billion in deferred tax assets at Dec. 31, 2021. Audited financial statements of these reporting issuers.

12 The big five Canadian banks have significant deferred tax assets. This means Alberta taxable income is concerned their benefit from future lower taxes is less than rates were higher. In other words, there are losers from the AJCTC.

13 Based on research by Dahlby and Ferede (2019).

14 In 2016, the NDP lowered the small business tax rate to two per cent from three per cent.

PUBLIC SECTOR

CHAPTER 12

CHOICE OVER RIGHTS: THE UCP'S ULTRA-RIGHT VISION FOR EDUCATION

Bridget Stirling

"If anyone, on the other hand, assuming a democratic, progressive position, therefore argues for the democratization of the programmatic organization of content, the democratization of his or her teaching — in other words, the democratization of curriculum — that person is regarded by the authoritarian as too spontaneous and permissive, or else as lacking in seriousness."

— Paulo Freire

IN LATE MAY 2016, as the Conservative Party of Canada met for their annual policy convention, Jason Kenney sat down for an interview with Ezra Levant. At the time, Kenney was still a member of Parliament — he would not announce his intent to resign and seek the leadership of the Progressive Conservative Party of Alberta until July — but he almost certainly was already contemplating entering into the provincial leadership race with the intention of merging Alberta's two major conservative parties. The interview covered a wide range of topics, one of which was why millennials were less likely to vote Conservative. Kenney blamed the public education system, claiming young people had been "hardwired with collectivist ideas, with watching Michael Moore documentaries, with identity politics, from their primary schools to universities" (*Rebel News*, 2016). Kenney stated this was a cultural problem for conservative parties and it was necessary for them to "break that nut" (*Rebel News*, 2016).

This was not the first time Kenney had commented on education. As early as 1993, Kenney made comments targeting the Alberta Teachers' Association (ATA) (Serres, 1993). However, as executive director of the Association of Alberta Taxpayers, president of the Canadian Taxpayers Federation, and then a federal member of Parliament, while he had a substantial platform, his direct power over education was limited. His comments to Levant foreshadowed his government's intense focus on the ideological content of curricula and his decision to make war on entities in the education system, in particular the ATA.

In 2016, education in Alberta was already the site of heightened ideological contestation around the NDP's passage of the Act to Amend the Alberta Bill of Rights to Protect our Children (2015), which came into force in June 2015. This legislation followed demands by sexually and gender diverse youth and their allies to strengthen protections for LGBTQ2S+ students, in particular the right to gay-straight alliances (also commonly known as queer-straight alliances or gender and sexuality alliances). The bill added explicit language protecting students' ability to establish GSAs in their schools as well as adding stronger anti-bullying requirements to the School Act (2012) and the (then-unproclaimed) Education Act (2012). It also removed a contentious requirement (introduced in 2009) that parents or guardians be notified when sexual orientation was to be discussed in classes. The debate over the legislation, more commonly known as Bill 10, revived longstanding divisions in Alberta education between groups claiming to be defending religious and parental rights and those claiming to represent children's rights and the rights of groups historically marginalized in schooling — in this case, LGBTQ2S+ students.

This drive in Alberta education toward parental rights comes from a convergence of two forces in Alberta politics: a libertarian orientation towards property rights (see Harrison, Chapter 5) — Alberta is notable as a province that has not one but two pieces of legislation specifically oriented to protecting property rights — and a socially conservative interpretation of children as property of their parents (rooted in Roman ideas of *patria potestas* that became entrenched in British and then Canadian understandings of parental rights via canon law). These two forces combine to create a strong orientation on parental rights in Alberta. These forces converged with a third, global political force, neoliberalism, with its focus on free markets and individual choice. These forces came together to shape Alberta's conception of parental rights in education, particularly during the Klein era beginning in the 1990s (Harrison and Kachur, 1999). This orientation on parental rights sat increasingly in contention with the understanding of children's rights as distinct from (although interconnected with) the rights of their parents or guardians reflected in the United Nations Convention on the Rights of the Child (1989) and emergent in Canadian jurisprudence (Blokhuis, 2020). Notably, Alberta did not endorse the convention until 1999, and then only reluctantly, making it the last Canadian jurisdiction to do so.

We Will Put that Curriculum Through the Shredder

In this milieu, the NDP took power in Alberta following the 2015 provincial election. The party had long supported LGBTQ2S+ rights and stood alongside sexually and gender diverse youth in their demands for Bill 10 to offer strong protections for the right to form a GSA. NDP Education Minister David Eggen's efforts to ensure legislative protections were put into practice quickly came into conflict with a group of religious private schools — primarily Christian, but also including conservative Sikh, Jewish, and Muslim schools — that refused to implement protections.

By the summer of 2016, the issue of GSAs and antibullying policy protections for LGBTQ2S+ youth was on full boil, with Eggen demanding school authorities comply with the law or (he implied) risk losing public funding if they refused to allow GSAs (Thomson, 2016). The battle lines were drawn: religious private schools and religious conservatives, aligned with parental rights advocacy groups, on one side, with sexually and gender diverse students, families, and teachers and their allies, including the Alberta Teachers Association, lined up on the other.

As the UCP leadership campaign began, Kenney took advantage of the opportunity to align himself with religious conservatives and parental rights groups. One organization in particular became central to Kenney's campaign: Parents for Choice in Education (PCE), an advocacy organization founded by Kenney's parliamentary colleague Garnett Genuis in 2012.

From its inception, PCE was engaged in opposing LGBTQ2S+ protections in schools, as well as advocating for issues of parental choice and parental rights. One of their earliest campaigns involved lobbying trustees to vote against a motion calling for policy protections against homophobia, transphobia, and heterosexism at the Alberta School Boards Association's fall general meeting in 2012. Although it was active in the 2013 school board election, the organization was not in the public eye until 2014, when Dr. Nhung Tran-Davies promoted a petition opposing "discovery math." Its real rise to prominence, however, came in organizing opposition to Bill 10 (Lamoreux, 2016). The PCE used Tran-Davies's petition to generate a substantial mailing list through which it promoted its anti-LBGTQ2S+ messaging.

The organization's growing base made it attractive to Kenney, and he quickly formed an alliance with the group, meeting with PCE communications advisor and board member Theresa Ng in the spring of 2017. Ng also met with then-Wildrose MLA and 2022 UCP leadership candidate Leela Aheer and was featured in a photo posted to Aheer's Facebook page (Press Progress, 2017). That fall, Kenney supported the "Students First" slate for the Calgary Board of Education (Klaszus, 2017) whose messaging aligned with some of PCE's positions. In Edmonton, although meetings with candidates and PCE involvement in campaigns were rumoured, Kenney did not publicly intervene. However, interference by activist groups including PCE became an issue during the campaign (French, 2017).

In the meantime, Kenney's support from PCE grew, receiving its endorsement in both the Progressive Conservative and subsequent United Conservative Party leadership races. PCE hosted booths at the UCP's annual general meetings, and policies reflecting its positions appeared prominently as motions at those AGMs. It became clear PCE and Kenney were in substantial policy alignment, with Kenney using PCE talking points in his campaign messaging (Jason Kenney Leadership Campaign, 2017).

Kenney's alignment with social conservative and parental rights education groups was a key element of his campaign strategy for leader and continued as part of the UCP's two-year campaign leading up to the 2019 general election. Star candidates engaged closely with PCE and affiliated groups, most notably Red Deer's Adriana LaGrange, the future UCP education minister, who hosted an event for PCE, as did PCE-affiliated candidate Tunde Obasan.

Despite public outcry over the UCP's opposition to Bill 24 (An Act to Support Gay-Straight Alliances [2017]), Kenney gambled that support from social conservative groups, combined with language that framed the issue as one of parental rights, would outweigh the electoral risk of opposing the bill (Banack, 2017). The depth of his relationship with these groups was subsequently evidenced in the UCP's 2019 education platform.

In the meantime, beyond their stance on GSAs and commitment to other equity issues in education, such as adding more Indigenous content across the curriculum, the NDP largely continued in the direction established in 2008 with the Inspiring Education consultations and subsequent report (Government of Alberta, 2010). Alberta Education and school jurisdictions across the province continued to function in alignment with the 2013 Ministerial Order on Student Learning. While Eggen announced in 2016 a comprehensive curriculum review, this work was largely built on previous curriculum consultations and developments initiated under the previous Conservative government (CBC News, 2016). Despite previously criticizing funding for private and charter[1] schools, the NDP in government continued the same funding model established in 2008, when the PCs under Ed Stelmach increased private school funding from 60 per cent to 70 per cent. It was largely business as usual, although the NDP made one notable move on charter schools late in their term when they announced infrastructure funding for a Calgary charter school. The announcement represented a significant shift, as public infrastructure funding would be transferred into the hands of non-governmental, albeit non-profit, school authorities.

Despite this largely unchanged direction in Alberta's primary and secondary education system, the UCP opened a second front in the ideological battle over schooling by launching an attack on the curriculum review process. At the party's first AGM in 2018, Kenney promised a UCP government would "put through the shredder" any NDP curriculum that "tried to smuggle more of their politics into the classroom" (Kenney, 2018). Of course, allegations of politicizing the curriculum were already a major part of the UCP's messaging on education.

Kenney was thus already signalling his party's intent to scrap the curriculum process, which in 2018 was built on a decade of consultation and development. The K-4 draft curriculum that was to be piloted in 2019 began with PC Education Minister David Hancock and had carried on through three subsequent ministers, although with some shifts to reflect changing expectations over those 10 years, and with input from thousands of participants — teachers, parents, education and subject matter experts, school authorities, members of the public, and non-governmental actors in both not-for-profit and corporate worlds — in the consultation process. Despite this, the UCP opposition alleged secrecy in the consultation process, demanded the release of the names of all parties involved in writing the curriculum, including hundreds of teachers, and accused the NDP of ignoring parents' voices. Once again, the UCP under Kenney was building the narrative that parents were being left out of education.

2019: The Shredding Begins

The full scope of the UCP's plans for education became clearer with the release of the party's 2019 platform document (United Conservative Party, 2019). While some areas seemed innocuous or even positive, much of the party's platform across primary, secondary, and post-secondary education indicated a substantial shift in direction that would restructure much of the system, including changing teacher qualifications and standards of evaluation, expanding privatization, reorienting education to focus on job skills, including a return to vocational and academic tracking, and entrenching parental rights in the Education Act. At the news conference launching the platform, Kenney also reignited the battle for GSAs, suggesting a UCP government would roll back protections included in Bill 10 and remove the strengthened antibullying policy and privacy protections introduced under Bill 24.

Despite sizeable protests in support of sexually and gender diverse youths and efforts by groups such as the ATA and school boards trying to make education an election issue, education was not a key issue in the 2019 campaign. The UCP's tactic of targeting socially conservative voters through the rhetoric of parental rights mobilized an important element of their base, but education was not a ballot box issue for the average Albertan struggling in the face of collapsing oil prices and a recession. Under new Premier Kenney, the UCP wasted no time after its election victory in putting their education plans into action, beginning with the introduction of Bill 8, which amended the Education Act. Although the Progressive Conservatives passed the Education Act through third reading in 2021, they had never sought royal assent for the legislation. At the time, a substantial outcry from education stakeholders delayed the legislation, so the PCs shelved it and the School Act (the previous legislation) remained in force. In the interim, education bills under the PCs were structured to amend both the School Act and the Education Act, ensuring parallel changes were made to both.

The most significant public opposition to Bill 8 concerned the issue of revoking the privacy protections and other protections on GSA access brought in by the NDP's Bill 24. Technically, Bill 24 would not be revoked; however, unlike previous amending legislation passed by the PCs, it only amended the School Act. This left the protections introduced in Bill 24 vulnerable to a sort of legislative sleight of hand when the Education Act came into force, replacing the School Act. The UCP successfully eliminated Bill 24's protections without the need to pass legislation explicitly revoking the legislation.

Other changes contained in Bill 8 furthered the UCP's planned education reforms. Also eliminated was the NDP's restriction on school bus fees. Added were new powers for school boards, including being able to provide alternative programs outside their geographic boundaries if the local board was not willing to offer them. It also added the ability of boards to fire trustees who breach board codes of conduct and opened up public school board elections to Catholic candidates. Finally, as a first step toward expanding the role of privatized education, the legislation lifted the cap on the number of charter schools permitted to operate in the province.

The end of the first session was followed quickly by the cancellation of the planned pilot year for the 2018 draft K-4 curriculum and the appointment of a Curriculum Advisory Panel to "review and enhance" the curriculum and draft a new ministerial order on student learning (Weisberg, 2019). The panel featured both industry representatives and academics, one of whom had previously endorsed the UCP's education platform.

Also included among the researchers was Ashley Berner, an American education reform advocate whose work was funded by a substantial grant from the right-wing Charles Koch Foundation (Press Progress, 2019). Berner's Institute for Education Policy had also published a report by another panel member, Amy von Heyking, a historian from the University of Lethbridge[2]. In a June 2019 commentary for the Thomas B. Fordham Institute, an ideologically conservative education think tank, Berner pointed to Alberta's system of funding private as well as public schools as a model for American education reform and praised aspects of the extant Alberta curriculum for focusing on content, rather than skills (Berner, 2019).

Notably, none of the panel members were active teachers, although the committee chair was former Edmonton Public Schools superintendent Angus McBeath, who retired from the board in 2005 after serving as acting superintendent during the 1994 Klein cuts and then as superintendent during Alberta's most recent teachers' strike in 2002. The appointment, combined with the shutout of the ATA from the panel, was an early indicator that the UCP under Kenney would follow through on his 1993 suggestion that the teachers' union should be the first target in a broader attack on public sector unions.

As the panel began its work, school authorities waited anxiously for the 2019 budget, which was not introduced until Oct. 24 — 24 days after the enrolment count on Sept. 30, when school divisions could normally solidify their budgets for the school year based on the province's spring budget. Uncertain of their options, boards opted largely to hold the line on spending beyond enrolment growth, anticipating the UCP's commitment to maintain or increase education funding would mean maintaining the per-student allocation from the previous year, although divisions also tried to predict which additional grants might be cut (Giovannetti, 2019). While the MacKinnon Report (Blue Ribbon Panel, 2019) suggested cuts were coming to the public sector, school authorities were left guessing where they might fall.

When Budget 2019 came down, however, school divisions across Alberta saw massive cuts. The deepest cuts hit Alberta's fastest-growing school divisions, mainly in urban areas and surrounding communities with Edmonton's public and Catholic boards reporting cuts of $79 and $52 million respectively and the Calgary Board of Education reporting a drop of $32 million. But cuts also affected school divisions in growing communities across the province, from Fort Vermillion to Lethbridge. Divisions scrambled to find funds, including announcing staff cuts in the middle of the school year, something not seen since the drastic cuts in the early years of the Klein era. Alberta parents who, until then, had assumed their children's schools would be protected from cuts as promised in the UCP platform, suddenly saw classes shuffled and students going without supports as teachers and EAs were laid off to compensate for the shortfall.

This began a pattern of funding cuts resulting in a continuing reduction of per-student funding year-over-year through to the 2022-23 school year, alongside cuts to infrastructure funding that left school divisions ever more unable to address the chronic problems of deferred maintenance that were a legacy of Klein's drive to pay off provincial debt rather than fund the upkeep of public infrastructure.

Some of the students most affected were students with disabilities and/or additional learning support needs. Deep cuts to Program Unit Funding, which supported children with delays through pre-kindergarten and kindergarten programs by increasing their readiness for Grade 1, saw access to those supports reduced or lost for thousands of Alberta children. Layoffs following cuts also disproportionately affected children who needed extra support as educational assistants were frequently the first to be laid off in many schools.

It became apparent the UCP was serious when they signalled massive changes to come in Alberta's schools. As with many of his statements, Kenney's promise to maintain or increase education funding needed to be parsed carefully — the UCP held the line on the overall budget allocation for education, but with 15,000 new students arriving in Alberta's schools in the fall of 2019, that pie was being sliced into smaller and smaller pieces.

Through the fall of 2019 into early 2020, education unions, alongside school divisions and their provincial bodies, primarily the Alberta School Boards Association, raised the alarm and launched campaigns to mobilize parents to take action in support of their children's schools. While the ensuring COVID-19 pandemic brought these campaigns largely to a sudden halt, it did not stop UCP efforts to restructure Alberta's K-12 education system regarding curriculum, educational choice, and parental rights.

Curriculum Battles

The UCP's curriculum review received perhaps the most media coverage of all its education reforms. The review process began with the review panel's appointment in the summer of 2019 followed by the public release of the panel's report on Jan. 29, 2020. Accompanying the report was a draft Ministerial Order on Student Learning replacing the 2013 order with a consultation window set to end on Feb. 24, giving the public less than four weeks to read the order and provide their feedback. Just 8,500 Albertans participated in that consultation process, as compared to initial curriculum consultations conducted by the NDP in 2016 that included more than 32,000 participants (Peck, 2020).

The finalized order, which came into effect on Aug. 6, 2020, represented a substantial shift in the direction of Alberta education from a focus on competencies in the 2013 order to a focus on "foundational, subject-specific content." The change presumed the development of a broad base of knowledge through learning facts would lead students to become strong communicators, problem solvers, and critical thinkers (Government of Alberta, 2020) — something pedagogy experts refuted, describing the UCP's approach as ineffective in equipping students to engage meaningfully with factual information, not just memorize and repeat it (Peck, 2020). The order also included language that read as coded references to a Eurocentric orientation on Western civilization and a "great works" curriculum focused on the traditional canon. Finally, while the focus was toned down in the final draft, the initial ministerial order was strongly focused on workplace readiness as the main purpose of education. Additionally, student outcomes focused on self-reliance, entrepreneurship, and individual responsibility, with minimal focus on the learner's relationships within their community and society beyond their rights and responsibilities within it.

In all, the order signalled a substantial divergence from the trajectory of Alberta's curriculum and learning model over the previous 12 years. While the conversation for more than a decade had been focused on what 21st century learning might look like, the UCP's plan for a back-to-basics curriculum seemed to set the clock back to the early part of the 20th century, ignoring decades of research in curriculum, pedagogy, and educational psychology.

Following the release of the final version of the ministerial order, the minister and two advisory panel members appeared at a news conference announcing the commencement of the curriculum development process. The conference had moments of absurdity (Peck, 2020) as Education Minister LaGrange appeared not to understand the difference between curriculum (what is to be taught) and pedagogy (how it is taught) and struggled to point to examples in the existing or draft curriculum that reflected the ideologically leftist content or bias that she claimed the new curriculum development process would correct (LaGrange, 2020). The minister stated that ministerial staff would be working over the coming months to develop the new curriculum but did not mention the appointment of a curriculum advisory panel or the identities of the panel's members; the existence of the panel and the members' names were only revealed a few days later in response to media inquiries.

Initially, eight panel members were identified. Notably, all were men, and only two of them, both professors in the Faculty of Education at the University of Alberta, had any expertise in teaching children. The minister's press secretary, Colin Aitchison, identified the other panel members as subject matter experts. The names of two members in particular stood out immediately.

William French, a lawyer and translator was appointed to advise on arts and literature. French's expertise seemed to be as a board member of Calgary's Shakespeare Company and the owner and author of Conservative, a website and blog on which he expressed his admiration for anti-modernist conservative philosopher Eric Voegelin, who advocated for the adoption of a "great books" curriculum centred on the European literary canon. The appointment of French, whose ideological positions were not readily evident, went largely unnoticed in the shadow of another appointed member.

Chris Champion, a former CPC staffer and Kenney advisor and editor of *The Dorchester Review*, was appointed as the adviser for the social studies curriculum. Champion's participation received significant scrutiny when his writing in *The Dorchester Review* and other venues came to light. Some critics cited as racist a 2019 article published under his byline in the review that called the inclusion of Indigenous content in curricula a fad and described the Kairos Blanket Exercise as brainwashing children (Bruch, 2020). He continued expressing similar views following his appointment to the review committee, including publishing an article and tweets denying Canada's genocide of Indigenous people and mocking stories of abuse brought forward by residential school survivors (Braat, 2021).

Under pressure from the public over the narrowness of the panel and the lack of advisors with expertise in areas such as Indigenous knowledges or Francophone literature and French-language literacy, the panel was eventually expanded to include 19 members by February 2021. However, the bulk of the initial draft curriculum was written under the direction of the eight initial appointees, with a draft completed and provided to a group of teachers and education professors to review in December 2020.

University of Alberta professor Carla Peck, who participated in the December 2020 review, has noted this group was only given two to eight days to complete their review and feedback, and that at this point, a full draft of the curriculum had been completed without the participation of experts in curriculum (Peck, 2022). Additionally, all participants in that stage of the process were required to sign a non-disclosure agreement, preventing them from discussing the contents of the draft materials. The rest of Alberta would have to wait until March 2021 to see the full draft K-6 curriculum, which the minister expected to see piloted in September 2021 — just six months away.

When Alberta teachers, families, and school boards read the draft curriculum, many responded with shock. Almost immediately, Alberta education Twitter exploded with comments pointing out factual errors throughout the document, including asking students where to find gravity on a globe and teaching students to use a map scale by calculating the distance from Regina to Duck Lake using a map of Alberta. Others noted the emphasis on the history, culture, and politics of the United States, such as the inclusion of a section on the Declaration of Independence in Grade 6 social studies and the removal of the Canadian Charter of Rights and Freedoms, part of the existing Grade 6 curriculum.

Several critics pointed out the draft curriculum's content was at times racist and xenophobic. Indigenous knowledges and worldviews and the perspectives of Alberta francophones had been almost entirely left out of the draft. They also noted it was highly gendered in presenting a very narrow view of women, and that LGBTQ2S+ perspectives were also absent. Health and wellness experts also expressed concerns about the physical education and wellness curriculum and its potential impacts on student mental health.

The inclusion of so much American content in the draft curriculum immediately drew attention and accusations of plagiarism. Sarah Eaton, a University of Calgary education professor with expertise in academic misconduct, analysed the curriculum and identified several instances of plagiarism from a variety of sources (Eaton, 2021).

Others likewise found substantial similarities to content from a curriculum sequence produced by CoreKnowledge, a publisher based in Charlottesville, Virginia.

Organizations and experts began publishing their criticisms of the curriculum, many of which were gathered on the Alberta Curriculum Analysis website (alberta-curriculum-analysis.ca), an initiative supported by the Alberta Deans of Education. (Disclosure: I worked on this website in the summer of 2021 as a research assistant to Carla Peck, one of three members of the steering committee). While much of the curriculum was panned, some aspects were noted as positive, such as the inclusion of phonological awareness and phonics in the language arts curriculum, although critics noted this only provided one narrow slice of the set of skills and competencies that make up literacy (Northern Alberta Reading Specialists' Council, 2021, June 10). Perhaps most damningly, however, the Northwest Territories, which had used Alberta's curriculum for more than four decades, decided instead to use British Columbia's recently redesigned curriculum.

It became increasingly clear the draft was nowhere near ready to pilot, let alone implement. When surveyed by the association, 91 per cent of ATA members said they were unhappy with the draft curriculum, with 93 per cent of those in leadership roles saying they did not feel confident in moving ahead. Fifty-six of Alberta's 61 school boards announced they would not participate in piloting the draft curriculum, with the majority of the remaining divisions opting only to pilot parts of the curriculum.[3] No school division opted to pilot the entire draft in any classroom in the 2021 school year, and only 360 teachers ended up piloting any part of the curriculum in their classrooms, representing just 7,800 of Alberta's more than 335,500 students in kindergarten to Grade 6 — just slightly more than two per cent (Peck, 2021).

Alongside feedback from the piloting teachers, Alberta Education over the following year initiated an engagement process with education stakeholders in an attempt to make up for minimal participation by the majority of Alberta school authorities. In April 2022, final versions of the English language arts and literature, mathematics, and physical education and wellness curricula were released, followed by updated drafts of the science, French first language and literature, and French immersion language arts and culture curricula in May 2022. Curricula for the fine arts (music, visual arts, dance and drama) and social studies — those areas largely developed by French and Champion, respectively — were still not released as of September 2022.

For the 2022 school year, despite continuing concerns over the updated final versions, school authorities were instructed to implement the K-3 English language arts and literature, K-3 mathematics and K-6 physical education and wellness curricula, with the option to implement Grades 4-6 in language arts and math if desired. The three updated drafts began a second year of piloting, with the yet-unreleased fine arts curriculum to be piloted at some point during the 2022-23 school year. Implementation of the full curriculum, with the exception of social studies, was planned to go ahead in 2023, with full implementation in 2024.

With an election slated for May 2023, the future of the UCP's draft curriculum remains uncertain. The NDP has stated that, if returned to government, it will scrap the UCP curriculum and begin a new round of consultations for another set of drafts. Additionally, however, none of the six UCP leadership candidates who attended the ATA's 2022 summer conference committed to moving ahead with implementation as planned (CBC News, 2022).

Choice in Education

A second key area of the UCP's transformation of Alberta's education system was an expansion of school choice, which primarily meant the expansion of charter and private schools and increased supports for homeschooling.

Functionally, Alberta offers seven forms of choice in education. In the category of publicly funded and publicly governed school authorities, the most prevalent are public schools. Alongside public schools, the province has publicly-funded Catholic schools that operate similarly, although only Catholics may serve on Catholic school boards, and schools may choose to deny registration to non-Catholic students — a rule that also applies for Catholic students wishing to attend a public school in areas where Catholic education is available, Francophone school boards are a third form of public school authority, offering both public and Catholic school programs for children who qualify under the minority language rights of Section 23 of the Charter of Rights and Freedoms. A fourth form of choice under this model, available across both public and Catholic systems, are known as alternative programs. Alternative schools complicate the question of what qualifies as a public school. These programs operate under publicly funded boards in the public and Catholic systems, and as such, are funded in the same way as are community public schools. Some education advocates, such as Support Our Students, have noted that public alternative programs are also a contributor to privatization, with entrance lotteries, high fees, and geographic locations creating barriers to entry that create school segregation within a public system.

Alberta also has three private or quasi-private systems. Fifth, and unique among Canadian provinces, is Alberta's charter school system. While they receive operational funding like public schools, charters until 2019 did not receive infrastructure funding. While charters are funded as public schools, their assets and governance are operated as independent entities under either the Societies Act or the Companies Act. They are not owned by, or accountable to, the public as a whole. The sixth group, private schools, still receives some public funding, receiving 70 per cent of the public grant, although they do not receive infrastructure funding. This funding is only available to accredited private schools, although non-accredited schools may operate in Alberta. Private schools do not have to admit all students and may have restrictive selection criteria as well as tuition fees. There is no cap on tuition. Finally, families in Alberta may choose homeschooling, with public funding available depending on whether the instruction is teacher or parent led.

The UCP's efforts at expanding school choice began with a package of amendments to the Education Act contained in Bill 8. The amendments lifted the requirement for charter schools to demonstrate "significant community support," meaning a charter could now be launched without needing to show the minister pre-existing demand. The amendments also removed the cap on the number of charter schools, previously set at 15. At the time, there were only 13 charters operating in Alberta (some of them multi-campus), meaning the legislative cap was lifted despite there not being demand up to that cap. This change was explained as a measure to stimulate more charter schools, a curious explanation that makes sense only in the context of removing the need to demonstrate community support and changes that came next in the Choice in Education Act.

The Choice in Education Act made additional amendments to the Education Act. It expanded further the potential for charter schools and opened up what can be considered an eighth choice in Alberta education. The act eliminated the requirement to request an alternative program from a local school board before launching a charter school. A group could now directly request establishment of a charter school without showing an existing demand or trying first to establish a public alternative program. The act also added a specific category of vocational charter schools, no longer subject to the same non-competition rules set out for other types of charters, meaning that vocational charters could duplicate public school programs in the same geographic area.

Charters were also now explicitly exempted from municipalities' joint-use and planning agreements. While an obscure change, this has potentially major implications for the privatization of public resources. Generally, when schools are built by publicly funded boards, they are built on municipal reserve lands or land owned by a board, both of which are public land. The building thus becomes a public asset. When a school building is no longer needed by a division, it reverts to the control of the public. Joint-use and planning agreements for municipalities generally have restrictions on what happens to that property, usually requiring it to continue to be used for public benefit. However, now that provincial funding for infrastructure is available to charters, charters can receive public funding to build private school buildings; that is, the non-profit society or corporation operating the charter school owns the building. The exemption from joint-use and planning agreements means the property can be sold into private hands, even though its original funding was public. In the U.S., many charter schools have amassed substantial real estate holdings and real estate investment as a significant part of their operations. As of this writing, it is too early to know for certain the long-term consequences of school privatization in Alberta, but it could have substantial effects on the transfer of public wealth into private hands.

Parental Versus Children's Rights

The Choice in Education Act also introduced a new form of choice: unsupervised homeschooling. Prior to its introduction, parent-directed homeschooling in Alberta had to be conducted under the supervision of a certificated teacher who evaluated the student's progress twice a year. While the child's education plan must still be submitted annually, approval of the plan is no longer required. This change concerned advocates for children's rights. While some homeschooling parents may be excellent teachers, with no checks on how the program is conducted, this cannot be assured; children could be vulnerable to receiving little to no meaningful education — a right that is well established in both the Universal Declaration of Human Rights (UDHR) and the United Nations Convention on the Rights of the Child (UNCRC). Under these agreements, children have the right to an education that supports their full potential. Additionally, the state has an obligation to ensure that such right is protected. Children's advocates raised concerns that the absence of accountability in unsupervised homeschooling settings risked leaving some children behind as parental rights were allowed to override the rights of children to a proper education.

The final legislative change that emerged from the Choice in Education Act may seem trivial to some, but it represents the legislative entrenchment of an ideological direction that has dominated Alberta's education landscape for decades. Anne McGillivray notes that "Childhood has bent the adjudication of rights in the Supreme Court of Canada" (2011); however, this effect extends beyond the decisions of the Supreme Court; it shapes how Canadian legislators, policy makers, and everyday citizens understand and interpret rights questions when it comes to the lives of children. These interpretations — that the Canadian Charter of Rights and Freedoms does not apply to children, that parental rights override children's rights, that children are not entitled to rights in education because they are subject to the school's power of *in loco parentis*, that privacy legislation does not apply to children (whatever the risk) because of the parent's rights over the child — these are rooted deeply in the historical and legal foundations of children's rights constructions that are frequently at odds with the principle that all human beings are born equal in dignity and in rights. Those roots are particularly deep in Alberta, the last province in Canada to endorse the UNCRC.

The Choice in Education Act enshrines Article 26.3 of the Universal Declaration of Human Rights into the preamble to the Education Act. Section 26 as a whole is about the right to an education. The third article, 26.3, reads: "Parents have a prior right to choose the kind of education that shall be given to their children." This section alone was added to Alberta's core education legislation.

To many, adding rights protections to legislation sounds like a positive step, but there is a subversion of children's rights in education in the decision to explicitly recognize 26.3, which is about parental rights, without the other two sections, which are focused on the recipient of that education — the child.

In full, the article reads:

> 1. Everyone has the right to education. Education shall be free, at least in the elementary and fundamental stages. Elementary education shall be compulsory. Technical and professional education shall be made generally available and higher education shall be equally accessible to all on the basis of merit.
> 2. Education shall be directed to the full development of the human personality and to the strengthening of respect for human rights and fundamental freedoms. It shall promote understanding, tolerance and friendship among all nations, racial or religious groups, and shall further the activities of the United Nations for the maintenance of peace.
> 3. Parents have a prior right to choose the kind of education that shall be given to their children.

The choice to protect only the third section without the balancing language of the first two, as well as the lack of recognition of the education protections in the UNCRC, is troubling. To focus only on the parents' rights over the child and not the child's rights to that education was profoundly ideologically driven.

This ideological addition to the Education Act reflects the foundation of Alberta's orientation toward education choice. If a society believes strongly in the idea that parental rights are the primary rights in education, and they believe that individual choice within a market is the best way to organize a society, then they will arrive at the place of ever-increasing educational choice where government's role is to act as a service provider to individuals, rather than government as something in which we all participate as citizens whose role is to provide for and protect the public good.

"Kids Don't Get Sick"

The UCP's libertarian orientation significantly affected its educational policy decisions in handling school safety during the COVID-19 pandemic. By March 2020, the seriousness of the pandemic was becoming increasingly apparent and Alberta's school authorities increasingly worried. Past influenza pandemics had shown that schools were a significant driver of transmission and that school closures were a recommended intervention to reduce the spread of a pandemic respiratory disease.

While other public infrastructure was being rapidly shut down across the province, Minister LaGrange and Alberta's chief medical officer of health, Deena Hinshaw, announced instead only sanitation measures for schools. People working in classrooms immediately recognized, however, the near impossibility of implementing these measures.

Trustees in Edmonton and Calgary began discussing school closures with their board colleagues and others within the Alberta School Boards Association. Behind the scenes, trustees pushed for closures, meeting by phone with local Alberta Health Services medical officers of health and their teams. Some began discussing the holding of special board meetings to make school closure decisions locally. Despite trustees' increasing alarm, however, Hinshaw announced schools would remain open. Finally, trustees were allowed to participate in a phone call directly with Hinshaw on a Sunday afternoon. Trustees and division administration again stressed how impossible the directives Hinshaw had given were to implement in schools.

That afternoon, trustees waited anxiously for a decision to come out of an emergency cabinet meeting. Finally, late on Sunday, Hinshaw announced that schools would immediately close for in-person schooling (Johnson, 2020). On March 16, schools across Alberta shut their doors to students, who would not return to in-person schooling until September of that year. For the remainder of the year, students would complete their classes remotely, most of them online.

The relief trustees felt, as strong public health measures were brought in to protect students and school-based staff, vanished quickly as the minister announced the province would claw back millions in funding for non-teacher support staff, including educational assistants (EAs), administrative assistants, bus drivers, and many others whose work supported students. Some divisions scrambled to juggle finances so that educational assistants would not need to be laid off and could continue working with students with additional learning support needs as they moved online. Despite their efforts, LaGrange directed schools to proceed with layoff notices (Omstead, 2020). In the end, 25,000 education workers were given notices, the majority of them EAs. It was the largest layoff in Alberta history. Children with learning support needs once again disproportionately experienced the effects of funding cuts.

In May 2020, LaGrange announced $250 million in funding for shovel-ready infrastructure projects for Alberta schools. This funding was part of the UCP's overall billion-dollar spending on infrastructure meant to stimulate the provincial economy during the global crash caused by the pandemic and get tradespeople back to work. At the time, the minister explicitly stated the funding was not intended for COVID-19 mitigation projects to prepare for students' eventual return to classes (Johnson, 2020), though she later (and repeatedly) contradicted her statement, retroactively describing this allocation as provincial COVID-19 funding for schools.

In the meantime, Alberta allocated very little to COVID-19 mitigation for schools beyond approximately $10 million to purchase hand sanitizer and two masks for each student and teachers returning to class in September 2020. LaGrange subsequently faced accusations of a conflict of interest when it was revealed that she had personal connections to one of the two selected suppliers, and that senior ministry staff might have intervened in the selection process. While the ethics commissioner eventually ruled there was insufficient evidence to prove a violation of the Conflict of Interest Act, questions lingered about the purchase (Toy, 2021).

Ironically, the largest support for COVID-19 safety in Alberta schools came not from the provincial government, but rather from Kenney's frequent target, the Government of Canada. Following a motion by Edmonton NDP Member of Parliament Heather McPherson, which received unanimous consent in the House, $2 billion was allocated to provinces to provide for safety measures for the return to school in September 2020. In total, $262 million was provided to Alberta, of which $250 million was distributed to school authorities (including charter and private schools).

Over the summer, school divisions continued to fight with the minister over questions of COVID-19 safety, coming to a head with the release of the initial school re-entry plan in July 2020. In the initial version of the plan, masking was not mandatory in schools, despite strong evidence by that point that masks were a useful part of a layered approach to mitigating the pandemic's impact. Masking was in fact part of recommendations of the Alberta Health Services COVID-19 Scientific Advisory Group Rapid Evidence Report on the topic of school safety, which advised cloth masking for older children and teens. Once again, trustees found themselves in conflict with the minister, with some divisions preparing to announce mask mandates independently. In response to the pressure, LaGrange made an early morning announcement on Aug. 4, 2020, ahead of scheduled news conferences later that day at which some boards had planned independent announcements on masking, along with the release of their local re-entry plans (Joannou, 2020).

Throughout the 2020-21 school year, despite urging from health professionals and community advocates, schools largely stayed open, although individual closures rolled across divisions as schools experienced outbreaks. While the winter break was extended for an additional week, and schools closed again for a two-week period in May 2021, by the summer of 2021 — labelled the "Best Summer Ever" by Kenney — the UCP government appeared to be determined to return to normal as quickly as possible. Throughout, the constant refrain was that children (other than those in high-risk groups) were not at significant risk from COVID-19 and rarely became severely ill. This narrative left out school-based education workers, who were in close contact with students throughout their workday, as well as children's families, and the risks to children as evidence mounted that children might be vulnerable to the effects of long COVID, even if their initial illness was mild. Children were treated as independent actors, not as people who by their nature live in close contact with adults in their communities.

In February 2022, the mask mandates in schools were removed, and school divisions were told they did not have the authority to defy the order. Critics speculated the decision to lift nearly all of Alberta's remaining COVID-19 restrictions was guided by political interests and not scientific evidence. A lawsuit launched by the Alberta Federation of Labour and a group of families with immunocompromised students revealed internal documents showing that Hinshaw had, in fact, informed cabinet that it was too early to remove restrictions, but the UCP overruled her advice. Documents also revealed that evidence in Alberta confirmed what advocates for school restrictions repeatedly said: Schools were a driver of community outbreaks in Alberta. Once again, the UCP's desire to appeal to libertarian elements in the party's base was the primary motivator for their decisions (Alberta Federation of Labour, 2022). In October 2022, King's Court justice Dunlop found the removal of the restrictions was improper and unreasonable and violated the Public Health Act in making a cabinet decision that overrode the decision-making powers of the CMOH. Additionally, Dunlop determined that the minister had overstepped her authority over school boards' local decision-making regarding masking in schools (Franklin, 2022). In response, in November 2022, Premier Danielle Smith announced regulatory changes barring school authorities from implementing mask mandates or closing schools to in-person learning, regardless of the severity of local outbreaks.

"If the Government Backs Down on This One"

In 1993, as president of the Alberta Association of Taxpayers, Kenney explicitly stated that education cuts were necessary to undermine the ATA, stating that if the government did not win in the fight with teachers, other public sector unions would become "more confident and more militant" (Serres, 1993). By the time Kenney became premier in 2019, he clearly saw the ATA as a central keystone in Alberta's public sector unions, as well as a formidable obstacle to the UCP's aims of remaking Alberta's education system.

Teachers as a profession, and in particular, their unions, have long been framed as enemies by both austerity-minded governments looking to cut public services and privatize education systems, and social conservatives who view teachers as significant actors supposedly indoctrinating children. The latter tend to see teachers (and the faculties of education that train them) as dangerously progressive. Their status as professionals with the ability to make pedagogical decisions grounded in their professional expertise, rather than as simply deliverers of curriculum content, is seen as threatening by those holding an authoritarian view of parental rights in education.

The UCP's attempts to gain control over areas of teachers' working lives can be understood as an attempt to gain power over both their labour and the futures of Alberta's children. The UCP began waging war on teachers and their association on multiple fronts, including shutting out the ATA from the curriculum process and refusing to engage with the association on COVID-19 policy.

The consolidation of central bargaining through the Teachers' Employer Bargaining Association in 2016, which negotiates the majority of items in teachers' collective agreements[4], significantly expanded the Government of Alberta's authority over teachers' conditions of employment, including salaries. As the UCP took office and made clear their intent to cut public sector wages across the board, teachers became an immediate target (although this would expand with the passage of the Public Sector Employers Act, giving the minister more ability to direct public sector wages across publicly funded institutions).

Teachers continued, however, to have autonomy in a number of other key areas; this was the case, in particular, for teachers who are part of the ATA, which functioned as both a labour union and professional association. The UCP's attempts, immediately after taking office, to take power over ever-increasing aspects of the teaching profession, and therefore over teachers themselves, began with efforts to take over management of the Alberta Teachers Retirement Fund (ATRF), placing it under the control of the Alberta Investment Management Corporation (AIMCo). Teachers opposed the transfer of their pensions not only because it shifted the fund's control to a body that had shown less success in investment management than had the teachers' own fund management, but also because teachers would have no authority over the fund's direction. Teachers eventually wrested back control through the courts, but the UCP's moves to assert power over the profession continued.

Education Minister LaGrange also began stripping away other areas of teachers' professional self-governance through the ATA as well. Along with its function as a labour union, the association supports teachers in areas such as professional development and professional practice. The minister set her sights on taking over control of teacher discipline.

Teacher disciplinary processes, as with many professions, are complex[5], with separate roles for the employer, the ATA as a professional body, and the minister of education in their power to grant or revoke teaching certificates. Many members of the public are understandably unfamiliar with these roles, a situation the minister took advantage of when a teacher was charged for a sexual assault that had occurred decades earlier. LaGrange accused the ATA of failing to inform police following a disciplinary hearing and blamed the association for not revoking the man's teaching certificate (French, 2021). Although it quickly came to light that the ATA had informed the minister at the time and that it was the ministry that had failed to act, the incident became an excuse to begin stripping away the ATA's self-governance powers and undermine confidence in teachers.

Following the passage of the Education Statutes (Students First) Amendment Act, Alberta Education launched a publicly accessible database of all persons holding teaching certificates, going back to 1954. Critics immediately began to identify significant issues with the database, including the listing of long-dead teachers, whose names were expected to show up as inactive, but frequently did not. Other concerns raised were the database exposed the former names of teachers, potentially identifying transgender teachers for harassment and putting abuse survivors at risk.

LaGrange also announced a commissioner would be appointed to oversee teacher discipline and develop a new, province-wide teacher code of conduct. Alberta already has three forms of provincial regulatory standards: Teaching Quality Standard, Leadership Quality Standard, and Superintendent Quality Standard. These were in addition to the ATA's own Professional Code of Conduct. While the minister stated this was to create one unified code and disciplinary system, the province's extant disciplinary process for private and charter school teachers was significantly less transparent than the existing ATA process. The ATA's code also included elements that, given the track record established in the curriculum process and the elimination of Bill 24, were potentially at risk, including rules protecting student confidentiality.

In September 2022, a new commissioner was appointed and a consultation process initiated to develop a new code. Notably, the appointed commissioner had no substantial experience with education law or policy (Johnson, 2022). The survey, as part of the announced process, included concerning questions about student confidentiality protections and the possibility of opening up complaints that would go beyond the scope of an established code. The trajectory was clear: To undermine the profession, to strip the power of teachers' representation as workers, and to further shift the balance of power and authority away from education experts toward a parent-as-consumer model of education.

Conclusion

On May 18, 2022, after scraping by a leadership review vote by just 51.4 per cent, Kenney announced he was resigning as premier, staying on until the party selected a new leader. While Albertans might have hoped for a substantial backing down from the drastic overhaul of education launched in 2019, it became clear as the leadership race wore on that, despite their stances on the curriculum process, the seven candidates were largely committed to expanding Alberta's choice model.

At an event hosted by none other than Parents for Choice in Education, candidates floated various proposals to expand education choice. Notably, Danielle Smith — who would months later win the leadership — endorsed Arizona's voucher model (Parents for Choice in Education, 2022). Her remarks on that occasion made clear she did not understand that Alberta's system already functions almost identically to Arizona's voucher model (and, in fact, private school parents there were receiving 64 per cent of the public per-student grant, as compared to Alberta's 70 per cent). The re-emergence of the voucher conversation, however, suggests a continued path toward increasing funding for private schools and the opening of doors to things like low-quality online charters, as seen in the US. These proposals are central to the UCP's discourse on education funding despite being unpopular with the majority of Albertans, whom polls consistently show as believing too much public money is given to private schools, with about two-thirds wanting that money redirected back to the public system. Smith's choice to keep LaGrange on as education minister after becoming premier signalled she had no intention of backing down. As with the UCP platform under Kenney in 2019, Smith is counting on education choice proposals mobilizing her socially conservative base, while education stays largely off the radar of the majority of Alberta voters. However, Albertans' anger over the UCP's direction on education, especially with the new curriculum, suggests that education will be a significant issue in the 2023 election.

References

Alberta Education (2010). *Inspiring Education: A Dialogue with Albertans.* Government of Alberta. https://open.alberta.ca/dataset/45370ce9-3a90-4ff2-8735-cdb760c720f0/resource/2ee2452c-81d3-414f-892f-060caf40e78e/download/4492270-2010-inspiring-education-dialogue-albertans-2010-04.pdf

Alberta Federation of Labour (2022, July 25). *Court documents reveal how Kenney prioritized his own interests over Albertans' when he removed masking in schools*

[Press release]. https://www.afl.org/court_documents_reveal_how_kenney_prioritized_his_own_interests_over_albertans_when_he_removed_masking_in_schools

An Act to Amend the Alberta Bill of Rights to Protect our Children, Alberta (2015). Retrieved from https://canlii.ca/t/53lcs

An Act to Support Gay-Straight Alliances, Alberta (2017). Retrieved from https://canlii.ca/t/53l43

Banack, C. (2017, Nov. 10). How Bill 24 may hurt the NDP and help Jason Kenney. *Folio*. https://www.ualberta.ca/folio/2017/11/commentary—how-bill-24-may-hurt-the-ndp-and-help-jason-kenney.html

Berner, A. (2019, June 25). Why Alberta is a model for education reform. *Thomas B. Fordham Institute*. Retrieved from https://fordhaminstitute.org/national/commentary/why-alberta-model-education-reform

Blokhuis, J. C. (2020). The Supreme Court of Canada and the Convention. In *A Question of Commitment: The Status of Children in Canada* (2nd ed., pp. 161–178). Wilfred Laurier University Press.

Blue Ribbon Panel on Alberta's Finances (2019). *Report and Recommendations*. Government of Alberta. Retrieved from https://www.alberta.ca/mackinnon-report-on-finances.aspx

Braat, T. (2021, June 17). Alberta UCP curriculum advisor denies Kamloops burial site part of Indigenous genocide. *CityNews*. https://edmonton.citynews.ca/2021/06/17/alberta-ucp-curriculum-advisor-denies-kamloops-burial-site-part-of-indigenous-genocide/

Bruch, T. (2020, Aug. 27). Calls grow to remove Alberta curriculum advisor over "racist" First Nation remarks. *CTV News*. https://calgary.ctvnews.ca/calls-grow-to-remove-alberta-curriculum-advisor-over-racist-first-nation-remarks-1.5082528

CBC News (2016, June 15). Alberta begins six-year overhaul of education curriculum. *CBC News*. https://www.cbc.ca/news/canada/edmonton/alberta-begins-six-year-overhaul-of-education-curriculum-1.3636519

CBC News (2022, Aug. 10). Alberta teachers grill UCP leadership candidates about new curriculum, education plans. *CBC News*. https://www.cbc.ca/news/canada/calgary/ucp-leadership-race-alberta-teachers-conference-2022-1.6547655

Convention on the Rights of the Child, United Nations (1989). 1577 U.N.T.S. 3.

Eaton, S. (2021, April 3). Analysis of Plagiarism in the Draft Alberta K-6 Curriculum. *Learning, Teaching, and Leadership*. https://drsaraheaton.wordpress.com/2021/04/03/analysis-of-plagiarism-in-the-draft-alberta-k-6-curriculum/

Education Act, Alberta (2012). Retrieved from https://canlii.ca/t/55j5j

Franklin, M. (2022, Oct. 27). Province overruled Alberta's CMOH in masking order, ruling finds. *CTV News*. https://beta.ctvnews.ca/local/calgary/2022/10/27/1_6128122

French, J. (2017, Oct. 6). Special interest groups wade in to school trustee elections. *Edmonton Journal*. https://edmontonjournal.com/news/local-news/special-interest-groups-wade-in-to-school-trustee-elections

French, J. (2021, Dec. 9). Alberta Teachers' Association to lose disciplinary role, province announces. *CBC News*. https://www.cbc.ca/news/canada/edmonton/alberta-teachers-association-discipline-1.6279621

Gerein, K. (2019, July 9). UCP had much to celebrate, lots to regret in gruelling first legislative session. *Edmonton Journal*. https://edmontonjournal.com/news/politics/keith-gerein-ucp-had-much-to-celebrate-lots-to-regret-in-gruelling-first-legislative-session

Giovannetti, J. (2019, Sept. 6). Guesses guide pre-emptive school spending cuts in the absence of an Alberta budget. *Globe and Mail*. https://www.theglobeandmail.com/canada/alberta/article-guesses-guide-pre-emptive-school-spending-cuts-in-the-absence-of-an/

Government of Alberta. *Department of Education Ministerial Order #001/2013: Student Learning*. Retrieved from https://open.alberta.ca/publications/education-ministerial-order-001-2013

Government of Alberta. *Department of Education Ministerial Order #028/2020: Student Learning*. Retrieved from https://open.alberta.ca/publications/ministerial-order-on-student-learning-2020

Harrison, T. W., & Kachur, J. L. (eds.). (1999). *Contested Classrooms: Education, Globalization, and Democracy in Alberta*. University of Alberta Press/Parkland Institute.

Jason Kenney Leadership Campaign (2017). Defend school choice in Alberta. *jasonkenney.com*. Retrieved from https://archive.ph/roClv

Joannou, A. (2020, Aug. 4). Alberta to require school staff, grades 4 to 12 to wear masks in common areas; Edmonton Public Schools mandates masks in classrooms. *Edmonton Journal*. https://edmontonjournal.com/news/local-news/back-to-school-alberta-covid-19

Johnson, L. (2020, March 16). COVID-19: Alberta cancels all school classes, closes licensed daycares. *Edmonton Journal*. https://edmontonjournal.com/news/local-news/covid-19-alberta-closes-all-schools-daycares

Johnson, L. (2020, May 20). Schools to get $250 million from Alberta's capital plan for repairs, projects. *Edmonton Journal*. https://edmontonjournal.com/news/politics/alberta-education-plan-lagrange

Johnson, L. (2022, Sept. 23). Alberta education minister appoints first commissioner in charge of regulating teachers. *Edmonton Journal.* https://edmontonjournal.com/news/politics/alberta-education-minister-appoints-first-commissioner-in-charge-of-regulating-teachers

Kenney, J. [@jkenney]. (2018, May 5). *If the NDP tried to smuggle more of their politics into the classroom through their curriculum, we will put that curriculum through the shredder. #ucpagm* [Tweet]. Twitter. https://twitter.com/jkenney/status/992961143374200835

Klaszus, J. (2017, Oct. 1). Students count trustee slate faces scrutiny. *The Sprawl.* Retrieved from https://www.sprawlcalgary.com/students-count-trustee-slate-faces-scrutiny-735e00b2a371

LaGrange, A. [@adrianalagrange]. (2020, August 6). *LIVE: With a new Ministerial Order on Student Learning, we are keeping our promises to Albertans to consult more openly with parents and educators on the curriculum, and put an end to "discovery" learning in the classroom. #abed* [Video attached] [Tweet]. Twitter. https://twitter.com/adrianalagrange/status/1291403583091380224

Lamoureux, M. (2016, May 14). *Protest groups clash over Bill 10 at Alberta legislature grounds. CBC News.* https://www.cbc.ca/news/canada/edmonton/bill-10-rallies-for-against-edmonton-1.3583002p

McGillivray, A. (2011). Children's rights, paternal power and fiduciary duty: From Roman law to the Supreme Court of Canada. *The International Journal of Children's Rights* 19(1), 21–54. https://doi.org/10.1163/157181810X527996

McIver, R. [@ricmciver] (2019, July 6). *It feels like school is out after 7 weeks in the legislature. New MLA's started a new "tradition" of a walk in the fountain. A little fun before we head to our families & constituencies for summer. #ucp.* [Tweet]. Twitter. https://twitter.com/RicMcIver/status/1147486938661707776

Northern Alberta Reading Specialists' Council (2021, June 10). *The Northern Alberta Reading Specialists' Council Response to the Alberta draft English Language Arts and Literature Kindergarten to Grade Six curriculum.* Alberta Curriculum Analysis. https://alberta-curriculum-analysis.ca/the-northern-alberta-reading-specialists-council-response-to-the-alberta-draft-english-language-arts-and-literature-kindergarten-to-grade-six-curriculum/

Omstead, J. (2020, March 28). *Alberta education cut expected to lay off thousands during pandemic. CBC News.* https://www.cbc.ca/news/canada/edmonton/funding-reduction-alberta-k-12-covid-1.5513803

Parents for Choice in Education (2012, Nov. 20). ASBA defeats sexual orientation motion. *Parents for Choice in Education.* Retrieved from https://www.parentchoice.ca/asba_defeats_sexual_orientation_motion

Parents for Choice in Education (2022, Aug. 10). 2022 PCE's Education forum for UCP leadership candidates [Video]. YouTube. https://youtu.be/FZS4fdJ49wo

Peck, C. (2020, Aug. 13). Reflections on the ministerial order on student learning released by Minister Adriana LaGrange on Aug. 6, 2020. Carla L. Peck Blog. https://carlapeck.wordpress.com/2020/08/13/reflections-on-the-ministerial-order-on-student-learning-released-by-minister-adriana-lagrange-on-august-6-2020/

Peck, C. (2022, Jan. 1). The absurd UCP curriculum. *Alberta Views*. https://albertaviews.ca/the-absurd-ucp-curriculum/

Press Progress (2017, Sept. 1). *Homophobic advocacy group tells supporters to vote for leader of Alberta's new conservative party*. Press Progress. https://pressprogress.ca/homophobic_advocacy_group_tells_supporters_to_vote_for_leader_of_alberta_new_conservative_party/

Press Progress (2019, Sept. 18). *Kenney government appointed foreign, Koch-funded researcher to rewrite Alberta's education curriculum*. Press Progress. https://pressprogress.ca/kenney-government-appointed-foreign-koch-funded-researcher-to-rewrite-albertas-education-curriculum/

Rebel News (2016, May 30). Ezra Levant's FULL interview with Jason Kenney at #cpc16 [Video]. https://youtu.be/OC4au3rUasE?t=559

School Act, Alberta (2012). https://canlii.ca/t/53j4d

Serres, C. (1993). The ATA plays chicken with Ralph. (Cover story). *Alberta Report / Newsmagazine 20*(48), 6.

Thomson, S. (2016, Sept. 4). Education minister demands assurances from private Christian schools on GSAs. *Edmonton Sun*. https://edmontonsun.com/2016/09/03/education-minister-demands-assurances-from-private-christian-schools-on-gsas

Toy, A. (2021, Aug. 16). LaGrange cleared of conflict of interest with procuring masks for Alberta schools but questions remain. *Global News*. https://globalnews.ca/news/8117369/lagrange-cleared-of-conflict-of-interest-with-procuring-masks-for-schools-unanswered-questions-remain-ethics-commissioner/

United Conservative Party (2019). *Alberta Strong and Free: Getting Alberta Back to Work* [platform document]. Retrieved from https://albertastrongandfree.ca/wp-content/uploads/2019/04/Alberta-Strong-and-Free-Platform-1.pdf

Weisberg, N. (2019, Aug. 22). New panel will review Alberta's revamped curriculum. *CTV News*. https://edmonton.ctvnews.ca/new-panel-will-review-alberta-s-revamped-curriculum-1.4561128

NOTES

1 While Alberta Education describes charter schools as a form of public school, they are operated by non-profit societies and function much like private schools (with the exception of being prohibited from offering religious education) but receive 100 per cent of the per-student grant. This quasi-public model is an example of the privatization of education introduced via neoliberal education reforms.

2 Von Heyking later criticized the UCP's curriculum process for not being developed collaboratively or consistently.

3 Lloydminster's public and Catholic school divisions use the Saskatchewan curriculum, although they are included in Alberta's 61 public and separate school divisions.

4 While some items are still bargained at local tables, those items are limited to local matters, and so most of the consequential bargaining now takes place at the central TEBA table.

5 School authorities, as the employer, hold authority over human resources processes such as hiring and firing, in which the ATA may play a role as a teacher's union representation in employer processes. The ATA, in its function as a professional body, is responsible for the Professional Code of Conduct and can conduct investigations into complaints, which may result in a hearing with potential outcomes including fines, suspension from the organization, and/or a recommendation to the minister to suspend or revoke a teaching certificate. The Minister of Education is responsible for teaching certificates, which allow the holder to teach in Alberta, and can suspend or revoke a certificate.

CHAPTER 13

ALBERTA IS OPEN FOR BUSINESS: THE RENEWED PUSH FOR HEALTH CARE PRIVATIZATION

John Church

"We could have gone further and deeper into health reform had it not been for COVID. And I think Canadians are now waking up to the reality that we do need fundamental health reform."

— Premier Jason Kenney, September 2022

DURING THE 1990S, the Government of Alberta under the leadership of Premier Ralph Klein pursued an overt policy of increasing the presence of the private sector in health care. The push for increased privatization was embedded within a larger neoliberal political agenda of deficit and debt elimination precipitated by precarious government spending during the 1980s and a significant downturn in the oil and gas market. In health care, deficit and debt elimination was pursued through the creation of 17 regional health authorities and a legislative framework to allow the new regional health authorities to contract for the delivery of surgical services with private corporations.[1] Government also sought savings through the contracting out of ancillary services (e.g. laundry, food and laboratory). While organized medicine publicly opposed cuts to the physician services budget, privately, it collaborated with government to create a framework to facilitate privatization.

Following the election of the United Conservative Party (UCP) in 2019, the Government of Alberta has once again embarked on an agenda of cost cutting, including the privatization of various aspects of public services. In particular, the government is pursuing a multi-pronged strategy to increase the role of private corporations in the health-care system. The major components of the strategy involve a legislative framework to facilitate privatization of service delivery beyond surgical care through direct contracts with private corporations, contracting out for the delivery of surgical services, re-privatization of laboratory services, the delivery of ancillary services and increasing use of e-health. Additionally, the government has pursued legislative changes that affect the governance of health care, especially the role of Alberta Health Services (AHS) and the College of Physicians and Surgeons. It also legislated authority to end the collective agreement with physicians without completing the established collective bargaining process. If successful, the net result of these various policy initiatives will be to shift the nature of the health-care marketplace to realign the balance of power among bureaucratic and professional interests in favour of private market solutions.

Dominant and Challenging Interests and the Realignment of the Health Care Market

In his seminal work on health-care politics, Robert Alford (1975) described a competition in health care among dominant and challenging interests. Physicians are dominant in the Canadian health-care market because of the state-sanctioned occupational monopoly established when universal health-care coverage was introduced, first through publicly-financed, universal coverage of hospital services (1958) and second through the introduction of publicly-financed, universal coverage of physician services outside of hospitals (1968). Increasingly, this monopoly has been challenged by health-care managers (public and private) who are primarily interested in controlling costs through the increased management of the organization and delivery of health-care services. Examples of this can be found in the introduction of regional governance structures and the associated changes in the organization of physician practice. In general, physicians are opposed to the growth of institutionalized medicine (whether public or private) because for them it leads to a restriction on individual clinical autonomy and a resulting decline in the quality of care (Tuohy, 1988; Freddi, 1989, pp. 5-6; Church and Barker, 1998).

Traditionally in Canada, this competition has been characterized by occasional conflict and accommodation between dominant and challenging interests in health care. For the most part, citizens are spectators to this competition. Over time, this has resulted in a gradual weakening of the control that organized medicine has exercised over the institutional context of health care in exchange for continuing exclusive control over the content of medical practice (clinical decision making) (Lomas and Barer, 1986; Tuohy and O'Reilly, 1992, pp. 278-279). Within this context, increasingly, Canadian governments have been looking to private corporations as managerial partners with a shared interest in improving the efficiency and reducing the costs of service delivery in health care (Fuller, 1998).

Despite popular belief, the existing federal legislative health-care framework does not prevent private corporate involvement in the delivery of health-care services (Boychuk, 2008). To the extent that private corporations employ cheaper labour substitution and more efficient service delivery modalities or are perceived to do so by political decision makers, they are in alignment with the major efficiency goals of government. In essence, governments have been intensifying their efforts to shift the balance of power among political/bureaucratic, professional and private markets[2], which has traditionally favoured a mutually beneficial relationship between political/bureaucratic and professional labour markets in Canada to increase the role of private labour markets at the expense of professional labour markets. In short, governments are interested in weakening the power of health professionals in the marketplace, especially physicians, in favour of third-party corporate interests.

Nowhere is this more apparent than in a renewed policy agenda for increased health care privatization in Alberta that is reminiscent of a similar agenda put forward during the 1990s (Freidson, 1990; Church and Smith, 2006). Simultaneous efforts by successive governments to create a legislative framework for increased privatization of the delivery of services while centralizing governance away from local communities have positioned Alberta to become the vanguard for health care privatization in Canada.

Privatization of Health Services Delivery

Historically, Alberta favoured a significant role for the private sector in health care (Marchildon, 2016). Having said this, the political popularity of compulsory public health insurance and the infusion of federal money was too great for Alberta to resist. Subsequently, the philosophical preference for private market solutions in health care was overwhelmed by the political benefits of expanding a highly popular public service fuelled by significant revenues from the oil and gas sector. However, by the late 1980s, economic downturn and repeated deficit spending led to a renewed focus on the role of government in the delivery of public services (Church and Smith, 2008). As part of a neoliberal political narrative, the themes of personal responsibility and choice would resonate throughout the 1990s. However, added to this would be a more general push for an increased role for the private sector in the delivery of health services[3] (Denis, 1995; Renouf, 1995).

For example, in a government report entitled Starting Points, "putting the customer first" was followed by the statement: "Underlying this goal is a clear direction that the customer must be at the focus of all decisions about the delivery of health services" (Alberta, 1993, p. 12).

The report also identified maximum access and maximum choice as key criteria in the reformed system (Alberta, 1993, p. 13):

> Consumers will be able to access the health system through a range of providers, institutions and locations; and...consumers will be able to choose from an array of health services and determine who will provide them and in what location....

> However, consumers will recognize that they will need to pay for services considered *non-essential* [their emphasis] under a newly created definition of basic health services.

Finally, the report encouraged newly created health regions "to work with non-profit associations and the private sector to establish joint venture (sic) or autonomous facilities" (Alberta, 1993, p. 21).

The government pursued this policy approach in several ways. First, it developed a set of principles (see Appendix 1) around which to negotiate with the federal government for clarity on the limits of privatization under the Canada Health Act. This negotiation occurred within the backdrop of a federal government interested in enforcing the provisions of the Canada Health Act, particularly the provisions relating to private clinics, after having taken a historically "passive stance" to enforcement of the legislation (Choudhury, 2000, pp. 51-53; Armstrong, 1999; Church and Smith, 2013, p. 488).

Despite this, resolution of the issue was crucial to enabling an expanded role for the private sector under the new regional health system (discussed below) that was introduced in Alberta in 1994 (Plain, 2000, p. 7; Health Canada, 1995; Church and Smith, 2008). From the point of view of the Alberta government, private non-hospital surgical clinics were seen as a potential option for contracting out of service delivery by regional health authorities operating in a financial climate where public resources would be significantly reduced. In particular, reductions in acute care expenditures in Calgary resulting in closure of three hospitals by 1997, including the demolition of one and sale of the other two to for-profit corporations, meant the capacity of the public health-care system to deliver surgical services was significantly reduced, resulting in increasing wait times, a major political issue (Taft and Steward, 2000, p. 29).

Second, Alberta Health collaborated with the Alberta Medical Association (AMA) to develop guidelines to inform government policy on the role of the private sector in health care within the boundaries of the CHA. The AMA developed a set of principles to aid in determining core services (what services would be publicly funded), including supporting an expanded private market for health services: "patients will have the right to purchase, and physicians will have the right to provide, non-core health services"[4] (AMA, 1994).

To achieve this, the AMA proposed Clinical Practice Guidelines (CPGs) that, in theory, would allow the government to realize its goal of reducing physician services expenditures through de-insurance while allowing physicians to bill patients privately for de-insured services. In effect, this represented a process of "passive" de-insurance (Health Law Institute, 1995, p. 61).

The government and the AMA also negotiated a principle related to the timeliness of access to MRIs that allowed patients to pay privately for an MRI if they could not access it publicly in a clinically acceptable time frame. Thus, through both CPGs and the timeliness principle, physicians continued to play a major role in what services would be publicly covered, what could be paid for privately, and the conditions under which MRIs (specifically) could be paid for privately (Plain, 2000, pp. 25-27).

Both the CPGs and timeliness principle were reflected in Principle 11 (revised) of the set of principles developed in negotiation with the federal government (cited in Plain, 2000, p. 10)::

> "All medically necessary services are insured services. A service is non-insured when deemed to be not medically necessary in that it does not meet a Clinical Practice Guideline (CPG) which would include criteria of medical condition, appropriate time frame, etc., or is otherwise determined not medically necessary through a medical condition... In the CPG situation, the patient pays the full cost of the procedure provided the patient is informed why the particular service does not meet the CPG and that the service would be covered if it met the CPG.

While this Alberta approach was confirmed in 2000 with the passage of the Health Facilities Act (Alberta, 2000), which created a framework for regulating private surgical clinics and a process for approving contracts with regional health authorities (RHAs), the federal government continued to have concerns about some aspects of the legislation that allowed for the sale of enhanced services and the potential for queue jumping, both of which might violate the CHA principle of accessibility. In correspondence between federal Health Minister Allen Rock and Alberta Health Minister Halvar Jonson, Rock noted that: "To permit for-profit facilities to sell enhanced services in combination with insured services would create a circumstance that represents a serious concern in relation to the [medicare] principle of accessibility" (Health Canada, 2000, p. 2).

While this intergovernmental battle ensued, the newly created Calgary Regional Health Authority (CRHA) negotiated contracts with a number of private surgery clinics, specifically in the surgical areas of ophthalmologic and orthopedic surgery (Steward, 2001; Armstrong 2000). Some of these initiatives came about as a direct result of the government's decision to demolish or mothball existing public health-care facilities and involved physicians who arguably were in a conflict of interest because of their direct involvement with the CRHA. The contracts were 10 per cent higher than services provided through public facilities. Although the public-private eye surgery market would survive, events were unfolding that would cause the market opportunity for one high profile orthopedic initiative, the Health Resources Group

(HRG), to evaporate. HRG went bankrupt in 2010, leaving the provincial government and taxpayers on the hook for $2.8 million (Derworiz, 2010; Taft and Steward, 2000, p. 48; Steward, 2001; Gibson and Clement, 2012; Lynch, 2013).

As a follow-up to the Health Protection Act, the government reinforced the push for increased contracting out through the release of the Mazankowski Report (2002) and a new policy framework, "The Third Way" (2005), that softened the previous restrictions on the sale of enhanced goods and services and increased the opportunity for Albertans to use private health-care insurance to pay for an expanding number of de-insured services (see Appendix 2). Of particular interest are principles seven and eight, which call for expanded consumer choice through the private sector (Alberta, 2006, pp. 2-3).

In this enabling environment, private membership clinics that offer both publicly insured and privately insured health-care services, but require a membership fee, began to emerge. As with the previous forays to increase the role of the private sector in health care, these initiatives attracted the attention of federal and provincial officials because of their potential to violate the CHA. However, repeated reviews of these clinics have failed to clearly determine whether or not violations have occurred[5] (Graff-McRae, 2017).

The overall legacy of the Klein-era policies was a significant increase in the number of private surgical clinics in the province. Between 1993 and 1999, private clinics increased from 36 to 49 within an enabling legislative framework (Armstrong, 2000). Klein's successor, Ed Stelmach, continued to de-insure services including podiatry, home care and pharmaceutical coverage for seniors and air ambulance, physiotherapy, chiropractic and massage therapy more generally (Graff-McRae, 2017, p. 10). Neither Allison Redford nor Jim Prentice, who succeeded Stelmach as premier, reversed this trend. However, the growing blurring of the lines between private and public provision of health-care services did result in a public inquiry into queue jumping during Redford's short time as premier (CBC, 2013; Alberta, 2013).

The political landscape in Alberta underwent a major shift in 2015 with the historic election of the New Democratic Party (NDP), a centre-left political party.[6] This marked the end of a 44-year conservative political dynasty. However, this proved to be only temporary, with the Conservatives returning to power again in 2019 under the banner of the United Conservative Party (UCP). As was the case during the early 1990s, Alberta's economy was sputtering because of a chronic downturn in the oil and gas industry. While the unstable leadership vacuum that occurred after Klein's departure and a booming economy derailed the privatization agenda, the election of the contemporary UCP on a platform of fiscal restraint has revitalized privatization as a solution to Alberta's fiscal woes.

While in opposition, UCP leader Jason Kenney signaled the UCP would support increased privatization in health care as a way to reduce wait times and save money (Bennett, 2019). During the election campaign, he evoked the political rhetoric of the 1990s, insisting that Alberta had a spending as opposed to a revenue problem and that significant expenditure reductions were the best means to return Alberta to fiscal prosperity (Markusoff, 2016). Part and parcel with the expenditure reduction agenda was the idea that private market solutions would lead to cheaper and more efficient delivery of services. In health care, privatization was seen as a solution to the chronic issue of wait times (Kanygin, 2020). The UCP's party platform identified several avenues to save money in health care including: greater use of private clinics to address the growing waitlists for a variety of procedures, greater use of nurse practitioners to increase access to services, contracting out of AHS laundry services, cancelling the public consolidation of laboratory services, and conducting a comprehensive review of AHS.

After being elected, the UCP moved aggressively to implement the expenditure reduction agenda. Following through on its health-care pledges, the government has taken a multi-pronged approach to reform. During its first legislative foray in the newly constituted provincial legislative assembly, the UCP passed legislation giving it the power to enter into contracts with third-party corporations for the provision of health-care services. This change was supported by two major external reviews.

The first, Report and Recommendations of the Blue Ribbon Panel on Alberta's Finances (Alberta, 2019a) recommended contracting out of surgical and other health-care services to reduce wait times and increase access (p. 3). The report also recommended greater use of licensed practical nurses (LPNs) as a means to improving access to services and saving money (p. 29). Finally, the report recommended moving away from paying physicians through fee-for-service (Recommendation 4), including legislative changes to achieve this goal (p. 6).

The second external review, Alberta Health Services Performance Review, was conducted by Ernst and Young (2019). That review confirmed the findings of the Blue Ribbon Panel report by recommending an increased use of private surgical facilities and the increased use of LPNs and nurse practitioners (pp. 28, 44, and 60).

In 2020, the government passed The Health Statutes Amendment Act that allows the government to enter directly into contracts with physicians or other third-party service providers. (Health Statutes Amendment Act, 2020, p. 6) The stated rationale for the provision is to allow more physicians to shift from fee-for-service to salary as a method of payment. (Doherty, 2019; Bellefontaine, 2020) In theory, this weakens the position of the AMA as the sole collective bargaining unit for all practicing physicians in the province. It also puts the government in control of decisions about contracting health services provided through AHS (more on this below).

Additionally, the government indicated it will double the percentage of surgeries provided by private for-profit clinics from 15 per cent to 30 per cent over the next three years [7](Parsons 2020). This directly contradicts legislation proposed by the previous NDP government that would have expanded existing capacity to provide surgery through public facilities (Russell, 2019). In addition to the current range of ophthalmalogic and major joint surgeries, the government is also interested in expanding privately provided surgeries to encompass mastectomies, hernia repair and some gynecological procedures (Belfontaine, 2020b). Included in this was an investment of $100 million to upgrade existing public health operating rooms to offer surgeries in underutilized facilities outside of Edmonton and Calgary and contract with private surgical facilities (Hardcastle, 2020, p. 2). Finally, the government announced approximately 11,000 health-care workers in nursing, laboratory, linen and cleaning services will be laid off, largely through outsourcing of service delivery (Russell and Rusnell, 2020).[8]

Again, following through on an election promise, the UCP cancelled plans to build a new $595 million centralized "superlab" public laboratory facility to consolidate all laboratory services under the control of AHS. The project was based on a review of laboratory services by the province's Health Quality Council and was championed by the previous NDP government. In essence, the government rejected creating a consolidated, public laboratory service for a service that is publicly funded, but privately operated (Trynacity, 2019). As noted by then UCP opposition leader Kenney during the election campaign, "The NDP priority is a government monopoly in everything including back-end lab services, which are already being done effectively by Albertans through an Alberta private company" (CBC, 2019). Because the initiative was already two years into implementation, the government lost $23 million that had already been spent on the project, plus another $12 million to restore the construction site (Cook, 2020).

Within the context of this emerging policy environment, health-sector actors have responded in various ways. In April 2021, AHS awarded an 11-year contract to K-Bro Linen Systems for the provision of laundry services for all AHS facilities, both urban and rural (previously only urban facilities were serviced) (Alberta Health Services, 2021a). It also contracted out for the provision of (public-facing) retail food services for Edmonton and Calgary facilities to move from 40 per cent to 100 per cent private delivery (Alberta Health Services, 2022).

In April 2021, AHS launched the Alberta Surgical Initiative (ASI) to increase the system capacity to provide surgery through contracts with private clinics (Alberta Health Services, 2021b).[9] According to AHS spokesperson Dr. Francois Belanger, "We already collaborate with partners to perform a number of surgical procedures at our

CSFs [Chartered Surgical Facilities]," adding: "These are publicly-funded procedures performed in a private facility. It's a way for AHS — and many other health systems across Canada — to expand surgeries and reduce wait times" (AHS, 2021).

At the same time, the government announced new contracts with two private companies, Holy Cross Surgical Services and Vision Group Canada, to provide 20,000 ophthalmology procedures in Calgary and at least 10,000 cataract surgeries in Edmonton chartered surgical facilities. In commenting on these new contracts, Health Minister Jason Copping argued, "We need to build both public and private surgical capacity and reduce wait times. Utilizing private clinics in Calgary and Edmonton for ophthalmological surgeries, particularly cataract surgeries, will free up space for other surgeries in AHS facilities" (Alberta, 2022).

In response to the ASI, two new proposals have emerged. The first, a $200-million proposal to build the largest state-of-the-art private surgical facility in Alberta history has been proposed by a group of Edmonton orthopedic surgeons, a land developer and lobbyists. If approved, the facility will provide potentially an additional 10,000 non- emergency orthopedic surgical procedures for the greater Edmonton area. The proposal, which would involve the use of non-unionized staff, would increase the number of annual surgeries by 60 per cent (Rusnell and Russell, 2020). The second proposal, which was approved by the government in June 2022, is a partnership between the Enoch Cree Nation and a private-sector company, Surgical Services Inc., to provide up to 3,000 uncomplicated hip and knee surgeries annually to Indigenous Albertans within an institutional environment that is sensitive to the cultural and health needs of Indigenous Peoples.

In July 2021, the government announced it will fund the integration of nurse practitioners (NPs) into Primary Care Networks.[10] While the Alberta Association of Nurse Practitioners are pleased by this development, they continue to advocate for the integration of NPs as independent health practitioners to boost the capacity of primary care beyond the existing complement of family doctors (Goulet, 2021; Alberta, 2021). The proposed labour reduction of 11,000 health workers has been postponed indefinitely due to severe workforce fatigue related to the pandemic and a resulting labour shortage.

Centralization of Governance

During the 1990s, Alberta actively pursued a policy to consolidate governance and encourage privatization in health care. In December 1993, the government laid out its vision for reforming Alberta's health-care system in a document (Starting Points) that described a new governance model in which the ministry of health would distribute resources to regional health authorities (local decision makers), which in turn would make decisions about how to provide health services within their geographic boundaries.[11] While it was pitched as a decentralization of power away from the ministry of health and back to local communities, in fact, it involved devolving some responsibilities away from the ministry while at the same time centralizing decision making away from the existing 200 local boards to regional boards with broader responsibilities (Church and Smith, 2008). The logic behind creating these regional boards flowed from New Public Management theory, which basically recommended government be run more like the private sector. In essence, the Alberta government wanted to create regional health authorities that would make service delivery decisions within fixed budgets and thus be able to better control costs in the overall health-care system. In practice, RHAs developed their own service delivery priorities and incurred annual deficits in pursuit of these priorities (Osborne and Gaebler, 1992; Aucoin, 1995).

Building on this initial restructuring, the government further consolidated decision making in 2004 when it collapsed the existing 17 regional health authorities into nine geographically larger health authorities. The rationale for this move was recognition that the current boundaries were not realizing potential economies of scale, something that had been highlighted through reviews in 1996 and 1997 of service patterns in the newly created health regions. In short, rural regions were sending significant numbers of patients into the two major urban regions (Calgary and Edmonton) because they didn't have sufficient capacity, and some regions were mismanaging their budgets by running deficits and overspending. The existing funding model did not allow the money to follow the patients when they crossed health region borders (Plain, 1996; Plain, 1997).

Klein's successor, Stelmach, further consolidated the nine regions into a single provincial health authority, Alberta Health Services (AHS), in 2008. As then health minister Ron Leipert (2009, p. 1) explained it:

> Many reasons for this particular decision have been cited, but in my view the purpose was to reverse the siloed and fragmented approach to the delivery of health care that had developed in Alberta — not by any devious means, but by evolution. This created barriers to good care that various entities had originally been created to provide. These silos did not just develop between cancer, mental health and other areas, but also between regions. Initially, the desire was to promote healthy competition between regions in their coordination of the delivery of acute care, public health and continuing care. And individual regions did become leaders in certain fields of delivery. But what struck me in taking over this portfolio … was the reluctance to adopt winning strategies developed in other regions.

The inaugural CEO of AHS, Stephen Duckett, offered this broader explanation (Duckett, 2010, p. 156):

> An inevitable consequence of regional autonomy is regional variability, labeled as 'regional responsiveness,' is among the arguments for regionalization. The positive side of this is regional innovation and harnessing the benefits of competition...The downside is inequity. And this is what happened in Alberta, in both clinical and non-clinical service choices.

In particular, Duckett characterized the competition between Edmonton and Calgary as a "medical arms race" that did not benefit the province as a whole (Duckett, 2010, p. 156).

Although all of this may have been true, health economist Cam Donaldson's analysis of the nine health regions' performance suggested "Alberta's health regions were already performing well... relative to other provinces and where changes have occurred since, they cannot necessarily be attributed to AHS" (Donaldson, 2010, p. 22). One noteworthy bright spot in all of this is that Alberta has the lowest administrative costs for health care compared to other jurisdictions in Canada. There is little doubt that the consolidation of governance has played a major role in this[12] (CIHI, 2022).

Despite heightened expectations that AHS would resolve many of the problems identified above, continuing conflict between government and AHS decision makers has impeded its ability to fulfill its mandate, punctuated by the firing of Duckett in 2010. This was the beginning of four years of leadership instability for AHS, further exacerbated in 2013 with the firing of the board for failing to comply with the government's concern about the payment of executive bonuses. Subsequently, Duckett's successor, Vicki Kaminski, resigned in 2016 because of excessive government interference in AHS decision making (Church and Smith, 2022, p.p. 174-177).

With the hiring of Verna Yiu as CEO in 2016, AHS would enjoy a new level of stability. Unfortunately, this leadership stability only lasted until the spring of 2022, when Yiu was fired by the government for undisclosed reasons, although AHS's failure to respond adequately during the pandemic was mentioned. Having said this, the coincidence between Yiu's firing and the government's announcement that it was proceeding with further contracting out of health services (previously discussed) is noteworthy. Underpinning all of this has been repeated claims by AHS leadership that the government is meddling in their legislative authority to make operational decisions.

Related to this issue, the government amended the existing legislation governing the roles and responsibilities of AHS as "final decision maker" in 2019 and replaced it with a new accountability framework that clearly recognizes government as the final decision maker on all matters related to health care. In essence, decision making in health has been centralized to the Ministry of Health and possibly the premier's office.

In October 2022, Danielle Smith replaced Kenney as leader of the UCP and became premier of Alberta. During the leadership race, Smith blamed the Chief Medical Officer of Health and AHS leadership for the perceived failure of the response to the pandemic. In November, she fired the Chief Medical Officer of Health and the entire AHS board, part of her intended restructuring of the health system.

The Frontal Assault on Doctors

In tandem with the moves to privatize health services delivery, the government passed three major legislative changes related to the governance of physicians. The Ensuring Fiscal Sustainability Act (Alberta, 2019b) changed the criteria used for issuing physician billing numbers to include geography. In essence, this gives the minister of health the power to require new or currently inactive physicians requesting a billing number to practice in underserved areas of the province. It also empowers the minister of health to cancel existing agreements with the Alberta Medical Association, including the collective agreement (Master Agreement). Finally, the government has also pursued changes to the governance structure of the College of Physicians and Surgeons of Alberta mandating that 50 per cent (previously 25 per cent) of the board of the college must comprise members of the public appointed by the minister of health.

In the wake of these legislative changes, the government cancelled the collective agreement with the AMA in February 2020 and imposed significant funding cuts to the physician services budget. The AMA responded in April 2020 by launching a lawsuit against the government for $250 million in damages. Subsequently, Alberta physicians rejected a proposed new collective agreement in March 2021, even though the minister of health had apologized and backpedaled on a number of key changes (Belfontaine, 2020a; Dormer, 2020; Bennett, 2021).

As for some of the other government actions, a settlement was reached with physicians in late September of 2022 in which the government has committed significant new money to physician services, although compensation is in line with other labour groups. Perhaps of greater significance is the government has effectively walked back some changes it made to collective bargaining and compensation reduction, although the legislation allowing the government to rip up the collective agreement and to enter into direct contractual relationships with groups of physicians remains in place (Bennett, 2022).

Conclusion

Since the 1990s in Alberta, the main strategy pursued by government for health care reform was the streamlining of administration and governance through the elimination of hospital boards and the divestment of direct program responsibilities from the ministry of health. Changes to governance and legislation also facilitated a limited expansion of the role of the private sector in health services delivery. This was facilitated by the creation of a legislative framework to allow contracting out for surgical services.

Efforts to reduce the physician budget were largely abandoned because of an effective public relations campaign by the Alberta Medical Association. The prospect of waging a political war on multiple fronts (hospitals and physicians) was more than the government was willing to risk. In fact, after the dust had settled, both parties, along with the newly created health regions, had established structures for the collaborative management of major health-care issues. This collaborative approach would endure until the election of the UCP in 2019. In essence, the ideological interests of government and the self-interest of physicians came together to develop a set of principles to govern the development of public-private partnerships within the boundaries of the CHA. The creation of AHS in 2008 represented the culmination of a 15-year effort to restructure administration and governance of the health-care system.

The move by the current UCP government to effectively strip AHS of final decision-making authority ensures it does not interfere with the government's plans for privatization of surgical services. As a lobbyist for the Edmonton orthopedic initiative (described above) noted, "their focus as lobbyists was on ensuring the surgeons controlled the facilities operation, something they couldn't do within the public health-care system under Alberta Health Services (AHS)" (Rusnell and Russell, 2020).

In effect, this eliminates any potential legislative barrier to moving forward with further privatizing Alberta's health-care system. As health law expert Ubaka Ogbogu notes (2020):

> While these changes may seem subtle, they affect AHS operations in potentially significant ways. Given that AHS is a party to contracts with private surgical facilities, these changes may be intended to ensure that AHS does not act as a barrier to the government's goal of doubling private delivery of health services.

This has left the government free to pursue a political agenda to expand the role of the private sector in health care.

Where the Conservative government of the 1990s was limited in what it could do by effective opposition from civil society actors backed by an activist federal government, this coalition between Alberta's civil society actors and the federal government does not currently exist. Having said this, various civil society actors including the AMA, the United Nurses of Alberta, the Health Sciences Association of Alberta, the Alberta Union of Public Employees and the Canadian Union of Public Employees and the Official Opposition have criticized the government for its plans to lay off large numbers of staff and contract out service delivery, especially in the shadow of the pandemic (Nash, 2020). However, the current government is more ideologically driven than the Klein government of the 1990s, so the efforts of civil society actors have been less successful.

As was the case during the 1990s, the current interaction between government and organized medicine is best characterized as conflict and accommodation between dominant and challenging interests, including the re-emerging pattern of government retreating from its original position with organized medicine (Tuohy and O'Reilly, 1992; Bennett, 2021). The recently ratified new collective agreement represents an accommodation of the collective interests of physicians.

What is different about the current period is the government appears to be moving forward on its privatization agenda without the collective support of physicians. Whereas during the 1990s physicians collectively did not oppose the government's agenda with some individually (a select few) supporting privatization, during the current period, government has attempted to do an end run around the AMA. Some physicians have publicly expressed concern about the adverse effects that may result from the current legislative agenda. While the arrival of COVID temporarily delayed the government's efforts to move forward on its privatization agenda, the government has not indicated it will abandon legislative changes related to increasing the role of the private sector in health care. If anything, the adverse effects of the pandemic on wait times has caused the government to double down on its commitment to further privatize health service delivery.

During an announcement about funding for privately delivered surgeries on September 2022, Copping re-affirmed the government's commitment to an increased private market presence in health care (quoted in Melgar, 2022):

> Adding more surgery capacity in central Alberta and communities south of Calgary means more Albertans can get their hip and knee replacements and other types of surgeries sooner and live a better life with less pain and limitations. We are funding more surgeries at chartered surgical facilities and hospitals to fulfil our promise to bring down wait times to the waiting period recommended by medical experts.

A growing number of physicians, particularly in rural Alberta, have also indicated their intentions to leave the province. Evidence emerging through the current tracking of physician resources suggests that this is occurring. Having said this, some individual physicians (e.g., Dr. Brian Day) have continued to push for increased access to private health care through private for-profit clinics (College of Physicians and Surgeons of Alberta, 2021; Ramsey, 2020; Dryden and von Scheel, 2020).

Beyond this, the conditions associated with COVID-19 have limited the commitment of the federal government to support local civil society groups in their current fight with the Alberta government over budget cuts and increased privatization. In short, parochial health-care policies are not currently on the political radar screen of the federal government, beyond pandemic response.

One other previous barrier that no longer exists is the lack of technological infrastructure to support an increased role for the private sector. Since the 1990s, the private sector has advanced technological solutions. For example, in 2020, the Alberta government contracted Telus Health to supply an app that allows Albertans to access information relevant to COVID-19, to check symptoms and to book appointments with family doctors and specialists, get prescriptions filled and get referrals for diagnostic imaging. Elsewhere in Canada, Telus has been buying up electronic emergency medical record solutions (EMR). Concerns about the privacy and monetization of patient data have been raised by the Alberta College of Physician and Surgeons and other health policy experts ((Fayerman, 2018; CBC, 2020; Turnbull, 2020; Hardcastle and Ogbogu, 2020).

More generally, in 2016, the U.S. health care giant McKesson Corp. purchased the walk-in clinics (Medicentres) and Rexall drugstores from the Katz Group in Alberta, a group that had previously created a vertically integrated network (primary care and drugstores) operating in parallel to the publicly-funded, physician-controlled, primary care market. McKesson is also offering to buy existing physicians' practices in Alberta. Primacy Property Management Inc. has also entered the primary health-care market by offering medical office space inside of or in close proximity to Loblaws grocery stores (Global News, 2011; Rexall, 2016; Primacy, 2021).

So, in light of the private-sector developments that have occurred since the 1990s, changes by the government to health legislation have created a governance structure and legislative framework that will provide significant opportunities for companies such as Telus and McKesson and other entrepreneurs to expand their market share in Alberta's health-care market beyond surgery, especially in the area of primary care (McKesson, 2022). The current government initiative to expand surgeries through the private sector will further facilitate an expansion of the private-sector surgical market. For physicians (collectively), who have traditionally dominated and controlled the health-care market through provincial medical associations and colleges, the emerging legislative framework allows private corporations to challenge this traditional central power position.

Additionally, if health spending accounts are introduced[13], the government will be able to further expand the private market through incremental delisting of services under the guise of creating greater choice. It will also essentially privatize the delivery of services provided by family physicians by tying health spending accounts directly to these services towards the larger stated goal of shifting costs from the public to the private market.[14]

Perhaps the most striking difference between the 1990s and 2021 is that, where large corporations previously lurked in the background, sometimes as investment partners, in the current environment, large corporations are now front and centre with technological and service delivery solutions. Depending on how many physicians are willing to sell their private practices, perhaps exacerbated by an increased outflow of physicians due to the current acrimonious environment, this could permanently shift the health-care market from a bureaucratic-professional marriage to an amicable union between bureaucratic and private sector corporate partners, not unlike what is the norm in the U.S. And perhaps most importantly, this can occur without violating the Canada Health Act. If this occurs, arguably, it will constitute a victory for challenging interests (including the private sector) in Alberta's health-care system that may have a ripple affect across Canada.

References

Alberta. (1962). *Submission to the Royal Commission on Health Services.*

Alberta. (1989). *Premier's Commission on Future Health Care for Albertans, The Rainbow Report: Our Vision for Health, Volume 1.* Edmonton: Queen's Printer.

Alberta. (1993). Health Planning Secretariat, *Starting Points: Recommendations for Creating a More Accountable and Affordable Health System*. Edmonton: Queen's Printer.

Alberta. (2000). *The Health Facilities Act.*

Alberta. (2002). *A Framework for Reform: Report of the Premier's Advisory Council on Health.*

Alberta. (2006, Feb.). *Health Policy Framework.*

Alberta. (2013). *Preferential Access Inquiry.*

Alberta (2019a). *Report and Recommendations Blue Ribbon Panel on Alberta's Finances.*

Alberta. (2019b). *Ensuring Fiscal Sustainability Act.*

Alberta. (2020). *Health Statutes Amendment Act.*

Alberta. (2021, July). *Primary Care Network Nurse Practitioner Support Program*

Information. Version 2. https://open.alberta.ca/dataset/f6e3828a-b05c-4956-a48e-d8223a9edbfe/resource/d42e5f36-5860-4804-b2c6-3487dcb38ddd/download/health-pcn-nurse-practitioner-support-program-information-2021-07.pdf

Alberta. (2022, April 22) *Moving Surgical Recovery into High Gear*. https://www.alberta.ca/release.cfm?xID=8240731BCB85A-ED2C-41E5-D54730000A077AE8

Alberta Health (1995, Oct. 16). *Public/Private Health Services: The Alberta Approach*.

Alberta Health Services (2021a, July 26). *AHS shares plans to transition linen services*. https://www.albertahealthservices.ca/news/releases/2021/Page16075.aspx

Alberta Health Services (2021b, April 13). *AHS seeks to provide more publicly-funded surgeries for Albertans*. https://albertahealthservices.ca/news/releases/2021/Page15861.aspx

Alberta Health Services (2022, Aug. 17). *AHS selects vendors to provide retail food services in Calgary and Edmonton*. https://www.albertahealthservices.ca/news/releases/2022/Page16787.aspx

Alberta Medical Association/ Alberta Health (1994). *MRI Taskforce*.

Alford, R. R. (1975). *Health Care Politics: Ideological and Interest Group Barriers to Reform*. Chicago: University of Chicago Press.

Armstrong, W. (2000). *The Consumer Experience with Cataract Surgery and Private Clinics in Alberta: Canada's Canary in the Mine Shaft*. Consumers Association of Canada (Alberta).

Aucoin, P. (1995) *The New Public Management: Canada in Comparative Perspective*. Montreal, PQ: Institute for Research on Public Policy.

Bellefontaine, M. (2020a, July 6). *New health omnibus bill clears way for privatization, doctors' contracts*. CBC News.

Belfontaine, M. (2020b, Jan. 21). *Mastectomies, hernias, possibilities for private surgical delivery*. CBC News.

Bennettt, D. (2019, Feb. 20). *UCP Leader Jason Kenney wants to explore private health-care options*. Canadian Press.

Bennett, D. (2021, March 23). *Alberta and its physicians move to end ugly feud over fees with new tentative deal*. Canadian Press.

Bennett, D. (2022, Nov. 1). Alberta NDP says B.C. doctor deal a wake-up call to Smith to knock off pseudo-science. *Globe and Mail*.

Boychuk, G. (2008, December). *The Regulation of Private Health Funding and Insurance in Alberta Under the Canada Health Act: A Comparative Cross-Provincial Perspective*. The School of Policy Studies. SPS Research Papers, The Health Series 1(1). Calgary: University of Calgary.

CBC (2013, Aug. 21). *Queue-jumping a fact, Alberta medical inquiry finds*. CBC News.

CBC (2019, March 11). *UCP government would cancel AHS "superlab" in Edmonton*. CBC News.

CBC (2020, March 20). *Telus virtual health-care app touted by Alberta government sparks outcry from physicians*. CBC News.

Choudhry, S. (2000) Bill 11, The Canada Health Act and the social union: The need for institutions. *Osgoode Hall Law Journal* 38(1), 39-76.

Church, J., & Barker, P. (1998), Regionalization of health services in Canada: A critical perspective, *International Journal of Health Services* 28(3), 467-486.

Church, J., & Smith, N. (2006). Health reform and privatization in Alberta, *Canadian Public Administration*, 49(4), 486-505.

Church, J., & Smith, N. (2008). Health care reform in Alberta: The introduction of regional health authorities. *Canadian Public Administration* 51(2), 217-238.

Church, J., & Smith, N. (2013). Health reform in Alberta: Fiscal crisis, political leadership and institutional change within a single-party democratic state. (Chapter 3). In H. Lazar, P.G. Forest, J Lavis, and J. Church (Eds.), *Paradigm Freeze: Why it is So Hard to Reform Health Care in Canada*. McGill-Queen's University Press.

Church, J., & Smith N. (2022). *Alberta: A Health System Profile*. Toronto: University of Toronto Press.

CIHI (2022). Corporate Services Expense Ratio details for Alberta. https://yourhealthsystem.cihi.ca/hsp/indepth?lang=en#/indicator/041/2/C20018/

College of Physicians and Surgeons of Alberta. (2021). *Physician Resources in Alberta Quarterly Update: Jan 01, 2021 to Mar 31*.

Cook, S. (2020, Aug. 21). *Whatever happened to the Alberta superlab?* CBC News.

Denis, C. (1995). The new normal: Capitalist discipline in Alberta in the 1990s. In G. Laxer and T. Harrison (Eds.), *The Trojan Horse: Alberta and the Future of Canada* (pp. 86-100). Montreal: Black Rose Books.

Deveau, S. (2005, July 12). Alberta's Third Way. *Globe and Mail*.

Derworiz, C. (2010, May 12). Clinic rescue costs $2.8M. *Calgary Herald*.

Doherty, B. (2019, Oct. 30). Medical Association slams provincial government over proposed changes to doctor's pay. *Star Calgary*.

Dormer, D. (2020, April 9). *Alberta doctors launch $250M lawsuit against the province*. CBC News.

Dryden, J., & von Scheel, E. (2020, Oct. 1). *More than 160 Alberta doctors were considered "high risk" to withdraw service, documents reveal*. CBC News.

Ernst and Young (2019, December). *Alberta Health Services Performance Review*.

Fayerman, P. (2018, August 3, 2018). Telus jumps into business of health clinics, buys national chain. *Vancouver Sun*.

Freddi, G., & Bjorkman, J.W. (1989). *Controlling Medical Professionals: The Comparative Politics of Health Governance*. Sage.

Freidson, E. (1990). The centrality of professionalism to health care. *Jurimetrics* 30(4), 431-445.

French, J. (2020, April 9). *Doctors association suing Alberta government for alleged breach of charter rights.* CBC News.

Fuller, C. (1998). *Caring for Profit.* Ottawa: Canadian Centre for Policy Alternatives.

Gibson, D., & Clements, J. (2012). *Private Matters: The High Costs of For-Profit Health Services in Alberta.* Edmonton, AB: Parkland Institute. April 10. https://www.parklandinstitute.ca/delivery_matters2

Global News (2011, April 25). *Katz Group sees profits in doctors, not just drugs.* Global News.

Goulet. J. (2021, July 14). More nurse practitioners add a boost to primary care in Alberta. *Lethbridge News Now.*

Graff-McRae, R. (2017). *Blurred Lines: Private Membership Clinics and Public Health Care.* Edmonton, AB: Parkland Institute. Nov. 28. https://www.parklandinstitute.ca/blurred_lines

Hardcastle, L. (2020, March 10). *Recent Health System Reforms: Sustainability Measures or a Push to Privatization?* ABlawg. Calgary: University of Calgary.

Hardcastle, L., & Ogbogu, U. (2020, March 26). Opinion: Alberta's virtual health-care app plagued with problems. *Edmonton Journal.*

Health Canada. (1995, Jan. 6). *Correspondence from Minister of Health Dianne Marleau to Provincial/Territorial Ministers of Health.*

Health Canada. (2000, April 7) *Correspondence from Minister of Health Allan Rock to Alberta Minister of Health Halvar Jonson.*

Health Law Institute. (1995, November). *Health Reform in Alberta: A Health Law Perspective.* Edmonton: University of Alberta.

Kanygin, J. (2020, March 4). *Province details $500M plan for utilizing private clinics to reduce surgical wait times.* CTV News Calgary.

Lazar, H, Forest, P-G., Lavis, J. N., & Church, J. (2013). *Paradigm Freeze: Why It Is So Hard to Reform Health Care in Canada.* McGill-Queen's University Press.

Lavis, J. N. (2002). *Political Elites and their Influence on Health-Care Reform in Canada.* Discussion Paper No. 26, Commission on the Future of Health Care in Canada (The Romanow Commission), 2002. http://publications.gc.ca/collections/Collection/CP32-79-26-2002E.pdf

Lee, J. (2020, July 8). *Alberta physicians express alarm over proposed health-care bill.* CBC News.

Liepert, R. (2008). *Changes to provincial health structures in Alberta.* Retrieved from 14 *http://www.healthinnovationforum.org/article/recentchanges-to-provincial-health15structures-in-alberta/*

Lynch, T. (2013). For-Profit Healthcare: A Lesson from Canada, *Pannon Management Review* 2(2), 11-32.

Marchildon, G. (2016). Douglas-versus-Manning: The ideological battle over medicare in postwar Canada. *Journal of Canadian Studies* 50(1), 129-149.

Markus, J. (2022, Sept. 24). *Jason Kenney says he never intended to "be in this gig for a long time."* CBC News.

Markusoff, J. (2016, July 6). "Kenney vies to become Ralph Klein, reincarnate." *MacLean's Magazine.*

McKesson (2022, March 29). *The future of healthcare: How convenient care clinics are disrupting the primary care market.* https://mms.mckesson.com/resources/patient-care-management/the-future-of-healthcare-how-convenient-care-clinics-are-disrupting-the-primary-care-market

Melgar, A. (2022, Sept. 9). *Alberta government to contract out surgeries to reduce wait times.* CityNews.

Nash, C. (2020, Oct 16). *Alberta unions say public health layoffs are the latest attack on labour movement.* Rabble.ca. https://rabble.ca/labour/alberta-unions-say-public-health-layoffs-are-latest-attack-labour-movement/

Ogbogu, U. (2020, Aug. 17). *Alberta eroding independence of key health institutions.* Healthydebate. https://healthydebate.ca/2020/08/topic/alberta-key-health-institutions/

Osborne, D, & Gaebler, T. (1992). *Reinventing Government.* Boston, MA: Addison-Wesley.

Parsons, P. (2020, Feb. 27). *Budget 2020: Health budget sees doubling of private clinic surgeries, deductibles on seniors drug plan.* CBC News.

Plain, R. (1996). *Working Together: The Reform of Regional Health Authority boundaries within Alberta.* Presented to the MLA Committee on the Review of Regional Boundaries.

Plain, R. (1997). *Meeting the Needs: Inter-regional and Intertemporal Differences in Health Status and Health Service Characteristics Among Regional Health Authorities Within Alberta.* Edmonton: Health Economics Research Centre, University of Alberta.

Plain, R. (2000, March). *The Privatization and the Commercialization of Public Hospital Based Medical Services Within the Province of Alberta: An Economic Overview from a Public Interest Perspective.* Edmonton: University of Alberta.

Primacy. (2021). https://www.primacy.ca/about-us/contact/)

Ramsey, C. (2020, July 10). *42% of Alberta physicians considering leaving the province for work: AMA survey.* Global News.

Rexall. (2016, Dec. 28). *Sale of Rexall Health to McKesson Now Complete.*

https://www.rexall.ca/newsroom/view/19/Sale-of-Rexall-Health-to-McKesson-Now-Complete
Renouf, S. (1995). Chipping away at Medicare: "Rome wasn't sacked in a day." In Gordon Laxer and Trevor Harrison (Eds.), *The Trojan Horse: Alberta and the Future of Canada* (pp. 223-238). Montreal: Black Rose Books.
Rusnell, C., & Russell, J. (2020, Aug. 10). *Proposed $200M private orthopedic surgical facility would be largest in Alberta's history*. CBC News.
Russell, J. (2019, March 14). *New legislation would curb health-care privatization, ban extra billing at private clinics*. CBC News.
Russell, J., & Rusnell, C. (2020, October 13). *Health Services to lay off up to 11,000 staff, mostly through outsourcing*. CBC News.
Steward, G. (2001). *Public Bodies, Private Parts, Surgical Contracts and Conflicts of Interest at the Calgary Regional Health Authority*. Edmonton, AB: Parkland Institute. March 19. https://www.parklandinstitute.ca/public_bodies_private_parts
Smith, D. (2021, June). *Alberta's Key Challenges and Opportunities*. University of Calgary, School of Public Policy. Calgary: Alberta.
Taft, K., & Steward, G. (2000). *Private Profit of The Public Good*. Edmonton, AB: Parkland Institute.
Taft, K., & Stewart, G. (2000, March). *Clear Answers: The Economics and Politics of For-Profit Medicine*. Edmonton: University of Alberta Press.
Trynacity, K. (2019, June 20). *It's official: Alberta government cancels Edmonton superlab*. CBC News.
Tuohy, C. J. (1988). Medicine and the State in Canada: The Extra-Billing Issue in Perspective. *Canadian Journal of Political Science* XXI(2), 267-296
Tuohy, C, & O'Reilly, P. (1992). Professionalism in the Welfare State. *Journal of Canadian Studies* 27(1), 73-92.
Turnbull, S. (2020, March 31). *Doctor warns of risks to patients' privacy amid surge in virtual care*. CTV News.
Rexall. (2016, Dec. 28). *Sale of Rexall Health to McKesson Now Complete*.

Appendix 1:

Public/Private Health Services: The Alberta Approach*

1. Ensure reasonable access to a full range of appropriate, universal, insured services, without charge at the point of service.
2. Alberta retains the authority and responsibility to manage the publicly funded health care system in the province.
3. Recognize the demands from both the public and health professions for an approach to health services that is consistent with long term sustainability and quality.

4. Ensure a strong role for the private sector in health care, both within and outside the publicly funded system.
5. The public and private sector should work together to provide patient choice, quality of service, and effective outcomes as the first priority.
6. Regional Health Authorities assess health needs in their regions and be funded to provide appropriate health services in accordance with the health needs assessment.
7. Consumers have the right to voluntarily purchase health services outside assessed need.
8. Maintain the restrictions on the role of private insurance, while introducing measures to expand the opportunities for the private sector to deliver services within the single-payer envelope.
9. Private clinics should have the option of becoming completely private (patient pays), or allowing them to enter into a variety of funding arrangements with the public sector to cover the full costs of insured services.
10. There is a place for medical training in both public and private settings, however, care must be taken to ensure there is no deterioration in the world class training physicians currently have.
11. Recognize that physicians can receive payment from both the publicly funded system and fully**
12. The province must at all times be able to demonstrate "reasonable access" to insured health services with no fee at point of services or penalties would apply. An understanding is necessary on the mechanisms to determine and measure "reasonable access."

* Alberta Health, Edmonton, Alberta, October 16, 1995. The original principles are cited in full in Plain (2000, p. 10).

** Principle 11 was subsequently reworded to read as follows: "The same physician can practice in both the public and private systems if he/she is offering insured services which are fully paid for by the public system and the non-insured services which are paid for privately" (as cited in Plain (2000, p.10).

Appendix 2:

Health Policy Framework

1. Putting Patients at the Center

New directions and strategies for the continuous improvement and development of the health system will give priority to the interests of the people being served. Albertans will be encouraged to play an active role in maintaining and improving their own health and in deciding on appropriate care and treatment.

2. Promoting Flexibility in Scope of Practice of Health Professionals

Alberta's new legislation governing the health professions provides greater flexibility in terms of the scope of practice of the various professions as a means of promoting greater innovation. Alberta will continue to work with health professions to take full advantage of the opportunities that the Health Professions Act has made possible. Team approaches to care will be promoted in areas that provide the greatest benefit. Mechanisms will be developed to determine the most responsible caregiver in each particular situation and to ensure that care is seamless and meets high standards of proficiency and quality.

3. Implementing New Compensation Models

Alberta will develop alternative compensation structures and models that provide incentives for quality of care, efficiency and inter-professional collaboration. These compensation models should be based on the achievement of measurable health outcomes and compliance with indicators of quality. New compensation models should also contribute to recruitment and retention strategies and take into account changes in clinical practice resulting from such initiatives as telehealth and the Electronic Health Record. Successful implementation of alternative compensation models also will require consultation and negotiation with professional associations.

4. Strengthening Inter-regional Collaboration

Regional health authorities and Alberta Health and Wellness will collectively plan for the delivery of health services and the establishment of shared service networks throughout the province. Building on the successful implementation of the heart institute and the bone and joint institutes, Alberta will continue to support the establishment of such institutes so as to combine leading edge research with advanced clinical care.

5. Reshaping the Role of Hospitals

The collective planning process between regional health authorities and Alberta Health and Wellness will include options for changing the role of urban and rural hospitals so as to provide better and more responsive service to Albertans. Changes may include shifting some day surgery and ambulatory care services to community settings, delivering more services through private surgical facilities, linking some rural hospitals to urban hospitals for the provision of less complicated acute or follow-up care and converting small rural hospitals to centres of multi-disciplinary primary care. If necessary, regulatory changes will be made.

6. Establishing Parameters for Publicly Funded Health Services

Public funding will still be used for the essential kinds of health services. Alberta will build on the work already completed to implement the recommendations of the Expert Advisory Panel to Review Publicly Funded Health Services. Services and benefits which are discretionary, are not of proven benefit or are experimental in nature may not qualify for public funding.

7. Creating Long-Term Sustainability and Flexible Funding Options
Alberta will closely examine how various alternative funding models for health-related benefits and services, such as prescription drugs, continuing care, dental care, allied (nonphysician) health services and non-emergency acute care services, would work in this province. Appropriate legislative changes will be introduced to ensure long-term sustainability and flexible funding options.
8. Expanding System Capacity
Where it makes sense, in response to identified needs, Alberta will expand system capacity and consumer choice in both the public and private delivery systems. The supply of skilled health professionals will be increased by providing onsite training opportunities in both public and private facilities in cooperation with professional teaching institutions.
9. Paying for Choice and Access while Protecting the Public System
Service providers will be encouraged to find innovative ways of providing improved consumer choice — provided that these innovations do not adversely affect the provision of essential services through the public health system. Mechanisms will be required to closely monitor the impact of the private system on the public health system to enable corrective action to be taken to safeguard the public system if necessary. Prohibitions in the Alberta Health Care Insurance Act that prevent physicians from "opting-out" in certain circumstances will be replaced with provisions that will not adversely affect the delivery of services within the public system.
10. Deriving Economic Benefits from Health Services and Research
Alberta will capitalize on its world class health service and research facilities and will encourage and assist in the national and international marketing of intellectual property and innovations developed in Alberta.
Source: Alberta (2006, February). Health Policy Framework.

NOTES

1 In essence, the government sought to centralize decision making away from the approximately 200 local health boards (128 hospital, 25 public health and 40 long-term care).

2 Friedson (1990) characterizes bureaucratic labour markets as being driven by bureaucratic norms and hierarchy with an emphasis on reliability and quality. Within these markets consumers can choose only between a limited range of services determined by governing bureaucrats at a predetermined and standardized price. Health providers compete to provide services based on professional credentials and delivery and quality standards. Professional labour markets are determined through occupational monopoly (protected markets) governed by professional collegial standards. Emphasis is placed primarily on public service (and quality) and secondarily on price, which is often standardized by the professions. Finally, private labour markets are determined through competitive pricing and require full consumer knowledge of the tradeoffs between different services and providers with an emphasis on profit maximization.

3 In tandem with the publicly-funded system, the province has facilitated a steady growth since the 1980s in the number of private clinics providing some form of health service with 60 clinics performing surgeries outside of hospitals by 2012. To maximize their economic success, these clinics have frequently tested the boundaries of the Canada Health Act (CHA), the federal legislation meant to prevent providers from directly charging patients for publicly funded services.

4 As reported in Plain, 2000, p.11.

5 Private membership clinics have been the subject of a series of legal challenges to the CHA brought by Dr. Brian Day, a British Columbia physician. So far, the courts have upheld the validity of the CHA.

6 While the NDP government planned to pass legislation to prevent the 10 private membership clinics operating in the province from charging membership fees, the legislation did not pass before it was defeated in the election (Russell, 2019).

7 This pledge was effectively delayed by two years because of the focus on the pandemic.

8 The strain of the pandemic on health human resources and the subsequent shortage of health-care workers has led the government to postpone most of these cuts.

9 AHS, https://www.albertahealthservices.ca/assets/news/rls/ne-rls-2021-04-13-bkg-ophthalmology-chartered-surgical-facility-rfp-faq.pdf
According to AHS, there are about 79 private clinics in Alberta of which 43 offer insured/publicly-funded surgeries and 36 offer non-insured surgeries, such as cosmetic surgery. CSFs are required to meet all safety standards to be accredited by the College of Physicians and Surgeons of Alberta to perform both insured and non-insured surgical services in Alberta.

10 First launched in 2003, PCNs are partnerships between Alberta Health Alberta Health Services and physicians to facilitate comprehensive team-based community care. There are currently 40 geographically-based PCNs providing services to 70 per cent of Alberta's population involving 80 per cent of Alberta physicians.

11 This echoed the recommendation of a previous provincial commission on health reform (Alberta, 1989).

12 Part of this long march to a single authority also involved the consolidation of the governance and delivery of ambulance services (EMS). Traditionally controlled by local municipalities and a source of security and employment, ambulance services have been centralized under AHS, resulting in increasing ambulance response times.

13 Premier Smith has indicated that she wants this to happen and has included it in the mandate letter of the current Minister of Health.

14 Smith (2021) has clearly indicated that she wants to shift costs away from the public sector to the private sector though employer and employee co-payments.

CHAPTER 14

GROUND ZERO IN CANADA'S HIGHER EDUCATION SHOCK DOCTRINE: UCP ALBERTA'S POST-SECONDARY SECTOR

Marc Spooner

"What we are witnessing is an attack on universities not because they are failing, but because they are public. This is not just an attack on political liberty but also an attack on dissent, critical education, and any public institution that might exercise a democratizing influence on the nation."

— Henry A. Giroux (2016)

LIKE A TIME-LAPSE VIEW of an old-growth forest being logged, tree by tree and grove by grove, watching the damage being done to post-secondary education (PSE) in Alberta is distressing and demoralizing. As a preview of what may be in store for the entire post-secondary sector in Canada, it is horrifying.

Worse still, this attack on PSE is only one front in the much wider culture war being waged by authoritarian and populist actors against education itself. It is no longer alarmist to conclude that these forces seek to control what is taught, what is studied, and even what is known.

Our neighbours to the south are currently engaged in nothing less than a reprise of the McCarthy era. This time the new bugbears of "woke" and "equity" substituted for the 1950s catch-all dog-whistle — communist — and the hysterical loyalty oaths and public interrogations of the post-war decade are now echoed not merely in attacks against individuals, but in sustained campaigns targeting entire areas of study.

Certain fields have become particular magnets for the manufactured fury of this populist inquisition; at last count, at least 35 states have passed bills to ban, or are considering banning, critical race theory (CRT) in name or in spirit (Alfonseca, 2022). Though there may be disagreements between scholars on the finer points of CRT, for the most part, as Prof. Kimberle Crenshaw of the Columbia Law School, explains,

> It is a way of seeing, attending to, accounting for, tracing and analyzing the ways that race is produced.... The ways that racial inequality is facilitated, and the ways that our history has created these inequalities that now can be almost effortlessly reproduced unless we attend to the existence of these inequalities (quoted by Fortin, 2021).

Each of these bills aims to curtail or outlaw entire areas of scholarship, with several of the bans extending through K-12 and into higher education. In some cases, the bills take aim at entire theoretical approaches or frameworks to knowledge creation and production — theoretical lenses upon which to view, observe patterns, make meaning of, and act upon the social worlds we inhabit. By going beyond individuals and extending into entire areas of historical and contemporary established fact, these authoritarian limits go beyond anything McCarthy ever envisioned. To properly measure the extent of the recent victories made in the present war on (certain kinds of) knowledge, it would serve us to recall that previous attempts at similar bans rightly died on the vine; for example, attempts to curtail Middle Eastern studies post 9/11. Close followers of American politics might recall, in 2003, House Resolution 3077, the International Studies in Higher Education Act, which intended to surveil and control Middle East Studies Centres throughout American campuses in order to "balance" course materials considered too anti-American (Doumani, 2006); the resolution was passed by the House of Representatives, but never by Senate.

The current CRT bans, and similar attempts to suppress a variety of related scholarly areas that seek to advance diversity, equity, and inclusion goals are therefore, at least in the post-war democratic West, exceptional and unprecedented in their ubiquity and boldness.

In fact, beyond mere McCarthyism, they hearken back to nothing less than the example of Galileo, where the mediaeval church infamously kept absolute control of the pursuit of empirical fact lest it threaten its own authority. Such state actions represent a form of epistemicide — an attempt to quash entire ways of knowing — while re-directing higher education away from its aspirational ideals and diminishing the ability of the university to fulfil its promise as an institution that seeks truth, justice, and the public good.

It would be the foolhardiest of optimism to imagine the Canadian post-secondary education sector is immune to this fevered anti-intellectualism. We face many parallel forces and pressures from the misguided initiatives authoritarian populists such as Pierre Poilievre seek to impose on Canadian universities under the guise of "free speech guardians" (Lévesque, 2022); to provincial legislation that conflates freedom of expression with academic freedom and in doing so transferring the ultimate arbitrary power to the minister of higher education (Friesen, 2022); to Alberta's most recent mandate letter issued to the minister of advanced education (Government of Alberta, 2022a) which has as one of the five main goals to "ensure our post-secondary institutions are adequately protecting the academic freedom and free speech of students and faculty."

However, nothing has threatened harm to the very foundation of Canadian PSE like the enormity of cuts and restructuring the Alberta post-secondary sector has been subjected to under the United Conservative Party (UCP). A party that recently considered its own K-12 resolution (Policy Resolution #5) seeking to similarly ban anti-oppressive K-12 education and related concepts. The proposed, but not passed, policy resolution read (UCP, 2022, p. 21):

> h) halt the practice of any student being taught that by reason of their ethnic heritage they are privileged, they are inherently racist or they bear historic guilt due to said ethnic heritage or that all of society is a racist system. Further any differential treatment practiced by any educator due to said ethnic heritage will be halted.

Instruction of these concepts will not take place whether it is advanced under the title of so-called critical race theory, intersectionality, anti-racism, diversity and inclusion or some other name.

Alberta is the frontline for imported American and global neoconservative ideology. Contrary to what some Canadians may believe, the province is no laggard, but rather, Canada's conservative and neoliberal avant-garde (Darts and Letters, 2021). Alberta is at the forefront. It is the northern staging ground for such ideology and its operational implementation. A foot in the Canadian door, the bellwether province. Our early distant warning. For this reason, what happens in Alberta post-secondary, is, or ought to be, of vital interest to all Canadians who believe and support a robust public sector as one of the great levelling mechanisms in a just society.

A Case Study of Alberta Post-secondary

Barely disguised behind a screen of mercenary consultants and high-powered consultancy firms, including Janice McKinnon, who (with support from KPMG) authored the government's Blue Ribbon Report (Government of Alberta, 2019), McKinsey & Company, and, in the University of Alberta's case, the NOUS Group (2022), the UCP's intent to rein in, redirect, and drastically restructure Alberta's post-secondary sector was easy to discern. Any doubts it was explicitly setting out to repurpose Alberta's universities were removed in September 2022 (Herring, 2022):

> It's the latest element of the United Conservative government's plans to overhaul higher education in Alberta, which places a premium on trades and jobs-focused programs.
> "[Jason Kenney] railed against university liberal arts programs, which he said had "modest" or "very poor" employment outcomes, and said government funding for universities should align with labour market demands."
> "That's why we're trying to retool the education system."
> "[Kenney] said gov funding for universities should align with labour market demands."

The UCP's *how to* playbook is not new — it is the same international restructuring playbook that has been employed in Australia, New Zealand, the United Kingdom, the United States, and elsewhere. To operationalize neoliberal ideology and dismantle the public sector, governments force New Public Management upon it by: (a) adopting private-sector management practices; (b) introducing market-style incentives and disincentives; (c) introducing a customer orientation coupled with consumer choice and branding; (d) defunding and devolving budget functions while maintaining tight control through auditing and oversight; (e) outsourcing labour with casual, temporary staff; and (f) emphasizing greater output performance measures and controls in the name of efficiency and accountability (Chomsky, 2018; Lapsley, 2009; Lorenz, 2012; Ward, 2012).

It is truly difficult to over-emphasize how dramatic and turbulent the times have been for Alberta's PSE sector under the UCP. But how does one prepare the public for the drastic repurposing, major restructuring, and vindictive de-funding that was to be meted upon the Alberta post-secondary sector? Experts, experts, and everywhere we turn, more experts for hire to set the battle stage.

The goal? Full-on shock and awe for those operating within the system, while the reports produced by the consultants incrementally re-frame and acclimatise the public to the government's predetermined objectives. These firms provide a patina of independence and disinterested objectivity. But with the government's narrow scope in its request for proposal (RFP) calls, and pre-specified review parameters, these costly public relations and blame shifting exercises were predetermined to provide recommendations that would further calls for major funding cuts, tuition hikes, and the deeper implementation of New Public Management within Alberta universities.

The management consultancy firms employ a type of shock doctrine to the system as well as to those who find themselves working within it, meanwhile the reports reframe the issues while appearing "objective" and disinterested but providing the predetermined direction and "answers" initially sought by government all wrapped up nicely in the technocratic accountancy-speak of consultancy expertise obscuring the ideological and partisan thrust behind them. The effect is to soften the public, directing their attention to third-party reports and expertise, deflect blame away from the government, and create the illusion that the trajectory is a fait accompli while the retooling sequence is initiated. Gradually, and cumulatively, they raise the stakes in a manner that makes it difficult for the public to notice the proverbial water reaching boiling (more on this later) all under the guise of the difficult to argue with efficiency and accounting banner.

The next few pages follow the predictable consultancy trajectory. First, from the broadest in scope, the final report of the MacKinnon-led Blue Ribbon Panel that examined Alberta's finances and budgeting, moving next to the McKinsey & Company Report focusing more specifically on the Alberta post-secondary sector, and ending with the NOUS Group's tighter focus on the University of Alberta as a case study.

The Blue Ribbon Panel

The Blue Ribbon Panel was presented and sold as an independent panel of experts who were to straightforwardly review Alberta's finances and provide recommendations to balance the budget (Government of Alberta, 2019). As the first and broadest of these external consultancy reviews, the Blue Ribbon Panel on Alberta's Finances chaired by Janice MacKinnon was tasked with balancing Alberta's budget. However, as Ascah, Harrison, and Mueller (2019, p. 3), point out:

> While the panel's terms of reference seemed at first broad in tasking the panel to provide "an independent review of the province's finances," the details — advice on balancing the budget, but concentrating only on expenditures, and with no increased taxes — immediately suggested the panel's examination would be something less comprehensive. Critics argued the panel's mandate meant it would not address bigger issues of balance or long-term fiscal sustainability. Indeed, it seemed intentionally designed from the outset to fall short of providing the government and Albertans with the information necessary to make sound financial decisions about the province's current situation or to plan for a fiscally sustainable future.
>
> The release of the MacKinnon report bears out these concerns, while raising others. Given the panel's mandate, its members were constrained to look only at government expenditures.

Marc Schroeder (2022) zeroes in on the problem with this arrangement: it "is consistent with the UCP's tendency to commission panels and reports that parrot predetermined aims."

What did the Blue Ribbon report mean for Alberta PSE? Under the heading "Addressing the spending challenge" in the section detailing specific recommendation for Alberta's advanced education, the framers of the Blue Ribbon report highlight that Alberta's post-secondary funding structure "...doesn't link funding to the achievement of specific goals or priorities for the province such as ensuring the required skills for the current and future labour market, expanding research and technology commercialization, or achieving broader societal and economic goals" (p. 42). They go on to recommend "...that the future funding model ensure a link between provincial macro goals and outcomes to be achieved by post-secondary institutions. The government should assess whether the current governance model can address the challenges facing post-secondary institutions in Alberta by exploring alternative models..." (Government of Alberta, 2019, p. 42). It thus set the stage for attacks on existing collegial governance structures, and harbingering the adoption of performance-based funding as two mechanisms by which to provide the government tighter controls and a bigger stick to coerce desired direction and behaviour from its institutions and faculty.[1]

It is difficult to see The Blue Ribbon report, with its recommendations for using indicators that sharply narrow the meaning of "performance" to labour-market, industry and economic outcomes, as anything but an ideologically-based attempt to redesign the fundamental mission of Alberta's universities. The call to link funding to achievement redirects universities away from fostering critical, creative, and well-rounded citizens — while performing research in the public interest — and instead toward a drastically retooled, narrowly conceived economic "outcome" focused on serving the current labour market and performing corporate-styled research and development. In this struggle, what is at stake is nothing less than the heart and soul of Alberta's universities.

While the KPMG/Blue Ribbon report mapped the terrain, McKinsey & Company provided the battalion of restructuring engineers to lay down the sight line and prepare the earthwork for the new roads and pathways. As noted by Adkin et al. (2022, p. 26),

> Another piece of the UCP government's strategy for restructuring the PSE sector... was its commissioning of a review of the PSE sector called Alberta 2030: Transforming Post-Secondary Education. The global consulting company, McKinsey & Company, was awarded a $3.7 million contract in June 2020 to conduct this review. The Request for Proposals stated that the government was looking for ways to implement the recommendations of the MacKinnon Report within the limits of the government's fiscal plan.

In the context of significant budget cuts to higher education, it is important to recognizethere have always been competing philosophical visions for education. On one side, critical educational scholars have long believed education should encourage democratic citizenship, support community engagement and social activism, and foment an appreciation for the intrinsic and emancipatory value of knowledge. On the other side, there are those who wish to see an increasingly instrumentalized university that primarily serves corporate and economic interests. Albertans face a stark choice between these competing philosophical perspectives, but they have largely been left in the dark as policies and governing structures are dramatically shifted.

McKinsey & Company

It is important to take a moment to put "The Firm" in proper context.

A February 2019 article in *Current Affairs* carries the following description of McKinsey & Company by one anonymous individual familiar with the firm (Current Affairs, 2019):

> The biggest, oldest, most influential, and most prestigious of the "Big Three" management consulting firms, McKinsey has played an outsized role in creating the world we occupy today. In its 90+ year history, McKinsey has been a whisperer to presidents and CEOs. McKinsey serves more than 2,000 institutions, including 90 of the top 100 corporations worldwide. It has acted as a catalyst and accelerant to every trend in the world economy: firm consolidation, the rise of advertising, runaway executive compensation, globalization, automation, and corporate restructuring and strategy.

In fact, McKinsey & Company is the very same consultancy firm that has been implicated in the now infamous Loblaws bread price-fixing scandal (though Loblaws denies McKinsey was involved) (Morrow, 2019).

A fact no one denies is The Firm's scheme to help Purdue Pharma turbocharge sales of OxyContin as reported in *The New York Times* (Bogdanich and Forsythe, 2020):

> Documents released last week in a federal bankruptcy court in New York show that the adviser was McKinsey & Company, the world's most prestigious consulting firm. The 160 pages include emails and slides revealing new details about McKinsey's advice to members of the Sackler family, Purdue's billionaire owners, and the firm's now notorious plan to "turbocharge" OxyContin sales at a time when opioid abuse had already killed hundreds of thousands of Americans

And in the unlikely event plans to turbocharge OxyContin sales were not enough to sway your opinion of this company, in "How McKinsey Helped the Trump Administration Carry Out Its Immigration Policies," *The New York Times* reveal how "newly uncovered documents show the consulting giant helped ICE [Immigration and Customs Enforcement] find 'detention savings opportunities' — including measures the agency's staff sometimes viewed as too harsh on immigrants" (MacDougall, 2019).

As Duff McDonald (2013), author of the bestselling book *The Firm*, once wrote for a *National Post* piece,

> ...it is also hard to overlook the mounting number of instances in which McKinsey advisers have behaved no better than mercenaries, collecting huge fees for work of dubious worth. At their most craven, they can be recruited to provide a high-gloss imprimatur of objectivity that is in reality mere cover for executives.

So, what did McKinsey and Company's consultation do for Alberta PSE? The Alberta 2030: Building Skills for Jobs website (see Government of Alberta, 2021a; 2022b) promises:

> Alberta 2030: Building Skills for Jobs is a transformational vision, direction and new way of working for our province's higher education system. The Alberta 2030 initiative will develop a highly skilled and competitive workforce, strengthen innovation and commercialization of research, and forge stronger relationships between employers and post-secondary institutions.

What initially was under the rubric Alberta 2030: Transforming Post-Secondary Education, was itself transformed into the more narrowly focused Alberta 2030: Building Skills for Jobs. The McKinsey & Company's full background report on Alberta's post-secondary education sector comes in the form of a 217-page slide deck (Government of Alberta, 2021c). The terms *citizenship* and *democracy* do not appear on any of the 217 slides; on slide 36, the slide deck report promises, "The system will be highly responsive to labour market needs and through innovative programming and excellence in research, contribute to the betterment of an innovative and prosperous Alberta." The government also published the companion 32-page report: "Alberta 2030: Building skills for jobs —10-year strategy for post-secondary education" (Government of Alberta, 2021a).

This document delineates six goals: improve access and student experience; develop skills for jobs; support innovation and commercialization; strengthen internationalization; improve sustainability and affordability; and strengthen system *governance*.

Table 14.1: Exceptional Tuition Increase (ETI)

FACULTY	Program	Approved Tuition 2021-22	Approved Tuition 2022-2023	Fall 2022 Proposed ETI (%)
UNDERGRADUATE				
Business	BComm	$8,012.48	$9,774.24	22.00%
Engineering	BSc Engineering	$7,309.44	$9,099.36	24.50%
Law	JD	$11,701.48	$15,094.84	29.00%
Medicine	BSc Radiation Therapy	$6,091.20	$7,309.20	20.00%
Medicine	BSc Med Lab Sciences	$6,091.20	$7,125.60	17.00%
Medicine	DDS	$23,109.16	$32,352.76	40.00%
Medicine	AP DDS	$57,093.40	$66,337.00	16.19 %
Pharmacy	PharmD	$11,431.68	$16,460.80	44.00%
GRADUATE				
Business	MBA	$14,380.80	$24,624.00	71.23%
Engineering	MEng	$7,345.20	$9,033.60	22.99%
Education	Master Counselling Psych • Course	$4,268.88	$8,573.76	100.00%
Education	Master Counselling Psych• Thesis	$4,192.80	$8,573.76	104.49%

Source: Adapted from Joannou (2022)

Let's take a closer look at each of the report's six stated goals.

1) Improve Access and Student Experience. It seems contradictory to set about to improve access by approving a seven per cent increase in tuition cap (Government of Alberta, 2021).

In addition to the seven per cent, in exceptional cases, tuition was permitted to as much as double. As reported in the *Edmonton Journal*, for instance, in at least 12 programs at the University of Alberta alone tuition rose at a far greater rate than the stated seven per cent cap (Joannou, 2022) (see Table 14.1, above).

2) Develop Skills for Jobs. Few would oppose developing skills for jobs. However, when such a goal is contained in a report that does not once mention *citizenship* nor *democracy*, and mentions *critical thinking* only in the service of careers, it is cause for concern. Given that goal 2 is contained in a report that is similarly titled with a focus on jobs, it reframes high education in reductionist, instrumental terms as if being solely about job training, thus erasing many other vital aims and goals of a sound, well-rounded, post-secondary education; for instance, developing critical and creative citizens capable of fully participating in a modern democracy. When viewed in this light, it is quite alarming.

Moreover, given the above, it is no surprise the UCP government would advocate for greater "micro-credentialing" which further offloads the cost, risk, and responsibility for workplace training and development onto individuals and away from its traditional industry responsibility. Micro-credentials also carve off and privilege a subset of narrow, context-specific skills, away from well-thought-out programs of study that feature the development and fostering of critical, creative skills and democratic habits of mind typical of degree programs (Wheelahan and Moodie, 2021a; 2021b).

3) **Support Innovation and Commercialization.** Here the report highlights that "both basic and applied research are important and intellectual property rights must be valued to ensure the appropriate incentives," (p. 14); though not everyone is motivated by intellectual property rights which implies a profit motive, when for many academics in a wide variety of fields, discovery, furthering our collective quest for truth, knowledge, and justice, is motivation enough.

Additionally, any indicator tying public funding to private funding would further incentivize commercialization of university research. It also elides important considerations including non-disclosure agreements, potential for delayed dissemination of findings, and questions surrounding access and use of generated data. Moreover, it rewards targeted industry research in the corporate interest, and concomitantly downgrades research in the public interest. An emphasis on commercialization also impacts society by devaluing less costly but no less important scholarship, including risky, yet innovative research; community-engaged research; and other valuable research endeavours that cannot easily be measured or reflected by a simple financial or corporate calculus. Rather than uncovering ground-breaking new ideas, following uncertain but innovative paths that become potential game changers, or working in the service of the communities in which their universities are situated, faculty are encouraged to seek out corporate research contracts instead.

Under this same goal, the report observes "Collaboration and communication are key. Barriers to collaboration among institutions, industry and government include differing goals, ineffective administrative processes, misaligned planning cycles and organizational culture" (p. 14). Yet, with a performance-based funding system as promised in this same report, universities would be all but compelled to favour inter-institutional competition over collaboration as they compete for the same limited pool of dollars.

4) Strengthening Internationalization. This goal seems contradicted by the government's own Alberta Tuition Framework: Version 2.1 which grants institutions "...the authority to set tuition for international students at their own discretion. Similarly, the rates at which international student tuition increase are entirely at the discretion of institutional boards (Government of Alberta, 2021b, p.14). Permitting, if not encouraging, the province's universities to treat international students as cash cows, while often not providing them with adequate and appropriate levels of support.

5) Improve Sustainability and Affordability. Under this heading, the report states, "We heard our stakeholders urge an increase in funding predictability and affordability for students, at the same time that they acknowledge Alberta is facing extreme fiscal challenges (p. 16). As in point 1, this is contradicted by the allowable tuition increases. And as covered in point 3, the report states "...encouraging industry investment in research and commercialization offer significant growth opportunities" (p. 16). Under this heading the report also suggests "eliminating duplication"; though what kind of duplication is not clear; should this refer to program duplication, it would actually have the effect of reducing access counter to the report's first stated goal as some students would have to go even farther afield to study.

6) Strengthen System Governance. Here it seems easy to agree with the suggestion for change that includes "... reducing the level of government control over factors such as institutional finances and board appointments, and ensuring boards have diverse perspectives and skillsets. It is important that appointments be timely, not political, and board members be experienced and highly competent" (p. 17). Though any agreement is short lived when one considers how the UCP government has stacked the university boards of governors (BOG) with political appointees, including appointing McKinnon herself to the University of Alberta BOG (Adkin et al., 2022). In comparing the 113 BOG appointees made by the NDP government, with the 121 appointed by the UCP government, they found (p. 3),

> while both parties disproportionately appointed individuals from business backgrounds, the NDP appointed significantly more individuals working in public or non-profit sector administration, and fewer individuals working in private sector management than the UCP. Our occupational analysis showed that PSEI[2] boards largely exclude representation from the working class. Specializations are highly skewed toward management, corporate law, accounting and finance.

Later the report calls for the establishment of "...a new governance framework with clear mandates and accountabilities for system- and institution level outcomes in teaching, research and innovation, and collaboration" (p. 30). The report further clarifies: "This strategy will result in a new governance structure that brings our institutions in sync to achieve greater success, and an outcomes framework that aligns performance-based funding and shifts incentives and accountabilities to ensure we are future-focused and achieving improved results" (p. 31).

When situating the Alberta 2030: Building Skills for Jobs report into the current Alberta landscape, it is difficult to portray it as anything more than a costly public relations scheme to deflect blame and attention toward a supposedly objective expert third party, as the report itself further presents and reframes post-secondary education in strictly economic terms.

How can one otherwise reconcile the report's stated goal to improve access and affordability, when juxtaposed with Alberta's ongoing 18.8 per cent cut to the post-secondary sector over the last four years? As Adkin et al. (2022, p. 1) note,

> ... the United Conservative Party (UCP) government has, from 2018 to 2022, cut the operating support budget for Alberta's PSEIs by 18.8 per cent, resulting in a trail of destruction across the province's universities, colleges, and technical institutes: thousands of employees laid off, increased workloads for remaining staff, teaching contracts cut, academic programs axed and, ultimately, tuition rising beyond the reach of many and growing student indebtedness.

On the heels of these cuts, the government has raised the allowable domestic tuition cap to seven per cent, permitted certain "exceptional" programs to double in cost, while completely deregulating international tuition fees. As the report calls for greater commercialization and industry-led research with more opportunities for securing intellectual property rights, it is difficult see it as anything but a plan to accelerate the university's drift away from the public interest and further down the road towards a more fully corporatized university.

Moreover, the performance-based funding systems malware it seeks to introduce into the system represents yet another mechanism to coerce the reprogramming of universities toward a narrow labour-market and industry focus with students recast as *Learner Economicus.* In so doing, these systems rob students of the possibility of experiencing higher learning as being about something more than getting a high-paying job, of being part of something beyond a market.

Students are further encouraged to view themselves as atomized entrepreneurs, to be upskilled at their own expense and risk via narrow career paths and clusters of stacked micro-credentials, and not as creative, caring, and engaged citizens of the province, the nation, and the world. Ironically, a properly functioning university should fail against such metrics as they are in essence attempts to deform the university away from its aspirational ideals.

Defunded, destabilised, and demoralised, the Alberta post-secondary sector finds itself twisting in the wind, its traditional rudder battered, as it is left mastless by the sustained violence of the storm visited upon it. In this tempest, no Alberta university bore a greater proportion of the cuts than the University of Alberta, setting the stage for the final consultancy firm to be presented in this exposé: the NOUS Group (appropriately pronounced, "noose").

NOUS Group Shock and Awe

Established in 1999, the Australian-based NOUS Group (2022) is an international management consultancy firm with offices in New Zealand, the United Kingdom, Ireland, and since 2019, Canada. They are an extreme solution for the problems that are always duly manufactured by its willing clients.

Echoing former premier Kenney's reckless and combative language around "retooling the university system to align with labour market demands" whilst simultaneously and erroneously maligning liberal arts programs for their supposed lack of employment outcomes, next in the procession of expert consultants, the NOUS Group are equally boastful of leveraging a type of shock doctrine to force massive restructuring upon vulnerable university systems. The shock doctrine, one might recall (Klein, 2007), is at its most succinct the "...the rapid-fire corporate reengineering of societies still reeling from shock."[3]

To put the magnitude of the operating cuts in proper perspective, for the University of Alberta alone, they represented "a 34 per cent reduction in provincial funding — a $222-million cut — that had to be reconciled in 24 months" (Flanagan, 2022). University of Alberta president, Bill Flanagan (2022) characterised it as "...larger in size and scale than ever undertaken by a major university in Canada."

The international management consultancy firm even has a dedicated case study featured on its website about how they "helped" the University of Alberta's strategic transformation in two years.[4] The site provides a helpful text box offering lessons learned from the Alberta experience where they extol the benefits of external shocks, including the "...burning platform that unite stakeholders behind fundamental change" (NOUS, 2022):

What You Learn from the University of Alberta

- External shocks can provide a burning platform that unite stakeholders behind fundamental change.
- Committed executive sponsorship and good governance structures such as a Program Management Office are essential to keeping a transformation program accountable and on track.
- Data comparing cross-function and cross-institution performance can enable targeted cost-reduction strategies and better discussions with faculty, staff and the executive.
- Large beneficial changes can occur when an organization's executive and staff lean into the challenge.

It reads like an open resumé targeting any and all like-minded conservative governments in Canada and should send chills down anyone's spine who even remotely cares about Canada's public universities as it cheerfully advertises its services and the *how-to* blueprint.

This could be viewed as an exaggeration or mischaracterization; however, the same website features an article entitled: "The environment for Canadian universities is changing. How can governance structures adapt?" (Rowe and Ashkanasy, 2022). In the article, they outline: 1) New Requirements from Government; 2) Intensifying Competition Among Universities, Locally and Globally; 3) Rankings And Metrics; 4) Competition For Students; and 5) Opportunities Through Digital as the five forces they see as challenges for university governance. Of the risks, they write,

> The consequences of governance structures that are not fit for purpose can be great, from exposure to legal risk to financial challenges and the inability to achieve the mission. On the other hand, a high-performing board and senate can help steer the university to achieve the mission and mitigate financial and other risks.

The article ends with the warning "Universities ignore the need for robust governance at their peril," and a pitch as they offer to help, "Get in touch to explore how we can support you to rethink your university's governance arrangements."

The Consultancy Merry-Go-Round

With that, the perpetual consultancy machine completes its cycle. It began with the UCP government seeking to weaken the public sector by operationalizing neoliberal policies through the adoption of New Public Management techniques. The consultancy process was first initiated by the limited scope of the Blue Ribbon report's predetermined review of the province's finances arriving at the predestined conclusion to make sweeping cuts across the public sector. From there, McKinsey & Company was inserted into the desired sector, regardless of its lack of field-specific expertise. McKinsey then sets in motion the engineered restructuring, while NOUS is contracted by the specific institution seeking to execute the plan. In this consultancy trickle down, each firm functions as an added shield to the public relations phalanx that third-party mercenaries are paid to provide. Having reached the end point in this cycle, the completed causation feedback loop idles while it awaits the next batch of ideologically driven public-sector raiders.

With each iteration of the cycle, society becomes all the more immersed in these neoliberal logics to such a degree that the short-term economic benefits become the only calculus with which to judge a system's value. As the narrative is repeated enough times, it becomes so normalized that any alternative appears alien. It is precisely such logics, and folly, that transform education into a commodity, or an old growth forest into lumber.

Message in a Bottle

The repeated onslaught of mutually reinforcing messaging, consultant after consultant, is effective and disorienting. Employing its full roster of high-powered management firms for hire, the Government of Alberta has successfully degraded the terms of reference for assessing the province's post-secondary sector. The lens through which higher education is viewed has been distorted, and in its new disfigured image labour market and economic outcomes are the foreground, while all else is faded to the background. In this upside-down, inside-out world it is the biggest victims of the UCP's slash-and-burn policies that seem to be its loudest cheerleaders. They may well be suffering from Stockholm Syndrome, though a less generous reading could conclude that ass kissing is a parroting reflex that appears to predominantly afflict ambitious ladder climbers. Still another take might be these are simply the words of a neoliberal wolf wrapped up in a sheep's rhetoric. Take for instance, the University of Alberta president's piece in the *Hill Times* where he uncritically champions the government's messaging as if he's been supplied these very talking points:

> Funders of all stripes — including governments, industry partners, and other private donors — are also demanding more. As finances are stretched in countless directions, there are higher expectations that universities make the most of every dollar. Employers are increasingly joining this chorus, especially amid labour market constraints. If tax dollars are going to universities, they

> reason, then graduates ought to have the right critical skills required for the jobs of today and tomorrow.

No mention of the university's larger purpose of serving the public good and society as a whole. The president's article is wholly framed in the narrative of industry, labour, and economy. No mention of the university's role in seeking and speaking truth, or fostering critical thinking, democratic habits of mind, justice, seeing value in non-economic terms, education's collective benefit, contributions to healthy communities, role in mitigating the great global environmental and democratic crises, or any other aspects of the university's aspirational mission.

As Schroeder points out, "The University of Alberta has fared especially poorly. And yet University of Alberta President Bill Flanagan has proven himself to be one of AB 2030's most ardent cheerleaders, echoing and amplifying UCP government talking points directly." (Schroeder, 2021). Sadly, this has been the public position of much of Alberta's post-secondary administrative sector, as noted by Adkin et al. (2022, p. 34),

> ...university presidents, vice-presidents, and other senior administrators — along with the UCP-appointed public members of the boards of governors — have remained silent, acquiesced to, or even endorsed the UCP's restructuring rationale without publicly questioning its wisdom or necessity. It appears that the UCP has succeeded in bringing the boards and the PSEI executives into alignment with its agenda.

Now, juxtapose President Flanagan's response with those of his presidential peers in Manitoba facing the similar prospect of their universities being reprogrammed by a conservative government to be more aligned with labour and industry needs through performance-based funding and other mechanisms.

> Manitoba's plans to change the way it finances universities and colleges is facing opposition from presidents of some of those institutions. In separate letters to the government, University of Manitoba president Michael Benarroch asked the province to refrain from tying funding to data, while Brandon University president David Docherty warned the metrics contemplated by government could come at the expense of students from disadvantaged backgrounds (Froese, 2022).

Subsequent to Froese's article, the *Winnipeg Free Press* reported the Manitoba government had taken a step back with regard to performance-based funding for its universities (Macintosh, 2022). The presidents' warnings were not the only element involved in the government's decision to pause its plans, but certainly one contributing factor. Only time will reveal if the pause becomes permanent or simply a regrouping strategy, but what is evident is the noticeably different administrative response that is generated between a collaborator versus a resistor. Not surprising perhaps, given Alberta's

> ...decades-long steeping of academic administrators in neoliberal rationality combined with a university governance structure that makes the president directly accountable to a UCP-appointed board means that senior management cannot be relied upon to mount any meaningful defense of their institutions despite being their nominal leaders (Shroeder, 2021).

There is another significant difference in the two responses. The resisting presidents help to create an on-message circuit breaker that permits a counter narrative to disrupt the government paid-for and produced narrative. By doing so, a key space for the public to view the issue from more than one repeated perspective is opened. Bear in mind, the greater the government's illiberal tendencies, the greater the importance of critique, dissent, and organizational opposition for the democratic health of the state.

Conclusion

In *On Tyranny*, Timothy Snyder (2017, p. 22) warns us that "institutions do not protect themselves. They fall one after the other unless each is defended from the beginning. So, choose an institution you care about— a court, a newspaper, a law, a labor union — and take its side." Here, I would most certainly add the university in all its aspirational ideals. While keeping in mind Giroux's (2016) observation that "in this case the autonomy of institutions such as higher education, particularly public institutions are threatened as much by state politics as by corporate interests."

Whether by riot or by occupation, the attempts in 2021 and 2022 to overthrow the democratically elected governments in the United States and Canada have clearly shown what a fragile construct democracy really is. Against this backdrop, the acquiescence shown by high-ranking officials in the PSE sector should give us all pause. They have forgotten — or have chosen to ignore — that the state's commitment to the independence of its universities is a key pillar of our democracy that joins the free press, the legislature, and the judiciary; each being mutually reinforcing and interdependent (Spooner, 2022).

Perhaps all of us would do well to commit President Flanagan's total capitulation to memory: "This is the new paradigm for post-secondary education in Canada. The University of Alberta is proud to be at its forefront" (Flanagan, 2022).

We can all go down his road, smiling at our aggressors, and hoping they turn their attention to someone else's budget, some other faculty's resources, some other university's core mission, or we can band together and fight them with everything we have.

In the end, the balance for the type of society our universities help shape is in our hands, my friends. It truly is as stark a contrast as a clearcut, barren landscape versus the fecundity of a lush old-growth forest.

References

Adkin, L., Carroll, W., Chen, D., Lang, M., & Shakespear, M. (2022). *Higher Education: Corporate or Public? How the UCP is Restructuring Post-Secondary Education in Alberta*. Edmonton, AB: Parkland Institute.

https://assets.nationbuilder.com/parklandinstitute/pages/1974/attachments/original/1661762422/Higher_Education_report.pdf?1661762422

Alfonseca, K. (2022, March). *Map: Where anti-critical race theory efforts have reached.At least 35 states have passed or considered legislation on race education.* ABC News. https://abcnews.go.com/Politics/map-anti-critical-race-theory-efforts-reached/story?id=83619715

Ascah, B., Harrison, T., & Mueller, R. E. (2019). *Cutting Through the Blue Ribbon: A Balanced Look at Alberta's Finances*. Edmonton, AB: Parkland Institute. https://d3n8a8pro7vhmx.cloudfront.net/parklandinstitute/pages/1712/attachments/original/1567981307/cutting_through.pdf?1567981307

Bogdanich, W., & Forsythe, M. (2020, Nov. 27). McKinsey proposed paying pharmacy companies rebates for oxycontin overdoses. *The New York Times*. https://www.nytimes.com/2020/11/27/business/mckinsey-purdue-oxycontin-opioids.html

Chomsky, N. (2018). An interview with Dr. Noam Chomsky on neoliberalism, society, and higher education. In M. Spooner, & J. McNinch (Eds.), *Dissident Knowledge in Higher Education*, (pp. 55-62). Regina, SK: University of Regina Press.

Current Affairs. (2019, February). *McKinsey & Company: Capital's willing executioners*. https://www.currentaffairs.org/2019/02/mckinsey-company-capitals-willing-executioners

Darts and Letters. (2021, Aug. 5). EP4: *The Conquest of Bread*. Podcast. https://dartsandletters.ca/2021/08/05/ep4-the-conquest-of-bread-rebroadcast/

Doumani, B. (Ed.), (2006). *Academic Freedom after September 11*. NY: Zone Books.

Flanagan, B. (2022, Sept. 21). Post-secondary transformation key to navigating the future. *The Hill Times*. https://www.hilltimes.com/ht_author/bill-flanagan/

Fortin, J. (2021, July 27). Critical Race Theory: A Brief History. *New York Times*. https://www.nytimes.com/article/what-is-critical-race-theory.html

Friesen, J. (2022, April 6). Quebec bill on academic freedom says no words are off-limits in classrooms. *The Globe and Mail.* https://www.theglobeandmail.com/canada/article-quebec-bill-on-academic-freedom-says-no-words-are-off-limits-in/

Froese, I. (2022, Nov. 8). *Presidents of 2 Manitoba universities wary of funding higher education using certain outcomes.* CBC Manitoba. https://www.cbc.ca/news/canada/manitoba/university-presidents-manitoba-performance-based-funding-1.6643810

Giroux, H. A. (2016, Sept. 15). *Neoliberal savagery and the assault on higher education as a democratic public sphere.* Café Dissensus. https://cafedissensus.com/2016/09/15/neoliberal-savagery-and-the-assault-on-higher-education-as-a-democratic-public-sphere/

Government of Alberta. (2019, August). *Report and Recommendations Blue Ribbon Panel on Alberta's Finances.* https://open.alberta.ca/dataset/081ba74d-95c8-43ab-9097-cef17a9fb59c/resource/257f040a-2645-49e7-b40b-462e4b5c059c/download/blue-ribbon-panel-report.pdf

Government of Alberta. (2021a). *Alberta 2030: Building Skills for Jobs: A 10-Year Strategy for Secondary Education.* April. https://open.alberta.ca/dataset/24e31942-e84b-4298-a82c-713b0a272604/resource/b5a2072e-8872-45f9-b84d-784d0e98c732/download/ae-alberta-2030-building-skills-for-jobs-10-year-strategy-post-secondary-education-2021-04.pdf

Government of Alberta. (2021b). *Alberta 2030: Building Skills for the Future.* Analysis of Stakeholder input. January. https://www.alberta.ca/assets/documents/ae-ab2030-analysis-and-stakeholder-input.pdf

Government of Alberta. (2021c). *Alberta Tuition Framework* (v2.1). December. https://open.alberta.ca/dataset/e934cad0-06a0-4e7b-a462-74f9423fed61/resource/ef91890d-186c-4c7f-8e8e-de6b9aa22892/download/ae-alberta-tuition-framework-version-2-1-2021-12.pdf

Government of Alberta (2022a). *Premier's Mandate Letter to the Minister of Advanced Education.* Office of the Premier. Nov. 15. https://open.alberta.ca/dataset/71ebe02e-bda3-46f3-8ddd-6bf3a0d3d7ca/resource/e9c06b90-21b2-4cc2-8d37-157c57b0a8a3/download/ae-mandate-letter-advanced-education.pdf

Government of Alberta. (2022b). *Alberta 2020: Building skills for Jobs.* Website. https://www.alberta.ca/alberta-2030-building-skills-for-jobs.aspx

Herring, J. (2022, Sept. 6). Alberta taps industry heads to advise on post-secondary needs. *Calgary Herald.* https://calgaryherald.com/news/ politics/alberta-taps-industry-heads-to-advise-on-post-secondary-needs

Joannou, A. (2022, March 9). University of Alberta students facing steep tuition hikes — as much as double — in a dozen programs. *Edmonton Journal.* https://edmontonjournal.com/news/politics/ledge-post-4

Klein, N. (2007). *The Shock Doctrine: The Rise of Disaster Capitalism.* Torontono: Knopf Canada.

Lévesque, C. (2022, June 21). Poilievre promises to protect freedom of speech on campus, appoint a "free speech guardian." *National Post.* https://nationalpost.com/news/politics/poilievre-promises-to-protect-freedom-of-speech-on-campus-appoint-a-free-speech-guardian

Lapsley, I. (2009). New public management: The cruellest invention of the human spirit? *Abacus* 45, 1–21.

Lorenz, C. (2012). If you're so smart, why are you under surveillance? Universities, neoliberalism, and new public management. *Critical Inquiry* 38, 599–629.

McDonald, D. (2013, Sept. 24). Cult of McKinsey. *National Post.* https://nationalpost.com/opinion/duff-mcdonald-cult-of-mckinsey

MacDougall, I. (2019, Dec. 3). How McKinsey helped the Trump administration carry out its immigration policies. *The New York Times.* https://www.nytimes.com/2019/12/03/us/mckinsey-ICE-immigration.html

Macintosh, M. (2022, Nov. 5). Province takes step back from proposed post-secondary funding formula. *Winnipeg Free Press.* https://www.winnipegfreepress.com/breakingnews/2022/11/15/province-takes-step-back-from-proposed-post-secondary-funding-formula

Morrow, A. (2019, Dec. 12). Pete Buttigieg's campaign denies his involvement in bread price-fixing scheme during work at Loblaws. *The Globe and Mail.* https://www.theglobeandmail.com/world/us-politics/article-pete-buttigiegs-campaign-denies-his-involvement-in-bread-price-fixing/

NOUS Group. (2022). *New Operating Model Helps a Canadian university to Realize More than $100 Million in Savings.* Downloaded Nov. 14. https://nousgroup.com/ca/case-studies/operating-model-university-savings/

Rowe, K., & Ashkanasy, Z. (2022). *The Environment for Canadian Universities is Changing. How can Governance Structures Adapt?* NOUS Group. https://nousgroup.com/ca/insights/canada-universities-governance/

Schroeder, M. (2021, May 28). *Alberta 2030: A Yoke for the University.* Guest post, Mount Royal/University of Calgary. https://artssquared.wordpress.com/2021/05/28/alberta-2030-a-yoke-for-the-university-guest-post-by-marc-schroeder-mount-royal-university-of-calgary/

Schroeder, M. (2022, May). Commentary: Alberta Is showing Canada how to destroy education. *CAUT Bulletin*. https://www.caut.ca/bulletin/2022/05/commentary-alberta-showing-canada-how-destroy-education

Spooner, M. (2022, Feb. 8). *Universities: The often overlooked player in determining healthy democracies*. The Conversation. https://theconversation.com/universities-the-often-overlooked-player-in-determining-healthy-democracies-175417

United Conservative Party (2022, October). *Policy and Governance Resolutions*. Annual General Meeting. https://www.unitedconservative.ca/wp-content/uploads/Plenary-Agenda-2022.pdf

Ward, S. C. (2012). *Neoliberalism and the Global Restructuring of Knowledge and Education*. New York: Routledge.

Wheelahan, L. & Moodie, G. (2021a). Analysing micro-credentials in higher education: A Bernsteinian analysis. *Journal of Curriculum Studies* 53(2), 212-228.

Wheelahan, L. & Moodie, G. (2021b). Gig qualifications for the gig economy: Micro-credentials and the "hungry mile." *Higher Education*, Aug. 3. https://doi.org/10.1007/s10734-021-00742-3 Downloaded Nov. 5, 2021.

NOTES

1 For an in-depth, meticulous examination of these attacks on university governance structures that is beyond the scope of this chapter, see Adkin et al. (2022).

2 Post-secondary Education Institutions.

3 https://tsd.naomiklein.org/shock-doctrine.html

4 https://nousgroup.com/ca/case-studies/operating-model-university-savings/

CHAPTER 15

SUBSIDIZED RENTAL HOUSING AND HOMELESSNESS UNDER THE UCP

Nick Falvo[1]

"Homelessness is not a choice. It is a lack of other choices."

— Louise Gallagher

THROUGHOUT CANADA, government supports housing affordability by seeing to it that rent levels in some housing units are kept at or below a certain threshold (often at or below 30 per cent of a household's income, with precise rent scales varying by program and jurisdiction). This role — while certainly not the only one played by government in housing — is crucial, largely because the private market alone does an inadequate job of creating and sustaining housing that is affordable for low-income households. This is the case both for households with limited labour market attachment and those receiving means-tested income assistance.

In Canada, all orders of government play important roles in housing, with the Government of Canada (GoC) often leading, and provincial and territorial governments responding to invitations to cost-share new initiatives by contributing funding. In some instances, staff support services are required by tenants and funded by federal, provincial and territorial governments. Municipalities play important roles pertaining to regulation and modest funding.

The Government of Alberta (GoA) has a regulatory, funding and leadership role in the provision of affordable housing. Alberta enters into funding agreements with the GoC and partners with municipalities and housing providers to deliver housing supports to Albertans. The GoA also plays an important role with respect to homelessness and harm reduction.

This chapter begins with an overview of the GoA's precise role with affordable housing and homelessness. It then discusses the United Conservative Party (UCP) government's various spending announcements and examines the following themes: GoA co-operation with federal housing initiatives; Alberta's Affordable Housing Review Panel; Stronger Foundations (Alberta's 10-year housing strategy unveiled in 2021); and homelessness policy under the UCP. The chapter closes with a discussion and conclusion.

This chapter is based largely on the author's own tracking of housing and homelessness developments in Alberta since the UCP formed government, including the author's ongoing budget analysis and review of publicly available documents. It also relies on results of 16 semi-structured interviews conducted via Zoom in February and March 2022 (see Appendix for the interview guide). Interviews were conducted with persons who have strong knowledge of Alberta's housing and homelessness sectors, and their identity is being protected. An early draft of this chapter was circulated to all interview subjects, one well-placed GoA staff person and various other subject matter experts in June 2022. Each person was given one month to provide the author with comments.

Subsidized Housing and Homelessness Policy in Alberta

Under former premier Jason Kenney, Alberta's Ministry of Seniors and Housing was the province's lead ministry for the design, administration and funding of affordable housing policy and programs (but in October 2022, housing became subsumed under the newly designed Ministry of Seniors, Community and Social Services). Its specific roles included: negotiating bilateral housing agreements with the GoC; providing program oversight of existing units under affordable housing agreements with private non-profits; and funding new units and repairs to existing units of public housing for low-income households. The Alberta Social Housing Corporation owns approximately 48 per cent of the province's affordable housing portfolio. Service Alberta is responsible for the Residential Tenancies Act and the landlord/tenant dispute board.

Some affordable housing in Alberta is administered by non-profit entities, some by public entities, and some by for-profit landlords. When it comes to housing on reserve, the Alberta provincial government typically plays no role at all — though it could if it so chose. Off reserve (and under Kenney), Alberta Seniors and Housing administered the Indigenous Housing Capital Program that provides capital funding to increase the supply of off-reserve, non-market, affordable rental units for Indigenous people in need.

The Alberta Housing Act is the regulatory framework for most subsidized housing in Alberta. It stipulates how rents are calculated and what kind of reporting must be done by operators. Alberta's Housing Management Bodies (HMB) — which manage most subsidized housing units in the province — are regulated by the act. As of April 1, 2022, there were 88 HMB in Alberta. Most HMB have boards that are partly or wholly appointed by municipalities. In most cases, the members appointed by municipalities are municipal elected officials. Outside of Calgary and Edmonton, HMB boards consist entirely of elected officials.

Most of Alberta' subsidized housing units are Rent Geared to Income (RGI) units, meaning rents are very low in those units. However, on a per capita basis, Alberta has far fewer subsidized housing units than the rest of Canada. Alberta's rate of social housing is just 2.9 per cent; for Canada as a whole, the figure is 4.2 per cent. This discrepancy is believed to exist for two main reasons: incomes have historically been higher in Alberta than the rest of Canada, resulting in less need for social housing; and Alberta's political climate is more conservative than the rest of Canada, meaning there has been less public appetite for social housing.

The percentage of Alberta households in core housing need has been rising steadily over the past three census periods, from 10.1 per cent in 2006 to 10.7 per cent in 2011 and 11.4 per cent in 2016. This refers to households either paying more than 30 per cent of their income on rent, are living in housing requiring major repairs, or are living in housing that has too few bedrooms for the household in question. In 2016, more than 164,000 Alberta households were in core housing need.

When Alberta's provincial government does fund new subsidized units, the process lacks transparency. This has been the case for decades (including under the previous NDP government). Provincial housing funding for capital is not allocated via a formal grant program through which non-profits (e.g. community housing/non-market housing providers) can apply for funding. Such a process has not been in place in Alberta since 2012. According to one key informant interviewed for the present chapter:

> They need to have open calls for funding initiatives. There's no program to apply for. People just lobby government. When provincial funding is provided for housing, we don't even know where the money comes from. We don't see funding in the budget for new builds, yet there will be announcements for new programs.

Another key informant made a similar comment with respect to permanent supportive housing:

> With the GoA, there's no clear application process for permanent supportive housing. They can meet with the minister and sometimes get approved. But there's not a transpareznt application process. Where's the portal? Where's the application process spelled out? It was similar under [former Alberta premier Rachel] Notley.

The GoA also lacks a clear, public reporting structure for subsidized housing. For example, most Albertans — including well-placed sources in the affordable housing sector — do not know: which projects have received funding; which types of households (e.g. singles, seniors, etc.) have been targeted; or in which municipalities the units are located. This lack of transparency makes it challenging for key actors in the non-profit housing and homeless-serving sectors to plan; it has also made it challenging for stakeholders to have a democratic dialogue about the appropriate allocation of public funding.[2]

Housing for seniors. Alberta has two seniors housing programs: the Seniors Lodge program and the Seniors Self-contained Housing program. The Seniors Lodge program is for semi-independent seniors and includes room and board. Unique to Alberta, the program was created in 1958 to free up spaces in auxiliary hospitals that were housing seniors who did not require such high levels of care. Municipalities have been a partner in this initiative from the beginning (GoA, 2015). Facilities that are part of the lodge program typically have large dining rooms and onsite recreational programs. Light housekeeping and home care are typically provided. Approximately 80 per cent of lodge units are located outside Calgary and Edmonton. The Seniors Self-contained Housing program provides apartment-style housing to seniors who are able to live independently, with or without the assistance of community-based services.

Rent supplements. In 2021, under the UCP, the GoA released a newly designed financial assistance framework for renters. The Rental Assistance Benefit (RAB) is similar to a program previously operated; it is a long-term subsidy intended for those in highest need (identified through social housing waitlists). The Temporary Rental Assistance Benefit (TRAB) is a shallower, time-limited subsidy (up to two years) intended for people with stronger labour market attachment. To be eligible for TRAB, households must either be currently employed or have been employed in the previous 24 months and cannot currently be receiving social assistance. The TRAB benefit is provided on a first-come, first-served basis. It was initially available only in Alberta's seven largest municipalities (Calgary, Edmonton, Red Deer, Lethbridge, Fort McMurray, Medicine Hat and Grande Prairie). However, on Aug. 5, 2022, the GoA announced TRAB's expansion to include a total of more than 80 communities (GoA, 2022a). In Calgary, both RAB and TRAB are administered by the Calgary Housing Company, the city's largest provider of affordable housing. In Edmonton, both programs are administered by Civida, a non-profit housing provider (Falvo, 2021).
Homelessness. In 2007, then-premier Ed Stelmach and Housing and Urban Affairs Minister Yvonne Fritz announced the GoA would embark on a 10-year initiative to address homelessness throughout Alberta. In January 2008, the GoA announced the creation of the Alberta Secretariat for Action on Homelessness, which was given a mandate to develop a 10-year provincial strategic plan to co-ordinate and end homelessness. The secretariat was established as an agency of the GoA,

> intended to not only develop but also to lead implementation of the provincial plan. To this end, the Secretariat was instructed to develop and coordinate new initiatives to address homelessness, such as prevention strategies, research programs, and the creation of a homeless information management system. The Secretariat was instructed to work with municipalities and communities throughout the province, and to support the development of community plans for action on homelessness (Alberta Secretariat, 2008, p. 3).

As such, in October 2008, the secretariat released *A Plan for Alberta: Ending Homelessness in 10 Years.* The plan placed great emphasis on Housing First, which means persons experiencing homelessness should be provided with permanent housing and all appropriate social work support, irrespective of whether or not they are deemed "ready" to maintain such housing.

However, according to one well-placed key informant, the GoA "slowly started losing interest" with changes in provincial leadership. The same key informant noted:

> The change in leadership with the Tories was as much a factor as elections. In a very short period we had Stelmach – Redford – Hancock – Prentice along with several cabinet changes. The funding kept coming but the leadership was lost after Stelmach and we entered a phase of benign neglect. When the NDP took over I think they dropped it altogether but didn't replace it. The money kept coming for the CBOs [e.g., Homeward Trust Edmonton, Calgary Homeless Foundation, etc.] but again there wasn't leadership.

The Secretariat was quietly disbanded well before the 10-year mark.

Today, when it comes to absolute homelessness (e.g. persons living in emergency shelters or sleeping rough) Alberta's seven largest cities have homeless-serving systems of care, each of which benefits from both federal Reaching Home funding and provincial Outreach and Support Services Initiative (OSSI) funding. Reaching Home is the Government of Canada's main funding vehicle for homelessness, while OSSI is a provincial program run through Alberta Human Services (OSSI funding is generally used for operating costs associated with Housing First).

In Calgary, Edmonton and Medicine Hat, both Reaching Home and OSSI funds are disbursed to large non-profit entities (namely, the Calgary Homeless Foundation, Homeward Trust Edmonton and the Medicine Hat Community Housing Society), which in turn disburse them to smaller non-profits providing direct service to persons experiencing homelessness. The province disburses the funds in the other four large cities to their local municipal governments, which in turn disburse them to small service providers.

Across Alberta there is a lack of publicly available reporting with respect to how either Reaching Home or OSSI funds are used. This makes it challenging to undertake critical analysis. It would be helpful, for example, to know the following for each funding source, for each city and for the most recent fiscal year:

- Which programs and organizations benefit from funding?
- What types of programs receive how much funding (e.g. prevention versus outreach versus wrapround supports)?
- Which age groups are targeted/prioritized?
- How does each system-planning organization (e.g. Homeward Trust Edmonton, City of Grande Prairie, etc.) allocate their administrative share of each fund — e.g. for their own internal use?

Throughout Alberta, provincial funding is also provided by the GoA directly to emergency shelters. However, this funding is inadequate. Provincial funding is not indexed to inflation and some emergency shelters have not received funding adjustments since 2008. This results in low wages for staff and high turnover.

There is very little publicly available, province-wide reporting on emergency shelters. It would be helpful, for example, to know the following about each of Alberta's emergency shelters:

- What percentage of each shelter's beds are occupied each night?
- What is the average length of stay in each shelter?
- What are each shelter's desired outcomes, as articulated in that shelter's core service agreements with the GoA (for example, are shelter officials encouraged and expected to help place residents into permanent housing)?
- How many new intakes and placements into permanent housing take place on a monthly basis? Of the housing placements, how many are assisted with rent supplements or housing allowances, and how many offer staff support once the person is housed?
- What efforts are made by shelter officials to provide follow-up services or to monitor recidivism once clients are placed into housing?
- How many FTE staffing positions does each shelter have?
- How many days of training does each new staff person receive, and what does that training consist of?
- What is each shelter's staff turnover rate?

Without such information, it is challenging for stakeholders of all types — including researchers and program evaluators — to engage in critical debate about how to improve programming and planning for persons experiencing homelessness.

Finally, during the last few years, the GoA has provided winter emergency funding to several emergency shelter sites in rural areas. In 2021-22, this included the eight rural communities of Cold Lake, Drayton Valley, Edson, Lac La Biche, Leduc, Peace River, Slave Lake and Wetaskiwin.

The Rachel Notley Years

In its 2016 budget, the Notley government announced its intent to nearly double annual provincial spending on housing (albeit on a time-limited basis). This represented a total of $892 million in new funding, initially spanning a five-year period;[3] approximately $13 million of this was earmarked for new units for vulnerable subpopulations, including for persons experiencing absolute homelessness.

In July 2017, Alberta's provincial government released their provincial affordable housing strategy, titled *Making Life Better*. Most of the new funding committed in this strategy was allocated to public bodies rather than to non-profits that operate at arm's length from government. Further, the funding was not allocated via a formal grant program through which non-profits (e.g. community housing/non-market housing providers) could apply for funding. As discussed, such a process has not been in place in Alberta since 2012. The provincial government initially claimed this would result in the creation of 4,100 new units of housing over five years through a combination of new builds and repairs. However, a lack of public reporting has made it challenging to critically assess these claims.

The UCP Years

During successive budgets, the UCP made several housing-related announcements, while generally taking a status quo approach with respect to homelessness funding. Meanwhile, UCP co-operation with federal housing initiatives was generally decent. The UCP struck an Affordable Housing Review Panel in July 2020, and that group's work was followed by the launch of a 10-year housing plan in November 2021. That same month, a provincial homelessness task force was announced. What follows is a discussion of the UCP's various housing-related initiatives.

Provincial budgets. The first three budgets introduced by the UCP included housing-related cuts, with some budgets containing modest increases. No major changes to funding for homelessness programming were announced in any UCP budget. Each UCP budget will now be discussed in turn.

The 2019-20 UCP budget announced that operating budgets for HMBs would be reduced by an average of 3.5 per cent. The budget also announced a 24 per cent reduction to the Rental Assistance Program that provides financial assistance for low- to moderate-income households to help with monthly rent payments for up to one year. This 24 per cent reduction was to begin in 2020 and take full effect within three years (see below for further discussion of the Rental Assistance Program). The budget also announced the following income assistance programs for low-income households would no longer be indexed to inflation: the Assured Income for the Severely Handicapped; the Alberta Seniors Benefit; Income Support; Special Needs Assistance for Seniors; the Supplementary Accommodations Benefit; and the Seniors Lodge Assistance program.

The 2020-21 budget included a 32 per cent cut to housing maintenance over three years. However, in the fall of 2021 (after the budget), some provincial stimulus funding was earmarked for affordable housing via the Municipal Stimulus Program (MSP). The MSP had a $500-million budget for roads, bridges, water and wastewater systems, public transit and recreation. Some of it was used for housing. For example, the City of Edmonton used $15.8 million from the MSP to support a Rapid Housing Initiative project[4] (specifically, the Westmount supportive housing site) and to assist several non-profit housing providers to undertake rehabilitation of existing housing units (Kjenner, 2022). The City of Calgary used $15 million from the MSP to renovate and repair affordable housing (Toy, 2020).

Budget 2021-2022 included a five per cent nominal increase in operating funding for provincially owned social housing units (though no new funding for the Seniors Lodge program). It also provided a $16-million annual increase in financial assistance to low-income households that rent primarily from for-profit landlords — a reversal of the 25 per cent cut to the Rental Assistance Program, announced by this government in its first budget (this funding is believed to have been claimed by the GoA as its provincial contribution to the Canada Housing Benefit initiative, discussed below). This budget also announced a six per cent nominal cut to capital maintenance for subsidized housing over the following fiscal year.

The 2022-23 budget announced $118 million in capital funding over three years for affordable housing in general. According to the budget document, this is intended "to begin implementation of Stronger Foundations: Alberta's ten-year strategy to improve and expand affordable housing." But three caveats are in order. First, this was financed in part by $90 million from the sale of provincially owned assets. Second, most of this funding was back-ended: $20 million for 2022-23, $39.9 million for 2023-24, and $58.1 million in 2024-25. And third, the Capital Investment budget for the Alberta Social Housing Corporation (e.g. provincially owned housing) for 2022-23 saw a 50 per cent reduction compared to the previous year. The new capital funding for housing was to include a modest increase, specifically for Indigenous people — namely, $20.7 million over three years for the Indigenous Housing Capital Program. A modest increase — $25 million over three years — was also announced for the operation of existing housing. This was to be back-ended as follows: $1.8 million for 2022-23; $4.2 million for 2023-24; and $19.3 million for 2024-25. Much of this was to go toward rent subsidies for low-income households.

Co-operation with federal initiatives. As discussed in this chapter's introduction, the GoC typically leads on affordable housing policy, with provincial and territorial governments cost-matching many of the federal initiatives.

Federal leadership on housing and homelessness in Canada saw a rebirth of sorts in 2017, when the GoC announced the National Housing Strategy (NHS). Alberta signed its bilateral agreement with the Government of Canada in March 2019, securing the maximum federal amounts available.

Several initiatives from the strategy required cost-sharing from provincial and territorial governments. What follows is a discussion of those initiatives in the Alberta context.

The Canada Community Housing Initiative (CCHI), unveiled as part of the NHS, focuses on preserving existing units of social housing across Canada until 2028. This entails $4.3 billion of federal funding over a decade and requires 50:50 cost-matching from provinces and territories. This is precisely the amount of federal funding that was set to expire over the course of the decade on existing social housing units, meaning this is about expiring operating agreements. Canada's approximately 500,000 social housing units that are administered by either provincial or territorial authorities and have rent-geared-to-income subsidies are eligible for this. This funding assists with repairs, helps keep rents affordable, and provides mortgage assistance for the operators.

As part of the NHS, the Investment in Affordable Housing (IAH) program was rebranded across Canada; in Alberta, it became known as the Alberta Priorities Housing Initiative and began on April 1, 2019 (after IAH ended). This bilateral program, requiring 50:50 cost-sharing with provincial and territorial governments, continues until 2028.

The Canada Housing Benefit (CHB) is a new federal initiative unveiled as part of the NHS consisting of financial assistance to help low-income households afford rent. It requires 50:50 cost-sharing with provincial and territorial governments. When it was announced in 2017, the GoC stated it expected the average beneficiary would receive $2,500 in support per year. As mentioned above, the 2019-20 provincial budget announced a 24 per cent reduction to the Rental Assistance program. This 24 per cent reduction was to begin in 2020 and take full effect within three years. According to a December 2019 *Canadian Press* article, "In negotiations over the funding arrangement, Alberta officials have sought to have their existing spending count towards the cost-matching approach instead of increasing funding as other provinces have said they would" (Press, 2019). Then, in July 2021, the GoA announced it had signed the Canada-Alberta Housing Benefit agreement, paving the way for the CHB to flow in Alberta. According to one well-placed key informant interviewed for this chapter:

> It's likely that the provincial cut announced in the 2019-20 provincial budget was later redirected toward the Canada Alberta Housing Benefit, and that there was no net increase in provincial funding for rent supplements [as expected under the federal initiative]. The GoA likely used that as their matching provincial contribution.

The Rapid Housing Initiative (RHI) is a federal initiative providing funding for newly built modular housing, the acquisition of land, the conversion of existing buildings into affordable housing, and the reclamation of closed or derelict properties. (Table 15.1, below).

Table 15.1: RHI funding announcements: Alberta

Date	Municipality	Stream	RHI funding	Units	Operator(s)
15 Dec 2020	Edmonton	Major Cities	$17.3M	80	Homeward Trust
17 Dec 2020	Calgary	Major Cities	$24.6M	178	HomeSpace, Horizon, Silvera
17 Mar 2021	Edmonton	Projects	$24.8M	130	Homeward Trust, GEF
6 Jul 2021	Edmonton	Major Cities	$14.9M	68	?
28 July 2021	Calgary	Projects	$16.6M	82	?
12 Jan 2022	Edmonton	Major Cities	$14.8M	125	Mustard Seed, Niginan
25 Jan 2022	Hinton	Projects	$2.3M	8	Town of Hinton, Evergreens Foundation
TOTAL			$115.3M	671	

Notes. Date refers to date of official CMHC announcement. The first five announcements in this table were made during Round 1, while the Jan. 12, 2022 Edmonton announcement was made as part of Round 2. Total figures in the bottom row are based on known CMHC announcements.
Source. Most of the data in this table was provided to the author by the Alberta Seniors & Community Housing Association on June 6, 2022.

Approximately $24.6 million in RHI funding was approved for Calgary through the RHI's Major Cities Stream. Edmonton received $17.3 million via the Major Cities Stream and $24.8 million via the Project Stream. In total, 210 units of modular housing will be delivered across five buildings in Edmonton. The GoA is providing a capital contribution of $16.3 million to the Edmonton projects via the Municipal Stimulus Program. In sum, RHI funding has been secured for projects — mostly for Calgary and Edmonton — but none have received provincial operating funding. This is problematic insofar as such operating funding must be in place for these units to support vulnerable tenants over the long term. Having said that, consultations by CMHC with provincial and territorial governments on the RHI before its roll out was extremely limited.

Affordable Housing Review Panel. In July 2020, Kenney struck the Affordable Housing Review Panel. As part of this process, Seniors and Housing Minister Josephine Pon appointed 10 experts to conduct a review with the view of providing "recommendations to transform affordable housing" (Alberta, 9 July 2020). The panel's final report, released in December 2020, included the following findings:

- The GoA owns almost half of Alberta's subsidized housing stock, with 60 per cent of units operating under a strict regulatory structure constraining the ability of providers to redevelop or partner with other entities (both non-profit and for-profit).
- Much of this stock is rather old — e.g., major repairs are needed in the near future.
- According to the report: "Because the government owns the assets, operators cannot leverage the properties that they operate to finance new development or reinvest in existing units" (p. 13).
- Many of Alberta's HMBs are small. "For example, 52 per cent operate fewer than 100 units and 34 per cent only manage one building. This means many operators lack operational and development expertise, have limited capacity to develop their portfolio and are not able to achieve economies of scale" (p. 13). This makes amalgamations appealing in principle.
- An important recommendation from the panel's final report was to ensure all proceeds from any transfer of assets be maintained within Alberta's affordable housing system. Specifically, the panel recommended "that the proceeds derived from the transfer of assets be held in a dedicated investment fund, with the income derived to be reinvested into affordable housing initiatives" (p. 15).
- The panel also recommended that the GoA's role change to "be as a regulator,

> policy maker, planner, funder and enabler of the sector rather than as an owner of affordable housing assets" (p. 14).

The panel's recommendations helped inform *Stronger Foundations: Alberta's 10-year strategy to improve and expand affordable housing*, a 28-page document unveiled in November 2021. The strategy also sought to directly address the panel's first recommendation, namely for the GoA to develop a clear long-term housing plan. *Stronger Foundations.* Stronger Foundations announced the intention of the GoA to undertake the following initiatives over a 10-year timeframe:

- *Reduce the GoA's role in property ownershi*p. According to the strategy, all proceeds from any real estate asset sales would be reinvested into the broader affordable housing system (keeping in mind that the Alberta Social Housing Corporation owns a considerable amount of aging stock). Having said that, real estate assets sold in one community will not necessarily be returned to that same community, especially if there is greater demand in a different community. The GoA was expected to start by transferring land (more so than buildings) for nominal sums. In cases where units are transferred, it is the province's preference to transfer them to their current operators.
- *Increase housing developments with mixed-income options.* According to the strategy: "Initiatives are planned to enable mixed-income developments through operating agreements with HMBs and new partnerships with the non-profit and private sectors" (p. 20).
- *Create 13,000 new units of affordable housing.* The strategy proposed to "bring Alberta closer to the national average of affordable housing supply without putting the entire burden on taxpayers" (p. 5).
- *Provide demand-side assistance to an additional 12,000 households.* The strategy proposed to "double the number of households receiving rent supplement (an increase of 12,000)" (p. 22). It is expected that such rent supplements might be channeled to communities with higher rental vacancy rates.

The strategy also committed to: shifting to competency based HMB boards; streamlining the income verification process for assisted households; increasing seniors' housing in line with population growth; enhancing the Seniors Lodge program (e.g. increasing the practice of co-locating continuing care beds in lodges so seniors do not have to move when they begin requiring care); investing in the Find Housing online tool; and establishing three-year targets for programs and new housing developments based on current and projected community needs. Bill 78, the strategy's enabling legislation, received royal assent on Dec. 8, 2021.

Homelessness policy under the UCP. Early in its mandate, the UCP funded the Herb Jamieson Shelter operated by Hope Mission in Edmonton. The first new emergency shelter funded in Edmonton in over a decade, it is a large shelter the UCP committed to funding in its election platform. According to one key informant: "Hope Mission is tight with the local faith community. When the UCP funded this, they appeared to be playing to the base of UCP supporters."

Similarly, the UCP intends to provide operating funding for a new emergency shelter in Edmonton operated by The Mustard Seed (another faith-based organization). This may provide between 40-50 new spaces.

In addition to the above, the government has made several pandemic-related funding enhancements related to homelessness; in most cases, the funding flowed directly to emergency shelters.

- In March 2020, $25 million in provincial pandemic-related funding was announced for homelessness. This was intended for "overflow homeless shelters and spots for people who need to self-isolate" (Bennett, 2020).
- On Aug. 5, 2020, the GoA announced an extension of this in the form of another $48 million for shelters and community organizations. According to a CBC News article, when this was announced, the Minister of Community and Social Services Rajan Sawhney indicated "there [were] no plans to reactivate emergency satellite shelters at convention centres in Calgary and Edmonton that wound down earlier [in the] summer. They were too expensive and the government wants a more affordable solution..." (French, 2020).
- In November 2021, the GoA announced it would provide $21.5 million for additional beds and isolation sites at emergency homeless shelters and emergency women's shelters until March 2022. About $13 million of the money would aid 14 shelters to expand space and provide meals, showers, laundry services and access to addictions and mental health services. Another $6.5 million is to be used to open about 285 isolation spaces in 10 communities, and $2 million would support emergency women's shelters (Mertz, 2021).

During the pandemic, there were some improvements in physical distancing at emergency facilities compared to the pre-pandemic period; however, this was inhibited by a lack of resources. There was a public health requirement of just one metre between persons in emergency shelters throughout the province — an exception to the province-wide two-metre requirement for the rest of the population. A two-metre requirement was in place at specific emergency shelters only during COVID-19 outbreaks — e.g. when there was one active case or more at the shelter in question (Falvo, 2020).

Harm reduction. Harm reduction focuses on reducing harm caused by drug use without requiring total abstinence. Harm reduction approaches include the distribution of condoms, clean syringes, safe inhalation kits and supervised consumption services. There is evidence that harm reduction approaches: reduce risk-taking behaviour; reduce the risk of transmission of blood-borne diseases; prevent overdoses; reduce crime; and increase contact with other supports, including health-care supports (Pauly et al., 2013). Recent research on supervised consumption services in Calgary has further found that: overdoses decrease steadily over time with such programming; and each overdose that is managed results in approximately $1,600 in cost savings resulting primarily from reduction in the use of emergency departments and pre-hospital ambulance services (Khair et al., 2022). Most harm reduction programs across Canada tend to target persons experiencing homelessness.

According to one key informant interviewed for this chapter, Alberta had rather robust harm reduction programming beginning in the 1980s, despite successive social conservative governments: "Activists worked quietly and effectively." For example, 1.4 million clean needles were being distributed in Edmonton annually even before the NDP formed a government in 2015. Alberta was also one of the first provinces in Canada to distribute naloxone kits (beginning in 2005), and one of the first provinces to distribute safer inhalation supplies (beginning around 2008).

The same key informant noted the UCP has since made harm reduction a wedge issue with voters. For example:

- Alberta Health Services (AHS) had a comprehensive organizational policy on harm reduction brought in under the NDP. It stipulated that people could not be excluded from health care due to substance use, and it covered all clinical programs covered by AHS and all AHS staff. The UCP changed this, limiting its scope and making it optional.
- Identification is now required to access supervised consumption services in Alberta. Patients must provide their personal health number. According to one key informant: "I'm not aware of anywhere else in the world that mandates this."
- Lethbridge had the largest supervised consumption service in North America, but it was closed by the UCP in 2020, and replaced with a two-booth mobile site. The previous initiative used to serve 20 people at a time; the two-booth mobile site today serves just two people at once.
- Edmonton lost Boyle Street, the city's largest supervised consumption site. The UCP had made an election promise about not having three sites in downtown Edmonton.
- The UCP closed injectable opioid agonist treatment (iOAT). AHS had operated

one in Calgary and one in Edmonton. This was long-term treatment for people with severe opioid challenges (e.g. persons who had tried other approaches without success). Patients were given one year to transition to oral medication — though most had not done well on it previously.

- The UCP plans to close Calgary's supervised consumption services (injection only) at the Sheldon M. Chumir Health Centre and intends to replace it with two shelter-based sites (*Canadian Press*, 2022).
- Medicine Hat was on the brink of getting supervised consumption. A building had been purchased, and then the UCP cancelled its funding.

According to one key informant who has strong familiarity with harm reduction in Alberta: "There was in effect quiet diplomacy for many years, but that's changed under the UCP, and people in the sector are now demoralized."

Admittedly, the UCP has made some abstinence-based treatment options more accessible to persons experiencing homelessness. Specifically, the UCP has reduced the need for fees for residential treatment programs. According to one key informant: "No one is supposed to pay out of pocket now to attend residential treatment." These programs are typically 30 days long, with some lasting up to one year. Some participants go from residential treatment to after care where they are still supported (but it is more like transitional housing). The UCP has also announced they will build five "recovery communities" for long-term residential treatment, some of which may now already be built.

The lack of harm reduction options has the potential to exacerbate homelessness, including amongst older persons. According to one key informant: "There's no seniors' housing for an alcoholic. Lodges don't take alcoholics. So those persons end up in a homeless shelter. They also don't fit well into long-term care."

In June 2022, the Legislative Assembly of Alberta's Select Special Committee to Examine Safe Supply released its final report. All NDP members of the committee had resigned four months earlier, calling it a "political stunt" (Smith, 2022). The report "was released online, without a government announcement or media conference" (Amato, 2022). To no one's surprise, the committee's final report did not recommend Alberta emulate British Columbia's exploration of safe supply. Rather, it recommended Alberta continue to focus on abstinence-based approaches (Select Special Committee, 2022).

Coordinated Community Response to Homelessness Task Force. Under the UCP, efforts were made to modify the funding model for emergency shelters. According to one key informant, the GoA (specifically, Community and Social Services) was developing

> a logic model for shelters. It [was rolled out] as of April 1, 2022 with the new funding cycle for shelters. Shelters will all be expected to have a logic model in place. Government is developing it and I think shelters will have some opportunity to give feedback on a draft.

Another key informant noted:

> Funding for our emergency shelter used to be based on shelter bed nights. Heads on mats and numbers of meals. I pointed out that this was not the right way to incentivize; so now [under the UCP] we have housing outcomes in part of our core service agreements. I was given the liberty of writing those into mine.

Possibly with the goal of building on such efforts, in November 2021, the UCP government announced the creation of a new homelessness task force. According to a government news release:

> The task force will look at how communities are affected by homelessness. It will also look at developing a model for responding to people with complex needs. Additionally, it will make recommendations that will help create an action plan on homelessness for the province (GoA, 2021b).

According to the GoA's website: "The Coordinated Community Response to Homelessness Task Force was established to find innovative ways to end recurring homelessness in our province and find long-term solutions to help those in need" (GoA, 2021a). The website reports the task force was also to:

> look at how communities are impacted by homelessness; conduct a thorough review of access to services, including shelter, food, financial assistance, health and recovery supports; and develop a coordinated and community-based model that responds to the individual and complex needs of vulnerable Albertans (GoA, 2021a).

Task force recommendations were to "inform an action plan for province-wide implementation and an evaluation framework" (GoA, 2021a). This work resulted in a Sept. 30 2022 report titled *Recovery oriented housing model: Report of the Coordinated Community Response to Homelessness Task Force*. The report itself consisted of little more than vague platitudes (it did not cite a single body of academic research). However, on Oct. 1 2022, the UCP government announced $63 million over two years

Table 15.2: Breakdown of $63 Million in Homelessness Funding

Action plan item	Funding 2022-23	Funding 2023-24
Additional funding for Edmonton community-based organization	$12 million	$12 million
Winter shelter demand	$9 million	$9 million
Expanding shelters to 24-7 service	$4.5 million	$9 million
Piloting the service hub model	$2.5 million	$5 million
		TOTAL: $63 million

Source. GoA, 2022c.

in new funding for homelessness, ostensibly in response to the task force report (GoA, 2022c). This new funding, which the UCP referred to as its Action Plan on Homelessness, consisted of the following:

- Equalizing funding between community-based organizations in Edmonton and Calgary (Calgary had previously been receiving more homelessness funding on a per-capita basis).
- The expansion of the number of emergency shelter spaces for the winter months.
- The conversion of all provincially funded emergency shelters to "24-7 access" (many had previously required that residents leave during the day).
- The piloting a new "service hub model" at emergency shelters in Edmonton and Calgary "to connect clients directly with supports and services such as recovery, housing and emergency financial support."

This new funding is summarized in Table 15.2 (above).

Also on Oct. 1, 2022, alongside the above announcements pertaining to homelessness, the UCP announced $124 million in new funding over two years for addiction services ($70 million in capital funding and $54 million in operating funding). The breakdown of this new funding is as follows shown in Table 15.3 (next page).

Discussion

While this chapter's focus is twofold (housing and homelessness), it effectively covers three areas: subsidized housing; homelessness; and harm reduction (largely due to how enmeshed harm reduction is with homelessness). The UCP government handled each of these areas differently, and as such deserve a different assessment.

Table 15.3: Breakdown of $124 million in addictions funding
Announced on October 1, 2022

Item	Funding level
Recovery Communities*	$65 million
Hybrid health and police hubs	$28 million
Therapeutic Living Units in provincial correctional facilities	$12 million
Medical detox	$11 million
Harm reduction and recovery outreach teams	$8 million

Notes. All budget items include both capital and operating funding, with the exception of "Harm reduction and recovery outreach teams," which consists only of operating funding. All funding is over two years.
* There are four recovery communities currently under development in Alberta, one in each of the following communities: Red Deer, Lethbridge, Gunn, and on the Blood Tribe First Nation.
Source. GoA, 2022b.

Subsidized rental housing: Insufficient funding, but a move toward better policy
From a budgetary standpoint, there were ups and downs with subsidized rental housing. The first three UCP budgets announced cuts to specific programs, although some programs saw modest increases.

The de-indexation of various income assistance programs (announced as part of the 2019-20 budget) has had a significant impact on housing affordability across Alberta, especially in light of high inflation.

The use (e.g. reprofiling) of previously allocated provincial rent supplement funding to count as the GoA's "matching contribution" toward the CHB initiative was disappointing for many affordable housing advocates. Federal initiatives such as these are intended to induce provincial and territorial governments to spend more, not to repackage existing funding.

A lack of provincial operating funding puts all RHI units in jeopardy across Alberta. All RHI projects will have great difficulty serving marginalized tenants without new operating funding from the GoA.

Both the Affordable Housing Review Panel and the Stronger Foundations strategy have been positive developments from the standpoint of housing operators across Alberta, especially as they relate to the transfer of ownership, the move toward a more diverse income mix among tenants, and commitments toward new units. Indeed, the recommendations appear sensible to most stakeholders, notwithstanding the fact that most recommendations depend on new provincial funding (not yet announced).

Homelessness: Additional funding and bias toward the faith community

UCP budgets were very status quo with respect to homelessness. Having said that, not everything has been status quo on the provincial homelessness front, as is outlined below.

The de-indexation of various income assistance programs in the face of high inflation will almost certainly lead to evictions and new homelessness across Alberta.

Pandemic-related funding for homelessness allowed for improvements in physical distancing across Alberta's homelessness sectors. Having said that, the fact that most shelters were unable to create two metres of physical distancing between individuals throughout the pandemic was revealing and disappointing.

It is not clear that preferential funding for faith-based organizations in the homelessness sector is good public policy. It likely is not, and reflects a willingness on the part of this government to play to its core supporters.

The work of the Coordinated Community Response to Homelessness Task Force resulted in important funding enhancements.

Harm reduction: A wedge issue and a move away from evidence

While not a central focus of this chapter, harm reduction is closely related to homelessness; most harm reduction initiatives have targeted persons experiencing homelessness, and the UCP government has been quite active on this file.

Harm reduction has been a wedge issue for the UCP, with the provincial government using its action on this file to distinguish itself from the NDP. The UCP has reduced access to harm reduction services, which has likely resulted in poorer health outcomes for vulnerable persons and has almost certainly resulted in the premature loss of lives. However, this government has shown an interest in providing funding enhancements for abstinence-based treatment, and some even for harm reduction (as announced in October 2022).

Conclusion

On the affordable housing front, this government's release of a 10-year strategy signalled the potential for improved policy. The strategy appears to open the door for better program design and more subsidized units. Having said that, the bold objectives articulated in the strategy have not yet been supported with appropriate funding, effectively making the strategy a North Star in search of a budget (though admittedly some of the strategy's recommendations may not require additional budgetary authority).

Under the NHS, it is commendable the GoA secured the maximum available federal funding amounts. However, many affordable housing advocates in Alberta are disappointed the GoA reprofiled existing provincial funding to count as its matching funding for the new CHB initiatives (rather than use *new* provincial funding). And while federal RHI funding has flowed in Alberta, supported units are still awaiting word on whether the GoA will provide operating funding.

No major funding changes for homelessness were announced during budgets, but a $63-million funding enhancement announced on Oct. 1, 2022 was welcome news for many. This government also signalled an interest in improved oversight of emergency shelters, with greater focus on the flow of people out of shelters into housing (but without additional funding to support successful transitions into permanent housing). Having said that, the inability of most of the province's emergency shelters to create two metres of physical distance between shelter residents during the pandemic revealed the extent to which emergency shelters were indeed under-resourced during this government.

Under the UCP, a major disappointment for practitioners, advocates, and researchers has been harm-reduction policy. The UCP appears to have used it as a wedge issue with voters, reducing access to harm-reduction initiatives in favour of abstinence-based approaches. This has likely resulted in a deterioration in health outcomes for both the vulnerably housed and those experiencing absolute homelessness. It has also quite likely resulted in the premature loss of lives of vulnerable Albertans.

References

Alberta (2020, July 9). *Expert panel formed to review affordable housing*. Retrieved from https://www.alberta.ca/release.cfm

Alberta Secretariat for Action on Homelessness. (2008). *A plan for Alberta: Ending homelessness in 10 years*. Retrieved from https://open.alberta.ca/

Alberta Seniors and Housing. (2021, November). *Stronger foundations: Alberta's 10-year strategy to improve and expand affordable housing*. Retrieved from https://open.alberta.ca/

Amato, S. (2022, June 30). *"'Rigged political circus': NDP, experts pan Alberta report on 'safe supply' of drugs."* CTV News. https://edmonton.ctvnews.ca/rigged-political-circus-ndp-experts-pan-alberta-report-on-safe-supply-of-drugs-1.5970133

Bennett, D. (2020, March 20). "'Unique vulnerability': Alberta to provide more aid for homeless in coronavirus crisis." *The Globe and Mail*. https://www.theglobeandmail.com/canada/alberta/article-unique-vulnerability-alberta-to-provide-more-aid-for-homeless-in/

Canadian Press, The. (2022, March 2). *Alberta in talks to open overdose prevention sites.* CBC News. https://www.cbc.ca/news/canada/calgary/alta-overdose-prevention-1.6370129

Falvo, N. (2020). *Isolation, physical distancing and next steps regarding homelessness: A scan of 12 Canadian cities.* Report commissioned by the Calgary Homeless Foundation.

Falvo, N. (2021). *Innovation in homelessness system planning: A scan of 13 Canadian cities.* Report commissioned by the Calgary Homeless Foundation.

French, J. (2020, Aug. 5). *Alberta government announces $48M to support homeless during pandemic.* CBC News. https://www.cbc.ca/news/canada/edmonton/alberta-government-funding-homeless-pandemic-1.5675211

Government of Alberta. (2015). *Seniors Lodge Program: Advisory Committee Final Report.* Retrieved from https://open.alberta.ca/

Government of Alberta. (2016). *Alberta's Affordable Housing System.* Retrieved from https://open.alberta.ca/

Government of Alberta. (2021a, Nov. 17). *Coordinated Community Response to Homelessness Task Force.* https://www.alberta.ca/

Government of Alberta. (2021b, Nov. 17). *Task force to develop homelessness action plan.* Press release. https://www.alberta.ca/release.cfm?xID=80397D793FD03-03BD-2650-6CF06810BAB22779

Government of Alberta. (2022a, Aug. 5). *Taking action on homelessness with $63 million in new funding.* Press release. https://www.alberta.ca/release.cfm?xID=84744C6B4DD90-939C-9D6E-8D42B2DE7839ED4A

Government of Alberta. (2022b, Oct.1). *$124 million in new funding to expand Alberta's response to addiction crisis.* Press release. https://www.alberta.ca/release.cfm?xID=84743C5D14A4A-A859-461B-4B6B0E7063E0ADDE

Government of Alberta. (2022c, Oct. 1). *Helping more Albertans pay rent: Temporary rent assistance is expanding to more than 80 communities across Alberta.* Press release. https://www.alberta.ca/release.cfm?xID=8436754CA2D7F-923B-0B6E-384023D950EC1893

Khair, S., Eastwood, C. A., Lu, M., & Jackson, J. (2022). Supervised consumption site enables cost savings by avoiding emergency services: A cost analysis study. *Harm Reduction Journal, 19*(1), 1-7.

Kjenner, C. (2022, May 9). Email correspondence with the author.

Mertz, E. (2021, Nov. 17). *Alberta provides $21.5M to extend COVID-19 supports to emergency homeless shelters.* Global News. https://globalnews.ca/news/8380929/alberta-covid-19-emergency-homeless-shelters/

Pauly, B. B., Reist, D., Belle-Isle, L., & Schactman, C. (2013). Housing and harm reduction: what is the role of harm reduction in addressing homelessness? *International Journal of Drug Policy*, 24(4), 284-290.

Press, J. (2019, Dec. 26). *Feds, Alberta set to clash over cash for new rent supplement.* CBC News. https://www.cbc.ca/news/canada/edmonton/alberta-ottawa-housing-benefit-rent-supplement-funding-1.5408759

Select Special Committee to Examine Safe Supply. (2022, June). *Select special committee to examine safe supply: Final report.* Retrieved from https://www.assembly.ab.ca/

SHS Consulting. (2020, December). *Final report of the Alberta affordable housing review Panel.* https://open.alberta.ca/

Smith, A. (2022, Feb. 4). *NDP members resign from Alberta safe drug supply committee, calling it a 'political stunt.'* Global News. https://globalnews.ca/news/8595760/alberta-ndp-resign-safe-drug-supply-committee-political-stunt/

Toy, Adam. (2020, Sept. 16). *Calgary to seek $152M in provincial stimulus money for municipal projects.* Global News. https://globalnews.ca/news/7339349/calgary-provincial-stimulus-money-alberta/

Appendix: Interview guide

Hello,

I'm writing a book chapter on the UCP government's performance on affordable housing and homelessness. The book in question is being co-edited by Trevor Harrison and Ricardo Acuna.

Here are some questions I'd like to ask you about your respective realm of expertise (I'm assuming you're more knowledgeable in one of the two areas, and not necessarily both). Please note that this interview would be non-attributable, meaning you would not be identified by name in what I write.

1. Let's assume you're talking to someone who knows the area in question (i.e., affordable housing or homelessness) but not the Alberta context. What are 3-4 Alberta-focused contextual factors such a person should know about in order to understand the policy area in question?
2. In the policy area in question, what are 3-4 significant things that this government has done?
3. What significant things has this government done relating to housing/homelessness and the overdose crisis?

14 Many would say that this government tried to shift GoA's focus from harm reduction to treatment. To what extent would you agree with that? To what extent were they effective in doing this? And to what extent were newly-developed treatment options accessible to persons experiencing homelessness?

5. What significant things has this government done relating to housing/homelessness and Indigenous peoples?
6. Of the things they've done in your area, which ones are you happiest with? In other words, describe why these things were good, and why this government deserves credit for doing those specific things.
7. What are some specific things this government has done in the area in question that you're unhappy with? What should people understand about how this government has let the sector down?
8. To what extent has the homelessness sector been tightly-aligned with the housing sector?
9. What about the alignment between the homelessness sector and the emergency shelter sector? Are those two systems working together cohesively?
10. The following types of roles are very important: ministers; backbench MLAs; senior public servants; mid-level public servants; and junior-level public servants. To what extend would you say that, under the UCP, these various players were all "on the same" page with each other?
11. To what extent did the various players listed above listen and dialogue respectfully with housing providers and senior homelessness officials in the community sector (e.g., CBOs)?
12. To what extent did they listen and dialogue respectfully with municipal governments?
13. Under this government, did you feel you could freely talk to your Minister and their staff? When you did try to dialogue with your Minister and their staff, to what extent was there a senior public servant (or several) trying to interfere or undermine those conversations?
14. What should the next government do to make positive changes in the policy area in question? In other words, what would be the 3-4 policy or program asks on your wish list, for the sector as a whole, for the next provincial government?
15. Do you have advice for the next government in terms of which types of people should be in key staffing roles in the senior public service?
16. Is there anything else you'd like to say about what we've discussed today?
17. Would you be willing to review a draft of this chapter?

NOTES

1 Trevor Harrison and Ric Acuna worked diligently to create the anthology in which this chapter finds itself. They provided inspiration, structure, and guidance to all authors. Sixteen research participants gave generously of their time toward this chapter. They all provided invaluable insight; many also provided detailed feedback on an early draft. The identity of participants is being protected due to the sensitive nature of the subject matter. Gary Gordon, Shaun Jones, and Steve Pomeroy provided feedback on an early draft. Others who provided comments requested their identity be protected. Susan Falvo and Jenny Morrow provided helpful proofreads of this chapter before it was submitted to the editors.

2 Unfortunately, this same criticism can be directed at most provincial and territorial governments, as well as at the Canada Mortgage and Housing Corporation.

3 This was eventually extended to seven years.

4 The Rapid Housing Initiative is a federal program discussed later in this chapter.

CHAPTER 16

EARLY LEARNING AND CHILD CARE: THE ROLE OF GOVERNMENT UNDER THE UCP

Susan Cake

"Early childhood education is a personal and social right, whose realization is essential for the integral development of the human being, the accomplishment of all other rights and the construction of a full citizenship, from birth."

— World Organization for Early Childhood Education, 2017

THIS CHAPTER EXAMINES Early Learning and Child Care (ELCC) in Alberta under the UCP from 2019 to 2022. During this time, several major changes were made, including cuts to government funding for diverse child-care arrangements, a review of and changes to Alberta's Child Care Licensing Act, the elimination of the Alberta Child Care Accreditation Standards, and the signing of three provincial-federal ELCC deals. In addition to all these legislative and budgetary changes, this chapter also considers the impacts of COVID-19 on the ELCC sector. Overall, the UCP has changed policies to reflect ELCC as a service meant to be bought and sold in a free market rather than a public service structured by government.

Background

Government investment in formal, high quality ELCC results in improvements to overall population health, helping women's equality especially around employment, and improving child development and future economic gains (OECD, 2018). Still, in many communities, caring for young children is seen as a private family responsibility, often left to mothers and women. This has been the case in Alberta, where most provincial governments have taken a hands-off approach and resisted calls to be more active in the delivery of care services.

Alberta currently has a mix of child-care arrangements that includes formally licensed facility-based care (e.g. daycares, out-of-school care, preschools), and home-based care providers who are licensed through family dayhome agencies. Family dayhome agencies are licensed by the provincial government to ensure, through support and monitoring, that dayhomes are meeting applicable standards and regulations. In total, there were 143,469 licensed spaces in Alberta in 2020-21 (Alberta, 2021a). There is also informal and unlicensed care, such as nannies and dayhomes not associated with a licensed dayhome agency. The number of unlicensed spaces in Alberta is unknown. As of 2021, there were approximately 530,000 children up to the age of nine in Alberta (Statistics Canada, 2021). As of March 2021, there were only 84,226 children enrolled in the approximately 144,000 licensed spaces, approximately a 58 per cent enrollment rate. Historically, enrollment has been around 80 per cent for licensed spaces in Alberta and the drop in 2021 is likely due to the COVID-19 pandemic (Alberta, 2021a).

In 2017, the Alberta New Democratic (ND) government launched the Early Learning and Child Care Centre Pilot, more commonly known as $25/day child care. With a later injection of federal funding, the NDP expanded the program to over 120 locations across Alberta. With this pilot program the NDP dramatically shifted away from the early learning and child care (ELCC) funding models previously used in Alberta that focused on income-based subsidies for families and a hands-off approach by government when it came to base staff wages, professional development, or curriculum. Many families described the pilot program as life-changing. Quickly after being elected in 2019, the UCP government cut the pilot program and returned once again to income-based subsidy funding focused almost exclusively on parental fees. In 2021, the Government of Canada announced an ambitious plan to fund ELCC across Canada which, in turn, required a more hands-on approach from provincial and territorial partners. The differences in governing approaches to ELCC highlights fundamentally different governing philosophies between the provincial UCP and the federal Liberals.

The funding for ELCC in Alberta has shifted since the UCP came into power and entails a patchwork approach with multiple streams designed to support specific aspects of ELCC. This approach to funding ELCC continues under the new Canada-Alberta agreement, which also has a heavy focus on subsidies. For example, there are parental subsidies to lower the parental fee. There are also grants that go directly to providers to offset costs or to incentivize specific kinds of care (such as infant care or overnight care). Generally, ELCC providers have been free to set their parental fees as they see fit within this mix of government funding.

Early childhood educators (ECEs) are the workers in the ELCC sector. The qualifications to enter the ECE field are quite low, and technically a person can start in the field and then begin training. There are three levels of certification, beginning with a 45-hour course at Level 1 and ending with a two-year diploma at Level 3.

Generally, ECEs receive relatively low pay and few benefits, although this can fluctuate with employers. The average ECE worker's annual wage is $34,691 as of 2019 (ALIS, 2020) for both certified and uncertified and workers in licensed and unlicensed facilities. The provincial government introduced a wage top-up for certified ECEs in the early 2000s based on certification level and had last been increased in 2008 to $2.14/hour for a Level 1; $4.05/hour for Level 2; and lastly $6.62/hour for Level 3. In late 2022 the UCP increased the top-ups up to $2.00 more per hour starting January 2023. Since 2016, the number of ECEs in Alberta has grown. However, the impact of COVID-19, along with some funding changes, have resulted in a shortage of ECEs at the same time as provincial and federal governments are focused on expanding ELCC (Dubois, 2021).

The UCP did not focus on ELCC in its platform in the 2019 election. The UCP platform criticized the NDP's promise to further expand the Early Learning and Child Care Centre Pilot, more commonly known as $25/day child care. The UCP claimed the NDP plan excluded "rural families, shift workers, parents who work at home, and those who prefer less formal kinds of child care" (UCP, 2019, para. 4). Subsequent to the election, the UCP made several changes to ELCC. Some changes were at the UCP's own initiative while others were in response to pressure and support from the federal government. These include:

- Funding Cuts: The UCP implemented several funding cuts focusing on what kind of care was supported by government funding for parents, operators, and early childhood educators. Since the 2019-2020 budget there have been decreases for ELCC: approximately $14.5 million in 2020-2021, and then another $0.5 million cut in 2021-2022 (AECEA, 2021a).
- Legislative and Policy Changes: The UCP discontinued the Alberta Accreditation Program and loosened aspects of Alberta's Child Care Licensing Act.
- Federal-Provincial Agreements: The UCP signed two agreements with the federal government to receive funding for ELCC in Alberta.

Overall, the changes the UCP has implemented are consistent with the approaches previous conservative governments in Canada have taken when it comes to ELCC (Friendly and Prentice, 2009). The UCP has attempted to limit government involvement in ELCC by scaling back different sources of funding and government oversight. At the same time, when the UCP has been forced to engage with ELCC, its interventions were often symbolic gestures that prioritize parental choice in a free-market system and claim to fund families most financially in need.

Funding Cuts

In its first year of office, the UCP government initiated several cuts to the patchwork of ELCC funding available in Alberta. These funding cuts were far-reaching, impacting children, families, care providers and their communities, and the cuts were often carried out with little to no consultation or warning from the government.

In November 2019, the UCP announced the elimination of the Kin Childcare Subsidy. This funding was geared toward families who had a non-parent family member caring for children. The subsidy provided $400/month for children who were not yet attending school and $200/month for children in school up until Grade 6. At the same time, the UCP also cut the Stay-at-Home Parent Subsidy, which was targeted to parents who attended school or worked part-time and helped cover the costs of preschool care. This subsidy provided $1,200 per year for each child. Both subsidies were cut without consultation. As a result, families were required to suddenly find new funding or new care arrangements. In some cases, families had recently received funding approval letters from government only to find out the subsidies had been cut before they were even able to access the funding (Edwardson, 2019).

Eliminating the Kin Childcare Subsidy affected the families receiving the funding and the non-parent family member who received the subsidy as income. For the Stay-At-Home Parent Subsidy, the loss of funding affected preschool operators, as families had less money for this type of care. The cancellation of these two subsidies in the last three months of 2020 resulted in $5.3 million in cost savings for the UCP (Alberta, 2020, p. 26).

Parent Link Centres, which support children and parents with free resources and supports focused on early childhood development, parenting skills, mental health supports, and ensuring family caregivers were connecting to community resources as needed, also had their funding changed and cut. In March 2020, the UCP cut funding to Alberta's Parent Link Centres and forced them to apply to the new Provincial Family Resource Networks program. The cuts to the Parent Link Centres were in combination with cuts to home visitation services, family resource centres, and other government prevention programs focused on children and families. Now Family Resource Networks are forced to compete for less overall funding and must serve children 0-18 in addition to families. These cuts in total amounted to approximately $12 million annually (McEwan, 2020).

In April 2020, the UCP also cut the Benefit Contribution Grant and the Staff Attraction Incentive, both of which supplement ECE wages. ELCC centres used the Benefit Contribution Grant to offset an increase in the cost of employer contributions to programs such as the Canada Pension Plan, Employment Insurance, and vacation pay under the provincial wage top-up program. The cut forced programs to shoulder these costs. The Staff Attraction Incentive was designed to retain ECEs by providing a bonus of $2,500 after completing 12 months as an ECE and an additional $2,500 after 24 months.

Also in April 2020, the UCP cut the Northern Living Allowance, which provided ECE workers in northern Alberta communities with an additional $1,000/month. For some ECEs, this cut represented about one-third of their monthly income, making it difficult to afford to continue to work. The UCP MLA for Fort McMurray, Tony Yao, openly stated the cut to the Northern Living Allowance would likely lead to increases in ELCC fees, which would disproportionately impact low-income families in the riding (Beamish and McDermott, 2020).

Legislative, Policy, and Accreditation Changes

Following these funding cuts, the UCP turned to legislative changes. The main ELCC legislation in Alberta is the Childcare Licensing Act. The act sets minimum standards for the health and safety of children in care. Another important policy was Alberta's Child Care Accreditation program, which was the only certification focused on quality ELCC in Alberta.

In March 2020, the UCP suddenly and without warning eliminated Alberta's Child Care Accreditation program in the name of cutting red tape. This optional, though highly subscribed, program allowed licensed providers to demonstrate and improve the quality of care in both dayhomes and centres. Accreditation standards focused on learning and developmental outcomes for children, evaluating how families and communities are supported by ELCC programs, and ongoing professional development and training for staff.

While in many ways the accreditation program had turned into an administrative exercise, accreditation was the only external verification parents and communities had for evaluating the quality of ELCC outside the basics of licensing. In fact, the Alberta government used program accreditation as a requirement for ECEs to receive provincial wage top-ups. With the removal of the Alberta Child Care Accreditation program, wage top-ups for staff began to flow to all licensed centres, regardless of quality.

When the UCP removed the accreditation program they indicated it would be incorporated into their review of the Childcare Licensing Act to ensure licensing addressed the quality of ELCC.

ELCC Licensing and Regulation Reviews

Despite the Child Care Accreditation being cancelled in March 2020, the UCP did not review the Child Care Licensing Act and Regulations until mid-April 2020, and the changes were not enacted until Feb. 1, 2021 — approximately a full year after the

UCP removed the accreditation program. The UCP reviewed the legislation under the general mantra of reducing red tape, improving quality and safety in the ELCC sector, and enhancing parental choice. The result was Bill 39: Child Care Licensing (Early Learning and Child Care Amendment Act) which, like its predecessor, covers care in facilities and dayhomes, as well as out-of-school care and preschools.

Since the previous voluntary accreditation program had been removed by the UCP with the promise they would incorporate quality standards into the act, it is important to consider how quality is addressed under the new legislation.

The term "early learning" was added to the title of the act and a preamble that includes Principles and Matters to be Considered sections was added as well. The Principles section focuses on the well-being and development of children, parental choice, and community engagement. The Matters to be Considered section includes vague statements on program planning such as: encouraging play experience to support development and learning; protecting children from physical, emotional and verbal abuse; and diversity in circumstances and backgrounds of children and the child's cultural, social, linguistic and spiritual heritage.

While the act now includes broad principles, there is no guidance at either the act or policy level for how to implement, assess, or enforce quality standards. These changes mean government does not have a strong role in enforcing quality in Alberta's licensed ELCC programs. The new broad sections of the act are likely far too vague to be enforceable except perhaps in the most egregious cases.

Some elements of quality ELCC are the curriculum used, the qualification of the staff, and the ratios of staff to children (Beach, 2020, p. 30). Mandated quality curriculum is a relatively easy way for provinces to promote quality ELCC that prioritizes optimum learning and development for children. Enhancing the qualification levels for early childhood educators has also been shown to improve the quality of care provided. Lastly, the ratio set between qualified staff and children is considered a key indicator of quality that interacts with both quality curriculum and staff qualifications; generally, the lower the ratios, the higher quality of care that can be provided, however staff with stronger educational backgrounds can provide quality care for higher ratios (Lesoway, 2020).

The UCP had the opportunity to mandate a curriculum for licensed ELCC operators that would have directly impacted quality. They did not take this step, even though Alberta does have its own early learning curriculum framework, the Flight curriculum. Other provinces, such as Ontario, have had a mandated curriculum since 2014.

There were no substantial changes made to the qualifications or professional development of the ECEs either. This is despite the educational standards in Alberta being on the lower side compared to other provinces (Beach, 2020). As of March 2021, approximately, 41 per cent of ECEs in Alberta were at a Level 1, 16 per cent were Level 2, and 44 per cent were Level 3 (Alberta, 2021a, p. 30). As mentioned before, these workers are not well compensated and go without various benefits that other workers may have access to, such as health and disability benefits, so it is difficult to retain ECEs. There is also limited time, funding, and access to further professional development for these workers.

The UCP did take the opportunity under this legislative review to change ratios for ECEs to children by introducing mixed ratios and increasing the ratios for dayhome operators overseen by a licensed agency. Recommended ratios for ELCC providers are generally set according to the age of the child, recognizing that younger children require more attention and, therefore, require lower ratios than older children. Generally, the UCP did stick with this recommendation. However, they also introduced mixed-age ratios, where the age of the majority of children is often the determining factor for which ratio level is applicable. While mixed-age ratios can be good for children's overall development, at the ratios set for operators, it would likely require careful planning and attention from a well-qualified ECE to be successful (AECEA, 2021b). The UCP also increased ratios in family dayhomes overseen by licensed agencies by no longer including the dayhome operators' own children in the ratio count. These two changes to ratios, in combination with leaving educational standards on the lower end, are unlikely to help increase the quality of care provided in Alberta.

One last change the UCP introduced was licensed overnight care. Prior to this change, overnight care existed in licensed agency dayhomes and in informal care, but licensed facilities were not able to be licensed to provide this care specifically. The UCP has not reported any official statistics on the rollout of overnight care beyond reporting that they used $10.3 million to create up to 1,600 spaces with 44 daycares and family dayhomes that included the creation of newly licensed care options, overnight care, and culturally and linguistically diverse care options (Alberta, 2021a, pg. 28-29). According to comments in the legislature, as of March 2022, there were only 97 new overnight spaces operating in Alberta (Alberta, 2022a, p. 590).

Federal-Provincial Agreements

The first federal-provincial agreement considered here is the Canada-Alberta Early Learning and Child Care Agreement under the Multilateral Early Learning and Child Care Framework, originally introduced in 2017. The second agreement is the Canada-Wide Early Learning and Child Care Agreements, (referred to as the Canada-Alberta Child Care Agreement), more commonly known as the $10/day child-care plan, announced on Nov. 15, 2021.

The former NDP government used the Canada-Alberta Early Learning and Child Care Agreement to expand their ELCC pilot project, known as $25/day child care. This pilot program, which covered 122 non-profit ELCC sites across Alberta, went far beyond implementing a fee cap for families. The program fee cap worked alongside the already established parent-fee subsidy in the province. This fee cap meant no family paid over $25/day for ELCC per child at a pilot centre. Families that also qualified for the fee subsidy were paying less than $25/day, and some families were even receiving free care. The program provided additional wage subsidy and professional development opportunities for workers in the sector. The program also included a made-in-Alberta curriculum (the Flight curriculum). Additionally, the pilot provided funding for ELCC centres to provide care during non-standard hours, such as late at night and on weekends, if enough families required this care. This pilot fundamentally shifted how ELCC was approached in Alberta, from being a product of the free market to being a service delivered by the non-profit sector and financially supported by government.

The UCP government rolled back the $25/day pilot program in phases. In phase one the UCP pulled all the provincial funding, impacting approximately 22 centres in June 2020. The second phase came with the expiration of the federal multilateral agreement that impacted an additional 100 centres in March 2021. Despite the generally positive reviews it received from the external review agency (R.A. Malatest and Associates, 2019; 2020), the UCP cancelled the program.

In July 2020, the UCP renegotiated and signed a new Multilateral Early Learning and Child Care Framework with the federal government for $45 million. They signed a second Multilateral Agreement in August 2021 for over $290 million in federal funding over four years. The UCP used the federal funding in three general areas: parental subsidies; funding for the workforce; and funding to address accessibility and quality in ELCC.

The UCP used the new federal funding to increase parental subsidies for the first time since 2008 (French, 2020). The new subsidy capped funding for families with annual incomes above $75,000, although it did increase the amount of subsidy families received, on average by 18 per cent (Buschmann, Fischer-Summers, Petit, Cameron, and Tedds, 2020). The increase to the subsidy only amounted to what subsidies would have been if they had increased with inflation (Buschmann et al., 2020). With the changed parental subsidy, the UCP claimed Alberta's 23,000 lowest income families would pay, on average, $13/day for care (French, 2020).

However, there were clear flaws with this subsidy program. Families with a yearly income that surpassed $75,000 experienced significant loss in financial support. A family with one infant in care saw a loss of $9,144. The loss increased with every additional child the family had in care (Buschmann et al., 2020). More generally, the parental subsidy continued the historical inequity wherein lower-income families end up spending a higher percentage of their earnings on ELCC than other families. According to Buschmann et al. (2020), a two-child family making $20,000/year would spend approximately 40 per cent of their income on ELCC with the subsidy, while a comparable family earning $60,000/year would only spend 19 per cent of their income on care with the subsidy.

The UCP later expanded the parental subsidy to include families with household incomes up to $90,000 starting Sept. 1, 2021 as part of the second Multilateral Early Learning and Child Care Framework. Shortly before this, the UCP also began including preschools as eligible for the parental subsidy. This offset some of the financial impact on families and operators caused by the elimination of the Stay-At-Home subsidy approximately a year-and-a-half earlier. However, without a redesign, the same general issues with the increased parental subsidy remained.

In Budget 2021, the federal Liberal government announced an ambitious plan to fund ELCC across Canada, commonly referred to as $10/day care. The pre-election timing of the announcement was an eerie re-living of the 2004 federal Liberal government announcement to introduce a national ELCC program under then-prime minister Paul Martin. The Harper government essentially dismissed the 2004 plan, after winning the 2006 federal election, by cancelling all the signed bilateral agreements. The Harper Conservatives then introduced Choice in Childcare Allowance that provided all families with $1,200 a year for each child under six years old, regardless of income or care arrangements.

The federal Liberals were able to win reelection in 2021 and carry on with signing bilateral agreements. The federal investment is significant at $30 billion from 2021 to 2026. The focus of the funding has been almost exclusively sold as lowering the fees for parents, first by reducing fees by an average of 50 per cent in the first year, then lowering fees gradually until achieving an average of $10/day in 2026. Included in the $30-billion investment is $2.5 billion for Indigenous ELCC (Alberta, 2022b).

On Nov. 15 2021, the Alberta and federal governments announced they signed the Canada-Alberta Childcare Agreement. This deal marks a federal investment of $3.8 billion in ELCC for children until kindergarten[1]. The funds are divided out as follows:

- $2.865 billion to lower fees for families
- $240.64 million to increase the number of regulated spaces available, predominately in not-for-profit and dayhomes
- $202.6 million to develop and fund care options for vulnerable and diverse populations, including children with special or additional needs
- $306.16 million to support licensed ELCC programs and certified ECEs

For-Profit

The current UCP government has been adamant that Alberta is unique when it comes to the role of for-profit providers in ELCC. The then-minister of Children's Services Rebecca Schulz switched from claiming Alberta's for-profit providers make up 60 per cent of current providers to 70 per cent but maintained this as one of the reasons the UCP could not sign a deal with the federal government sooner. While it is true that most licensed ELCC providers in Alberta are for-profit, that statement is also true for several other provinces, including British Columbia, Newfoundland and Labrador, P.E.I., Yukon, and Nova Scotia (Akbari, McCuaig, and Foster, 2020), most of whom signed deals months before Alberta.

Despite the UCP's assertion that the mixed-market provision of ELCC in Alberta was a barrier to an agreement, they were eventually able to sign a deal. Like the deals done with other provinces, all current licensed ELCC spots are eligible to receive the federal funding, including for-profit and not-for-profit. It becomes far less clear if newly created ELCC spaces will be funded in the for-profit sector. According to the federal government, the space creation aspects of the Canada-Alberta Childcare Agreement include the creation of at least 42,500 spaces in the not-for-profit and dayhome fields in ELCC, which would exclude for-profit private providers from any space creation funding from the federal government. However, the UCP has released its own document indicating that, while for-profit providers are not eligible to have new spaces funded under the current deal, with input from the newly formed Canada-Alberta Implementation Committee, the UCP "expect that for-profit providers will be invited to create new spaces near the end of 2023 to meet the demand for spaces throughout Alberta" (Alberta, 2021b, p. 2, para. 9). This foreshadowed an unexpected amendment to the Canada-Alberta Childcare Agreement.

Although the agreement between Alberta and the Federal government was signed in November 2021, neither government released the agreement text until June 2022. When the Federal Government released the Alberta agreement, they also released an Amendment to the Alberta deal signed in April 2022 that contained additional information unique to Alberta. This Amendment specially references Alberta's "mixed market system" and commits to a new yet to be negotiated For-Profit Expansion Plan, along with a cost control framework, in addition to creating 42,500 not-for-profit spaces. The Canada-Alberta Implementation Committee, made up of representatives from provincial and federal governments along with some select stakeholders, will conduct these negotiations for the new For-Profit Expansion Plan. It is gravely unfortunate this amendment was negotiated seemingly in secret from represented constituents, but also more importantly from stakeholders in the child-care sector in Alberta.

Operating Grants

The Canada-Alberta Childcare Agreement introduced operating grants that help offset ELCC user fees via supply-side funding. Generally, supply-side funding is preferable to what is called demand-side funding (funds targeted at parents through subsidies or tax breaks). Demand-side funding is linked to a market ideology and the idea that families looking for ELCC are consumers in a free and open market (Friendly and Prentice, 2009, p. 68). Under demand-side funding, parental choice, whether a practical choice or not, is prized above other aspects and parental choice is given through consumer power in the free market (Friendly and Prentice, 2009, p. 68). However, supply-side funding allows governments far more control in helping to develop an ELCC system, create sustainable infrastructure, and assure quality in the ELCC programs in ways parents as consumers cannot.

Under this new funding model in Alberta, ELCC providers must be licensed and committed to reducing their fees on average to $10/day by 2025-2026 to be eligible for an "affordability grant." All providers, regardless of their set up (for-profit, not-for-profit, and home-based) are currently eligible for any licensed spaces that already exist or have already been approved. Providers must also commit to keeping their annual fee increases to three per cent or lower in the coming years.

The UCP designed the grant based on an average of current parental fees across Alberta. The operating grant fluctuates depending on the ages of the children and the setting in which they are cared for — facility versus home-based care — ranging from $635/month to $300/month. Having the operating grant fluctuate with the age of the child recognizes that it costs more to care for infants because they require lower staff ratios. Many daycares adjust their fees downwards as children age to reflect the lower costs; however, the costs to parents are not often perfectly reflected in parental fees. Most often, care for infants is indirectly supported by all families in a centre to keep it at affordable levels. Some ELCC facilities and home-based care providers have also opted to charge a flat rate regardless of age. This nuance in how ELCC fees are set is not reflected in the current one-size-fits-all operating grants for either facility or home-based settings.

Averaging the current parental fees across Alberta means that, under the current formula, ELCC operators that have more expensive costs, perhaps due to higher rents or higher costs for food or equipment, will be unlikely to provide $10/day care in the future. This makes it unlikely that higher costs communities, such as Calgary and Edmonton, will benefit as much as lower costs ones (Buschmann, 2021).

Another notable gap in this form of funding is the lack of an overall parental fee cap. At the time of writing, there is nothing preventing ELCC operators from adding additional fees for what may be considered value-added services such as food, or even access to activities such as swimming or child yoga or music programs. There is no regulation that any value-added fee also needs to reflect the actual cost of the value-added program. Lastly, there is nothing preventing these value-added fees as being informally required to gain access to a ELCC program.

There are several other issues with the current design of the operating grants. These are not true operating grants but rather a different form of parental fee subsidies, as the grants are designed to lower parental fees and not to support and enable the provision of quality ELCC. The operating grants are not a reflection of the actual operating costs to care for children, but rather a reflection of the cost that parents were being charged, and the "savings" are to be passed along to the parents. Still, the grants allow the government considerable latitude to impact program delivery and development in ways the parental subsidy does not, such as limiting fee increases.

Parental Subsidies

One of the cornerstones of many Conservative approaches to funding ELCC is demand-side funding. This is a broad category of funding arrangements wherein government funding is targeted directly at families and includes parental subsidies, tax credits, and child benefits. Some of these approaches have historically been income-based, such as the ones in Alberta, while others have been universal with restrictions set according to the age of the child, such as the federal one implemented by the Harper Conservatives.

Conservatives often promote demand-side funding, claiming it supports "parental choice" (Friendly and Prentice, 2009). As mentioned before, the choice comes from parents having more consumer power in a free and fair market system, not from the funding increasing the availability of ELCC. As has traditionally been neoliberal practice, more than just funds are downloaded onto families: so too is the responsibility for keeping the ELCC system in check and to ensure providers are offering quality care. The expectation is that parents will avoid low quality care providers, and this will incentivize higher quality ELCC. The reality of the situation in most communities, though, is there are not nearly enough ELCC spaces available to have market competition. This is especially the case for families with children who have additional needs and families with multiple young children needing care at the same time. These dynamics, combined with the limited ability of many parents to properly evaluate ELCC (due to a lack of information, expertise, and time), raise questions about whether this is a suitable approach to regulating ELCC.

Alberta has had a parental-based subsidy available for families seeking ELCC for many years now, and the most recent version is funded jointly by Alberta and the federal government. This subsidy is income-based, and families must apply through a government website. The ELCC provider, which must be a licensed provider, receives the subsidy directly from the government on behalf of the family. The UCP set the subsidy for preschool children in 2021 to a maximum of $644 per month for children in facility-based programs and $516 per month for children in home-based daycares. Families who made under $50,000 received the maximum subsidy, which decreased in steps to zero when families made $90,000. The Canada-Alberta Childcare Agreement was announced alongside changes to the parental subsidy, which included expanding the income threshold and lowering the overall amounts received by families. Families who have children covered under the agreement can receive a maximum subsidy of $266 for families who earn under $120,000 per year. Families earning $120,000-$179,999 will receive a partial subsidy. The subsidy rates are equal for facility-based and home-based ELCC arrangements and do not fluctuate with the age of the child.

While the overall amounts for the parental subsidy have been lowered and the coverage expanded to higher incomes, what has not changed in Alberta is the hour qualification to receive the subsidy. To receive the full subsidy, children must be in care for at least 100 hours per month. This incentivizes families to send their children to ELCC even when it is not required or when children are sick in order to qualify for the subsidy. If a child is not in care for at least 100 hours per month, the family will not receive the full subsidy for that month and is expected to make up the difference[2].

Another issue with the parental subsidy is the lowest-income families are still often benefiting the least from the current funding arrangement (Buschmann et al., 2020). This outcome sits awkwardly with the UCP's insistence that it must target government spending to families the UCP has determined required the assistance (Alberta, 2021c).

An analysis by Buschmann (2021) focused on the new parental subsidy and operating grant found that although fees for all families will decrease, lower-income families (less than $50,000 per year) would not see the promised 50 per cent reduction in fees. In some cases, their reduction would only be 24 per cent for a low-income family in Edmonton and 13 per cent in Calgary. The variation for these fee reductions is largely the result of how the operating grant under the new program is calculated using an average from across the province that does not account for variations costs (e.g. higher rents, utilities) as well as additional services that some facilities provide (e.g. transport, food). This means, "the new system is more generous to more families overall, but is less sensitive to family need" (Buschmann, 2021, p. 2).

This situation could have been remedied if the UCP had approached the reduction in ELCC fees by implementing a fee cap for everyone and then used their parental-subsidy system to lower fees based on family income. Other options could have included a fully public system or making further increases for low- and modest-income families alone.

The Workforce

Under the Canada-Alberta Childcare Agreement there is just over $300 million dedicated to the ECE workforce. No announcements have been made to change the current wage-top-up program available to all certified ECEs working at a licensed program. There have also been no announcements on plans to train more ECEs to enter the field or to help current ECEs achieve higher certifications. One explanation for the UCP maintaining the current wage top-up is a desire to keep the general wages that have been set in the free market. Without an increase in the ECE workforce, it is doubtful the government will achieve its ambitious space creation goals.

Impact of COVID

In March 2020, the UCP government closed all licensed ELCC centres in Alberta to prevent the spread of COVID-19. Select dayhomes were permitted to continue providing care, although with limited capacity. Select ELCC centres were gradually opened, with priority going to children of frontline workers (Alberta, 2021a). Eventually, all ELCC facilities were able to reopen. ELCC providers continue to be under incredible financial stress and the workers faced unique occupational issues with the delayed rollout of a vaccine for children under five years of age and the difficulty for younger children to socially distance or wear masks in any useful manner.

COVID-19 has had a detrimental impact on staffing for ELCC facilities. In November 2021, a survey done by the Association of Early Childhood Educators of Alberta (AECEA) found approximately one in five programs reported at least one COVID-19 case among staff in August or September 2021. However, the risk of COVID-19 goes beyond the physical illness and has impacted staff morale, with about one-third of respondents concerned about staff leaving in the next few months (AECEA, 2021c). Since 2016, the number of certified ECEs in Alberta has increased every year from approximately 14,600 in March 2016 (Alberta, 2017) to a high of 18,818 in March 2020 (Alberta, 2020). The most recent report shows Alberta only has 14,984 certified ECEs as of March 2021, a loss of nearly 4,000 (Alberta, 2021a). According to the Children's Services Minister, as of December 2021, the number of ECEs has increased back to pre-pandemic levels (Johnson, 2022). Historically, the government tracks the number of ECEs based on the wage top-ups distributed. Now, the wage top-up is given to more ECEs, including preschool ECEs, and licensed operators have expanded to include those working at dayhomes that have recently joined a licensed agency. Given the changes to who receives wage top-ups, which is how the government tracks ECEs, it is disingenuous to claim the workforce is back up to pre-pandemic levels.

Another outcome from COVID-19 has been a drop in overall enrolment for ELCC centres. The drop in enrolment is caused by several factors that include parents no longer requiring care as their employment or education situation has changed, families no longer being able to afford care at these licensed centres, and centres struggling to keep stable staffing with ECEs leaving the field or needing to isolate. The drop in enrolment left many operators without their usual funding and few ways to make up the shortfalls. It is likely the lower parental fee structure introduced in January 2022 will cause enrolment with licensed providers to increase.

The provincial government has provided intermittent funding injections from provincial and federal funds to the sector in response to COVID-19, but this funding lacks consistency and predictability. The intermittent funding has included:

- $30.0 million through the Child Care Relief Program, which provided support to reopen;
- $72.9 million from the provincial/federal Safe Restart Agreement;
- $11.1 million in surplus funding left unspent from the Canada-Alberta Early Learning and Child Care Agreement in February 2021; and
- $15.7 million through a one-time Critical Worker Benefit payment (Alberta, 2021a).

The UCP used part of the unspent funding from the Children Services Ministry budget (due to lower ELCC enrollment overall) for the Working Parent Benefit that provided a one-time payment of $561 per child to families who paid for ELCC during the pandemic and had incomes below $100,000 per year. Approximately 21,000 families were able to receive this benefit (Alberta, 2021a). This funding likely did not actually support the ELCC sector, as it went directly to parents who had already paid for services months prior.

Overall, the sector continues to face pressure with under-enrolment, staff exhaustion, and lack of consistent and stable funding that supports ELCC operations stemming from COVID-19.

Assessment

The overall trends from the many changes made by the UCP are consistent with approaches used by other Conversative governments in Canada. The most notable approach the UCP has taken has been to consistently centre the free market as key to the provision of ELCC, and to repeatedly attempt to limit government interventions where possible. Although the UCP has not delivered on its core messages around ELCC, which were to ensure parental choice and that funding was directed to those they deemed most in need, it has succeeded in centreing the free market in the delivery of ELCC services — except where limited by the federal agreements which tend to emphasize supply-side funding.

The way the UCP has begun implementing the federal ELCC plan is by keeping the free market as involved as possible. This can be seen in their dedication to the parental subsidy, the calculation of the operating grant, their hesitancy to move away from wage top-ups to a provincial wage grid for ECEs, and how they have avoided setting an overall fee cap for ELCC in Alberta. The parental subsidy keeps demand-side funding as part of the provincial funding package. The UCP calculation of the operating grant is based on the market rate previously set by programs and does not necessarily reflect the cost of delivering quality care, and avoiding a provincial wage grid means ECE wages are still primarily set by the free-market.

Despite the federal goal of growing the not-for-profit and public parts of the sector, the UCP's insistence that for-profit ELCC providers be supported with government funding demonstrates the government's dedication to a free-market system. The UCP has repeatedly stated that for-profit providers are not expected to become non-profit, and the amendment to the federal agreement is evidence of their continued push for for-profit providers to receive expansion funding in 2023. The UCP often justifies this approach by claiming Alberta's mixed market of providers is unique in Canada, which it is not.

The UCP argues that supporting for-profit ELCC providers in Alberta supports women entrepreneurs, as outlined by an op-ed by then-minister Rebecca Schulz (2021). What the UCP often ignores here is that most ELCC providers, whether home-based, not-for-profit or for-profit, are led by females. Females are the vast majority of the workforce also. There is nothing unique about for-profit operations in terms of who owns them compared to any other providers in this sector, and to attempt to single out for-profit owners and praise them for being women-led appears disingenuous.

Lastly, it appears the UCP is doing little when it comes to ensuring quality ELCC in Alberta. The UCP removed the accreditation program, eliminated the previous government's ELCC program which mandated the Flight curriculum, did not required any peer-reviewed curriculum as part of licensing or government funding, and has not yet addressed ECE education or compensation. If quality was truly a goal of the UCP, they could do more in terms of improving staff qualifications and ratios, mandating ongoing professional development, improving operating funding, and including metrics for evaluation and development. The quality component of ELCC is incredibly important, as high-quality care can be beneficial for all children, but especially for low-income children (Friendly and Prentice, 2009). Further, the shortage of licensed ELCC spaces means most families have limited ELCC options and thus little ability to select a provider on the basis of quality. This means that the UCP's approach to relying upon market pressure to regulate quality is unlikely to be successful. These barriers for parents to assess and regulate quality, combined with the vulnerability of children, is key to why governments must have sound regulations and legislation that are backed by adequate funding.

Conclusion

Considering the UCP's performance, they have limited provincial government oversight and funding of ELCC while claiming to prioritize parental choice. Doing so, while simultaneously constraining parental choice by cutting off funding for specific ELCC options, such as the Stay-at-Home Subsidy, is impressive spin doctoring. In terms of affordability, the UCP has made ELCC generally more affordable for families accessing licensed care, although this is largely due to the federal government agreements. The UCP has done little to ensure quality care, opting instead to take actions like removing the accreditation program and not requiring licensed programs to offer a peer-reviewed curriculum. As programs become more affordable and the demand for ELCC increases, it is not clear how the UCP will deal with increasing the number of spaces and the number of ECEs required to staff those spaces. They have also been successful in presenting ELCC as a service that government should only ensure is offered to the lowest income families, despite not actually implementing any programs that accomplish this goal. Additionally, the lack of predictable and stable action in the face of COVID-19 left the ELCC sector to fend for itself, and aligns with a minimal role for government.

The influx of federal funding seems to have forced the UCP into greater direct participation in the setting up of provincial ELCC than perhaps it would have done on its own. Yet even here, the pushback by the UCP continues with its dedication to a free-market model. Ultimately, how successful the party will be in permanently shaping ELCC in Alberta is yet to be fully determined.

References

Akbari, E., McCuaig, K., & Foster, D. (2020). *The Early Childhood Education Report.* https://ecereport.ca/media/uploads/2021-overview/overview2020_final2.pdf

Alberta. (2017). *Children's Services Annual Report 2016-2017.* Edmonton.

Alberta. (2019). *Children's Services Annual Report 2018-2019.* Edmonton.

Alberta. (2020). *Children's Services Annual Report 2019-2020.* Edmonton.

Alberta. (2021a). *Children's Services Annual Report 2020-2021.* Edmonton.

Alberta. (2021b). *Canada-Alberta Canada-Wide Early Learning and Child Care Agreement Fact Sheet for Operators.* Edmonton.

Alberta. (2021c). *Making Childcare More Affordable for Families.* Edmonton. https://www.alberta.ca/release.cfm?xID=80488A8915525-D24A-753B-D41250B4B115874C

Alberta. (2022a). Legislative Assembly of Alberta, The 30th Legislature Third Session. Standing Committee on Families and Communities. Ministry of Children's Services Consideration of Main Estimates, Tuesday March 8, 2022. https://docs.assembly.ab.ca/LADDAR_files/docs/committees/fc/legislature_30/session_3/20220308_0900_01_fc.pdf.

Alberta. (2022b). *Federal-Provincial Child Care Agreement*. Alberta. https://www.alberta.ca/federal-provincial-child-care-agreement.aspx#:~:text=Parents%20per%20cent%2020will%20per%20cent%2020see%20per%20cent20reduced%20per%20cent20fees,they%20per%20cent%2020earn%20%20per%20cent20%20%20per%20cent24180%20%20per%20cent2C000%20%20per%20cent20and%20%20per%20cent20above

ALIS (2020). *Early Childhood Educator*. Alberta. https://alis.alberta.ca/occinfo/occupations-in-alberta/occupation-profiles/early-childhood-educator/#other-sources-of-information.

Association of Early Childhood Educators of Alberta. (2021a). *Statement on the Provincial Budget*. https://aecea.ca/sites/default/files/AECEA per cent20Statement per cent20Provincial per cent20Budget per cent20March per cent202021 per cent20FINAL.pdf

Association of Early Childhood Educators of Alberta. (2021b). *AECEA's Response to the Early Learning and Child Care Regulation Changes*. https://aecea.ca/sites/default/files/AECEA%20Response%20to%20the%20Early%20Learning%20and%20Child%20Care%20Regulation%20Changes%20FINAL2.pdf

Association of Early Childhood Educators of Alberta. (2021c). *A Rough Two Months and a Worrying Future: Results from a Survey of Alberta Early Learning and Child Care Programs During the Fourth Wave of COVID-19*. https://aecea.ca/sites/default/files/211122%20A%20Rough%20Two%20Months%20Full%20Report.pdf

Beach, J. (2020). *An Examination of Regulatory and Other Measures to Support Early Learning and Child Care in Alberta*. Edmonton, AB: Edmonton Council for Early Learning and Care and the Muttart Foundation. https://static1.squarespace.com/static/5f170b16bf7d977d587e43c4/t/5fbe82537acac6192a96998c/1606320730938/Beach+Report+Measures+to+Support+Quality+ELC+2020-11.pdf

Beamish, L., & McDermott, V. (2020, March 10). Workers, parents protest loss of living allowance for child care workers. *Fort McMurray Today*. https://www.fortmcmurraytoday.com/news/local-news/workers-parents-protest-childcare-cuts

Buschmann, R. (2021) *Still Unaffordable for Low-Income Families? In Alberta's New Child Care System, Out-of-Pocket Fee Reductions are Smaller for Lower-Income Families*. Edmonton Council for Early Learning and Care. https://static1.squarespace.com/static/5f170b16bf7d977d587e43c4/t/61bcb0777d9f5a2d09679750/1639755895178/Still+unaffordable+for+low+income+families+2021-12-14.pdf

Buschmann, R., Fischer-Summers, J., Petit, G., Cameron, A., & Tedds, L. (2020). *Fiscal Policy Trends: Analyzing Changes to Alberta's Child Care Subsidy*. University of Calgary, The School of Public Policy. (https://www.policyschool.ca/wp-content/uploads/2020/11/FPT-AB-child-subsidy.pdf).

Dubois, S. (2021). *Alberta lost one in five licensed early childhood educators during 1st year of pandemic, data shows*. CBC News. https://www.cbc.ca/news/canada/edmonton/alberta-chlid-care-educators-federal-covid-19-1.6139406

Edwardson, L. (2019, Nov. 8). *"It's tough": Alberta government to end 2 child-care subsidies at end of year*. CBC News. https://www.cbc.ca/news/canada/calgary/preschool-melanie-nickle-roots-and-wings-leanne-collings-1.5352556

French, J. (2020). *Federal funding to increase Alberta child-care subsidies for first time in 12 years*. CBC News. https://www.cbc.ca/news/canada/edmonton/federal-funding-to-increase-alberta-child-care-subsidies-for-first-time-in-12-years-1.5660687). Friendly, M. & Prentice. S. (2009). About Canada: Childcare. Fernwood Publishing.

Johnson, L. (2022, Feb. 28). NDP warns Alberta won't hit $10-per-day child care target without extra $200M from UCP. *Edmonton Journal*. https://edmontonjournal.com/news/local-news/ndp-warns-alberta-wont-hit-10-per-day-child-care-target-without-extra-200m-from-ucp

Lesoway, M. (2020). *Regulations Can Support Quality Early Learning and Child Care*. Edmonton Council for Early Learning and Care. https://static1.squarespace.com/static/5f170b16bf7d977d587e43c4/t/5fd2a41df800b4559815925f/1607640093402/ECELC+Brief+-+Quality+and+Legislation+2020-12-08.pdf

McEwan, T. (2020). *New networks offering services for families after funding cancelled for Parental Link Centres*. CBC News. https://www.cbc.ca/news/canada/edmonton/family-resource-network-parental-1.5651479

OECD. (2018). Poor children in rich countries: why we need policy action. *Policy Brief on Child WellBeing*, OECD Publishing, Paris.

R.A. Malatest & Associates. (2019). *Early Learning and Child Care Centre Pilot: Year one Evaluation Summary of Access to Quality Child Care at $25 Per Day*. Edmonton.

R.A. Malatest & Associates. (2020). *Evaluation of Early Learning and Child Care Centres: Annual Report 2018-2019*. Edmonton.

Royal Commission on the Status of Women in Canada. (1977). *Report of the Royal Commission on the Status of Women in Canada*. Ottawa: Information Canada.

Schulz, R. (2021, May 1). Opinion: A truly universal child-care plan would meet the needs of all families. *Edmonton Journal*. https://edmontonjournal.com/opinion/columnists/opinion-a-truly-universal-child-care-plan-would-meet-the-needs-of-all-families.

Statistics Canada (2021). Table 17-10-0005-01 Population estimates on July 1st, by age and sex

UCP. (2019). *Alberta Strong and Free: Getting Albertans Back to Work*.

NOTES

1 The original November 2021 announcement stated the funding would be provided to children "not yet in kindergarten." This was changed in January 2022 to include children in kindergarten, although it still does not include children in Grade 1 and above who require out-of-school care.

2 There is a temporary COVID-based measure where families can still qualify for the full subsidy if they are absent from care due to illness; all other reasons are reviewed on a case-by-case basis.

SOCIAL ALBERTA

CHAPTER 17

DYSFUNCTION IN THE FAMILY: PROVINCIAL-MUNICIPAL RELATIONS UNDER THE UCP

Ben Henderson

"Cities have the capability of providing something for everybody, only because, and only when, they are created by everybody."

— Jane Jacobs, *The Death and Life of Great American Cities*, 1961

I HAVE A BIAS. I served for 14 years as an Edmonton city councillor, a city now of one million people. I have felt for many years that provinces are a redundant order of government. I say this partly in jest, but the reality is Edmonton is larger in population than over half the provinces and territories, and is comparable in size to Saskatchewan and Manitoba. At an approximate annual budget of $4.5 billion, its expenditures are also comparable at least to the smaller provinces. Similarly, Calgary sports a population of nearly 1.5 million and an approximate annual budget of $5.5 billion. The populations of several other Alberta cities are also substantially large, with concomitantly sized budgets.

Yet municipalities have no constitutional status and are completely governed in Alberta by the province's Municipal Government Act (MGA). In fact, they are often referred to as the "children of the provinces," a term not said out of any feeling of parental fondness. Municipalities have a direct and close relationship to their constituents and provide a huge range of day-to-day services. They have natural boundaries and a direct relationship to their communities. Municipalities serve important functions different from, and not replicable by, their federal and provincial counterparts. But municipalities are often frustrated at being treated as junior forms of government, especially by provincial governments.

Of course, this frustration runs both ways. Provincial governments have their own agenda and are no doubt annoyed that another order of government — one they perceive as a *lesser* order of government — does not do what it is told; especially when that order of government plays by a similar, but not identical, set of rules. In Alberta, particularly, municipalities use a formal set of rules for debate, but do not operate in a traditional parliamentary and partisan system with formal parties, caucuses, governments and oppositions. Indeed, as I will detail, overt partisanship is not useful in good municipal governance. But there are also tensions created as municipalities — because they lack significant independent sources of revenue generation — are forced to seek provincial dollars for most of their infrastructure and various other needs, while the province is attempting to control its own spending.

Add to this mélange the expectations of a public served by a complex system of three orders of government — four, if you include school boards — whose nuanced responsibilities can only really be enjoyed, or understood, by political nerds. The average citizen does not understand these nuances nor, ultimately, do they care. They simply want their collective governments to provide the services they expect, at the quality they expect, for the lowest possible cost to their personal funds.

Aided by interviews I conducted with some key informants, this chapter journeys through the 14 years, six premiers, and significantly more ministers of municipal affairs, with whom I am most familiar, to examine the complex and changing relationship between Alberta's municipal governments, large and small, and the provincial government. As I will detail, this relationship has often been bumpy and uncomfortable, but has been particularly fraught since 2019, when the United Conservative Party (UCP) was elected to office. I begin with a brief historical perspective.

Traditional Conservative Governments Through the Rear View

While hindsight is often clouded by nostalgia, a look back is useful for comparison and context. Talking with my municipal colleagues, all have a fond spot for Premier Ed Stelmach's government (2006-2011). It was not perfect — there were some tough issues that needed to be addressed and were ignored — but, in general, they felt Stelmach both understood and appreciated what local governments brought to the table. The cuts of the previous Ralph Klein government left municipalities with virtually no funds to maintain their infrastructure, let alone to invest in growth. Stelmach's creation during his tenure of the Municipal Sustainability Initiative (MSI) represented a long-term and significant re-investment. The MSI has since been whittled down, and the province's commitment to get out of the Education Property Tax business as a way to replace the MSI, thus allowing for a stable source of municipal funding, never materialized. Nonetheless, the MSI program was critical in allowing municipalities to dig out of the infrastructure deficit created over the previous decade. Stelmach understood the value of municipalities and believed investment in them would benefit Alberta as a whole; that the province could not thrive if its cities, towns, counties and municipal districts did not also thrive.

And there was a good reason for this. Stelmach's government understood the value of municipalities and municipal governments because he had been a municipal reeve and much of his caucus had experience in local government. This contrasts with the current UCP government, of which only two or three caucus members have experience in local government, one of whom, Nathan Cooper, is legislative speaker and thus expected to be distanced from the government and the creation of policy.

Rightly or wrongly, the old PC party traditionally used municipal government as a kind of farm team for their own caucus. While inherently patronizing and undervaluing of the contribution of municipal servants, this was politically useful as provincial party candidates came ready-built with name recognition. And the benefit to municipalities was that many government MLAs understood and had dealt with the struggles of municipal government and could bring that perspective to the legislature. That connection currently seems detrimentally severed.

Barry Morishita, former mayor of Brooks and former president of the Alberta Urban Municipalities Association (AUMA), points out another, albeit perhaps minor, contributing factor to the changed relationship: access to government planes. Government planes, and the sense of entitlement they created, later became a political hot potato and were sold off by the Jim Prentice government. But while they existed, the planes allowed provincial ministers to fly *en masse* to geographically isolated areas for an evening event and still return to Edmonton for the next day's House sitting. They allowed a sense of connection to government outside the major centres.

As fondly as the Stelmach years are remembered, the short-lived Conservative governments that followed evoke fewer or less positive recollections. Communications with former premier Alison Redford were far less than with Stelmach; the sense of connection began to erode. There was some work on the ongoing subject of the city charters, a possible new governance and financial deal with the two big cities, about which I write more below. Redford apparently also made a specific commitment, asking her cabinet to find ways to support the mid-sized cities, though neither I nor Bill Given, then mayor of Grand Prairie who mentioned it to me, are clear why this particular commitment was important or ultimately significant to her. The Redford years did, however, see a substantial investment in the expansion of Edmonton and Calgary's light rail transit systems, the result of hard-fought battles with their two newly elected and re-elected popular mayors, Don Iveson and Naheed Nenshi, respectively.

The Dave Hancock and Prentice PC governments that followed were short-lived and hence largely unremarkable in their impact upon municipalities. An MOU was signed between Prentice and the large cities to continue to move forward the Big Cities Charter, but its execution continued to be absent. The relationship between Prentice and the Edmonton mayor's office was at least one of constructive respect, but those representing the smaller more rural municipalities felt Prentice viewed municipalities as unnecessary; indeed, he apparently said so. Whether Prentice's view would have changed with time is hard to know, but the story highlights the negatively evolving municipal-provincial relationship.

Prentice's political background is also instructive. Like Premier Kenney, he came from the federal government. He did not grow up politically through the provincial government scene, let alone the local government one. Relationships can develop and improve with time, but neither Prentice nor his government had the chance to exhibit a capacity to listen and adapt. In that absence, some might argue, the Prentice government's disconnection from local government hastened its downfall.

The First Changing of the Guard: A New Democrat Government

Alberta has a long history as a one-party state. To borrow, with some mischief, the old joke, Albertans change governments once every two or three generations whether we need to or not. For municipalities, this meant ingrained habits of getting to know, work with, and inform the provincial government; fostering relationships on the assumption this would assist in advocacy. Further, the Klein years made people fearful of being punished if they stepped too far out of line, and this included working too closely with the opposition. While the arrival of a new premier might indeed bring about some slight change, the constant for municipalities was to try to work with the existing provincial government, not against it. The only exception to this habit was Calgary, which, by nature of its size and historic predilection for voting PC, was more inclined to take an ornery tone if necessary. Edmonton, which traditionally did not vote PC, did not have the same luxury, and no other municipality alone had the collective clout to safely take on the province.

So, while the 2015 election of Rachel Notley's New Democrats replaced old relationships, the long habit of municipalities working and building relationships with the provincial government did not change. The necessity of getting to know and work with the new government to see where interests overlapped, remained.

There was once significantly different condition, however. With such a large number of new members elected (only four New Democrats had sat in the legislature prior to the election), the new government lacked experience and a relationship with the municipalities. Joe Ceci had been a Calgary city councillor, but he was asked to take on the ministry of finance, not that of municipal affairs. The one advantage to this was that much of the negotiation on the city charters for Edmonton and Calgary, which in large part focused on a new funding formula, went through his office. I am informed, however, the new government proved willing to learn, an approach that helped re-establish relationships with municipal governments, particularly outside the big cities.

The one caveat is the New Democrats did not entirely trust the municipalities, which they suspected of being part of the old Tory machine. This harkens back to my earlier point that municipal politics are largely absent partisanship. Getting a pothole filled does not beg a partisan response. But, when one lives in the legislature's partisan world, one comes to believe everyone and everything is partisan. I always understood that getting anything accomplished on my council of 13 meant getting six other members to agree with me. The next issue might require six different people. It was foolish to make permanent enemies. With 87 members, the provincial legislature, for good or ill, works in a very different way. The NDP's initial mistrust of municipal councils, believing them as populated by Tories, was a mis-reading they needed to overcome. Ironically, the UCP government has repeated this lack of trust. The current government appears to believe municipal councils — at least those in the major cities — are dominated by New Democrats and it is no more true now than it was then.

It is fair to recognize the mandate of provincial governments does not always align with the needs of the municipalities. And municipalities are not completely compliant, but why should they be? Like those before it, the Notley government, tried to show fiscal restraint at a difficult economic time. This restraint bumped up against municipalities' real needs for continued investments in infrastructure and such long overdue things as affordable housing. The negotiations on the big city charters funding formula and the new financial funding formula for all the remaining municipalities was understandably fraught and, in the latter case, ultimately unsuccessful.

The New Democrats did, however, take on what became the Intermunicipal Collaboration Frameworks that allowed for rural counties and municipal districts to work out relationships with surrounding towns. This framework created a way for regions to plan together, share resources, and, most contentiously, share revenue. The issue was less critical in Calgary, which had been allowed to amalgamate decades before, but Edmonton, which is surrounded by some large municipalities, such as St. Albert and the County of Strathcona, felt the pressure of providing services for a region despite lacking any real say on how the region was growing or who would pay for the services. As an example, there is only ever going to be one Commonwealth Stadium in the Edmonton region, but everyone from the region expects to use it and, additionally, get to it. Who should pay? If we were aiming to build more compact and thus affordable cities and keep expenses down, then everyone in the region had to play by an agreed upon set of rules. And in Edmonton, this was further exacerbated by the industrial tax base being unevenly distributed, whereby significant revenue might go to one player, but the expenses to another.

This was not just an Edmonton problem. It was an even trickier issue outside the large centres, where towns and mid-sized cities provided needed amenities for counties and municipal districts that, in some cases, due to linear assessments on industrial development, had significantly more access to revenue than did others. Previous PC governments, with their strong rural roots, had been reluctant to address the issue because it pitted two of their loyal constituencies against each other; though, to his credit, Stelmach paved the way with what has become the Edmonton Metropolitan Region Board. Less encumbered, the New Democrats took the issue on and worked out a framework that, for the most part, was acceptable to all parties. It rendered the work of counties more complex, as they needed to create multiple frameworks for the numerous towns in their midst, but people I interviewed seem OK with the result. For its part, the UCP seems to have stuck to that framework, but also took away one of its most useful tools: the ability to work out a revenue sharing formula that could be adopted by and agreed to by all parties. Without this ability to create a formula, it is no longer possible for municipalities to share risk and have a framework that adapts to changes over time.

One observer told me the New Democrats recognized municipalities as a useful tool and collaborator in economic development. This may in part have been because the issue was particularly important to them in trying to restart an economy suffering from a long slowdown in its traditional oil and gas economy. In part, it may also have been because the New Democrats felt vulnerable to the narrative that they did not care about, or had no experience and connections in, the economic/business field. Whatever the NDP's motivation, it clearly understood what the federal government was also coming to understand: that municipalities are a key tool in driving economic growth and can be useful partners in achieving their goals.

For some time, Alberta's two largest cities pushed for the creation of city charters. The idea had a two-fold rationale. It was part governance/policy and part financial. It was long apparent the MGA did not fit the needs of either Edmonton or Calgary, let alone the smallest summer village. Both the largest and the smallest communities were frustrated, but for different reasons. The size of both big cities required a governance document that could describe and define their relationship with the province in a different way, based also on, it must be noted, their own quite different priorities. The charters were not designed to pull the big cities out of the MGA; rather, to identify their different needs and accommodate for those differences. At the same time, the plan was to design a fiscal framework that could give some certainty to the fiscal relationship, either through new taxation powers for the cities or through some kind of reliable formula to replace the MSI. Municipal governments had been long frustrated at not knowing from year to year the level of provincial revenue support they could expect. This uncertainty made planning difficult, especially as municipalities moved to multi-year budgets dealing with capital/infrastructure spending.

Neither the Alberta Urban Municipalities Association (AUMA), representing the rest of the cities and towns, nor the Rural Municipalities Association (RMA), representing the counties and municipal districts, enthusiastically viewed the idea of the separate city charters, worrying they might create an unfair advantage. But they understood why the two cities wanted them and begrudgingly tolerated the effort. The mid-sized cities were particularly nervous, but also hoped that if Edmonton and Calgary's issues could be sorted out, then perhaps they could be next. Parallel to these discussions was work being done to overhaul the MGA itself, including a look at replacing the soon-to-expire MSI funding.

Work on the charters proceeded slowly throughout the New Democrat's term in office, and much of the effort from the provincial side of the table was driven by the bureaucrats and government officials. This probably affected the ultimate agreement. From the city's point of view, the ability to do a deal did look to create an opportunity for a political win for the province and always had, but the discussion had gone on inconclusively for most of a decade.

A large part of the governance focus of the charter was that, as Alberta's two big cities had grown, there were more and more areas where municipal and provincial responsibilities overlapped, but no good mechanisms for resolving how to work together on those files. As an example, one can see how a partnership model, where overlapping jurisdiction was sorted, could help with challenges like homelessness, the opioid crisis and mental health. COVID-19, of which I will say more later, also showed where clear collaboration between both orders of government could provide huge benefit. The original work focused on creating frameworks to facilitate both orders of government working together in a spirit of partnership, with the understanding that the real work of building out the details could happen later. In the end, while only some minor powers were transferred to the cities, the charters did establish an initial framework which still exists, but the harder task of fleshing out the details remains unfinished. There is little sense the UCP is committed to or understands the value of that partnership, the lone exception being during the COVID-19 pandemic. I am told the current government said it intended to return to charter discussions in the latter part of its mandate, but that commitment seems forgotten and time has run out.

The fiscal conversation was always going to be the trickier one, in part because where money goes, so also goes power. As mentioned before, the municipalities wanted some form of stable, predictable funding. There are two ways to achieve this. The first is to create a funding formula. The second is to give municipalities, in this case the two big cities, their own increased taxing powers.

Regarding the second option, Alberta's municipalities really only have two significant sources of revenue. One is the property tax, based on market value assessments; the other is user fees. There are some other smaller revenue sources, from fines and investments, but these are usually insignificant. The problem with property taxes is they are inelastic. Unlike a sales or income tax, property taxes do not grow or contract directly with the economy. There is no connection between revenues and expenses during times of growth. For a municipality to increase its tax base, it must change its mill rate; that is, raise taxes. By contrast, provincial and federal tax revenues increase as the economy expands or, in the case of Alberta, the oil patch heats up. Of course, it is harder for a province to adapt during a downturn, but in times of growth, when the demand for expenditures is greatest, the revenue is automatically present and there is no need to be seen raising taxes to meet that increased demand.

To deal with their unique fiscal problems, the cities argued either for access to some different tax measures, linked to what was actually happening in the world around them, which would relieve the province to a degree of having to deal with them when times were tough; or keep direct funding in place but tying it to the performance of the overall provincial economy. This latter choice was settled upon.

I am not sure why the province historically has been reluctant to give the cities more tax tools, as this would get the municipalities out of the province's hair. I suspect it is largely because provincial governments are loathe to give up power and influence, and credit for helping with various projects; few politicians want to forego the chance of cutting ribbons and having a photo-op. Surprisingly, while this attitude has existed with all the provincial governments, it has heightened under the UCP. Given the UCP's mantra of cutting the size of government and lowering taxes, one would think passing on tax responsibility to another order of government so that you can cut your own would seem attractive. But, and I am hypothesizing, this option is less attractive if your intent is also to get municipalities to shrink their own governments. I will say more about this shortly. This latter narrative would have been driven by years of advocacy against the charters by anti-tax groups like the Taxpayers Federation. They equated "charter" with "more tax." Governments, including the NDP, didn't want to be seen to be "increasing taxes."

A final agreement was struck between the two major cities and the NDP government just prior to the 2019 election. It represented yet another decrease in the base amount of the traditional MSI funding, but contained a formula for increasing the amount based on increases in provincial revenues. It was not ideal from the cities' perspective, but it did at least provide a predictable funding source. The Big Cities Charter was embedded in legislation supported by all parties in the House. The UCP committed to standing by the charter and the funding agreement as part of their election platform. It looked like years of effort had finally paid off.

Prior to the election, there was hope a similar agreement could be struck with the rest of the municipalities, but they were not offered the same deal and felt compelled to decline what was offered. In the end, it did not matter. After the election, the UCP tore up the big city financial framework. Having been left out of the deal, the smaller municipalities not surprisingly were muted in their sympathy, outrage, and general solidarity with the big cities, but the tearing up of the big cities' financial deal was just the start of what was to come.

And the Pendulum Swings Again: The UCP Government

This brings us to the UCP and the life of municipalities under the new government. In general, the municipalities once again entered the relationship in good faith. Irrespective of any partisan leanings during the election, municipal leaders continued to understand the need to work together and had hopes, too, that a sense of partnership with the province could continue. As always, there was a desire to find places where interests overlapped and the means by which both orders of government could continue to move forward. But, as confirmed in all my conversations with representatives of all sizes of communities, the relationship quickly went off the rails.

The Alberta municipal world comprises four forces: Edmonton, Calgary, the Rural Municipal Association (RMA), and The Alberta Urban Municipal Association (AUMA), with perhaps a subset of mid-sized cities in the latter. Alberta politics, when it takes the form of division, which it sadly too often does, pits these forces against each other. In traditional Alberta politics, the province might foster this division, playing one group against another in a strategy of divide and conquer. What no one expected was that the newly enthroned UCP government would pick a fight with all four municipal forces at the same time. No one I talked to has given a definitive answer as to why the government did what it did.

Two simple answers are possible, however. One is that the UCP government, and former premier Jason Kenney in particular, wanted by any means possible to impose everywhere their agenda of lower taxes, cutting back on government, and making way for business. The second explanation, not entirely separate from the first, is that the incoming Minister of Municipal Affairs Kaycee Madu simply believed he could run the municipalities better than they could. Numerous sources told me Madu saw himself as the "boss of the municipalities," and went out of his way to pick fights. (I have also been told, albeit second-hand, that certain ministers were indeed given instructions to intentionally do so.) This explanation is largely borne out by op-eds he wrote at the time suggesting, "My entire department is working to help empower municipalities like Edmonton and Calgary to get the job done for their residents, and I have no doubt we will be successful" (Madu, 2019). Ironically, the municipalities had for years been working on such issues as cutting red tape, greater efficiency in government, and economic development. But these efforts were ignored.

A couple of comparisons about the past two Alberta governments may be instructive. Prior to 2015, there had been no formal political change in Alberta for decades, hence no real need for a massive change of personnel. This was appropriately true of the bureaucracy, but it also applied to political staff. When the NDP and the UCP came into office, however, they had to hire their political staff from scratch. There is nothing inherently wrong with this, and in fact such change is a healthy and necessary part of democracy. But, in both cases, if for different reasons, the new governments heavily imported their staff from outside Alberta. For the New Democrats, it was largely because they needed people with previous political experience in governing. For the UCP, I think, the reason lay in the premier's political connections and experience in Ottawa. Previous PC governments had strong connection, through their caucus and political staff, to Alberta's grassroots. But this was not the case for the UCP, and was later exacerbated by constant staff turnover, particularly at the senior levels. In consequence, the UCP lacked strong staff connection with municipalities and Alberta's grassroots.

Like the NDP, the UCP came into office lacking experience in municipal matters. Unlike the former, however, they chose not to recognize or understand what they did not know. Chris Spearman, then mayor of Lethbridge, tells of two early contacts with their local MLAs. At the first, Grant Hunter, the UCP MLA from Cardston, attended a regular meeting of forty mayors from across southern Alberta at which he chose not to listen, but instead to lecture them on what should be the municipalities' priorities. The UCP, he asserted, had been elected on a mandate and it was the municipalities' job to support the provincial government. The second contact involved Nathan Neudorf, the newly elected UCP MLA for Lethbridge East, who early on came to speak to Lethbridge city council. Mayor Spearman characterizes that meeting as, "A guy with no prior electoral experience coming in and telling us our priorities were all wrong and that we were to support the priorities of the UCP government." In effect, the MLAs were telling municipalities their role as MLAs was not to represent the region to government, but to represent the provincial government and its desires to the region. Everyone was to fall in line. There was no recognition by the new government that they were dealing with other politicians who had their own mandate, and who also had a significantly deeper and longer-standing knowledge of their communities.

Whether the attitudes of Hunter and Neudorf prevailed among UCP members elsewhere is unclear, but it was the situation in Lethbridge. The same attitude was repeated when Economic Development Lethbridge's "Team Lethbridge" made their regular trip north to meet with the provincial government. This team is made up, not just of politicians, but a broad cross-section of the city, including the chamber of commerce, the university, and the police. But, again, the government would not listen, but told them of their mandate. The city's officials wanted to meet either the minister or the associate minister to discuss the critical issue of harm reduction and drug usage in Lethbridge, but were instead given a meeting with Joseph Schow, the MLA for Cardston-Siksika, who contradicted everything they put on the table. Mayor Spearman says they felt "bullied."

The big slap for municipalities came with the first budget. The UCP government employed nearly every mechanism it could to cut the flow of money to all municipalities, large and small. The city charters fiscal frameworks, discussed earlier, were torn up, despite an election promise not to do so, and further cuts were made to the MSI for all municipalities and to grants in lieu.

The latter requires some explanation. Grants in lieu are based on the amount of tax the province would usually pay on their own buildings. The premise is that, although they are exempt from property tax, they use — and should therefore pay for — services provided by municipalities, such as roads, fire departments, and snow clearing. By cutting the grants in lieu, the UCP government arbitrarily decided essentially not to pay their taxes for the use of those services. Additionally, the grants in lieu covered provincial affordable housing, which was also exempt from property tax, but which also used municipal services — a significant download once again of costs onto the municipalities.

I specifically reference this latter cut because it is instructive. For years, municipalities begged the province to build more affordable housing. Municipalities were picking up the costs of dealing with homelessness, but it was argued the province would garner the greatest financial benefit if more housing was built through savings in the form of lower costs for health care, policing, and other forms of social service support. Both former premier Ed Stelmach and the NDP made commitments, or at least talked about committing, to building additional public housing, but little progress had been made. The province's efforts to control their budget always won out. There was one hopeful sign, however: the federal government, after years of inaction, was willing to provide direct funding to the municipalities, albeit for capital costs only, if the provinces picked up the operating costs.

The UCP would not budge. I was at a meeting in Ottawa years ago with then prime minister Stephen Harper, Candace Bergen (who had just taken over the homelessness file), Denis Lebel (who was minister of infrastructure), and Jason Kenney. We were arguing for federal support for housing and a national housing strategy. Kenney argued then that housing was a provincial responsibility. His later response, as premier, suggests he either does not think there is an urgent need for additional housing or that it is now a municipal responsibility, despite municipalities being the only order of government with no constitutional responsibility in the matter. Outside of the emergency COVID-19 aid, to-date the only significant funds the City of Edmonton has received, related to housing, has been for Hope Mission's shelter program. Curiously, the latter was not part of any City of Edmonton ask, while the city's request for permanent supportive housing has gone unanswered. The city has used its own money, and that of the federal government, to start filling that need. It is unknown where the money will come from to operate facilities, though this clearly falls under provincial responsibility. The UCP government seems determined not to get into any ongoing operational funding relationships for housing, despite its relationship to mental health. Perhaps because it is more ideologically palatable to the UPC and its base, the government seems more comfortable with the idea of religious organization's running temporary shelter housing. Recently, money has finally been announced for housing, but it is not clear how much is new money or how the money will actually be spent. It is likely it will just continue to support the temporary shelter programs with still no real investment in long term solutions. This is coupled with a commitment from the province to sell a significant portion of its existing affordable housing stock. Little detail is known of this proposal, but it is hard to imagine how a choice such as this is driven by anything more than an ideological conviction that housing is a private sector problem with a private sector solution, or alternatively a problem best left to the charitable sector. And it is also hard to imagine how it will help.

Back to the budget. The clumsiest cut was perhaps the provincial government's taking of a much larger proportion of the traffic fine revenue. The province had always taken part of this revenue in return for processing the tickets, but the UCP's first budget saw a huge jump in this percentage. As some of the money did not go to the municipalities themselves, but to their police forces, the UCP was ironically the first government to essentially "defund" the police.

The irony does not stop there. In June 2022, Minister of Justice Tyler Shandro used the Police Act to demand the City of Edmonton immediately create a public safety strategy for its troubled downtown core and put more money into its police budget. But the police budget's earlier cut of $22 million did not come courtesy of the city; it was the result of fine revenues the province took. Edmonton did indeed transfer some of the money targeted for the police budget to its community wellness initiative; in essence to alleviate the root causes of crime, using social service tools and supports to those in need, rather than depending on police intervention that may not be the best tool. The city has begged the province to step up in their provincial jurisdiction to deal with mental-health supports, funding for supportive housing, and a real response to the opioid crisis, but to no avail. Ironically, some of the money withheld from the police increase has gone to these areas in the absence of the province taking responsibility. This makes Shandro's overtly political intervention even more galling.

But, back again to that first UCP budget. While fiscal restraint played a major and obvious role in the budget cuts, another motive may also have been in play. Much of what was done had the effect of starving municipalities. If you believe municipalities are inefficient and tend to over-spend, the best corrective solution lies in cutting back on their resources. The municipalities must then either reduce their expenses — the services they provide — or raise taxes. The former tactic theoretically creates smaller government, while the latter makes the provincial government, as tax cutters, look favourable in comparison.

But it may be the UCP's approach to municipalities was also based on misperceptions about the latter's wealth. Under the MGA, municipalities are not allowed to have budget deficits. They can borrow for capital infrastructure, but they cannot borrow for operating expenditures. If they go into a deficit in any given year on the operating side, they have three years to replenish the funds. To deal with the unpredictability of events, such as big snow years, and thus the imprecision of budgeting, most municipalities have stabilization reserve funds. In the case of the City of Edmonton, with which I am most familiar, there is a Stabilization Reserve Fund Policy that sets the fund's minimum and maximum amount based on a percentage of the overall operating budget. The city uses the reserved fund to cover the deficit in loss years and replenishes it in surplus years. If the fund is healthy, it can be drawn upon to cover one-time expenses, such as infrastructure or one-time programming. Using the fund to cover ongoing budget expenses could mean digging a future hole, however. While some municipalities have been tempted due to special circumstances to do this to keep tax increases down in a given year, it means doubling the tax increase the next year to replace the one-time money.

The UCP government seems to have seen these reserve funds as evidence the municipalities had plenty of money. Certainly, some wealthier rural municipalities may have been able to build up healthy reserves. This reflects the varying access rural municipalities have to industrial and linear assessments, already mentioned, and the tension it causes with their neighbouring towns and cities. But that was an issue that needed to be dealt with by revenue sharing mechanisms, and not necessarily evidence of vast amounts of hoarded wealth. Nevertheless, the UCP apparently felt it was evidence of available money. In the case of Greenview MD, for instance, when the province finally stepped in to fill a long-standing need to twin a major highway in the Grand Prairie region, the government insisted Greenview use $100 million of its reserves to pay for part of the road.

This was not the only way the UCP government went after what they considered excess wealth hidden in counties' and municipal districts' coffers. Arguing that oil and gas companies were more deserving of help, it refused to help the counties collect back-taxes from those companies and, without the participation of the RMA, started a review of the linear/industrial assessments. The resulting proposals would have meant a significant hit to many of the MD's and counties and a big benefit to the largest oil and gas companies.

I think the UCP believed helping the oil and gas industry would be a winner in rural Alberta and, in fact, on first blush it sounded attractive to many. But it is one thing to let companies off the hook for their taxes; it is quite another to say to the average citizen their taxes are going up significantly to pay for this generosity. In sparsely populated but large areas, the infrastructure costs to keep everything going meant that industry — the greatest beneficiary of that infrastructure — should pony up. It did not go over well with property owners when it became clear it was them, and not the oil and gas companies, who would have to pick up these costs.

Apart from misreading how this would play in rural Alberta, the UCP also did not count on the ability of local governments to rally the troops. Bill Given, the former mayor of Grand Prairie, points out municipal politicians have an ability to do political "micro targeting" that other orders of government can only envy. A friendly conversation with someone in the supermarket or at the coffee shop provides the opportunity to pass on good information quickly and effectively. And that is what the RMA did. Before long, government MLAs were being deluged with calls to the point they were calling the RMA, telling people to back off. In the end, the government blinked on the linear assessment review, but the damage was done, and the bond broken with a usually supportive constituency.

The municipalities still tried to work with the government. The City of Edmonton continued, as it always had, to seek further efficiencies. Everyone was on the same page in trying to "cut red tape." This was an issue where it felt the province and municipalities could work together. But even here it did not feel like a partnership. Morishita, who was then the president of the AUMA, provides an example. He tells of an offer the AUMA made to take over cutting cheques to the municipalities for the Federal Gas Tax Fund, money sent to each municipality per a federal formula. Other provinces had happily passed cheque-cutting on to their similar associations. Doing the same in Alberta would have relieved the province of another level of bureaucracy. But the AUMA's offer received no response.

As mentioned earlier, it is hard to know how much Minister Madu's negative attitude towards municipalities was his alone. Former mayor Given relates a meeting at which the minister, "talked about foreign oil coming into Canada, where the room wanted to hear about FCSS funding. He did not seem to care very much about our needs. He really did not know his audience." But perhaps Madu's attitude also reflected the government as a whole. In any case, it was with no small amount of relief when he was moved out of municipal affairs in August 2020. Everyone I spoke to felt the new minister, Tracy Allard, was a breath of fresh air. People sensed she was prepared to work with people, rather than lecture. But within months Allard was gone, caught in a scandal over a trip to Maui against the government's own rules on travel during COVID-19. How things might have played out if Allard had stayed and COVID-19 had not overwhelmed events, we will now never know.

Rick McIver became Allard's interim replacement as minister. He had previously been a City of Calgary councillor and thus was one of the few UCP MLAs with municipal experience. Like Allard, he too was an improvement over Madu. He took over the portfolio with a reputation as a "straight shooter." But, as an interim minister, and in the midst of COVID-19, while the relationship with municipalities may have been better, it did not really move things forward.

COVID-19 Changes Everything

Analysis of the UCP's response to COVID-19 could fill an entire book. Here I will examine how it changed the relationship, albeit temporarily, between the province and the municipalities. For reasons unclear, Kenney at some point early on changed his tune and suddenly recognized that collaboration with the municipalities could be beneficial in dealing with the pandemic. In particular, he struck up conversations with mayors Iveson and Nenshi, as well as Morishita. Phone numbers were shared and texted communications between the group became frequent. While no flow of significant health data seems to have occurred, there was clearly a co-ordinated strategic response between the different governments.

The first real test was over the question of mandatory masking and the various rules around shutdowns. These measures could be enforced either within provincial powers, or through local states of emergency called by the municipalities, or both. The premier apparently believed the measures desired in the large urban centres might not be as supported, or viewed as necessary, in more remote rural communities. The solution was to let the cities make their own rules, avoiding province-wide rules. Nenshi tells me he told the premier he would need the latter to publicly state that municipalities had this power for him to get council's support. The premier agreed.

There was a huge fear the cities' homeless populations were going to be particularly vulnerable to COVID-19. Through significant co-operation and provincial government support, the cities were able to set up both safe spaces for the homeless and places where they could isolate should there be an outbreak. Provincial support for shelter space continued through most of the pandemic response.

Those involved tell me that, while initially slow to get going, good communication and co-operation between the orders of government evolved. The mayors felt their perspectives were listened to, perhaps due to more regular personal communication and the development of personal connections.

A cautionary note, however. Texted personal communications can create their own problems, as staff, bureaucrats, and other political colleagues may get left out of the loop, their advice left out of personal channels. Government decisions being done through text messages also raises concerns. The irony is that these are subject to the Freedom of Information and Protection Privacy Act (FOIP). This is particularly true of municipalities. The provincial government was apparently surprised to learn city records, including text messages, were subject to FOIP regulations. Subsequently, the text channel went largely silent.

The experience of the two big cities was not that of everyone else. Former Lethbridge mayor Chris Spearman approached the premier for help dealing with what was essentially a regional divide. COVID-19 numbers were spiking in Lethbridge, and Spearman asked for help dealing with the tensions between the people of Lethbridge and the people from the surrounding hinterland who had very different ideas about the need for restrictions and vaccinations. Lethbridge is a centre for a large, rural population that uses city services, but that population was not particularly respectful of the rules the city and public health implemented. Quite apart from health concerns, tensions also emerged between the two populations. Early on, people from outside Lethbridge arrived in the city to demonstrate. Stories emerged of people standing in line for vaccines being harassed by those who were not. Spearman asked the premier for help, but was told, "half the people are in favour of these measures and half are opposed. What do you want me to do?"

As Alberta entered summer 2021 — the premier's "greatest summer ever" — collaboration on how to respond to COVID-19 began to fray, even within the big cities. The cities were inherently much more cautious in removing restrictions than was the province. In retrospect, I think the municipalities were proven right, but at the time they were frustrated the province was basing its choices on health data to which the municipalities did not have access. In consequence, it was hard to know the degree to which caution was warranted.

A final, interesting point. Public health, as opposed to medicine, had been historically the responsibility of municipalities. This is why, in Ontario, Toronto now makes decisions based on the recommendations of their own officer of public health. There are strong arguments for why this makes sense. Many of the tools for achieving public health — things like recreation, community support and local environment — rest with the municipalities. Likewise, municipalities also experience many of the effects and costs of poor public health. I would never argue public health be transferred back to municipalities without financial support, but given our experiences over the past few years with COVID-19, a re-localizing of public health is worth considering.

The opioid crisis is another place where there is a clash between the province and municipalities over public health. The municipalities, particularly the cities of all sizes, live daily with the consequences of this crisis. For the most part, the cities have been very supportive of harm-reduction strategies, like safe injection sites. Many of these sites were already in place, green lit by the federal government and the New Democrat government prior to the 2019 election. But the UCP opposed the sites and, from the beginning, though not easily able dismantle them entirely, made clear it had different ideas on how to deal with the crisis.

The UCP government created a task force, with a limited mandate and membership, to study the subject. The task force's mandate was to look at the effect of the centres on the community around it, but not at the efficacy of the centres themselves. Given this limitation, the report's lack of enthusiasm for the sites was not surprising. Following the report, the Lethbridge safe injection site was shut down using financial mismanagement as the excuse. A subsequent police investigation absolved the site's management of error, but by then the Lethbridge site had been shut down and has not been replaced.

The government's focus has been on treatment and recovery, but its policy choice seems based in ideology, fed by religious belief, and not on what effected communities say they need. No one argues against the importance of treatment and recovery, but the stand-alone approach has a long history of ineffectiveness. The problems of opioid deaths and the COVID-19 pandemic have taken place at the same time as Lethbridge and other cities have also seen a huge spike in homelessness. The cumulative result is a crisis for the downtown core of Alberta's cities. Unfortunately, despite the advice and desire of effected communities, the province has done little to address the issue, except for Minister Shandro's cynical intervention, mentioned earlier.

By the summer of 2019, any progress during COVID-19 to recreate a useful partnership between governments was reversed in threats to remove the power of municipalities to invoke masking bylaws. That power was useful to the UCP early on as COVID-19 allowed the cities to respond to the pandemic, without the premier having to do so province-wide. By 2021, however, as he faced internal caucus pressures and public anger from some of his supporters over restrictions (see Harrison, Chapter 4), Kenney found it politically expedient to be seen overruling the desire of municipalities to keep restrictions. The province went so far as to propose a bill to remove permanently the power of municipalities to establish local responses to the current crisis and future ones.

Besides politics, the UCP's inherent distrust of some municipal governments can be traced to another factor. Particularly in the two largest cities, the relationship with the federal government has been strengthening. In part, this relationship goes back to the 2008 financial crisis. Harper's government wanted quickly to move out the stimulus funding into the country. Due to its constitutional history, the federal government found its relationship with the provinces to be clumsy, slow, and ineffective in distributing the funding. By contrast, it found municipalities could respond quickly, effectively, and efficiently on projects when funding became available. But I suspect the federal government at some point figured out the municipalities were simply more fun to play with. In any case, traditional arguments against a direct relationship between the municipal and federal two orders of government fast disappeared and money flowed directly to the municipalities.

Prime Minister Trudeau's Liberal government, through the Federation of Canadian Municipalities (FCM), further strengthened the relationship around things like infrastructure, transit, housing, and the environment. In Alberta, much of this relationship was directly with the big city mayor's caucus of FCM, which was chaired by Iveson, while Nenshi provided another strong voice. Their arguments that cities could be an important partner for the federal government did not fall on deaf ears. But cutting out provinces in their role as constitutional intermediaries threatens provincial power and, in the case of Alberta, is exacerbated by the stated dislike by Alberta's UCP government for Ottawa's Liberal government. I suspect, in turn, that the largely successful relationship between the federal government and the two big city mayors did not help to build their relationship with the premier. It also led to a provincial bill banning municipalities from signing agreements with the federal government without provincial approval.

As noted, there were, and are, real differences between some of the priorities of the municipalities and those of the province. As Iveson pointed out to me, the City of Edmonton declaring a "climate emergency" certainly did not help to align the objectives of the two orders of government.

The fall of 2021 saw municipal elections held across Alberta. The elections saw significant changes in Alberta's two large cities, as Iveson and Nenshi declined to run again for mayor. The elections thus presented a pivotal point for establishing a new relationship with the province. Edmonton's incoming mayor, Amarjeet Sohi, subsequently reached out to the province, but none of Edmonton's asks were accommodated in Alberta's spring 2022 budget. I have already mentioned Minister Shandro's antagonistic response to many of these asks. To outside observers, not much seemed to have changed.

The UCP seems always to have believed all would be right in their world if friendly voices were elected into local governments. To this end, the government changed the local elections act, loosening the restrictions on election financing the previous NDP government had passed. The new rules allowed for PAC-style fundraising by third parties; lifted some of the limits on how much any donor could give, and to how many candidates; and removed the provision that donor lists should be published before the election. The government also piggybacked Senate elections and a referendum question on federal equalization payments onto the municipal election, actions suspected by many as intended to attract more conservative voters to the polls. Quite apart from the fact it muddied the purpose of holding municipal elections, the UCP's actions brought partisan politics into the mix, while changes to the election funding also brought the influence of bigger money back into play. The AUMA, in particular, took a strong stance against all these measures, but to no avail.

The impact of these actions is hard to measure. At least in Edmonton, the electorate treats with hostility any whiff of partisan politics in municipal affairs. As discussed earlier, the challenges faced by municipalities make partisan politics unworkable at that level. Nevertheless, the government's intervention in the municipal election did not help repair relationships.

There are a number of other ongoing clashes of note. The provincial government's decision to amalgamate ambulance response/911 dispatch into three large, essentially province-wide, regions was another significant area where municipalities were not consulted. Communities expressed concerns calls would be made to a regional dispatcher without knowledge of their community and living hundreds of kilometres away. Mayor Spearman pointed out to me, by way of an example, a specific incident in his city involving the report of a heart attack happening at Henderson Lake Park. It is a large and significant park in Lethbridge, but the called dispatcher did not recognize the city in which the incident was occurring, let alone how to dispatch help to a place lacking a specific address.

Spearman notes a central dispatcher lacks knowledge of local dispatch capacity. In the past, for instance, the local fire department could fill in if the ambulance service could not immediately respond. This still happens in the two largest cities, where that kind of link is possible, but is not possible in a province-wide system. And the recent sad story of the woman who died as a result of a dog attack in Calgary while the dispatch misunderstood the severity of what was being related to them and did not send an ambulance until half an hour later when the police showed up and asked for it to be dispatched shows the effect this has in even the largest cities.

If the central dispatch system creates problems for cities the size of Lethbridge and Calgary, consider its impact on smaller communities where the effective sharing of available resources is even more important to deal with problems in real time. I was told another sad story of a young woman who died in Raymond when the ambulance took fully half an hour to arrive. One week earlier, the ambulance had been moved from Raymond to McGrath, 20 minutes away. Requests for province-wide data about the issue have gone unanswered, but what seems clear is there are no longer set standards concerning speed of dispatch and waiting times. This is the sort of local issue municipalities understand in a very real way, but which gets lost in a large regional government. Local officials rightly fear the Raymond example will be repeated as their advice continues to fall on deaf ears.

The UCP's deafness to community concerns is replicated in the government's moves to replace the RCMP with a provincial police force, part of its symbolic efforts to "free" Alberta from Ottawa's control. For most municipalities, the RCMP are their only police force. Apart from the question of who a provincial police force would answer to, there are great and real fears that municipalities will bear the cost. Most municipalities fail to see any advantage in the change, and centralization gives them little comfort. But, once again, these concerns seem unheard by the UCP government.

Conclusion: So Where Does This Leave Us?

Something is broken in the current relationship between the province and municipalities. I want to go back to the chapter's beginning to address why the relationship seems broken. As early acknowledged, there are legitimate and expected differences between the different orders of government based on their respective mandates. Conflict is expected, but in the past both the municipal and provincial orders of government were generally able to come to a fruitful, if not harmonious, resolution. This changed under the UCP after 2019.

Municipalities, especially when you look at Alberta's cities, are not insignificant governments. Councillors are elected with just as much democratic agency as MLAs. In smaller centres, they probably have a greater connection to the micro desires of the population than an MLA. In larger centres, they have significantly larger mandates than the MLAs. Consider the Edmonton ward I represented. It was more than double the size of a provincial constituency. I shared constituents with three MLAs. I was elected by those constituents in exactly the same way and had an equal responsibility to those constituents. And to fulfill that responsibility, I had an equal onus to represent them.

Nenshi likes to point out he was elected by all the people of Calgary, a larger constituency by far than anyone else in the province, closely followed by Iveson. I am not sure the UCP understand this. Iveson shared with me his observation that the UCP government seemed to think municipal governments were like Edmonton's community leagues, involving local volunteers. But community league executives are not elected by their community to represent them. By contrast, municipal councils are legal bodies, duly elected to make decisions on behalf of their citizens. They have both responsibility and agency.

A recent comment by MLA Shane Getson (Gosselin, 2022) — "Municipalities are children of the province. If the children get not aligned, maybe it's time for someone to get spanked" — highlights everything wrong with the current relationship; patronizing, yes, given that many municipalities have as many or more constituents than his riding, but also insulting and ignorant in completely disregarding the responsibility of elected officials and the value municipalities can bring to the table. As one of the people I talked to observed, if it is true that municipalities are children, what does it say about the parent? What is the responsibility of those with greater power in a relationship? One would never blame a child for the dysfunctionality of the family, but that is what has repeatedly happened under the UCP. Such an abuse of power is the very definition of bullying.

The UCP's lack of respect is aided and abetted by misperceptions that municipalities are dens of progressive politicians, and that ideology reigns over their priorities. Given's comment, "There is no ideological way to fill a pot hole," is closer to the mark.

Because municipal governments are closer to the people, they see their challenges differently from those of the province. The specific choices made at the municipal level have immediate effect on the quality of people's lives. Local citizens expect service. They expect their garbage to be picked up; their snow to be plowed; their playing field lawns to be cut; their municipalities to be clean; their roads to be maintained; and there to be recreation centres and libraries. These are important things that citizens do not appreciate seeing compromised.

Decisions made by other orders of government can also profoundly affect citizens, but the line of sight between that decision and a real change in the quality of life for them or their community is often obscured or the effects delayed. Provincial and federal governments often fail to understand what citizens expect of their municipality. Municipalities often understand the need for regulation and austerity in a different way and it is not due to ideology, it is because their responsibilities are different. With little practical experience in, or connections to, Alberta municipalities, the UCP after 2019 proceeded with little or no consultation to enact, or threaten to enact, a host of policies that will impact directly the lives of citizens: budget cuts; the opioid crisis; the centralization of ambulance services; and the replacement of the RCMP by a provincial police force.

I mused earlier of provinces being a redundant order of government, but I want to turn that idea on its head. Imagine we had no municipal governments. Perhaps some provincial governments might wish this. In any case, it is very hard to picture how a regional government, driven by a parliamentary system, could understand and provide the level of services expected of municipal governments.

Municipalities are organic structures, their borders driven by geography and community, whereas provinces are artificial constructs. There is no particularly communal rationale to Alberta and Saskatchewan's current boundaries; indeed, they are wholly the result of a political decision enacted in 1905. Consider the example of Lloydminster, whose citizens must deal with both Alberta and Saskatchewan's rules and priorities, even as it struggles to address the common interests of its citizens. The ability of local governments to give voice to that shared interest is vital to democracy and to the prosperity and wellbeing of the whole.

I have written this chapter in hopes the relationship between municipalities and the province can once more become a partnership. That can only happen, however, if all the voices at the table are respected.

References

Madu, K. (2019, October 22). Madu: If the province can cut costs, why can't Alberta's big cities? *Calgary Herald*. https://calgaryherald.com/opinion/columnists/madu-if-the-province-can-exercise-spending-restraint-why-cant-big-cities

Gosselin, A. (2022, March 3). Vaccine-injured UCP MLA feared he would be barred from job. *Western Standard*. https://www.westernstandard.news/news/vaccine-injured-ucp-mla-feared-he-would-be-barred-from-job/article_b7134360-2d9d-5df8-b5a4-049b80803f36.html

CHAPTER 18

BACK TO THE FUTURE: THE UCP'S INDIGENOUS POLICIES, 2019-2022

Yale D. Belanger and David R. Newhouse

"In Alberta, one should expect that Indigenous people be returned to the margins of society, blurred from the past, present and future of Alberta."

— Creezon Iamsees, 2019

IN THE WEEKS following its resounding victory over the NDP in Alberta's May 2019 election, the Jason Kenney-led United Conservative Party (UCP) abruptly turned to dismantling the NDP's policies, and the Indigenous file was an immediate target.[1] Kenney's first serious foray into Indigenous policy making, nonetheless, initially appeared progressive as the UCP spoke of helping Indigenous students succeed and its impending financial investment of $1 billion in loan guarantees to foster Indigenous economic development. Discouraged at federal obstruction of Indigenous LNG pipeline construction goals, the UCP offered an additional $10 million to litigants looking to sue the federal government for affirmation of their Aboriginal rights to develop. Indigenous leaders expressed concern as plans began to circulate. Notably, the UCP's intentions to reduce Indigenous content in their proposed K-4 curriculum was unsettling, as was the exclusion from economic programming of communities unwilling to pursue energy, mineral and forestry projects. What wasn't mentioned was similarly distressing. Treaties were rarely discussed and Indigenous self-government and urban Indigenous issues were ignored. Pitting energy-producing Indigenous communities against those pursuing alternative strategies was judged a cynical divide-and-conquer tactic, while a $10-million litigation fund was construed as a blatant attempt to bypass the constitutional division of powers Kenney had earlier described as constraining Alberta's oil pipeline file. Finally, the premier's office took pleasure maligning the duty to consult and the United Nations Declaration on the Rights of Indigenous Peoples (UNDRIP) — both were framed to be questionable means of granting Indigenous leaders a veto over vital economic projects and threats to Alberta's economic autonomy.

Making sense of the UCP Indigenous policy's contradictory features is an onerous task. As an analytical starting point, we conducted a brief content analysis of major news media stories from 2010 to 2019 where Kenney commented. When debating Indigenous issues, he commonly referenced the constitution acts of 1867 and 1982 and mentioned "responsibility" to frame his observations. Actions such as condemning the UNDRIP and the Supreme Court of Canada-affirmed duty to consult offered more insights. The data pointed to a distinctive trend: Kenney embraced an originalist approach to Indigenous policymaking. Considered a conservative legal ideology, originalists do not believe constitutions require updating (Oliphant, 2015). Nor do they deem Indigenous peoples' claims to treaty and/or Aboriginal rights as contemporary concerns, or legally actionable (Borrows, 2012). Instead, complex dialogues concerning the precise responsibility for "Indians, and lands reserved for the Indians" guide conversations that in turn demand as corroboration past examples of Aboriginal practices that do not help to empower "present-day Indigenous claims" (Borrows, 2017, p. 116). As John Borrows reminds us, dependence on legal and political originalist interpretations inhibits Indigenous peoples from proving "contemporary rights to self-government, child welfare, *education*, *economic regulation*, and so on because the courts have found such claims do not have strong historical analogues at the moment of European encounter" (Borrows, 2017, our emphasis). As we illustrate, Kenney's originalist leanings extended to most facets of his political ideology, and thus the UCP's legal and political tactics for developing its Indigenous file.

This chapter proceeds as follows. We provide a brief biography of Kenney that focuses on his engagement with Indigenous leaders prior to 2019, which in turn acts as an interpretive lens to aid with our assessment of the UCP's approach to Indigenous policymaking. Kenney was essentially a one-man band, the UCP's key policy voice and source of public communications. Hence it is nearly impossible to disentangle Kenney's ideas from how the party pursued Indigenous policy. Consequently, when we discuss the UCP we are referring to Kenney and vice versa. We confine our analysis to the UCP's Indigenous economic policy, which emerged as a means of aiding Alberta's economic recovery; and the party's efforts to write Indigenous peoples out of the K-4 curriculum.

Jason Kenney and Indigenous Issues

Kenney's highly publicized anti-abortion advocacy while an undergraduate student at the University of San Francisco brought him public notoriety in the early 1990s. He dropped out before completing his bachelor of arts, re-surfacing to serve as the Canadian Taxpayers Federation's (CTF) first president until 1996 (Willmott and Skillings, 2021). In 1997, he was elected a Reform Party MP and helped facilitate the

Reform-Progressive Conservative Party merger into the Conservative Party of Canada (CPC) in 2003. He informally served as CPC leader Stephen Harper's "general" in the party's struggle to recalibrate Canadian conservatism (Harris, 2014). During his tenure as prime minister (2006-2017), Harper appointed Kenney to several high-profile cabinet posts (Indigenous affairs was not a major part of any of these postings). After serving two decades as an MP, Kenney abruptly quit in 2016 to spend the next two years criss-crossing Alberta in his quest to "Unite the Right." In October 2017 he won the leadership of the UCP, which had been founded the previous July when the Wildrose Party and the Progressive Conservative Association of Alberta merged. Embracing the spirit of Harper's partiality for concentrating governing authority, Kenney emerged as the UCP's face and its most prominent political voice.

Kenney's Indigenous affairs comprehension is difficult to discern, and we have little to go on prior to 2017 and in the years leading up to the 2019 provincial campaign. Two specific episodes do, however, help shed some light on his views. In September 2022, Kenney denounced future UCP leader Danielle Smith's Alberta Sovereignty Act. Declaring it an attempt to undermine the rule of law, he portrayed Smith's proposal to be a banana-republic denial of the authority of Canadian courts and the federal ideal. Describing himself unapologetically as a Canadian patriot, Kenney affirmed the rule of law "a fundamental conservative principle" (Melgar, 2022). The second episode occurred during his election victory speech in May 2019, where he described Alberta as an idea "of a society that believes deeply in the dignity of the human person, and in a great tradition of ordered liberty" that is "essential to human flourishing" and "what gives meaning to our choices" (National Post, 2019). It is through these public statements combined with the new media content analysis, that Kenney's ideological disposition begins to emerge.

Drawing liberally from Edmund Burke's political thought that animated the political careers of John A. Macdonald and Harper, Kenney proposed an Alberta "that values freedom: free expression, free inquiry, freedom of conscience, and freedom from intrusions by the state into our lives" (National Post, 2019). His goal, arguably, was to preserve an ordered liberty, which was closely aligned to Harper's blend of Burkean conservatism (social conservatives) and classical liberalism (economic conservatives). Kenney on the one hand emphasized "individual freedom" which "stresses private enterprise, free trade, religious toleration, limited government and the rule of law" (Wesley, 2022). On the other hand, his social order emphasis stressed "respect for custom and traditions (religious traditions above all), voluntary association, and personal self-restraint reinforced by moral and legal sanctions on behaviour" (Boessenkool and Speer, 2015). With these ideas in mind, Jared Wesley interpreted Kenney's brand of Alberta conservatism as linking "hard work and individual wealth to personal virtue and, in religious circles, the will of God. On the flip side, a lack of wealth is associated with lack of personal effort or worth" (Wesley, 2022). Added to this, Kenney's self-declared patriotism was grounded by his faith in the rule of law (constitution, public and private law) and federalism (division of powers assigned to federal and provincial governments vis-à-vis sections 91/92).

As the UCP would learn, Indigenous policy files are incredibly elaborate. They are developed according to well established policy ideas Kenney was conscious of, as our review of the relevant news media interviews and social media indicates. To begin, as founded by the Constitution Act, 1867, Canadian federalism divided powers between federal and provincial orders of government with the federal order of government (section 91) assigned exclusive legislative authority over "Indians, and lands reserved for the Indians" (subsection 24). Declared to be wards of the state lacking political agency, Canada's diversity of Indigenous peoples was collapsed into an administrative term, "Indians" (see our discussion of legibility below) that colonial and then federal officials employed to fashion policies aimed at civilizing some Indigenous peoples while (largely) ignoring others (Dyck, 1991). To act on the narrowest interpretation of its responsibility, and to give legislative backing to its civilization agenda, Canada enacted the Indian Act in 1876 that permitted the federal government to, among other provisions, define who is an "Indian" (and, therefore, which Indigenous peoples are excluded from the federal government's limited consideration of its responsibilities), create an "Indian" registry, confine "Indian" people to reserves, and replace traditional governing systems with imposed, foreign political models (Palmater, 2011; Grammond, 2009).

The federal government's civilization agenda evolved in the decades prior to Confederation, and one of its essential institutions was and remains self-government (Nichols, 2019). Then and now, Canada delimits the powers accessible to First Nations governments and the roles the federal government is expected to fulfil through laws, policies and self-government institutions (Borrows, 2016; Titley, 1992). Most First Nations continue to operate under the Indian Act, meaning they are delegated forms of government analogous to municipalities. Indian Act Indian bands exercise a limited form of self-rule with delegated authority (guided by Indian Act limitations) that is restricted to the reserves' confines, whereas self-governing First Nations have recognized decision-making authority within limits defined by modern treaties. The leaders of Indian Act bands challenge the Indian Act's ongoing prominence by claiming an inherent right to self-government, which they insist confirms their capacity to regulate their own laws, determine priorities and policies, and negotiate with federal and provincial governments on matters of law and public policy (Manuel and Derrickson, 2021). Our contention that Kenney interpreted these provisions as rule of law, in this way, is further supported by the UCP's indifference for urban Indigenous peoples. Urban Indigenous peoples are not mentioned in the Constitution Act of 1982, nor do they figure prominently in key policy such as the Inherent Rights Policy of 1995. In sum, using an originalist frame, Kenney and the UCP meticulously interpreted the wording of the Constitution Acts of 1867 and 1982 to fashion policies for, and responses to, Indigenous peoples.

Despite his general silence about Indigenous peoples, Kenney hinted that Canada remained burdened by an ongoing "Indian problem," which we describe as Indigenous peoples' rejection of the state's coercive attempts at economic and cultural transformation (Dyck, 1991). David Newhouse (2016) explains that dating to 1857, the Indian problem and the resulting bureaucratic desire to terminate the issue underscored the creation of all Indigenous policy until at least the 1970s:

> Public policymakers saw Indigenous peoples as a problem. As a result, Indigenous people endured more than a century of assault on their lands, economies, cultural practices, knowledges and identities. Canadian policy's fundamental role during this period was to "solve" the Indian problem by either moving Indians from their traditional lands and territories or removing "the Indian" from within them.

Seeking to upend this institutionalized narrative, Newhouse and Yale Belanger have argued that the "Indian problem is in fact a Canada problem" that should be more accurately considered an "ongoing colonial project that involves both the denial of [Indigenous] political sovereignty and the theft of [Indigenous] lands" (Newhouse and Belanger, 2019, p. 35). The UCP did not respect nor did it avoid interfering with Indigenous peoples' ability to live as sovereign nations, it failed to recognize the inherent rights and jurisdictions of Indigenous peoples, and it did not live up to its historic and contemporary treaty promises (Newhouse and Belanger, 2019, 35). The party correspondingly ignored the growing set of Supreme Court of Canada decisions that delimit Aboriginal rights.

Kenney's approach to Indigenous affairs reflected former prime minister Pierre Trudeau's White Paper (*Statement of Policy of the Government of Canada*) (Canada, 1969), which envisioned "Indians" as ordinary citizens with neither special status nor any entitlement to different administrative arrangements or legal relationships. The White Paper was underscored by liberal notions of equality and equity for all individual citizens, expressed in the overall goal of enabling Indian people "to be free to develop Indian cultures in an environment of legal, social and economic equality with other Canadians" (Canada, 1969). Trudeau's government argued the policy of a separate legal status for Indians "kept the Indian people apart from and behind other

Canadians," and that this "separate road could not lead to full participation, to better equality in practice as well as theory" (Newhouse and Belanger, 2010, p. 344). In other words, the White Paper proposed "Indians" would assume mainstream citizenship responsibilities by being accorded the same legal status and rights afforded all Canadian citizens, and that relations between the government and "Indian" people would move away from the policy regime of protection and special rights. Trudeau's policy in effect aimed to do away with "Indian" status and the unique constitutional relationship established through treaties.

The "Indian problem" surfaced in many of Kenney's comments dating to his time as minister of Citizenship, Immigration and Multiculturalism (2008-13) and of Employment and Social Development (2013-15). In 2011, for example, while promoting immigrant economic security, Kenney linked new Canadians with "Indian" assimilation by claiming that "social and cultural integration are essential. Integration was a politically incorrect word only a few years ago. It was considered synonymous with forced assimilation" (Babiak, 2011). On Aboriginal Day in 2016, in responding to criticism of his representing pre-contact Indigenous populations as Canada's first immigrants, a bewildered Kenney replied that his statement was not "really a point of contention" (Kay, 2016). Then there was his public show of support for Canada's first prime minister, John A. Macdonald, whom he argued had become an unfair victim of cancel culture. Ignoring Macdonald's documented attempts to starve western bands out of western development's pathway (Daschuk, 2019), Kenney represented Macdonald as a revered man of his era who eagerly embraced his role of nation builder that would regrettably force him to make difficult albeit necessary decisions (Kenney 2021). Later that year, he sponsored raising a statue in Calgary to commemorate former British prime minister Winston Churchill, who presided over the 1943 Bengal famine that left three million dead and who would decree that the Indigenous peoples in Australia and the United States had in fact not been mistreated, by virtue of the fact that a "stronger race, a higher-grade race" had "taken their place" (Markusoff, 2022).

Kenney's support for two historical figures known for their scandalous treatment of Indigenous peoples resonates on two levels. One, Macdonald endorsed ordered liberty, ideas that endure for Kenney, the principal of which is the need to defend individual rights and reward individuals embracing the Protestant work ethic. Two, Macdonald and Churchill's tactics underscore their belief that Indigenous peoples lived outside of civilization's norms (O'Neill, 2016). Civilization in this context originates with the French and the British, Canada's two founding nations that Indigenous peoples preceded to North America by several millennia. Kenney nevertheless proclaimed Indigenous peoples as analogous to post-Confederation immigrant populations, they needed "Canadianization" (i.e., assimilation), or the need to adopt middle-class Canadian social and moral codes and pro-capitalist values (Bohaker and Iacovetta, 2009).

Even though the specific details discussed offer us an imperfect gauge of Kenney's disposition toward Indigenous peoples, when read together they offer us a fuller picture of his and thus the UCP's orientation. Employing an originalist frame that relies on the rule of law and tradition as an interpretive frame, Indigenous peoples are not one of Canada's founding peoples. Rather, they are enumerated as "Indian," now First Nations, Metis and Inuit, in section 35 of the Constitution Act, 1982. The quickly growing urban Indigenous population is not mentioned and as such has no legal standing. First Nations are considered self-governing, administrative units vis-à-vis federal delegated authority. Consequently, First Nations and Metis communities are permitted to negotiate agreements with the Alberta government. As a BNA Act, 1867, Ss. 91(24) concern, Alberta is thus not responsible for "Indians, and lands reserved for the Indians," and should have no say nor should it be compelled to participate in public and social policymaking or adopt any of the associated costs. An originalist, we argue, regards assimilated Indigenous peoples as progressive, whereas traditional Indigenous communities and individuals are deemed antagonistic to Canadian social and political norms and mores. As Kenney himself stated, communities and individuals seeking advancement must socially and culturally integrate, and this entails adapting to middle class social and moral codes.

In sum, following in two of his heroes' footsteps, Kenney recognized that getting the job done (e.g. advancing Alberta's economic revitalization) would be a thankless task that would also necessitate his adopting unpopular strategies, inevitably placing him at odds with Indigenous and federal leaders. It is with these ideas in mind that we now turn to our analysis of the UCP's Indigenous policy.

UCP Emergence and Indigenous Policymaking

On April 16, 2019, two years after Kenny assumed leadership of the party, the UCP defeated the NDP in the Alberta general election. With 54.4 per cent of the popular vote, the UCP won 63 of 87 provincial seats. The UCP immediately promised to restore Alberta's prosperity that had been undone by steep oil price declines dating to 2015, and federal intransigence. During his victory speech, Kenney proclaimed Alberta "one step closer to a government focused on prosperity so that we have the means to be a compassionate and generous society" (Wesley, 2022). Notably, this "comment marked one of the few times a major conservative leader in Canada has said the quiet part out loud: that in neoliberal ideology, prosperity must come before compassion" (Wesley, 2022). Embracing Burke's ordered liberty and the rule of law laden with a healthy dose of the Protestant work ethic, Kenney's neoliberal beliefs presaged extreme cuts to governmental spending he claimed would be offset by anticipated increases in productivity resulting in a higher standard of living (e.g., Harvey, 2007). Indigenous peoples were expected to play a critical role in Alberta's revitalization, which would likewise benefit their communities. As Kenney stated, "Alberta's government will sit down with you in the spirit of the treaties, and of reconciliation, to develop real partnerships that can help move First Nations people from poverty to prosperity through resources" (National Post, 2019).

Prior to 2015, Alberta's Indigenous file was extremely limited, a trend former premier Rachel Notley helped to reverse. In addition to leading the province's first party to include an Indigenous mandate in its platform, Notley apologized for Alberta's role in both the Indian Residential Schools (IRS) system and the Sixties Scoop. The NDP proceeded to invest $120 million in affordable housing for Indigenous peoples and $100 million to improve First Nations drinking water, and publicly supported a public inquiry into missing and murdered Aboriginal women and girls. A job training program was established to help Indigenous women build skills as heavy equipment operators while ensuring Indigenous partners played an active role in the NDP's renewable energy plan (25 per cent of new capacity brought onstream had Indigenous equity participation) (Alberta Native News, 2019).The Climate Leadership Plan was an aggressive climate change strategy that included an economy-wide carbon tax and a 100 megatonne cap on oils sands GHG emissions, and also promised to return three per cent of revenues to Indigenous communities through the Indigenous Climate Leadership Initiative (ICLI). This amounted to an estimated $90 million in interest-free grants for Indigenous communities and organizations that provided short-term employment, created solar panel arrays, evaluated energy efficiency of buildings, retrofitted existing structures and supported new builds including schools and houses (Iamsees, 2019).

The UCP could have benefited from the NDP's policy and consultation infrastructure. Instead, the party kicked off its Indigenous policy talks by emphasizing Indigenous poverty, an idea that grounded most of the UCP's public statements. As Ponting and Voyageur argue, accentuating "the woes, conflicts, and other problems of First Nations and on their status as victim" is a common gambit politicians use to rationalize external interventions to resolve perceived problems (Ponting and Voyageur 2001, p. 277). Portraying Indigenous peoples as in need of economic aid allowed the UCP to link provincial prosperity with Indigenous prosperity. The UCP's efforts to end Indigenous poverty relied on four pillars. The first was the Indigenous Opportunities Corporation (IOC), a crown corporation that offered up to $1 billion in loan guarantees to "make financing more affordable and improve lending terms to create economic prosperity and social improvements in communities" (Alberta,

2022c). Secondly, the UCP established a $10-million litigation fund in support of First Nations seeking to sue the federal government and included Indigenous communities in its Stand Up for Alberta strategy. Thirdly, economic development rights were to be added to the Alberta Aboriginal Consultation policy's preamble to empower First Nations financial participation in major resource projects (Alberta Native News, 2019). Finally, the UCP would demand a federal consultation to clarify what precisely the duty to consult entailed to ensure clearer timelines and legal certainty for all project stakeholders working with Indigenous communities or on traditional lands. The UCP also ventured into education. Specifically, it intended to reproduce the Whitecap Dakota First Nation/Saskatoon Public Schools agreement (2014) to safeguard Indigenous student access to the provincial education system, which, significantly, would be subsidized with federal dollars.

On the surface, the UCP's policy approach appears benevolent in nature and progressive in scope. A deeper reading reveals some troubling issues. For the remainder of this section, we focus on the education and economic development streams.

Education. The Whitecap Dakota-Saskatoon model agreement is our starting point (Craig, 2014). Signed in September 2014, with provisions for renewal, it is supported by a companion funding agreement between the Saskatoon Public Schools and Indigenous Services Canada (ISC), which redirects federal funding for Whitecap Dakota Nation students to Saskatoon Public Schools. UCP members who were frustrated at redirecting municipal school fund expenditures (derived from local school taxes) to support Indigenous student attendance supported the Whitecap Dakota-Saskatoon model. Why? Because the federal government would contribute to Indigenous education costs. The UCP frequently suggested that restricting Indigenous student access to public schools was likely if the needed federal funding did not materialize.

From the UCP's perspective, Indigenous students were construed as First Nations members whose membership was territorially bounded to First Nations communities ("… lands reserved for the Indians"), which the UCP took to mean they were "Indians" by virtue of Ss. 91(24) of the BNA Act, 1867, and consequently a federal responsibility. The UCP failed to publicly admit however, that according to section 92, provincial governments are exclusively responsible for education, and that all students living in Alberta can attend public schools free of charge, provided they are Canadian citizens or permanent or temporary residents. Indigenous students are therefore legally entitled to attend provincial schools. The UCP challenged this certainty by employing an originalist frame to argue that by accepting exclusive responsibility for "Indians" in 1867, the federal government likewise accepted responsibility for the costs for urban Indigenous education. Kenney and the UCP petitioned the federal government to intervene and pay for urban Indigenous education. Internally, the party appeared minimally concerned with the fallout of using Indigenous students/provincial citizens as pawns in what was clearly a manufactured constitutional crisis designed to pressure federal officials to hand over additional funding the UCP intended to use to supplement the provincial education file. The UCP also strategically resurrected the antiquated theory that off-reserve Indigenous peoples are a social problem that can infiltrate provincial and municipal environs, (Indian) problems that can only be resolved through federal intervention (Newhouse, FitzMaurice, McGuire-Adams, and Jette, 2012).

Paradoxically, as he was threatening to block Indigenous students from school attendance, Kenney encouraged First Nations students to abandon substandard reserve schools and relocate to towns and cities to attend provincial schools. Here is where the Whitecap Dakota-Saskatoon model agreement is important, for it transfers federal funding from First Nations to the province to support school operations. The potential agreement was needed to safeguard Indigenous student access to the provincial education system, Kenney argued. Yet he failed to mention it was the UCP the students needed safeguarding from — it was, after all, the UCP's education policy that threatened urban Indigenous student access to provincial schools. Indigenous leaders discovered the UCP's interpretive tactics took on a trickster-like quality. Urban Indigenous students were deemed "Indians" whose education costs fell to the federal government. Prior to funding being allocated, the UCP seemed willing to sacrifice their education in an effort pressure a federal response. As these threats circulated, the UCP was simultaneously encouraging First Nations students to transfer into provincial schools once First Nations-municipal agreements were in place guaranteeing federal funding transfers. Though each subset of students was treated distinctively based on how the UCP interpreted the Constitution at any one moment, Indigenous student education outcomes was the cudgel used to draw federal funding to Alberta and to compel moving from First Nations and into cities. The latter strategy echoed historic rationales for transferring students to residential schools, which appears lost on the UCP.

Tactically, the UCP has been able to effectively deflect blame for poor Indigenous educational outcomes to the federal government. The simplicity of the UCP's interpretation masks several key trends that force us to question both its moral foundation and long-term sustainability. For example, the UCP has not discussed the substantial rates of urban Indigenous and off-reserve population growth, rates that almost double that of the non-Indigenous population. The Indigenous community also trends toward remaining younger (Canada 2021). The anticipated increase in Indigenous K-4 students will place greater demands on provincial officials to generate the required funding, which is not a federal responsibility.

For students considering and permitted to attend provincial schools, what can they expect from the curriculum? Dating to his rebirth as a provincial politician, Kenney was apprehensive about the "pedagogical fad" of including Indigenous history and the Indian Residential Schools' legacy in K-4 curriculum (French, 2018a). In his efforts to revise the NDP's program implemented in 2019, one of his first acts was to appoint Adriana LaGrange as education minister. A former Red Deer Catholic School trustee with limited practical education, LaGrange was perhaps best known for prompting First Nations, Metis and Inuit representatives to "tell positive stories of residential schools in the curriculum as well."[2] Kenney then hired former Parliament Hill staffer Chris Champion (2007-2015) to lead the curriculum review. Champion had achieved his own brand of notoriety after warning the inclusion of Indigenous perspectives in curriculum could brainwash non-Indigenous children (French, 2018b). Verifying these opinions are comments that appeared on the Champion-edited *Dorchester Review's* Twitter account, which announced Indigenous parents preferred their children attend residential schools; and that the Truth and Reconciliation Commission (TRC) was influenced by "politics and cashola" that was a source of "cash for anyone who spins a tale (Press Progress, 2021)." The ensuing public outcry and the calls for Champion's termination were ignored. With literally a blank cheque in hand, Champion set to work on his final review that led to the production of a wildly unpopular K-4 social studies curriculum that continues to generate severe opposition.

This overview of the UCP's Indigenous education policy is not meant to be exhaustive, but rather intended to briefly highlight how originalist interpretations have materialized as barriers to the enhanced Indigenous education Kenney claimed as a specific goal. In doing so, the UCP government demonstrated its willingness to manufacture a constitutional conflict that, in one brief policy phase, led Kenney and LaGrange to: (1) declaim responsibility for urban Indigenous peoples who were similarly denied political recognition; (2) portray First Nations education as substandard in relation to superior provincial education; (3) reject Indigenous citizenship by threatening to withhold provincial and municipal resources (despite students' parents contributing the appropriate taxes); while (4) encouraging Indigenous parents to transfer their children to provincial schools despite UCP efforts to limit Indigenous content, to ensure Kenney was not accused of succumbing to an ideological fad. We now turn to the UCP's economic policy.

Economic Development. Reflecting on Kenney's public antipathy for the NDP's K-4 curriculum, the UCP's Indigenous education policy was not unexpected. The UCP did, however, catch many off guard by touting Indigenous peoples' role in Alberta's economic transformation. During his victory speech, Kenney thanked "the growing number of progressive Indigenous leaders across Canada who want to be partners in responsible resource development" (National Post, 2019), despite the leaders having been long marginalized from the provincial economy. Indigenous participation made sense, the premier claimed, to end the endemic poverty threatening all Indigenous communities. "Alberta's government will sit down with you in the spirit of the treaties, and of reconciliation, to develop real partnerships that can help move First Nations people from poverty to prosperity through resources" (National Post, 2019).

In this regard, the Alberta Indigenous Opportunities Corporation (AIOC) was the UCP's key policy pillar. Established to administer sustainable Indigenous economic development by providing up to $1 billion in loan guarantees, AIOC funding access was nevertheless contingent on communities pursuing natural resource projects. The UCP's decision to exclude non-energy-producing Indigenous communities was in response to the recommendations of the Fair Deal Panel that, in December 2019, personally heard from Albertans to discern how they wished to move forward. Seeking to ascertain ways to help Alberta secure a better deal in the federation and to promote the province's economic interests, the panel's final report advised the UCP to secure "Indigenous participation in resource development," thus ensuring "the successful development and movement of Alberta's resource products to market" (Alberta, 2020b). The refusal to consider renewable resource projects confirmed the UCP's pledge to improve Alberta's fortunes vis-à-vis enhanced oil and gas production. It was a way, the AOIC board argued, to help Indigenous communities help themselves in ways that would also benefit Alberta.

The AIOC aligned with the UCP's Partners in Prosperity program, which was linked to the UCP's reconciliation model encouraging Indigenous nations, provincial government and private industry partnerships. Kenney's desire to capitalize on the Indigenous economy made sense. In 2019, the Indigenous economy in Alberta generated $6.74 billion, representing two per cent of provincial GDP (equivalent to the provincial agricultural sector) (MNP, 2021). Any future growth would help stabilize a contracting provincial economy suffering from the lingering effects of a recession dating to 2015-16. The AOIC was deemed the tool needed to unlock this potential by enabling Indigenous peoples to take charge of their economic future and "fund the development and social programs that will benefit" their communities (Alberta, 2019). Program eligibility extended to Indian Act bands, Metis Settlements Act communities, those approved by the minister, and entities fully owned by those entities (Alberta, 2022). A handful of Indigenous leaders initially were supportive of the AIOC. Enoch Cree Nation Chief Billy Morin celebrated the announcement while the Métis Settlements General Council president Herb Lehr indicated that his people now had "the opportunity to work with other people to develop the resources ourselves rather than trying to get into agreements with other energy players" (Bellefontaine, 2019). Within one year of its launch the AIOC announced a loan guarantee to a consortium of six First Nations enabling their purchase of an equity stake in the $1.5-billion Cascade Power Project (Alberta, 2020a).

Many Indigenous leaders were wary of the UCP's approach, specifically the AIOC's focus on liberating entrepreneurial skills, ideas intimately linked to institutional frameworks promoting strong private property rights, free markets and free trade. Targeted for its lack of Indigenous consultation, the AIOC was vilified as an extra community enterprise and little more than an extension of colonization (e.g. Rist, 2014). What particularly galled most Indigenous leaders was the UCP's refusal to observe emergent best practices guiding state-Indigenous economic partnerships. For example, the Assembly of First Nations (AFN) report Advancing Positive, Impactful Change argued Indigenous communities must be considered project owners rather than imminent barriers to development. The UNDRIP featured prominently in the AFN's report, specifically the requirement that states consult and co-operate in good faith with the Indigenous peoples concerned to obtain their free, prior and informed consent (FPIC) ahead of adopting and implementing legislative or administrative measures that may affect them. Consultation is obligatory prior to the state adopting legislation or administrative policies that affect Indigenous peoples (Article 19) and undertaking of projects that affect Indigenous peoples' rights to land, territory and resources, including mining and other utilization or exploitation of resources (Article 32) (United Nations, n.d.).

Reflecting on Kenney's victory speech comments and the news media content analysis, the UCP's policy, generally, and the AIOC more specifically, were structured to directly impact Indigenous communities. This intent should have been but was not accompanied by FPIC processes, which one critic argued was "counter to the overall goal and selling point of the initiative to empower Indigenous communities to partner in resource development" (Iamsees, 2020). The UCP's failure to consult with Indigenous leaders combined with its refusal to secure FPIC was deemed problematic. So was the party's decision to exclude non-oil and gas producing communities from AOIC consideration. But why would Kenney and the UCP chose a strategy designed to privilege some Indigenous communities over others? Especially after publicly trumpeting the important role Indigenous communities were expected to play in Alberta's economic recovery. Here we return to Kenney's originalist leanings, specifically his categorizing Indigenous communities as progressive (and not). Beginning with his victory speech when he thanked "the growing number of progressive Indigenous leaders across Canada who want to be partners in responsible resource development," the UCP's policy privileged Indigenous communities that supported energy production. Those Indigenous communities resisting resource development were considered at best passive or misguided and at worst, antagonistic to the UCP's economic revitalization objectives (see Alberta, 2021). Being positioned in the latter group punished Indigenous leaders for their choice to not pursue energy development. The punishment? The UCP would deny them access to capital. As prescribed by the UCP's policy, the choice was simple: become an oil and gas producing nation for which capital funding was available, or "choose" to be shut out of the AIOC program by exercising a like right to self-determination.

The UCP's approach is ideologically similar to the federal government's historic strategy of promoting Indigenous economic absorption into Canada based on the still popular conviction that Indigenous and non-Indigenous peoples emerged from separate historical trajectories that led to noteworthy economic imbalances becoming institutionalized (Tough, 2005). Post-Confederation-era officials believed Indigenous peoples needed to align their institutions with majority Canadian society to ensure success, ideas Kenney previously alluded to when discussing the need for immigrants' "economic integration" — recall the premier deems immigrants and Indigenous peoples as analogous. In their efforts to encourage economic integration, historic Indian Affairs officials privileged First Nations that embraced farming and ranching, and individuals who adopted wage labour, all in support of Canadian economic prerogatives (e.g. Carter, 1990; Belanger, 2022). Kenney similarly would privilege energy-producing Indigenous communities. Crucially, the UCP's strategy represented a calculated attempt to produce legibility. According to James Scott, legibility is a bureaucratic method the state employs to engineer a common citizenry from culturally complex communities, by encouraging everyone to disregard past differences in pursuit of common societal linkages. Legibility's success depends on standardized bureaucratic language and legal discourses to help policymakers arrange populations. Federal officials worked to institute legibility to simplify the task of Indigenous civilization (Tobias, 1976). Take the state's seemingly innocuous act of rebranding Indigenous peoples "Indians" (Scott, 2008). Manufacturing "Indians" in this way reduced Indigenous complexity, which inaugurated decades of Canada and the provinces rewriting Indigenous cultural inclusion to support larger societal economic and political desires (see Veracini, 2010).

The UCP's resistance to Indigenous self-government is a prime example of a provincial government embracing legibility. By virtue of the Inherent Rights Policy of 1995, the 643 First Nations in Canada (42 in Alberta) have "an inherent right of self-government guaranteed in section 35 of the Constitution Act, 1982" (Canada, 2022b). Comparable to the federal government in 1867 and beyond, the UCP employed legibility as a homogenizing scheme to erase political and cultural uniqueness and deny the federally acknowledged inherent Indigenous right to self-determination, which the UNDRIP claims comes with state obligations to ensure FPIC. Indigenous self-determination, the FPIC and the Supreme Court of Canada's affirmation that governments and businesses must consult and accommodate Indigenous communities, when read together, award Indigenous communities the legal standing needed to challenge projects intended to traverse bounded and traditional lands. Legibility is the first step that sets the stage for future provincial policy, for once the political and economic environment is simplified, the state may freely place constraints that privilege selected communities while compromising others' ability to develop according to internally determined standards (Sen, 1999).

As the Nobel Peace Prize winning economist Armatya Sen explains, a crown corporation such as the AOIC also acts as an external constraint by granting selected communities limited access to capital and a freedom to achieve other communities were likewise denied. What the UCP's policy failed to acknowledge for all Indigenous communities was their inherent rights to self-determination and FPIC. In fact, outside of the AOIC acting as a source of capital infusion, the UCP framed all Indigenous communities as special interest groups rather than self-determining nations engaged in diplomatic relations. For instance, Kenney's efforts to kill the federal government's plans to enshrine the UNDRIP as law were based on his fear that Indigenous groups would be granted a veto over provincial projects and thus represented a rejection of Indigenous self-determination (MacCharles, 2020). Pasternak (2020) refers to this as racial capitalism — where the UCP publicly maintained support for Indigenous peoples' market economy inclusion and promised to share in resource development's bounty, these desires were overtaken by settler colonial impulse to limit Indigenous nations' ability to act as sovereign decision makers within their own territories. In sum, even for those who were granted the financial means to pursue local development, the UCP refused to endorse Indigenous rights of self-determination.

The second set of constraints took on various arrangements within the UCP's administrative apparatus. One tactic was to simply avoid consulting with Indigenous leaders. Take Kenney's proposal to raise revenue by selling First Nations-adjacent crown land in Peace County. Treaty 8 First Nations of Alberta chief administrative officer Darryel Sowan publicly reminded Kenney of his people's inherent rights to the lands and that "we also signed a treaty. That's our land, that's Crown land, and you can't do that without proper consultation." Treaty 8 Grand Chief Arthur Noskey criticized the UCP's omission of Indigenous peoples as an expression of "blatant disrespect" and a form of assault linked to "oil, gas, diamonds, now farming. We're just whittled down where actually First Nations are kind of isolated" (Leavitt, 2018). When challenged, Kenney argued "It is our (Alberta's) land. It's always been our lands."

Seeking to relieve the tension, he noted the lands in question were not "in a reserve or treaty area, it's crown land and it's land that belongs to Albertans which is not being put to economic use." Invoking his prosperity doctrine, Kenney argued that,

> The whole province was built on converting crown land into productive economic use. The notion that all of Northern Alberta should suddenly be turned into a park, I think, runs contrary to our entire history as a people where we seek in a responsible way to allow for the development of our resources (quoted in Bellefontaine, 2018).

This episode highlights two important issues. One, in responding to the disapproval of his proposed land sale, Kenney asserted that through the Natural Resources Transfer Acts of 1930, the federal government conferred to Alberta ownership of provincially bounded lands and regional natural resources jurisdiction (Hall, 2006). Indigenous leaders continue to depict the NRTA as an unlawful territorial allocation that placed the prairie provinces "in a position of equality with other Provinces of Confederation" (Alberta, 2013) that compromised First Nations' pursuit of traditional livelihoods (Calliou, 2007; Tough, 2003). The second related issue: the UCP's originalist outlook discounts subsequent decades of legal, political and social developments that occurred after 1930, and that were calculated to offset decades of colonization's detrimental effects that continue to adversely impact Indigenous peoples. Originalism thus conceals colonialism and the attendant legal and policy advancements. As Kenney would discover, proceeding with his proposal to sell Crown lands would have inevitably triggered a duty to consult. What further irritated Indigenous leaders was the premier's ignorance — intentional or otherwise — of the Province of Alberta's Policy on Consultation with First Nations on Land and Natural Resource Management, whereby the

> Government of Alberta seeks to reconcile First Nations' constitutionally protected rights with other societal interests with a view to substantially address adverse impacts on Treaty rights and traditional uses through a meaningful consultation process. Alberta's management and development of provincial Crown lands and natural resources is subject to its legal and constitutional duty to consult First Nations and, where appropriate, accommodate their interests when Crown decisions may adversely impact their continued exercise of constitutionally protected Treaty rights (Alberta, 2013).

Upon further reflection, Kenney's proposal clearly fell into the category "strategic and project-specific Crown decisions that may adversely impact the continued exercise of Treaty rights and traditional uses," and related "to land and natural resource management is contemplated." Accordingly, any sale of Crown lands would not be possible prior to prescribed consultations occurring with the affected First Nations (Alberta, 2013).

Kenney refused to accept colonization's ongoing institutionalized impacts, which is further confirmed by his consistent return to historic era policies and legislation the federal government employed to assert hegemony over Indigenous peoples. Apologists such as Kenney (see his Macdonald defence) will reason we need to forgive these historical figures' political choices for they lacked the expertise we've developed in relation to legal and economic theory. Many will also argue the exigencies of nation building, for instance, needed to take precedence to ensure greater economic independence required to safeguard Canada from U.S. annexation and ongoing colonial oppression. As the last four decades of academic scholarship has revealed, however, Confederation-era politicians understood their actions defied international laws confirming Indigenous rights to self-determination and land ownership. In their efforts to foster Alberta's economic recovery, Kenney and the UCP adopted approaches similar to Canada's post-Confederation approach to Indigenous peoples. The UCP chose to loosely interpret the duty to consult and FPIC, aggressively opposed the UNDRIP, openly challenged reconciliation as conceived by Prime Minister Justin Trudeau and adopted by former premier Notley's NDP government, and discounted the Truth and Reconciliation's (TRC) Calls to Action (CTA), described as "actionable policy recommendations meant to aid the healing process" by acknowledging the full, horrifying history of the residential schools system" and "creating systems to prevent these abuses from ever happening again in the future" (Canada, 2022a).

The UCP's ideological resistance to these ideas is evident in political institutions. Dating to 2013, Alberta's consultation policy the UCP inherited sought to foster agreements between Metis and First Nations in Alberta that "provide a framework for continued collaboration," the goal being to "support meaningful discussion, information sharing, and the exploration of issues of mutual concern" (Alberta 2013a). As of 2022, three protocol agreements were in place that Alberta signed with three partners: (1) Blackfoot Confederacy (2019); (2) Stoney Nakota-Tsuut'ina Tribal Council (2020); and (3) Confederacy of Treaty Six First Nations (2022). The UCP also inherited the 10-year, Long-Term Governance and Funding Arrangements signed in 2013 with the Metis Settlements General Council (Alberta 2013b). Established to discuss assorted concerns, the agreements stated the signatories were to meet at minimum once a year to nurture collaborative working relationships. The agreement's provisions reveal surprising differences that offer insights of the UCP's Indigenous engagement. While the UCP signed the Stoney Nakota-Tsuut'ina Tribal Council and Blackfoot Confederacy agreements, each one was largely negotiated while the NDP was in office. Conspicuously, the UNDRIP and the TRC's CTAs act as guiding principles for those consultation tables, which in turn obligates Alberta to recognize the right to Indigenous self-determination, the Indigenous right to be recognized as distinct peoples, as well as an Indigenous right to FPIC. The TRC also reaffirmed First Nations, Inuit, and Métis peoples' treaty, constitutional, and human rights while reminding Albertans that, as treaty peoples, they were duty-bound to establish and maintain mutually respectful relationships.

Neither the TRC nor the UNDRIP is mentioned in the Confederacy of Treaty Six First Nations protocol agreement negotiated with the UCP. As noted, the UCP's policy does not mention the Indigenous right to self-determination or the need to recognize Indigenous peoples as distinct peoples. Nor does it confirm FPIC. Excluding those provisions from the agreement signals the UCP's originalist leanings and its refusal to seriously consider the post-1930s federal legal and policy landscape. Accordingly, the UCP's Indigenous economic policy avoids wrestling with the institutionalized colonial ideas that both influence policy and perpetuate economic imbalances. By avoiding any mention of colonialism's ongoing implications and frequently invoking colonial-era policies to set the contours of provincial-Indigenous interface, the UCP signalled its aversion to systemic change. Instead, it chose to delegate the duty for trust-building, fostering accountability, and transparency to an unproven Crown corporation. Surely the $1 billion in secured loans is a considerable investment. But the $1 billion is almost to the dollar the amount of revenue five on-reserve First Nations casinos have contributed to the provincial treasury since 2007 for the privilege of operating (Belanger, 2018). The UCP, in effect, reassigned $1 billion acquired from First Nations to the AIOC, for redistribution to Indigenous applicants in the form of secured loans.

The final pillar of the UCP's Indigenous economic policy was a $10-million litigation fund to support First Nations looking to sue the federal government to ensure pipeline development through Indigenous lands. The UCP's real target was Bill C-48 (Oil Tanker Moratorium Act), which received royal assent on June 20, 2019. Established to protect a remote region in northern British Columbia-Hectate Strait, Queen Charlotte Sound and Dixon Entrance from the risk of oil spills by prohibiting tankers carrying more than 12,500 tonnes of oil, Bill C-48 antagonized Kenney. In response, the UCP effectively offered to subsidize First Nations lawsuits to help Indigenous communities win right of pipeline access the UCP would otherwise be powerless to expedite. In November 2021, the Fort McKay and Willow Lake Metis nations were awarded $372,000 from the Litigation Fund to initiate a constitutional challenge of Bill C-48.

Despite their frustration with the federal barriers to pipeline construction, and the troubling nature of the litigation fund, Kenney and the UCP never suggested overtly ignoring federal legislation due to their belief in the rule of law. Rather, the UCP tried to entice First Nations to do their bidding. Unfortunately, by privileging Indigenous communities pursuing energy projects over non-oil and gas producing Indigenous communities the premier took to publicly mocking, the litigation fund unsettled already fragile relationships. Students of the premier's long political career recognized the divide-and-conquer tactics, which were not unexpected of an individual labelled a binary politician. As a scathing column by Graham Thomson concluded, "You are either with him or against him. He signals this every time he praises his allies and attacks his enemies, even if those enemies are fellow Albertans" (Thomson, 2019). Kenney took it upon himself to act as a self-styled ambassador for the 30 of 42 Alberta First Nations supporting the Northern Gateway project in his efforts to get the project off the ground, which included demanding the prime minister's office (PMO) place clearer limits on the duty to consult to permit construction. Oil and gas producing communities quietly defied the premier, for their leaders understood the duty to consult's capacity to safeguard diverse interests. Nevertheless, Kenney continued to press the PMO for change by emphasizing the duty to consult's anti-democratic nature — it was an institution that accepted a minority of Indigenous leaders could disrupt majority Indigenous desires.

It appears Kenney and the UCP were less concerned about Indigenous rights and more focused on promoting provincial development. According to this thinking, Indigenous rights for some may be protected, as may self-government. But for all, in various ways, it will be forsaken to guarantee economic growth.

Conclusion

The UCP policies that, on the surface, appear progressive and supportive remind us of the White Paper, a policy that was intended to force the assimilation of Indians into Canadian society by: (1) removing the legal distinctions between "Indians" and other Canadians; (2) dismantling the structures and processes that kept "Indians" separate; and (3) proposing to turn over all federal responsibility to the provinces. The UCP has reluctantly agreed that legal distinctions have become too entrenched to remove and that Indigenous distinctiveness is impossible to diminish. In response, as this chapter demonstrated, the UCP employed an originalist lens to create an Indigenous policy that fostered an Alberta-first philosophy that encouraged Indigenous peoples to contribute to provincial economic revitalization but denied Indigenous self-determination for fear of slowing Alberta's progress. The goal was to safeguard Alberta's autonomy while encouraging partnerships with *special interest* Indigenous communities.

From the start, the UCP refused to acknowledge colonialism's ongoing threat to Indigenous betterment while ignoring the UNDRIP, the duty to consult, reconciliation, and the TRC's final report. Discounting these institutions outright was not possible for a party that heralded law and order as a conservative tradition. The UCP turned to originalism to emphasize the Constitutional division of powers, early Indigenous policy, and the NRTA's ongoing relevance. "Indians, and lands reserved for the Indians" remained a federal responsibility and First Nations were self-administering units that provincial officials generously described as self-governments. Urban Indigenous peoples were not mentioned and consequently have no legal standing. Treaties were historic documents of limited contemporary value whereas the NRTA conferred to provincial ownership land and resources alike. Admittedly the UCP crafted fascinating policy angles. The education policy's originalist approach led the UCP to manufacture a constitutional crisis to declaim fiscal responsibility for Indigenous students and compel the federal government to pay for "Indians" attending provincial schools. The economic policy shrewdly employed originalism to help the UCP interpret the constitution and the NRTA, to fortify provincial land ownership claims and autonomy over economic development by denying Indigenous self-determination.

The policy of assimilation and civilization has not worked, but that will not stop the UCP from continuing to treat Indigenous peoples in ways they've been rejecting since the White Paper debates of the early 1970s. And it seems as though the new UCP leader and Alberta Premier Danielle Smith picked up where her predecessor left off. To date she has not mentioned Indigenous affairs, nor did she or any of the candidates for premier discuss Indigenous issues during the campaign prior to the Oct. 6 vote. Indigenous leaders, such as Chief Mel Grandjamb (2022) of the Fort McKay First Nation, remarked about the silence and challenged the candidates to "Tell us how — under your leadership — the province will work with Indigenous communities on land and water rights, wellness and education." The Métis Nation leadership raised the same set of issues with the previous UCP government and was greeted with a similar silence. Unlike Kenney, Smith appointed a special advisor on Indigenous Relations, specifically Chief Billy Morin — one of the earliest and loudest proponents of the AIOC and who has remained on the forefront promoting Indigenous oil and gas production. Though for some it may be too early to pronounce on Smith's Indigenous policy, it appears the status quo will prevail and she will deviate minimally, if at all, from the Kenney-influenced Indigenous policy framework. It seems that the UCP policy takes us back to the future.

References

Alberta. (2013a) *Government of Alberta-Metis Settlements General Council: Long-term Governance and Funding Arrangements.* https://open.alberta.ca/publications/metis-settlements-long-term-governance-and-funding-arrangements

Alberta. (2013b). *The Government of Alberta's Policy on Consultation with First Nations on Land and Natural Resource Management.* https://open.alberta.ca/dataset/801cf837-4364-4ff2-b2f9-a37bd949bd83/resource/88e318b9-8f1b-4110-9a85-73c7ade8c261/download/6713979-2013-government-policy-fn-consultation-2013-08-02.pdf

Alberta. (2019). *Alberta Indigenous Opportunities Corporation: Annual Report 2019.*

Alberta. (2020a). *Alberta Indigenous Opportunities Corporation Announces First Participation in Cascade Power Project.* https://www.theaioc.com/about/news/alberta-indigenous-opportunities-corporation-announces-first-participation-in-cascade-power-project/

Alberta. (2020b). *Fair Deal Panel Report: Government's response.* https://open.alberta.ca/dataset/735ca928-7938-41c0-ba33-a38a0d7e3251/resource/7153e491-4274-4ec7-bcbe-ecb56d663683/download/fair-deal-panel-report-government-response-june-2020.pdf

Alberta. (2021). *Report of the Public Inquiry into Anti-Alberta Energy Campaigns.* https://open.alberta.ca/dataset/3176fd2d-670b-4c4a-b8a7-07383ae43743/resource/a814cae3-8dd2-4c9c-baf1-cf9cd364d2cb/download/energy-report-public-inquiry-anti-alberta-energy-campaigns-2021.pdf.

Alberta. (2022a). *First Nations Agreements: Provide Frameworks for Exploring Issues of Mutual Concern, Information Sharing and Continued Collaboration.* https://www.alberta.ca/first-nations-agreements.aspx

Alberta. (2022b). *Alberta Indigenous Opportunities Corporation: Loan Guarantee Investment Program Guidelines.* https://www.theaioc.com/wp-content/uploads/2022/06/220413-AIOC-Loan-Guarantee-Investment-Program-Guidelines-Updated-Final.pdf

Alberta. (2022c). *AIOC: Alberta Indigenous Opportunities Corporation.* https://www.theaioc.com.

Alberta Native News. (2019, April 1). *Alberta NDP Unveils Election Platform: Reconciliation With Indigenous Peoples.* https://www.albertanativenews.com/alberta-ndp-unveils-election-platform-reconciliation-with-indigenous-peoples/

Babiak, T. (2011, March 19). Mural of bishop opens old wounds; First Nations people recall a different Canadian history. *Edmonton Journal.*

Belanger, Y. (2022). Permanent precarity? Racial exclusion, discrimination, and low-wage work and unemployment amongst Canada's First Nations. In J. Peters and D. Wells (Eds.), *Canadian Labour Policy and Politics: Inequality and Alternatives*. Vancouver: UBC Press.

Belanger, Y. (2018). First Nations gaming in Canada: Gauging its past and ongoing development. *Journal of Law and Social Policy* 30(1), 175-184.

Bellefontaine, M. (2018, Dec. 6). *Alberta Crown land proposed for sale not treaty land, Kenney says*. CBC News. https://www.cbc.ca/news/canada/edmonton/jason-kenney-alberta-crown-land-first-nations-1.4935913

Bellefontaine, M. (2019, Oct. 8). *Government plans to help First Nations, Métis groups get capital for resource projects*. CBC News. https://www.cbc.ca/news/canada/edmonton/government-plans-to-help-first-nations-métis-groups-get-capital-for-resource-projects-1.5313673

Bohaker, H., & Iacovetta, F. (2009). Making Aboriginal people "immigrants too": A comparison of citizenship programs for newcomers and Indigenous peoples in postwar Canada, 1940s–1960s. *Canadian Historical Review* 90(3), 427-462.

Boessenkool, K., & Speer, S. (2015). Ordered liberty: How Harper's philosophy transformed Canada for the better. *Policy Options*. https://policyoptions.irpp.org/2015/12/01/harper/

Borrows, J. (2017). Challenging historical frameworks: Aboriginal rights, the trickster, and originalism. *Canadian Historical Review* 98(1), 114-135.

Borrows, J. (2016). *Freedom and Indigenous Constitutionalism*. Toronto: University of Toronto Press.

Borrows, J. (2012). (Ab)Originalism and Canada's Constitution. *Supreme Court Law Review* 58, 351-398.

Calliou, B. (2007). Natural Resources Transfer Agreements, the Transfer of Authority, and the Promise to Protect the First Nations' Right to a Traditional Livelihood: A Critical Legal History. *Constitutional Studies* 12(2), 173-213.

Canada (1930). Alberta Natural Resources Act S.C. 1930, c. 3 [online] https://laws-lois.justice.gc.ca/eng/acts/A-10.6/FullText.html

Canada. (1969). *Statement of the Government of Canada on Indian Policy*. https://oneca.com/1969_White_Paper.pdf

Canada. (2021). *Projections of the Indigenous populations and households in Canada, 2016 to 2041*. The Daily. https://www150.statcan.gc.ca/n1/daily-quotidien/211006/dq211006a-eng.htm

Canada. (2022a). *What Are the Truth & Reconciliation Commission's 94 Calls to Action & How Are We Working Toward Achieving Them Today?* https://www.reconciliationeducation.ca/what-are-truth-and-reconciliation-commission-94-calls-to-action

Canada. (2022b). *Self-Government* [online] https://www.rcaanc-cirnac.gc.ca/eng/1100100032275/1529354547314

Carter, S. (1990). *Lost Harvests: Prairie Indian reserve Farmers and Government Policy*. Montreal: McGill-Queen's University Press.

Craig, M. (2014, Oct. 24). *Whitecap, Saskatoon public schools sign historic education agreement*. Global News. https://globalnews.ca/news/1634530/whitecap-saskatoon-public-schools-sign-historic-education-agreement/

Daschuk, J. (2019). *Clearing the Plains: Disease, Politics of Starvation, and the Loss of Indigenous Life*. Regina: University of Regina Press.

Dyck, N. (1991). *What is the Indian "Problem? Tutelage and Resistance in Canadian Indian Administration*. St. John's: Memorial University Press.

Elliot, J. (2019, June 21). *Why critics fear Bill C-69 will be a "pipeline killer."* Global News. https://globalnews.ca/news/5416659/what-is-bill-c69-pipelines/

French, J. (2018a, Aug. 18). *Alberta social studies curriculum adviser calls inclusion of First Nations perspectives a fad*. CBC News. https://www.cbc.ca/news/canada/edmonton/alberta-social-studies-curriculum-adviser-calls-inclusion-of-first-nations-perspectives-a-fad-1.5690187

French, J. (2018b, July 19). Draft curriculum documents "misinterpreted," education minister says. *Edmonton Examiner*.

Grammond, S. (2009). *Identity Captured by Law: Membership in Canada's Indigenous Peoples and Linguistic Minorities*. Kingston & Montreal: McGill-Queen's Press-MQUP.

Grandjamb, M. (2022, Sept. 19). When are UCP leadership candidates finally going to discuss Indigenous issues? *Calgary Herald*. https://calgaryherald.com/opinion/columnists/opinion-when-are-ucp-leadership-candidates-finally-going-to-discuss-indigenous-issues

Hall, D. (2006). Natural Resources Transfer Acts 1930. *Canadian Encyclopedia*. https://www.thecanadianencyclopedia.ca/en/article/natural-resources-transfer-acts-1930

Harris, M. (2014). *Party of One: Stephen Harper and Canada's Radical Makeover*. Toronto: Penguin Books.

Harvey, D. (2007). *A Brief History of Neoliberalism*. Oxford: Oxford University Press.

Iamsees, C. (2019). *Indigenous Policy in Alberta: Considering the First 100 Days of UCP Rule."* Toronto, ON: Yellowhead Institute. Aug. 28. https://yellowheadinstitute.org/wp-content/uploads/2019/08/indigenous-policy-in-alberta-brief.pdf

Iamsees, C. (2020). *The Alberta Indigenous Opportunities Corporation: A Critical Analysis*. Toronto, ON: Yellowhead Institute. Feb. 13.

https://yellowheadinstitute.org/2020/02/13/alberta-indigenous-opportunities-corporation-analysis/

Kay, B. (2016, June 29). History vs. narratives. *National Post.*

Kenney, J. (2021, June 2). Cancel John A. Macdonald and we might as well cancel all of Canadian history. *National Post.* https://nationalpost.com/opinion/jason-kenney-cancel-john-a-macdonald-and-we-might-as-well-cancel-all-of-canadia n-history

Leavitt, K. (2018, Nov. 30). Treaty 8 First Nations slam Jason Kenney's proposal to sell crown land in Peace Country. *Toronto Star.* https://www.thestar.com/edmonton/2018/11/30/its-bribery-treaty-8-first-nations-slam-jason-kenneys-land-auction-proposal.html

Manuel, A., & Derrickson, R. (2021). *Unsettling Canada: A National Wake-Up Call.* Vancouver: Between the Lines.

MacCharles, T. (2020, Feb. 24). Alberta pushes to kill Liberal plan to enshrine UN declaration on Indigenous rights. *Toronto Star.* https://www.thestar.com/politics/federal/2020/02/24/alberta-pushes-to-kill-liberal-plan-to-enshrine-un-declaration-on-indigenous-rights.html

Markusoff, J. (2022, Aug. 28). *Churchill statue is Jason Kenney's way of declaring what we must celebrate, warts be damned.* CBC News. https://www.cbc.ca/news/canada/calgary/winston-churchill-statue-calgary-jason-kenney-analysis-1.6564011

Melgar, A. (2022, Sept. 6). *Smith releases plan for sovereignty act; Kenney calls it "catastrophically stupid."* CityNews. https://calgary.citynews.ca/2022/09/06/smith-sovereignty-act-kenney-catastrophically-stupid/

MNP. (2021). *Opening the Door to Opportunity: Reporting on the Economic Contribution of Indigenous Peoples in Alberta.* https://www.mnp.ca/en/insights/directory/opening-the-door-to-opportunity

National Post. (2019. April 17). Read Jason Kenney's prepared victory speech in full after UCP wins majority in Alberta election. https://nationalpost.com/news/canada/read-jason-kenneys-prepared-victory-speech-in-full-after-ucp-wins-majority-in-alberta-election

Newhouse, D. (2016). *Indigenous Peoples, Canada and the Possibility of Reconciliation.* IRPP Insight 11. Montreal: Institute for Research on Public Policy.

Newhouse, D., & Belanger, Y. (2010). Beyond the "Indian problem": Aboriginal peoples and the transformation of Canada. In J. Courtney and D. Smith (Eds.), *The Oxford Handbook of Canadian Politics.* Toronto: Oxford University Press.

Newhouse, D., & Belanger, Y. (2019). The Canada problem in Aboriginal politics. In D. Long and G. Starblanket (Eds.), *Visions of the Heart: Canadian Aboriginal Issues*, 5th Edition. Toronto: Oxford University Press.

Newhouse, D., FitzMaurice, K., McGuire-Adams, T., & Jette, D. (Eds.) (2012). *Well-Being in the Urban Aboriginal Community: Fostering Biimaadiziwin*. Toronto: Thompson Educational Publishing.

Nichols, J. (2019). *A Reconciliation Without Recollection? An Investigation of the Foundations of Aboriginal Law in Canada*. Toronto: University of Toronto Press.

Oliphant, B. (2015). Originalism in Canadian constitutional law. *Policy Options*. https://policyoptions.irpp.org/2015/06/25/originalism-in-canadian-constitutional-law/

O'Neill, D. (2016). *Edmund Burke and the Conservative Logic of Empire*. Berkeley Series in British Studies, no. 10. Oakland: University of California Press.

Palmater, P. (2011). *Beyond Blood: Rethinking Indigenous Identity*. Saskatoon: Purich Publishing.

Pasternak, S. (2020). Assimilation and partition: How settler colonialism and racial capitalism co-produce the borders of Indigenous economies. *South Atlantic Quarterly* 119(2), 301–324.

Ponting, J. R., & Voyageur, C. J. (2001). Challenging the deficit paradigm: Grounds for optimism among First Nations in Canada. *Canadian Journal of Native Studies* 21(2), 275-307.

Press Progress. (2021, June 3). *"Reprehensible and Disgusting": Key Author of Jason Kenney's Education Curriculum Under Fire Over Tweets About Residential Schools*. https://pressprogress.ca/reprehensible-and-disgusting-key-author-of-jason-kenneys-education-curriculum-under-fire-over-tweets-about-residential-schools/

Rist, G. (2014). *The History of Development: From Western Origins to Global Faith*. London & New York: Bloomsbury Publishing.

Scott, J. (2008). *Seeing Like a State: How Certain Schemes to Improve the Human Condition Have Failed*. New Haven: Yale University Press.

Sen, A. (1999). *Development as Freedom*. Oxford: Oxford University Press.

Tharoor, I. (2015, Feb. 3). The dark side of Winston Churchill's legacy no one should forget. *Washington Post*. https://www.washingtonpost.com/news/worldviews/wp/2015/02/03/the-dark-side-of-winston-churchills-legacy-no-one-should-forget/

Thomson, G. (2019, Nov. 15). The way Jason Kenney treats Indigenous peoples. *The Tyee*. https://thetyee.ca/Analysis/2019/11/15/How-Jason-Kenney-Treats-Indigneous-Peoples/

Titley, B. (1992). *A Narrow Vision: Duncan Campbell Scott and the Administration of Indian Affairs in Canada*. Vancouver: UBC Press.

Tobias, J. (1976). Protection, civilization, assimilation: An outline history of Canada's Indian policy. *Western Canadian Journal of Anthropology* 6(2), 13-17.

Tough, F. (2005). From the "original affluent society" to the "unjust society." *Journal of Aboriginal Economic Development* 4(2), 26–65.

Tough, F. (2003). The forgotten constitution: The natural resources transfer agreements and Indian livelihood rights, CA. 1925-1933. *Alberta Law Review* 41, 999-1048.

United Nations. United Nations Declaration on the Rights of Indigenous Peoples. https://www.un.org/development/desa/indigenouspeoples/declaration-on-the-rights-of-indigenous-peoples.html

Veracini, L. (2010). *Settler Colonialism: A Theoretical Overview.* London: Palgrave MacMillan.

Wesley, J. (2022). The dramatic fall of Jason Kenney: A failure to unite. *Alberta Views.* https://albertaviews.ca/dramatic-fall-jason-kenney/

Willmott, K., & Skillings, A. (2021). Anti-Indigenous policy formation: Settler colonialism and neoliberal political advocacy. *Canadian Review of Sociology/Revue canadienne de sociologie* 58(4), 513-530.

NOTES

1 This chapter uses the terms "Indian," "Aboriginal," "First Nation," and "Indigenous" as appropriate to the historical context. In 2016, the Government of Canada moved to use "Indigenous peoples" to describe the original inhabitants of the continent, and this term is used throughout in speaking more generally of the constitutionally recognized Indians, Métis, and Inuit. The singular "Indigenous people" is used in reference to an agglomerate of people as individuals.

2 https://twitter.com/SeanDunn10/status/1514465238535442434

CHAPTER 19

SIMPLY CONSERVATIVE? RETHINKING THE POLITICS OF RURAL ALBERTA

Laticia Chapman and Roger Epp

"How come we don't have that strong rural voice that we thought we were going to have?

— A county reeve (quoted in Parsons, 2019)

THE PROPOSITION THAT "rural Alberta" — readily identifiable and *different* — is the populist-conservative "base" for the United Conservative Party (UCP) government is incanted almost as an article of faith in political talk. It serves as a convenient shorthand for journalists, a home-truth for political scientists, and a spectre that both haunts and excuses the urban progressive political corners of the province. Midway through the UCP's contentious term in office, for example, we read that Premier Jason Kenney's "segment of Alberta" was "heavily rural" (Markusoff, 2021); that it was represented by a "tough, often ornery bunch" of MLAs (Mason, 2021); and that if it was "rural dominance" that elected the UCP in 2019 (Dawson, 2021), rural impatience with pandemic restrictions had also exposed its weakness (Bratt and Foster, 2021). Indeed, it was said, "much of rural Alberta, with exceptions of course, is still stamped with this notion that COVID-19 is some kind of urban plot against the countryside," when, in fact, by spring 2021, "rural Alberta is more infected than the big cities, to the extent that rural patients now put serious pressure on the city hospitals" (Braid, 2021). In this picture, rural Alberta is represented by the Whistle Stop Cafe in the village of Mirror and the anti-lockdown rodeo outside of Bowden — sites of noisy defiance of public health regulations; by crude anti-Trudeau billboards on semi-trailers along the highway between Calgary and Edmonton; and by news stories of Nazi flags flying over acreage properties near Boyle and Breton, and a Ku Klux Klan hood worn into the post office in Grimshaw. Alternatively, rural Alberta is home, as Kenney noted, to "thousands of Alberta families who put food on the table because of the mining industry" — not like the condescending big-city critics of his government's quiet move to open the province to coal development; that is, those who "look down on those folks" (Weber, 2021). Once again, it is different. Except that ranchers and rural municipalities were prominent among the critics of the policy shift and the proposed Grassy Mountain project. Somehow rural Alberta's status as conservative base is most secure when it seems not to be, and the former premier was desperate to regain its support (Thomson, 2022). When, in the end, Kenney was pushed to resign after a leadership review in 2022, it was the "outsized share" of membership votes in the "heartland of disgruntlement" that decided the outcome (Markusoff, 2022).

Even before she was sworn in as Kenney's successor, Danielle Smith had announced that she would call and contest a byelection in Brooks-Medicine Hat, not Calgary, where a seat was open, as a signal that she intended to represent the "forgotten corners of the province" — places where the UCP base presumably lived. In some of those same corners, rural municipal leaders who were elected to speak for their communities quickly declared their opposition to a policy scheme for which Smith had voiced support as a lobbyist and then as premier: to give oil companies royalty credits to do the well-site cleanup already required in their drilling licences. The scheme, one rural leader said, amounted to a misuse of public funds; it was "exactly how a fox would design a henhouse" (Bennett, 2022). The lesson, however, remains: rural opposition might be enough to help check industry-friendly UCP plans for the countryside but not to complicate the convenient political caricatures.

Five Propositions

This chapter begins from other propositions. First, rural Alberta does not decide provincial elections — including the last one — in what is one of the most urban provinces in the country, especially now that the electoral map has converged incrementally toward that reality. While Calgary and Edmonton votes are still discounted, more than four in five ridings are either part of a metropolitan region or else anchored by the population of a small or mid-size city. Airdrie, Lethbridge, Red Deer, Wood Buffalo, and Sherwood Park are not rural, unless rural is a residual category that means everything outside of the two largest cities. Second, rural Alberta does not set the policy agenda. It has no conservative think tanks or schools of public policy or media venues to match those in Calgary. Third, it is increasingly difficult to say with precision where rural Alberta is and where it is not. While the provincial map, of course, is still dotted with small towns, rural municipalities, and First Nations reserves, rural Alberta is not a straightforward idea or a single place with a shared set of interests. It is northern and southern, forested and farmed, Indigenous and not. It is not one, large libertarian enclave. Fourth, in the past quarter century, rural places have experienced the provincial state's visible retreat, its consolidation of services at greater distances, its efficiency formulas, its relentless pruning of local authority, and its abandonment of an active role in rural and northern development. In the neoliberal era, the "practice of governing from a distance" is about "maintaining structures that facilitate capital extraction and increasing the opportunity costs of living/working in rural Canada, rather than supporting the development of rural places or peoples" (Hallstrom, 2018, p. 31). Communities that imagine alternative futures are mostly on their own. Fifth, rural places have experienced most intensely the land-use conflicts and the environmental risks that come with proximity to resource industries, whose development — mostly out of sight, out of mind — has long been a matter of public interest for the entire province. But the shiniest evidence of the prosperity of the Alberta Advantage has always been harder to find, say, in Falher or Edgerton or Empress. Already in 2001, the government was compiling internal documents that showed striking regional disparities, otherwise hidden in provincial aggregates: health outcomes, educational attainments, and per-capita incomes in the outer perimeter of the province, for example, were well below those on the inside (Alberta Economic Development, 2001; also, Conference Board of Canada, 2012).

This chapter is *not* interested, except indirectly, in the puzzle that has vexed progressive parties and thinkers from Alberta to Kansas: why do rural people vote the way they do, that is, against their own presumed interests? Instead, we propose reconsidering rural Alberta along different lines of inquiry. We follow the American sociologist Loka Ashwood in thinking about the limits of conservative or libertarian as reliable descriptors for the politics of rural places. Ashwood suggests the lived experience of near-statelessness — or else the experience of the state as handmaiden of large-scale resource extraction — informs a distinct politics that is not reducible to either simple conservatism or populist resentment, even when it is denominated in grievance and a sense of loss. It might contain traces of a community-based anarchism or self-organization. It is locally scaled. It does not put its primary political energies into electing the right people in Edmonton, even when communities have mobilized to convince the provincial government to stop something or save something. It is not simply about delivering what Ashwood calls a better state. In this chapter, we (1) rethink rural places in a different register, outside the realm of either identity politics or state-centric politics; (2) survey the UCP government's extraction-focused approach to the rural as well as the pushback, whether from doctors, surface rights associations, and municipalities; (3) relocate rural places in a trajectory that has eroded the possibility of democratic, community-level practices of responsibility; and (4) in conclusion, suggest what it might mean for left politics and a post-UCP government to take them seriously.

The Politics of Rural Spaces

In his book, *Rural* (2011), British geographer Michael Woods follows the important work of Keith Halfacree (2006) and, behind him, Henri Lefebvre, in locating rural space inside the conceptual triangle formed by three vertices. The first is represented by localities inscribed by practices of land use: production and consumption. Think ranching, pipelines, coal mining, or tourism. The second is represented by the framing of capital and state policy in a market economy: For what shifting purposes have rural spaces been colonized and commodified, bought and sold? What is rural for? The third is represented by the everyday lives of those who inhabit those spaces — a corrective to a simple materialist reading of capitalist accumulation — as they negotiate the pressures of work, livelihood, and community futures, and interpret their own fragmented histories and experience. Rural, in other words, is material, imagined, and practiced. Those three points of the triangle are connected, but the logics to which they refer, Woods argues, are not necessarily aligned or stable over time. Rather than marking a "congruent and unified rurality," they define the tensions at the centre of a politics "in which the meaning and regulation of the rural is the core issue of debate" (Woods, 2011, p. 11). That politics is often about land, what it is for, what it enables, tolerates, and precludes, and who has the authority to decide. In its most dramatic form, whether successful or not, the politics of the rural takes the form of a refusal. In Alberta, over the past half century, rural people in particular localities have said no, for example, to a sprawling coal mine on Class 1 farmland southeast of Edmonton; to massive hog barn complexes in counties like Flagstaff and Forty Mile; and to the risks of proximity to the oil and gas industry: sour gas plants, hydraulic fracturing, flaring, rusted pipelines, and water tables disrupted by seismic blasts. Typically, those fights have divided communities and exposed the provincial government as both regulator and promoter, with its own calculus of public interest (Epp, 2012). Moreover, there is nothing about them that is particular to Alberta. In outline, they can be found across rural North America (e.g., Catte, 2018; Griswold, 2018; Flowers, 2020).

Woods's conceptual triangle can also help orient us toward a significant political problem for rural places, the phenomenon of *rural burden*. As described by rural sociologists, rural burden is the historical and material processes of "[c]onsolidation of ownership over the means of production and the metabolic rift cultivated by rural resource extraction for largely urban consumption," persistent poverty, environmental degradation often connected to the presence of extractive and hazardous industries or the loss of economic opportunity, and decreasing demographic and political power (Ashwood, 2018, pp. 1-2). It is the sometimes hidden-in-plain-sight substrate to the relationship between rural places and the state, national or provincial. For Ashwood, it is important neither to overlook rural burden nor to assume that the standard political and ideological vocabularies are adequate for understanding rural places. The orientation of modern social science is "rooted in the classic, theoretical groupings of conservatism, radicalism, and liberalism, and more recently neoliberalism, which each assume the state as a starting point" (Ashwood, 2018, p. 720). From that starting point, it is easy to overlook the possibility that the state is not the central focus of all political articulations, even when the state has had a significant hand in creating local burdens. Ashwood's work to understand the anti-state and "stateless" politics of rural places is connected to her interest in strengthening rural communities' "participation in democracy" and, reciprocally, strengthening "democracy's accountability" to them (Ashwood, 2022). To put this in stark terms, rather than assuming that democracy basically works for everyone and that rural people can be judged as political actors within a functioning, impartial system, Ashwood has developed a body of research that shows how rural people, places, and lands have often served — and continue to serve — as sites of extraction, dumping grounds, and a legal and cultural "other."

How might assuming the state as a starting point result in distortions or misrepresentations of rural politics? What might constitute an alternative for researchers with an interest in the politics of rural places? Although conventional political analysis might contrast conservatism, liberalism, and radicalism, Ashwood groups the three ideologies together under the label "pro-statism." For the sake of space, we'll focus on Ashwood's differentiation between conservative and anti-state politics. While conservatism can support "anti-state rhetoric and some anti-state action," Ashwood notes that, like neoliberalism, "[i]n practice, modern-day conservatism is anything but stateless, imposing [through policy and other governmental means], for example, the corporate agribusiness agenda and resource extraction that leads to rural social and economic decline" (Ashwood, 2018, p. 720). Social conservatism is also compatible with pro-state politics, in that the "imposition of morality and traditionalism .... requires the imposition of a certain kind of state" (Ashwood 2018, p. 722). While the imposition of corporate agribusiness is often alienating to rural communities, there can be affinity between pro-state social conservatism and rural politics. This affinity, however, is not universal and may be more relevant when the moral society being envisioned is one that emphasizes community, kinship structures, and work ethics rather than individualization and competitiveness.

How does differentiating between conservative and anti-state politics help us to understand politics in rural places? The issue of academic and cultural puzzlement has special relevance here. Numerous books and articles from Canada, the United States, and western Europe published in the last decade testify to a sudden sense that something has gone wrong for (but more often with) rural people, especially if those people can be described as working class and white. Katherine Cramer's book *The*

Politics of Resentment (2016) is an ethnography of "rural political consciousness" in Wisconsin that, as the title indicates, invests heavily in the idea of resentment as a dominant affect. From France, novelist and essayist Édouard Louis's *Qui a tue mon père/Who Killed My Father* (2019) offers a difficult but sympathetic personal account of the social and physical toll of neoliberal policies on working class and poor rural French people and families and how such policies have created a foothold for Euroskeptic, nationalistic political rhetoric within these formerly left-leaning communities. And in Alberta, Clark Banack explicitly follows Cramer's methodology of "listening to rural people, in rural places" in an attempt to capture a sense of social identity and its intersection with political opinion (2021).

Each of these texts highlights rural attitudes of wariness, distrust, and sometimes outright hostility towards the state, as well as "urban elites." But they and others expend varying levels of effort on contextualizing rural "anti-state" attitudes. To greater or lesser extent, these publications testify to the "conservative" character of rural people and places, implicitly or explicitly calling on ideology to explain rural politics. But as we have seen, the label of conservatism tends to focus our intellectual energy on the state, instead of helping us to pay attention to the political issues that shape people's everyday experiences and concerns. One of the dangers of studies and explanations that emphasize identity over history is making rural political attitudes seem timeless, essential, and independent of people's experiences or social context, including the experience of rural burden.

Blue Trucks and Golden Eggs

In the months prior to the 2019 election, the UCP positioned itself as the champion of rural Alberta against a New Democratic government that was ideologically out of touch with the countryside. Its campaign famously featured Kenney in a blue pick-up truck, which, as a magazine profile innocently put it, "appeals to rural Albertans" (Stepanic, 2022). Alternatively, the truck might well have suggested how ill-at-ease the UCP leader was in small towns, on gravel roads, and at the gas pumps; it did not help him overcome a lack of trust that was present from the start in the party's electoral coalition. Nor did it deliver a robust rural policy agenda. For all that, the results on election night were decisive enough. In winning 63 of 87 seats, the UCP dislodged the NDP from every part of the provincial map except Calgary and Edmonton, Lethbridge and St. Albert; NDP candidates, including incumbents, were competitive in only a handful of other ridings outside the two metropolitan regions, among them Lesser Slave Lake and Banff-Kananaskis. The results, however, were met with little enthusiasm or sense of rural restoration to a rightful place at the governing table. That would not change over the UCP's term in office.

In the first place, Kenney appointed a Calgary-dominated cabinet. It included very few MLAs with rural addresses: Travis Toews from Beaverlodge in Finance, Jason Nixon from Sundre (though he was born, raised, and educated in Calgary) in Environment; Devin Dreeshan from Pine Lake in Agriculture and Forestry; and Rick Wilson from Wetaskiwin County in Indigenous Relations.[1] The previous NDP cabinet had about as many rural MLAs in major economic and social portfolios.

In the second place, the UCP government signalled from the start it would spare rural communities none of the austerity measures contained in its first budgets.[2] It did not make policy with a rural lens. It cut funding to regional economic development alliances. It brought an abrupt end to a long, diminishing history of provincial involvement in agricultural research and extension, eliminating some 250 positions and closing facilities in several locations. Not least, the UCP government commissioned and accepted a performance review of Alberta Health Services from an international consulting firm. The review, advised by a "global expert panel," consistently targeted "small/medium" facilities outside the major cities for "reconfiguration or consolidation" of emergency, acute, and obstetrics care, based on standardized measures of "clinical viability," volume, and cost-effectiveness, as well as the promise of virtual technologies. In short, it recommended that most of the province's rural hospitals become night-time ambulance stations, shuttling emergency patients down the road, and that acute care be phased out at almost half of them, with five facilities closed completely (Ernst and Young, 2019). The global pandemic pre-empted the implementation of the review's recommendations, meant to save $1.9 billion per year. But it did not stop then-health minister Shandro from imposing a new fee regime on the province's doctors in April 2020 that prompted a rural physicians' group to warn of a "looming crisis" in many communities as their colleagues chose to withdraw hospital services or simply close their practices (Rural Sustainability Group, 2020).[3] The effect over the next two years — compounded by a shortage of nurses — was regular disruption of care at thinly staffed hospitals from Athabasca to Pincher Creek, Elk Point to Rocky Mountain House. A modest plan to recruit new doctors to specific rural and remote places was launched in 2022 after Shandro was shuffled out of the health portfolio (Ferguson, 2022).

If the UCP government generally struck an adversarial tone in the first two years of its mandate, that was particularly evident in its rocky relationship with Rural Municipalities Alberta and its members. The premier put them on the defensive in his first speech to the provincial organization's fall convention, historically an important political event. Kenney focused on his government's efforts to return the oil and gas industry to profitability. To that end, he instructed them to do their part to help him attract new investment: cut red tape, match the "generous property tax breaks" available to the petrochemical industry "in places like Texas and Louisiana," and accept the budget cuts that were part of his government's determination to "stop running Alberta on a credit card." The MacKinnon Report, he noted, concluded that grants to Alberta municipalities were 20 per cent higher than in the rest of Canada (Government of Alberta, 2019). By the time of the convention, the UCP had already made deep cuts under the Municipal Sustainability Initiative, a major source of funding support for basic infrastructure renewal. It had also announced a new funding formula that off-loaded a larger share of policing costs onto rural municipalities.

Nothing, however, reflected the twin priorities of austerity and resource extraction more than the government's refusal to side with municipalities in their efforts to collect unpaid oil and gas industry taxes that had doubled to $173 million in the year 2019 alone (RMA, 2020a; 2020b). That figure would increase to $250 million by 2022.[4] Moreover, despite strong protests from rural leaders, the government imposed a set of changes to the municipal property tax regime to exempt new wells and pipelines, cut assessments on existing well sites, and eliminate a levy on well-drilling equipment (Graney, 2020; Varcoe, 2020). It had already cut the assessment on sour gas wells. As Premier Kenney said at the time: "if dozens of companies go bankrupt, then the goose that laid the golden egg is dead at the end of the day and there is no property tax revenue" (quoted in Johnson, 2020). But the net effect of the industry's tax revolt and the new assessment model was a massive transfer of wealth from the countryside and a challenge to the viability of some rural municipalities. Add to it the challenges faced by individual landowners in oil and gas country, whether in collecting surface rights leases or in ensuring proper remediation of well sites even after reclamation certificates have been issued, generally, as regulations permit, without an on-site inspection. Landowners can apply to the Surface Rights Board for eventual compensation for rent owed — a form of public subsidy to the industry the government rarely recoups (Riley, 2021); but they are often left to rely on their own vigilance when it comes to cleanup in a province where ownership can be transferred, responsibility is difficult to trace, and the track-record of the Alberta Energy Regulator in enforcement is uneven (Goodday and Larson, 2021; Morgan, 2021). The province acknowledges 170,000 abandoned and inactive wells (Government of Alberta 2022d). The environmental liability associated with those wells has been estimated at more than $250 billion, a liability that far exceeds the capacity of the industry-funded Orphan Well program, which was fortified with a provincial loan in 2017, and that was scarcely diminished by a $1-billion federal job-creation allocation to the province in 2020 (Egler, 2021; Boychuk, 2021). As the leader of one surface rights association, representing about 200 landowners in southern Alberta, told a journalist: "It seems like this government is in industry's back pocket and they'll do whatever they need to do to ensure these companies are allowed to drill. . . . There's a huge rural backlash coming and it does not appear that the government recognizes this" (quoted in Anderson, 2020).

All of this had been set in motion before the prolonged crisis of pandemic management, vaccinations, mask mandates, and border blockades seemed to some commentators to define the stark cultural-political gap between urban Albertans, who believed in science, public health, and the rule of law, and rural Albertans, who believed in personal freedom and responsibility. Indeed, provincial statistics at various intervals did tend to show rural-metropolitan differences in vaccination rates, notably in age ranges below 60 and especially below 18 (Government of Alberta, 2022c). But this is a more complicated story than ignorance or ideology. In some locations, it is also about convenient access for working people. In some northern regions, it is about lower uptake in Indigenous communities, which are imprecisely grouped in the province's "local geographic areas." In at least three of four of those areas where vaccination rates have been markedly lower — High Level/Mackenzie in the northwest, Two Hills, Taber, and Forty Mile in the southeast — the response also reflects the presence of conservative Mennonite populations with their own historically informed wariness of government. In rural communities generally, where anonymity is limited by scale, mandates and resulting employment losses in precarious sectors like health care have been a matter of significant local conflict rather than consensus (e.g., Baig, 2022).

What COVID-19 did serve to expose was the structural disadvantage of rural internet. In the work-from-home, study-from-home, do-business-from-home phase of public health restrictions, the lack of reliable broadband connections affected almost half a million Albertans in rural and remote locations. In this case, the federal government led the policy response with a major infusion of funds in 2019 for its national Connectivity Strategy. The province announced a $150-million matching commitment in July 2021. In a rare show of collaboration, the two orders of government agreed subsequently to raise their respective contributions to $390 million each, the total to be administered through the established mechanism of a federal fund (Government of Canada, 2022). For the UCP government, the investment became a feature part of its post-pandemic economic recovery strategy, but also gestured toward virtual health and support for automation in sectors like large-scale agriculture. Whether it reduces the structural gap, rather than, say, benefiting Shaw and Telus with public money in the name of rural development, will be a matter for future assessment.

For all COVID exposed in the countryside, however, it also might be said to have served as cover. In the first months of the pandemic, without public consultation, the government quietly rescinded a policy that since 1976 had restricted coal development in the Eastern Slopes of the Rockies, opened 1.4 million hectares to exploration leases, and granted 150 of them. The flashpoint became Grassy Mountain, an open-pit, metallurgical coal project proposed for an area north of Crowsnest Pass, on a former mine site that was abandoned in the 1950s and never fully protected under provincial policy. It was already under review by a federal-provincial joint environmental panel that held public hearings in October 2020. It had the support of the municipality of Crowsnest Pass, which anticipated well-paid jobs and an influx of families, and three First Nations within Treaty Seven that had signed long-term benefit agreements with the Australian coal company behind the project, Riversdale Resources. In the region, however, opposition to Grassy Mountain over its potential environmental impact in the Oldman River watershed had coalesced around the Livingstone Range Landowners group, municipalities, environmentalists, and Indigenous peoples upset at their own leadership. The government was clearly surprised by the backlash and the wider public sentiment that the mountains should be off limits to the kind of resource development it had invited. Its first response was to paint the opposition as mostly urban. But the opposition coalition stood its ground (Nikiforuk, 2021). The UCP's fall-back response was to cancel most of the new leases and strike a carefully constituted committee, including landowner and Indigenous (but not environmental-group) representation, to hold public consultations and recommend a new coal policy.

The government seemed resigned, or else relieved, when the joint federal-provincial panel ruled in June 2021 that the Grassy Mountain project was not, on balance, in the public interest. That did not mean it had necessarily abandoned its interest in coal, or that opponents had let down their guard. The much-awaited report from the policy committee, released in 2022, did not close the door on new mines, but recommended regional and sub-regional land-use plans involving Indigenous communities be completed first in order to rebuild public trust, provide "investment certainty," and determine where, if at all, development should happen along the Eastern Slopes (Coal Policy Committee, 2021). The government could regard the report as partial validation of its position that the old policy was obsolete. Meanwhile, a parallel report, "written by Albertans, for Albertans," and based on what was presented to the policy committee, took a stronger position: no new mines or exploration, "timely and effective remediation" of all lands disturbed by coal activity, and sufficient financial security, on the polluter-pays principle, to cover those existing liabilities. The report has been endorsed by an unlikely mix of ranchers, recreationists, and environmental organizations, as well as the Town of High River ("A Coal Policy for Alberta," 2022). The chair of the government's policy review committee, Ron Wallace, declared the community report to be "pretty much" consistent with its recommendations and likely to add momentum to the shaping of a different policy outcome (Derworiz, 2022).

The politics of land, water, and community futures are often politics for keeps in rural places; they concern what Woods calls the meaning and regulation of the rural. What is striking about the UCP government, "rural base" or not, has been its one-sided and open commitment to rural as an extraction site. To that end, it has also opened old-growth forest for logging near the Willmore Wilderness Park and native grassland in the south for oil and gas exploration leases. It has sold off Crown land and maintained a hands-off policy approach to the financialization of farmland, which has become an investment commodity priced beyond any realistic economic returns in agriculture and beyond the reach of a next generation (Aske, 2022). It has sought out strategic alliances with individual First Nations and the extraction-friendly Indian Resource Council; it created both a billion-dollar Opportunities Corporation to provide access to credit for "Indigenous-led investment in energy, mining, and forestry projects" (Government of Alberta, 2022a). At the same time, it can claim no signature achievement on behalf of rural communities and rural people. Those rural people who may — and do — disagree with each other about the desirability of particular resource developments, sometimes on the basis of their own working lives in those industries, are already living in places where local hospitals no longer deliver babies, schools close, municipalities are hard-pressed to provide basic services, and landowners cannot hold companies accountable.

The UCP's backbench MLAs have had little to say in public about any of this, not even in a late-term leadership race to replace Kenney in which rural is back in play, if only as a marker of cultural authenticity and identity: rural is the ability to ride a horse, not just wear a hat. Of the two MLAs who now sit as Independents for having challenged Kenney's pandemic leadership, Todd Loewen (Central Peace-Notley) cited failed negotiations with doctors, the rescinding of the coal policy, and the premier's neglect of his own caucus among his reasons (Bennett, 2021). The other, Drew Barnes (Cypress-Medicine Hat), floated a discussion paper, testing the waters for a new political movement that could be a "truly rural voice" (Barnes, 2021). The paper began from the premises that "rural marginalization is real," and that Kenney's UCP had abandoned rural voters, silenced their MLAs, spent recklessly, and taken a "sharp left turn" in order to chase urban votes. That left turn included a desire to "shut down rural Alberta's resource-driven economy." Rural Voice, he wrote, would be committed to listening, but Barnes already knew what people wanted: economic and social freedom, limited government, grassroots democracy, entrepreneurship, and an "embrace" of Alberta exceptionalism. Rural, in other words, as uncomplicated, boiler-plate conservative and as wholly different in its values. A year after the discussion paper was posted, Rural Voice had shown little sign of uptake.

Governed from a Distance

The questions of who speaks for the rural and how much its politics are actually oriented to the conventional theatre of elections and legislatures are both close to the heart of this chapter. So is Ashwood's interest in democratic accountability and in rural as history — a lived experience of the state, present and absent — rather than simply as identity or an oppositional consciousness. The post-war migration to the cities, the mechanization of farming, and the challenges of what came to be called rural modernization confronted governments across much of North America. In Saskatchewan, the Royal Commission on Agriculture and Rural Life in the 1950s delivered a 14-volume examination of issues such as farm income, education, local government, roads, and electrification. One good analysis describes the commission in terms of a tension between public participation and the technocratic influence of academics, bureaucrats, and business leaders, who preferred central authority and consolidation over local control (Skovron, 2011). In Alberta, which had become an urban province by 1951, what emerged at the same time was a pattern of patron-client politics enabled by energy revenues, which gave the provincial government the means to spend generously in select and visible areas — schools, hospitals, seniors' homes, roads — but also to withhold such benefits when a constituency might be tempted to vote for an opposition candidate. This bargain helped to solidify electoral dynasties for Social Credit and then Progressive Conservative governments. It also delivered tangible benefits to rural places, especially as oil prices rose in the early 1970s.[5] But patron-client politics came with costs too. One was an erosion of local autonomy, as governments tied municipal grants to specific program expenditures. Another was the need to accept the priority of oil and gas development as a matter of overriding public interest for the province, set in law, even though rural people often lived closest to the related health and environmental risks. A third, not least, was a political deskilling: a loss of the capacity to think, act, speak, and organize in defence of distinct interests that had marked an earlier generation of self-organization and office-holding (Epp, 2001).

It was Ralph Klein's government that ended its part of the relationship. Faced with public debt and declining energy prices, it consolidated school and health authorities, forced municipal amalgamations, shuttered regional planning commissions, and consolidated service delivery in larger centres. It took back municipal tax powers; once again, less government turned out to mean more centralized and less local government. For the province had its own ideas for what, from Edmonton, looked increasingly like an empty countryside. It proposed to reignite oil and gas production by reducing royalties, stripping any language of conservation from regulatory mandates, and introducing elements of industry self-regulation on a changing industry. It recruited global investors to build large, messy, export-oriented resource developments; it narrowed the scope for community intervention in public hearings; and it took the power to say no to intensive livestock facilities away from rural municipalities. Those were angry years in an increasingly divided, industrialized countryside (Epp, 2008;

2012). The anger culminated in a post-Klein firestorm over proposed provincial legislation to restrict development on land designated for long-term public infrastructure development, this after the government's own regulator was caught spying on rural landowners during hearings into a major North-South transmission line. The response was framed largely in the language of property rights (Nikiforuk, 2011; Stenson, 2015).[6] Then and now, no other available language has seemed as powerful or direct — a point of affinity with, or susceptibility to, conventional conservative ideological themes, but also a source of ideological confusion when exercised against Conservative governments in Edmonton. The other anti-establishment legacy of that period was a deepening distrust of expertise — the kind that discounted local knowledge in public hearings, offered soothing reassurances, and typically worked for those who could afford it. It was the 2012 provincial election that detached rural voters from the Progressive Conservatives for the first time in a generation and put their MLAs in the opposition benches with the Wildrose Party.

Told in this way, the story of rural Alberta aligns closely with Ashwood's account of rural America. The recent experience of the provincial state's retreat as well as its entanglement with large-scale resource development is typical. So is the steady erosion of local jurisdiction and democratic political spaces within which to act, with the result that rural people — with the possible exception of First Nations communities, as they reassert a meaningful right of self-determination — live increasingly under the domain of decisions made elsewhere. The Canadian political theorist Warren Magnusson calls this the "consolidation" and "upward transfer of functions…that integrates central and local government" (Magnusson, 2015, p. 67). He has cities in mind, not rural places, but some of his analysis might apply most acutely to the latter. Local government, he writes, is now a complex assemblage: not just municipal offices but also the institutions and agencies — hospitals, schools, social services and economic development — that represent other political jurisdictions in a particular location. In this mix, municipal governments have become more dependent on provincial financing, and at the same time have "been subjected to closer administrative supervision and legislative controls" (Magnusson, 2015, p. 88). This sense is compounded in rural places in Alberta where the local has become the regional, and where people live inside a Venn diagram of geographically disparate administrative districts and, often, inside a media desert, having lost the traditional sources of local news. The result is two-fold. The first is to frustrate and devalue a democratic conception of community politics as an assembly for the purposes of caring for local matters of public concern. The second, as Magnusson puts it, is to "distance authority from ordinary people" (Magnusson, 2015, pp. 112-13). The political options that remain for local people tend to be reactive, oppositional, and sometimes misdirected.

Conclusion: Beyond the Better State

In June 2022, on a rain-soaked spring night in Forestburg, a small town in grain-farming country in east-central Alberta, people filled the hall — a restored train station — for a pandemic-delayed celebration of the 10-year anniversary of the Battle River Railway co-operative. There was not a government MLA in sight. After music and a meal of beef on a bun, speakers, men and women, recounted the history of their efforts to save the Camrose-to-Alliance line. Like so many others in the rural West, it had been quietly marked for abandonment. But it was, they knew, the economic backbone of their communities and farms. Over several years, their efforts included the formation of the co-operative, the hard negotiations with CN Rail, the purchase of the line and an engine, then another, the regulatory hurdles, and the training for volunteers who, especially in the early years, did much of the work (Barney, 2013). Speakers recounted the leadership it took, and the time. But they also spoke of the friendships they formed, and how a new generation was now filling the volunteer board. They did not say how little support they received from provincial governments. They did not say how much the co-operative contributed to the province's gross domestic product. Instead, they spoke of small economic successes, jobs for smart young people, and tourism opportunities to showcase their region.

It may be a fool's game to predict the future in electoral politics, but the existence, success, and longevity of organizations like the Battle River Railway co-operative suggest the time is long overdue for politicians and political thinkers in Alberta to reconsider its rural communities. We want to close the chapter with some thoughts geared toward those who consider themselves to be on the left of the political spectrum, but they are hardly exclusively left-oriented. We could instead label them *democratically oriented*. We've found scholars like Woods and Ashwood helpful in orienting our thinking about rural places and democratic accountability because they challenge the orthodoxy of state-centric political analysis and because they show rural places as particular, politically and socially complex communities.

Neoliberalism is the political idiom of our time. The political conundrum for the left in rural places in neoliberal times is this: on the one hand, the proposed alternative to UCP-like governments looks like what Ashwood calls the better state, one that is unafraid to regulate, tax, and even own the productive assets of the economy. To be sure, a better state could make a difference in rural Alberta if, for example, it can stabilize rural health care in a post-COVID world, deal with environmental liabilities, support economic transition, and set limits on the commodification of land for the sake of both the next generation and other models of agriculture. On the other hand, the better state is a centralizing state. It, too, governs from a distance. It thinks in universals. It may not champion neoliberal community volunteerism to cushion the impact of public-sector cuts, but neither does it necessarily trust local authorities to deliver transformation or provide the equivalent measure of services to citizens, regardless of where they live. The better state can fill a gap but then reopen it at any time, or reorganize a system so that it can no longer be navigated or negotiated from below. It is unreliable. This has been the experience of the state in rural places.

The political conundrum for the NDP — the standard-bearer for the better state — is that the path to re-election does not need to take it very far outside Alberta's major cities. To engage people on the periphery is to waste strategic electoral resources; for everyone *knows* rural Alberta is the conservative heartland. In this chapter, we have offered challenges to identity-focused explanations of rural politics, moving instead to think about the histories and experiences that inform rural opinions about the state. What everyone knows offers no hope of political change. It does not strengthen the democratic accountability of the provincial government to its communities. Nor does it recognize the political creativity and persistence of groups like the Battle River Railroad co-operative or the Livingstone Range Landowners, or like the rural municipalities that have found a sharper, more sophisticated voice at the edges of their legislated jurisdiction.

But understanding doesn't necessarily win elections, and understanding doesn't alter the reality of the political map. Rural Alberta, as we said at the outset, does not decide elections. It has held symbolic importance to conservative parties in recent decades, but the benefit has been substantially in one direction, with rural communities increasingly losing much more than they gain at the hands of conservative governments. Our final question, then, is to ask why the NDP should care about rural Alberta. The answer lies partly in larger questions of food, climate, and environmental futures, but by themselves we suggest, those policy interests might just produce a different set of projections imposed on the countryside. Instead, they require an engagement with people who live in real rural places, and an interest in local manifestations of democracy. By itself, a social-democratic better state is not sufficient. The challenge is to balance those commitments with a contribution to better rural municipalities and, more generally, to the renewal of meaningful decision-making at the level of communities and watersheds, including the choice to say no to the next provincial grand design. In that case, rural Alberta and Alberta's political left might find themselves with new and unexpected things to say to each other.

References

A coal policy for Alberta. (2022). Accessed July 25, 2022, at https://www.acoalpolicy foralberta.com

Anderson, M. (2020, Jan. 28). Turned to stone: how quickly Jason Kenney betrayed rural Albertans. *The Tyee*. https://thetyee.ca/Analysis/2020/01/28/Jason-Kenney-Betrayed-Rural-Albertans-Turned-To-Stone/

Ashwood, L. (2018). *Rural conservatism or anarchism? The pro-state, stateless, and anti-state positions*. Rural Sociology 83(4), 717-748.

Ashwood, L., Diamond, D., & Walker, F. (2019). Property rights and rural justice: A study of U.S. right-to-farm laws. *Journal of Rural Studies* 67 (April), 120-29.

Ashwood, L. (2022). Loka Ashwood. Accessed June 29, 2022, at https://lokaashwood.com/

Aske, K. (2022). *Finance in the Fields: Investors, Lenders, Farmers, and the Future of Farmland in Alberta*. Edmonton, AB: Parkland Institute. June 13. https://www.parklandinstitute.ca/finance_in_the_fields

Baig, F. (2022, April 22). COVID-19 response strains relationships in northern Alberta county. *Globe and Mail*. https://www.theglobeandmail.com/canada/alberta/article-covid-19-response-strains-relationships-in-northern-alberta-county/

Banack, C. (2021). Ethnography and political opinion: Identity, alienation and anti-establishmentarianism in rural Alberta. *Canadian Journal of Political Science* 54(1), 1-22.

Barnes, D. (2021). *Making room for a rural voice: A discussion paper*. Accessed July 26, 2022, at https://www.drewbarnes.ca/blog-2-1/blog-post-title-one-xb848

Barney, D. (2013). "That's no way to run a railroad": The Battle River branchline and the politics of technology in rural Alberta. In John R. Parkins and Maureen G. Reed (Eds.), *Social Transformation in Rural Canada: Community, Cultures, and Collective Action*. Vancouver: UBC Press.

Bennett, D. (2021, May 13). *Internal dissent boils over into open call for Alberta Premier Jason Kenney to resign*. CBC News/Canadian Press. https://www.cbc.ca/news/canada/edmonton/alta-kenney-caucus-todd-loewen-1.6024857

Bennett, D. (2022, November 11). Alberta rural leaders blast oil well cleanup plan. *Globe and Mail. https://www.theglobeandmail.com/canada/alberta/article-alberta-rural-leaders-blast-oil-well-cleanup-plan/*

Boychuk, R., Anielski, M., Snow, J. Jr., & Stelfox, B. (2021). *The big cleanup: How enforcing the polluter pay principle can unlock Alberta's next great jobs boom*. Calgary: Alberta Liabilities Disclosure Project. Accessed June 29, 2022, at https://www.aldpcoalition.com/_files/ugd/6ca287_2ffe90ca7c354d3eac43b5f141b6ec8a.pdf.

Braid, D. (2021, May 17). Kenney tries to bust the rural myth that powers his dissidents. *Calgary Herald*. https://calgaryherald.com/opinion/columnists/braid-kenney-tries-to-bust-the-rural-myth-that-powers-his-dissidents

Bratt, D. & Foster, B. (2021, April 13). *UCP caucus revolt latest in a long history of splintering conservative parties in Alberta*. CBC Opinion. =https://www.cbc.ca/news/canada/calgary/road-ahead-alberta-conservative-parties-splinter-history-1.5984055.

Catte, E. (2018). *What You Are Getting Wrong About Appalachia*. Cleveland: Belt Publishing.

Coal Policy Committee. (2021). *Recommendations for the Management of Coal Resources in Alberta*. Report commissioned by the Government of Alberta.

Conference Board of Canada. (2012). *Alberta's Rural Communities: Their Economic Contribution to Alberta and Canada*. Report commissioned by the Government of Alberta.

Cramer, K. (2016). *The Politics of Resentment: Rural Consciousness in Wisconsin and the Rise of Scott Walker*. Chicago: University of Chicago Press.

Dahlman, R. (2020, Oct. 24). UCP job cuts in ag research deal major blow to Lethbridge, Brooks and Alberta's future. *Medicine Hat News*. https://medicinehatnews.com/news/local-news/2020/10/24/ucp-job-cuts-in-ag-research-deal-major-blow-to-lethbridge-brooks-and-albertas-futuree/

Dawson, T. (2021, April 8). Kenney's divided house: COVID pandemic widens urban-rural rift within Alberta government. *National Post*. https://nationalpost.com/news/politics/kenneys-divided-house-covid-pandemic-widens-urban-rural-rift-within-alberta-government

Derworiz, C. (2022, May 25). *Alberta town endorses community-developed policy saying no to coal mining in Rockies*. Global News/Canadian Press. https://globalnews.ca/news/8869402/alberta-high-river-no-coal-mining-rocky-mountains/

Egler, M. (2021). *Not well spent: A review of $1-billion federal funding to clean up Alberta's inactive oil and gas wells*. Edmonton, AB: Parkland Institute. July 7. https://www.parklandinstitute.ca/not_well_spent

Epp, R. (2001). The political de-skilling of rural communities. In Roger Epp and Dave Whitson (Eds.), *Writing Off the Rural West: Globalization, Governments, & the Transformation of Rural Communities*. Edmonton: University of Alberta Press.

Epp, R. (2008). Two Albertas: Rural and urban trajectories. In *We Are All Treaty People: Prairie Essays*. Edmonton: University of Alberta Press.

Epp, R. (2012). Off-road democracy: The politics of land, water and community in Alberta. In David Taras and Christopher Waddell (Eds.), *How Canadians Communicate IV: Media and Politics*. Edmonton: Athabasca University Press.

Ernst and Young. (2019). *Alberta Health Services performance review: Final report*. Report commissioned by the Government of Alberta. Dec. 31.

Ferguson, D. (2022, Feb. 2). Alberta moves to attract rural doctors. *The Western Producer*. https://www.producer.com/news/alberta-moves-to-attract-rural-doctors

Flowers, C. C. (2020). *Waste: One Woman's Fight Against America's Dirty Secret*. New York: The New Press.

Goodday, V., and B. Larson. (2021). *The surface owner's burden: Landowner rights and Alberta's oil and gas well liabilities crisis*. SPP Research Paper 14(1).

Government of Alberta. Economic Development. (2002). *Regional disparities in Alberta: Resource package*. March 4.

Government of Alberta. (2019). *Premier Kenney speech at the Rural Municipalities of Alberta convention - November 15, 2019*. Retrieved July 26, 2022, from https://www.alberta.ca/release.cfm?xID=6612007BB17DE-D64B-88FF-75B1F120688D9889

Government of Alberta. Alberta Indigenous Opportunities Corporation. (2022a). Retrieved June 29, 2022, from https://www.theaioc.com/

Government of Alberta. (2022c). *COVID-19 Alberta statistics*. Retrieved June 29, 2022, from https://www.alberta.ca/stats/covid-19-alberta-statistics.htm

Government of Canada. Innovation, Science and Economic Development. (2022). *Alberta and Canada expand partnership to improve access to high-speed Internet for Albertans*. Media release, March 9. https://www.canada.ca/en/innovation-science-economic-development/news/2022/03/alberta-and-canada-expand-partnership-to-improve-access-to-high-speed-internet-for-albertans.html.

Graney, E. (2019, Dec. 1). Alberta's property-tax break for gas producers leaves municipalities facing loss of revenue. *The Globe and Mail*. https://www.theglobeandmail.com/business/article-albertas-property-tax-break-for-gas-producers-leaves-municipalities/

Griswold, E. (2018). *Amity and Prosperity: One Family and the Fracturing of America*. New York: Farrar, Straus and Giroux.

Halfacree, K. (2006). Rural space: constructing a three-fold architecture. In P. Cloke, T. Marsden, & P. Mooney (Eds.), *The Handbook of Rural Studies*. London: Sage.

Hallstrom, L. (2018). *Rural governmentality in Alberta: A case study of neoliberalism in rural Canada*. Revue Gouvernance/Governance Review 15(2), 27-49.

Johnson, L. (2020, Oct. 7). Notley cuts down early UCP proposal that could cost rural taxpayers. *Edmonton Journal*. https://edmontonjournal.com/news/politics/notley-cuts-down-early-ucp-proposal-that-could-cost-rural-taxpayers

Louis, É. (2019). *Who Killed My Father* [*Qui a tue mon père*], trans. Lorin Stein. New York: New Directions.

Magnusson, W. (2015). *Local Self-government and the Right of the City*. Montreal and Kingston: McGill-Queen's University Press.

Markusoff, J. (2021, Feb. 11). This is not what Kenney came back for. *Maclean's*. https://www.macleans.ca/longforms/jason-kenney-alberta/

Markusoff, J. (2022, May 11). *This is where the Albertans deciding Jason Kenney's future live (think small)*. CBC News. https://www.cbc.ca/news/canada/calgary/jason-kenney-united-conservative-party-leadership-review-ridings-1.6448382

Mason, G. (2021, April 15). Jason Kenney, once considered invincible, is facing an uprising. *The Globe and Mail*. https://www.theglobeandmail.com/opinion/article-jason-kenney-once-considered-invincible-is-facing-an-uprising/

Morgan, G. (2021, Sept. 22). "Blindsided": Alberta farmers fret as regulator eyes moving bankrupt company's idle oil wells to new insolvent firm. *Financial Post*. https://financialpost.com/commodities/energy/oil-gas/blindsided-alberta-farmers-fret-as-regulator-eyes-moving-bankrupt-companys-idle-oil-wells-to-new-insolvent-firm

Nikiforuk, A. (2011, Feb. 4). Does Keith Wilson look like a revolutionary to you? *The Tyee*. https://thetyee.ca/News/2011/02/04/KeithWilsonRevolutionary/

Nikiforuk, A. (2021, June 18). Who saved Alberta's mountaintops and precious clean water? Albertans. *The Tyee*. https://thetyee.ca/Analysis/2021/06/18/Alberta-Mountaintops-Precious-Clean-Water-Saved-Albertans/

Parsons, P. (2019, Oct. 10). *New funding proposal for policing angers rural municipalities*. CBC News. https://www.cbc.ca/news/canada/edmonton/rural-policing-funding-alberta-1.5315674

Riley, S. (2021, March 9). Alberta covered $20 million in unpaid land rent for oil and gas operators in 2020. *The Narwhal*. https://thenarwhal.ca/alberta-oil-gas-land-rent-2020/

Rural Municipalities of Alberta. (2020a). *Municipal taxation and assessment*. Position Statement. Retrieved July 26, 2022, from https://rmalberta.com/wp-content/uploads/2021/01/Municipal-Taxation-and-Assessment-Position-Statements.pdf

Rural Municipalities of Alberta. (2020b). *Unpaid taxes owed from oil and gas companies to rural municipalities continue to increase*. Media Release, Jan. 20. https://rmalberta.com/news/unpaid-taxes-owed-from-oil-and-gas-companies-to-rural-municipalities-continue-to-increase/

Rural Sustainability Group. (2020). *Doctors warn of looming health crisis in rural Alberta*. Media release, April 20. https://www.albertadoctors.org/Media%202020%20PLs/press-release-doctors-warn-looming-healt.pdf

Skovron, W. M. (2011). *A people's commission?: High modernism, direct democracy, and the royal commission on agriculture and rural life, 1952-1957*. [Unpublished master's thesis]. Simon Fraser University.

Stenson, F. (2015, June). Landowner rights: How Big Oil trumps private and public good. *Alberta Views*, 28-33.

Stepanic, M. (2022, Summer). "I wanted to have a normal life." *EDify*, 16-20.

Thomson, G. (2022, Feb. 15). *Jason Kenney sees opportunity in political chaos*. CBC Opinion. https://www.cbc.ca/news/canada/calgary/opinion-alberta-jason-kenney-political-opportunity-chaos-1.6352249

Varcoe. C. (2020, Aug. 14). UCP stuck in middle of tax clash between oilpatch and rural municipalities. *Edmonton Journal*. https://edmontonjournal.com/opinion/columnists/varcoe-ucp-stuck-in-middle-of-tax-clash-between-oilpatch-and-rural-municipalities/wcm/07267095-8896-476c-a95a-871da142d65c

Weber, B. (2021, Feb. 4). Premier holds firm on coal-mining changes, says old policy was obsolete. *The Globe and Mail*. https://login.ezproxy.library.ualberta.ca/login?url=https://www.proquest.com/newspapers/premier-holds-firm-on-coal-mining-changes-says/docview/2485997385/se-2

Woods. M. (2011). *Rural*. New York: Routledge.

NOTES

1 Grant Hunter from Taber was initially appointed as associate minister for "red tape reduction." He was shuffled out of cabinet before he chose to side with the trucker protests at the U.S.-Canada border crossing at Coutts in winter 2022. Dreeshan was pushed to resign his portfolio in 2021. He was replaced by Nate Horner, a rancher from Pollockville in eastern Alberta and part of a multi-generational political family.

2 The exception might be the UCP's change to base K-12 education funding on a three-year rolling average of student numbers. The rationale was to make funding more predictable. While metropolitan school boards struggling to keep pace with growing student numbers complained this amounted to a subsidy to rural districts, the change could not reverse rural enrolment decline or prevent the school closures that continued in 2019 and 2020.

3 The minister did hastily rescind an additional cost-saving measure, identified in a memo, to de-list 141 communities from a Rural, Remote and Northern Program that supplemented physicians' funding.

4 In 2020 the Minister of Municipal Affairs introduced legislated measures — "a hammer" — in response to RMA demands, but a year later conceded they hadn't gone far enough. Among other things, the RMA has asked that the Alberta Energy Regulator refuse to approve applications from companies with outstanding debts to municipalities or landowners, and not rely on self-reporting, rather than simply "consider" those debts as "risk factors."

5 Curiously, when those governments made generational investments to bridge rural and urban living standards – electricity in the 1950s, natural gas in the 1970s — they chose to work through local co-operatives rather than private capital, a gesture toward an older rural culture of self-organization.

6 See Ashwood, Diamond and Walker (2019) on the experience in the U.S., where property rights, sometimes entrenched as right-to-farm laws, sometimes turn out to protect corporate agribusiness, but in some contexts, "can help rural people push back against the market economy in defense of their health and environmental rights when other political means falter" (p. 120).

CHAPTER 20

PEPPER SPRAY FOR ALL: THE UCP'S APPROACH TO COUNTERING RACE-BASED HATE IN ALBERTA

Irfan Chaudhry

"A more immediate need for Alberta's diversity activists is to confront the intolerance promoted by its own government."

— Dr. Darren E. Lund

THE ABOVE QUOTE, written by late Prof. Darren Lund in 2006, echoes strongly in present-day Alberta, where recent interventions (or lack thereof) by the UCP government toward hate-motivated crimes and incidents have fallen short of community expectations. During the COVID-19 pandemic, Alberta experienced sharp rises in reported and unreported hate-motivated violence toward numerous racialized communities, particularly Black Muslim women who observe the practice of wearing the hijab. Edmonton, in particular, has been host to a number of these violent attacks since December 2020 (Wakefield, 2021), prompting many advocacy groups, like the National Council for Canadian Muslims (NCCM) to put pressure on all layers of government to intervene and put an end to these hate-fuelled acts.

Oftentimes, governments try to utilize short-term funding as one approach to address social issues, hoping the money will make the problem go away. The UCP government, for example, introduced funding in 2021 to help increase security and support minor infrastructure improvements to facilities that serve communities at risk of hate or bias-motivated crimes or incidents. Such facilities could include mosques, gurdwaras, temples, churches, synagogues, LGBTQ2S centres, or other physical spaces where diverse communities gather. According to the Alberta Security Infrastructure Program Grant website (Alberta, n.d.), eligible organizations could apply for funds to support education, training, and security assessment, or they could apply to improve physical security infrastructure (such as gates, fences, security systems, etc.). While this funding opportunity might help support physical security enhancements, it does little to help deal with the ingrained and systemic racism, bias, and discrimination that lead to these spaces being targets in the first place. Although these hate-motivated forms of violence and intimidation are not unique to Alberta, recent responses from the UCP government have been unique (and not in a good way), highlighting a significant gap in awareness and understanding of how governments can meaningfully intervene and provide support to groups at a higher risk of being targeted by hate motivated violence — especially racialized Albertans.

The 2020 Statistics Canada report on police-reported hate crime shows race continues to be the most common motivating factor for those victimized by hate-motivated crime in Canada (Moreau and Wang, 2022). Alberta is no different in this regard, and a more sustained and community-focused approach is needed to adequately address hate-motivated violence and intimidation in Alberta. This chapter will explore how the UCP government has attempted to address race-based, hate-motivated violence and intimidation in Alberta and how these approaches have been out of synch with community, law enforcement, and other layers of government expectations. This chapter will also trace how previous provincial governments have supported the work to combat hate, and how these historical lessons can help our understanding of countering race-based hate in Alberta.

Alberta: A History of Race-based Hate

Alberta, like many other Canadian provinces, was built through a process of colonization and exploiting the labor of new arrivals to the country. This brought on emerging aspects of racism and discrimination. For example, a 1911 Edmonton Board of Trade petition pushed back against an influx of black settlers, whom they described as a "serious menace" (City of Edmonton, 2022). Federal legislation followed shortly after, preventing further immigration of black people into Alberta (City of Edmonton, 2022). The black community was not the only target of this race-based hatred. As a recent anti-racism report from the City of Edmonton (2022, p. 6) highlights:

> Alberta also has a long history of anti-Asian racism, which grew and spread as Chinese and Japanese immigrants began arriving in British Columbia in the late 1800s and early 1900s. During the Great Depression, Albertans openly discriminated against Chinese Canadians, even while showing greater acceptance of Japanese Canadians. This pattern reversed during World War II, when many Japanese Canadians were interned, and some forced to work on Alberta sugar beet plantations. In the 1950s and 60s, Alberta's sugar beet industry later relied on coerced labor from Indigenous people.

This form of industrialized racism formulated some of the foundational years for the province of Alberta. "In the pioneer colonial years of Alberta, there was a clear ethnic hierarchy in the work force. Skilled labour jobs were reserved for Anglo-Saxons, while central, eastern, and southern Europeans worked jobs labeled unskilled" (Finkel, nd). This form of racial and ethnic hierarchy made conditions ripe for the presence of certain white supremacist groups in Alberta, such as the Ku Klux Klan, more commonly known as the KKK (Baergen, 2000). While some have come to label Alberta as a hard-working, blue-collar type of community (due to the vast agriculture and oil and gas industries), others see Alberta as a "Bible-belt — the three hundred kilometer wide east-west band between Edmonton and Calgary" (Finkel, n.d., p. 11). It is this latter representation that provides insight into a darker side of Alberta's history, one in which the KKK was able to flourish in the 1930s with much popularity. As Baergen (2000, p. 12) explains in some detail,

> The Ku Klux Klan mentality — a religious and racial bigotry...emerged during Alberta's formative years when its foundation Anglo-Saxon stock felt its racial purity threatened by a wave of non-Anglo-Saxon immigrants. The Alberta Klan had its roots in the same soil as the Bible thumpers, but the emphasis on white supremacy was its own unique contribution to the Alberta psyche.

While alarming, it might not be surprising the ideology of the KKK took root in the province of Alberta during this timeframe, where fear and concern was permeating within the Bible Belt regarding the increase in immigration that would be a threat to the "Canadian" way of life. This sentiment echoes the heart of the Klan's purpose: to maintain white hegemonic control (Southern Poverty Law Center, n.d.). From 1921 to 1931, Alberta's population increased by 144,000 — mostly by immigration from European countries not named Great Britain. German, Polish, and Ukrainian immigrants all helped build Alberta's population in this era. As Baergen mentions, "in retrospect, such numbers do not appear alarming, but the perception at the time gave rise to wide-spread paranoia" (Baergen, 2000, p. 13), creating fertile grounds for the message of racial and ethnic purity from the KKK to take root and embed itself in the province. In fact, "Alberta is unique in that it is the only jurisdiction in Canada on record to have granted charter to the KKK — a charter that made its activities legal, even though one of its stated objectives was racial purity" (Bergen, 2000, p. 13). Non-British Europeans were not the only group to be impacted by hateful sentiment in Alberta. Muslims from Lebanon, for example, had been establishing roots in the province since the early 1900s. In fact, Canada's first mosque, Al Rashid mosque, was established in Edmonton in the mid-1930s. This presumably gave further rise to concerns about the potential replacement of white settlers, giving further fodder for groups like the KKK to exploit. This history of overt and covert racism impacted many racialized identities. Anti Semitism, anti-(non-white) immigration, and anti-Black racism all seemed to take root in the province through various processes and policies during the 1940s, '50s, and '60s (see City of Edmonton, 2022).

Interestingly, it is also around this same time Canada began to develop its hate speech laws under the criminal code, spearheaded by the creation of the Cohen committee to address the horrors conveyed to the global public during the Second World War. The disclosure of the systematic genocide of Jewish people by the Nazis during the war motivated a delegation from the Canadian Jewish Congress to appear before a 1953 joint committee of the House of Commons looking at revisions to the Canadian Criminal Code (Anand, 1998). Despite their vigorous attempts, and a perceived increase in neo-Nazi organizations distributing contentious literature across Canada, the Canadian Jewish Congress was unsuccessful in lobbying the government for legislation proscribing religious and racial hate propaganda (Janhevich, 1997; Anand, 1998). It was not until 1964, following an overt revival of Nazism in Canada, that strong lobbying for an anti-hate law began (Anand, 1998). Following a discussion in the House of Commons in 1964, then-justice minister Guy Favreau announced the appointment of the Cohen Committee — a committee with the mandate of studying problems associated with the spread of hate propaganda in Canada (Anand, 1998). The Cohen Committee released its report in November 1965, and while it found the extent of the problem in Canada was limited to a small number of persons, it stressed such activity created a climate of malice and destructiveness to the values of society (Ross, 1994; Janhevich, 1997; Anand, 1998). Despite the limited scope of the problem in Canada, the Committee (1966, pp. 24-25) ascertained that:

> ...the individuals and groups promoting hate in Canada constitute a clear and present danger to the functioning of a democratic society. For in times of social stress such hate could mushroom into a real and monstrous threat to our way of life. Nor does giving these hate promoters a radio or television platform serve any valid debating purpose...In the Committee's view, the hate situation in Canada, although not alarming, clearly is serious enough to require action".

A bill incorporating most of the recommendations from the Cohen Committee Report was tabled in Parliament in 1966 (Ross, 1994). Four years later, in June 1970, these recommendations were finally incorporated into the Canadian Criminal Code as sections 281.1, 281.2, and 281.3 (Ross, 1994). Because of a revision of the entire Criminal Code during the 1980s, these now fall under section 318 to 320 (Hate Propaganda

Laws). Under the new provisions, it was illegal to espouse religious, racial and ethnic hatred in Canada (Janhevich, 1997). The new hate propaganda laws prohibited the advocating of genocide, the public incitement of hatred, and the willful promotion of hatred (Ross, 1994).

While these federal legislative moves had an impact across the country, they did not slow down the emergence and prevalence of organized hate groups in Alberta. In fact, the province of Alberta further legitimized some of them through provincial incorporation. In 1972, "forty years after granting a charter to the first Klan, the province of Alberta reaffirmed its tolerance for intolerant societies by issuing a Certificate of Incorporation to the Confederate Klans of Alberta" (Baergen, 2000, p. 265). As Bashir Mohamed found, "this certificate of incorporation lapsed in the 1950s when events in the U.S made the group universally reviled. But the certificate was renewed in 1980, remaining in effect until 2003" (Bashir Mohamed, as quoted in McMaster, 2019). Although the likes of the KKK began to fizzle out in the province in the '60s and '70s (though, there were still pockets in certain areas of the province), "the Aryan Nations Movement has given one of its main tenets — racial purity — a new sense of life" (Baergen, 2000, p. 263). In the 1980s, Alberta would begin its notoriety as being home to one of the country's landmark cases related to hate propaganda, testing out our newly minted hate propaganda laws. In 1981, James Keegstra was the subject of multiple complaints from parents regarding him teaching students the Holocaust was a hoax (Boyko, 2006). As Boyko explains (para 3):

> They discovered that, for more than a decade, Keegstra had been teaching students that the Holocaust — in which Nazi Germany murdered six million European Jews — was a hoax. He had also referred to Jewish people using ugly, disparaging terms and claimed that Jews were trying to take over the world through internationally organized plots. Keegstra was warned to stop teaching his racist, anti-Semitic views and obviously false conspiracy theories, but he refused. On 7 December 1982, Keegstra's 14-year teaching career ended with his dismissal and expulsion from the Alberta Teachers Association.

This Alberta case was the first successful conviction under the Canadian Criminal Code's hate speech provisions and is often used as foundational law to assess current cases related to hate speech and hate propaganda cases.

Alberta has also been host to some lesser known race-based hate events. For example, in 1990, a white supremacist event called Aryan Fest took place in a farm near Provost, (300 kilometres east of Edmonton). This event was organized by factions of the Church of Jesus Christ Christian-Aryan Nations and had hate on full display, including cross-burnings, signs reading "KKK White power," and a swastika flag (Alberta Labour History Institute, n.d). While several people who witnessed the event made complaints to the Alberta Human Rights Commission, an inquiry board created to assess the event was only able to order the respondents (the event organizers) to refrain from the same or any similar public display of discriminatory signs and symbols in the future (Alberta Labour History Institute, n.d).

Alberta: A History of Addressing Race-based Hate

While the above section only provides a snapshot of Alberta's history of race-based hate, it includes suitable context to consider the provinces' historical approaches to addressing racial discrimination. Legislation guiding Alberta's commitment to addressing bias and discrimination was formally promoted in 1990 by way of the Alberta Multiculturalism Act (Commission, 1994), replacing the Alberta Cultural Heritage Act of 1984. As a report by the Alberta Multicultural Commission (Commission, 1994, p. 2) explains:

> In 1988, Albertans told the Government of Alberta during a series of province-wide meetings that they wanted to live in a society that allowed citizens to be true to themselves and their traditions. [Albertans] said multiculturalism policy should encourage respect for, and integration of, all cultures. It should also ensure every citizen has the same opportunities to achieve their hopes and dreams without racial or cultural prejudice.

The main objectives of the Alberta Multiculturalism Act were to encourage respect for and promote an awareness of the multicultural heritage of Alberta, and to foster an environment where Albertans can participate and contribute to the cultural, social, economic and political life of their province (Leman, 1999). The act established a Multiculturalism Commission to advise the government on policy and programs respecting multiculturalism, as well as a Multiculturalism Advisory Council to advise the commission on policy matters (Leman, 1999). A Multicultural Fund was also set up to finance programs and services. The Alberta Multicultural Commission was established in 1991 to introduce a three-year Multiculturalism Action Plan (MAP) to provide a roadmap for accomplishing the act's objectives (Commission, 1994). However, the Alberta Multiculturalism Act, and associated commission, was short-lived. As Leman (1999) explains, in the spring of 1996, as part of its efforts to streamline government programs and operations, the Conservative provincial administration introduced new legislation that merged the human rights and multiculturalism programs. This resulted in the Alberta Human Rights Commission taking over the duties of the former Multiculturalism Commission, and the provincial human rights act usurping many of the responsibilities of the Alberta Multiculturalism Act, making it obsolete. With this move, the Alberta Human Rights Act provides the most authoritative legislative response to bias and discrimination in the province. According to the Alberta Human Rights Commission website, "in 1972, Peter Lougheed introduced the Individual's Rights Protection Act and the Alberta Bill of Rights, which formed the basis of Alberta's human rights legislative framework today. This legislation was responsible for establishing the Commission, whose function is to administer the Act and uphold the principle that all persons are equal in dignity, rights, and responsibilities" (*Alberta Human Rights Information Service*, 2017).

Although authoritative, it is limited in scope. The act prohibits discrimination in the following five areas: statements, publications, notices, signs, symbols, emblems or other representations that are published, issued or displayed before the public; goods, services, accommodation or facilities customarily available to the public; tenancy; employment practices; employment applications or advertisements; and membership in trade unions, employers' organizations or occupational associations. The act protects against discrimination in the above areas, on the following 15 protected grounds: race, religious belief, colour, gender (including identity and expression), physical and mental disability, age, ancestry, place of origin, marital status, source of income, family status, and sexual orientation. It is important to note the act does not apply to criminal events motivated by hate, as these would fall within the scope of the Criminal Code, where section 718.2(a)(i)[1] and the other hate propagation laws mentioned above could apply to the situation. Outside of the Alberta Human Rights Act and the Criminal Code, there are limited legislative routes to address race-based hate and its root causes. As a result of this gap, various layers of government have created funding opportunities to try to support initiatives that address hate and bias from a broader context.

Show Me the Money

In the wake of 9/11 and following a reported increase in hate motivated violence towards racial and religious minorities, the federal government launched an action plan against racism called A Canada for All: Canada's Action Plan Against Racism (Canadian Heritage, 2005), a first of its kind for the Canadian government. There was strong commitment put toward the plan: the 2005 federal budget even included a five-year, $56-million investment to support its implementation (Government of Canada, 2005). A key aspect of this plan was a focus on strengthening the role of civil society in the fight against racial discrimination. As the action plan highlights (Government of Canada, 2005, p. 33):

> The Action Plan Against Racism supports the development and capacity-building of racial and ethnic groups by fostering community participation and by coalition building between communities and non-governmental organizations. Collaboration between government and civil society continues to focus on strengthened partnerships, capacity-building, and public education on multiculturalism and anti-racism issues.

Through this approach, the federal government made it possible for various civil society organizations to apply for funds to address forms of racial bias and discrimination. This form of funding was significant, as it was an intentional sum of money made available to combat racism and racial discrimination. As limited funding was available at various provincial levels (including Alberta) to counter racial hate, this federal fund was integral in supporting all provinces in their efforts to counter race-based hate, supplementing limited funding programs available at the provincial level.

In Alberta, for example, although it was short lived, a key contribution of the Alberta Multiculturalism Act was the establishment of the Multicultural Fund. When the Alberta Multiculturalism Act was abolished, the Alberta Human Rights Commission became the main distributor and regulator of the Multicultural Fund, which was rebranded to the Human Rights Education and Multiculturalism Fund. According to information available on a publicly archived website,

> the Human Rights Education and Multiculturalism Fund provides support for educational programs and services that promote an environment where all Albertans can participate in and contribute to the social, economic and cultural life of the province. A key function of the Fund [was] to provide grants to community organizations for projects that foster equality and reduce discrimination (Human Rights Education and Multiculturalism Fund, 2014).

This source became the main funding stream available to various organizations across the province with the goal of addressing various forms of bias and discrimination, including race-based hate, through different types of education and engagement activities[2]. Projects spanned from urban centres to rural areas of Alberta, empowering and building capacity for local stakeholders to address racial bias through programming, education, and action. Available funding was also significant. Between 2011 and 2019, for example, the total monetary amount of grants paid out through the fund was just under $8 million, supporting a total of 209 unique projects and initiatives to help support and build a more inclusive and hate-free Alberta. Table 20.1 (below) provides a breakdown of these figures.

Table 20.1: Human Rights Education and Multiculturalism Grant amounts and initiatives funded

Year	Amount	Initiatives funded
2011-2012	$ 1,082,155.00	25
2012-2013	$ 1,014,555.00	26
2013-2014	$ 895,190.00	19
2014-2015	$ 996,500.00	25
2015-2016	$ 1,095,085.00	33
2016-2017	$ 940,798.00	27
2017-2018	$ 868,696.00	25
2018-2019	$ 744,570.00	29
TOTAL	$ 7,637,549.00	209

Sourced from the Alberta Human Rights Commission website

By having dedicated provincial funding to address hate, the provincial government conveys a strong message to their constituents that they are committed to countering and combating race-based hate. Provincial funding provides a consistent stream of money that civil society can apply to in order to support initiatives aimed at addressing the issue. Unfortunately, under the UCP government, this funding was officially eliminated in 2019 (Yousif, 2019) in an effort to reduce government spending and redirect monies to other "higher priority" areas. While some may argue it was a required cut to ensure financial stability, Table 20.1 shows the annual financial cost of the grant was around $1 million, hardly making a dent in the overall provincial budget. More importantly, however, cancelling this funding opportunity sent a strong message to Albertans that the UCP government does not take the issue of race-based hate seriously. These cuts could not have come at a worse time. In 2020 and 2021, Alberta saw an unprecedented increase in race-based hate attacks to members of Alberta's Black Muslim community. According to information available on the National Council of Canadian Muslims' (NCCM) website, between December 2020 and May 2021 (a five-month time frame), there was a total of nine reported attacks on this community in Edmonton and Calgary alone. Alberta also saw an increase in police-reported hate crimes in 2020, with 211 reported hate crimes targeting victims due to their racial background (Statistics Canada, n.d). This is up almost 100 reported cases from the previous year. Table 20.2 (below) shows Alberta has seen steady increases over the past number of years in police-reported hate crimes targeting people due to their race.

Table 20.2: Number of police-reported hate crimes in Alberta where race was the main motive, 2009 to 2020

Year	Number of police-reported hate crimes where race is main motive
2009	74
2010	91
2011	74
2012	76
2013	83
2014	86
2015	108
2016	92
2017	124
2018	119
2019	106
2020	211

Source: Statistics Canada, Canadian Centre for Justice and Community Safety Statistics, Incident-based Uniform Crime Reporting Survey.

Addressing Race-Based Hate Under the UCP Government

As mentioned above, Alberta had the highest number of police reported race-based hate crimes in 2020, one year, coincidentally, after the Human Rights Education and Multiculturalism fund was cut by the UCP government. Although the UCP government did establish a new (albeit smaller) pool of funding to address equity and diversity issues through the Multiculturalism, Indigenous and Inclusion Grant Program (currently on hold at time of writing), other "unique" approaches were more aggressively pursued to address race-based hate in Alberta. The following section provides a high-level overview and summary of three specific initiatives established by the UCP government to counter race-based hate: the establishment of a provincial hate crimes co-ordination unit; the creation of and increased funding for the Alberta Security Infrastructure Program Grant; and seeking federal support to amend the Criminal Code and authorize individual use of pepper spray for self-defense.

The Provincial Hate Crimes Coordination Unit. Alberta saw an alarming increase in race-motivated attacks in 2020-2021. This compelled the UCP government to enact a handful of strategies to address the issue. On June 11, 2021, then-minister of justice Kaycee Madu announced his office would be creating a hate crimes co-ordination unit in the provincial government, utilizing already available funds within the province. According to Madu, "the Alberta unit will be made up of civilians and law enforcement with the tools to investigate, enforce and prosecute hate-motivated incidents" (Wakefield, 2021b). Interestingly, the previous NDP government proposed a similar unit in 2018 following a report commissioned after the 2017 attack on a Quebec mosque, but the unit was never created (Wakefield, 2021b). According to the UCP, the purpose of this newly established unit is to create better connections within the province's law enforcement community to address hate-motivated crimes in a more concerted manner. It is also an opportunity for better intelligence gathering and sharing regarding active hate groups in the province between the various police services in Alberta.

While this is a positive step in co-ordination and collaboration, it is something anti-hate advocates have been asking for years in the province. For example, a 2007 report by the Alberta Hate Crimes Committee — a volunteer led, non-partisan group committed to ending hate in Alberta — highlighted the creation of a province-wide hate crime unit as one of the main recommendations. According to the report, "forming an Alberta Hate Crime Team, under the umbrella of the Solicitor General will ensure consistent and professional responses to all Albertans in relation to the prevention, detection, investigation, and prosecution of hate crimes" (Stewart, 2007). The report suggests an Alberta Hate Crime Team could consist of a trainer, an investigator (from law enforcement), a prosecutor (Alberta Justice), and a civilian crime analyst. This robust and collaborative model would ensure a broad lens is applied to addressing and assessing hate motivated crimes in Alberta. While it is not clear how the current Alberta Hate Crime Coordination Unit is set up (due to limited open-source information), the structure of the team seems to be quite small, with one analyst and two police officers assigned to the unit for a predetermined period of time. While it is too early to determine the effectiveness of this collaborative unit, the average Albertan will likely notice minimal difference, as most of the changes are internal and within the government of Alberta.

Creation of and increased funding for the Alberta Security Infrastructure Program Grant. Shortly after the announcement of the Provincial Hate Crimes Coordination Unit, former premier Jason Kenney announced the establishment of the Alberta Security Infrastructure Program Grant. On June 12, 2021, the Province of Alberta made funds available for "religious and ethnic organizations that are at risk of being targeted by hate-inspired violence or vandalism. This includes places of worship, temples, synagogues, gurdwaras, community centres such as Indigenous friendship centres, ceremonial facilities, and monuments" (Gibson, 2021). This funding stream would allow for the purchase of security and technology improvements, as well as the hiring of security guards, improved security infrastructure, and other similar security enhancements. According to a news release, the province set aside $500,000 for the 2021-2022 fiscal year, with applicants eligible for up to $10,000 to assist with security assessments and training, as well as $90,000 for the purchase and installation of alarms, gates, motion detectors, and security systems (Gibson, 2021). As of May 2022, the program had been renewed, with application deadlines removed entirely, and total funding expanded to $5 million. Organizations are now only able to apply for up to $35,000 for the purchase and installation of security equipment, and to retain security training, instead of the $90,000 available in the first year of the program (Antoneshyn, 2022). According to media reports, roughly 100 applicants have been awarded a grant since its inception, totaling an overall amount of $1.2 million (Antoneshyn, 2022).

Although this type of funding opportunity is a step in the right direction, it is not a unique program, as the federal government has had a similar Security Infrastructure Program (SIP) for communities at risk since at least 2015, offering funding options for similar improvements to security infrastructure. Whether or not this funding is duplicating federal efforts, or supplementing a pre-existing program at the provincial level, there seems to be an interesting trend with the UCP government in armoring our buildings, or as the next section explains, armoring our citizens, in an effort to combat race-based hate.

Amend the Criminal Code and authorize individual use of pepper spray. In the wake of the increased attacks on predominantly Black Muslim women in Alberta, the UCP government, under the direction of Madu, put forward a bold recommendation with the hope of empowering citizens to take control of their own safety. On July 21, 2021, Madu publicly expressed his support for the Alberta government asking the federal government to amend the Canadian Criminal Code and allow people to carry and use pepper spray in self-defense. In a letter drafted by Madu's office, and addressed to the Minister of Justice and Attorney General of Canada David Lemetti and then-minister of Public Safety and Emergency Preparedness Bill Blair, Madu urges the federal government to take consideration and "allow individuals, including vulnerable persons, to carry capsaicin spray, commonly known as 'pepper spray,' for self-defence" (Taniguchi, 2021). According to Madu, his reason for promoting this idea was to protect vulnerable people, particularly women, from being attacked in a hate-motivated offense. As Madu explains:

> I think when you are a vulnerable woman who is faced with an attack by folks that you don't know, where they're coming from, the question is, what tool (is) out there, what type of empowerment can we provide to them that would help them ward off the attackers? I don't think anyone out there would doubt that pepper spray is an effective tool in helping them ward off these attacks" (quoted in Taniguchi, 2021).

Not surprisingly, this strategy was met with resistance and criticism from a variety of perspectives in Alberta. For example, Mike Elliot, president of the Edmonton Police Association, said this proposal was ill-advised and could lead to more crimes being committed by people using pepper spray (Leavitt, 2021). Additionally, one Alberta-based researcher who focuses on the study of hate crime prevention cautioned against the proposal, suggesting the idea is disconnected from what communities impacted by race-based hate are looking for as possible solutions (Leavitt, 2021). The federal government agreed. On Aug. 4, 2021, both Blair and Lametti said the pepper spray proposal would lead to further violence:

> When looking at this request, we have to be mindful that all weapons that are prohibited have been prohibited for a reason, as they are extremely dangerous when they fall into the wrong hands. When confronted with a problem, the solution cannot simply be to increase accessibility to prohibited weapons (quoted in Johnson, 2021, para 4).

Not happy with the federal government's response, Madu issued a statement shortly after, accusing the Liberal government of being "soft" on crime and siding with criminals rather than victims. As Madu posits:

> They [the Liberal Government] seem to take the shameful stance that Canadians themselves are responsible for not standing-up against hate¾instead of putting the blame squarely on perpetrators. They have no real solutions when it comes to stopping crime in its tracks; they rather leave Albertans empty-handed and vulnerable when faced with a potential assault or other related crime (quoted in Johnson, 2021, para 11).

The proposal to arm citizens with pepper spray was not well thought out and was an extremely short-sighted approach to an issue needing systemic and structural change. As discussed below, this strategy is not surprising, as it speaks to a broader Conservative-based ideology where a solution to emerging social problems and criminal issues is to take a more reactive and potentially, punitive approach.

Discussion

Based on the above UCP strategies to address race-based hate crimes, it is clear they have not considered long-term solutions, focusing rather on short-term, reactive-based approaches. The above strategies give the impression the only way to counter race-based hate is through more security and more armed citizens, at more cost. This draconian approach is not sustainable, and will not get at the root causes of race-based hate in society. Furthermore, the above strategies scream to the "tough on crime" approach championed under the Stephen Harper Conservative federal government. Perhaps this should not come as a surprise, as Kenney was a key player within the Harper government during this time, holding several influential ministerial portfolios' such as immigration and multiculturalism. The many critiques of the federal Conservative government's "tough on crime" approach can be just as validly deployed against the UCP strategies outlined in this chapter. Quite simply, the "tough on crime" approach is a failed play from an overused playbook. After coming to power in 2006, the Harper government implemented a host of legislative and policy changes designed to "tackle crime," "hold offenders accountable," and "make communities safer" (Cormack, Fabre, and Burgher, 2015). Almost 15 years later, the Alberta UCP government is echoing these same sentiments, with Madu on record stating his proposed strategies will hold offenders accountable and make communities safer (Johnson, 2021).

The "tough on crime" approach has been criticized often as not being evidence-based. In fact, a 2015 report assessing the impact of the "tough on crime" approach on frontline workers found that many viewed the approach to be more ideologically driven than evidence-based (Cormack et. al, 2015). As one respondent from the report highlights, "they're [the government] vocal about all the things they're doing to make the community safer, yet their policies actually create more harm and more danger to communities that they say they're so committed to protecting" (Cormack et. al., 2015, pg. 30). This quote also holds true for the UCP approach to countering race-based hate. The UCP government has been extremely vocal in announcing its solutions, without taking much consideration of the impact these policies will have on communities. As mentioned above, most rational thinkers found the pepper spray idea ill-advised, prompting the *Toronto Star* to run a feature story on the topic with the headline "Why critics are slamming Alberta's 'ill-advised' proposal to fight hate-motivated attacks with pepper spray" (Leavitt, 2021). To be frank, no community asked for pepper spray to be provided to them as an option to counter race-based hate in the province.

The security infrastructure program also misses the mark in terms of proactive and data-informed approaches to countering race-based hate. As longitudinal data from Statistics Canada highlights, from 2011 to 2020 only two per cent (124 instances) of police reported hate crimes deemed as a violent violation were located at a religious institution (Statistics Canada, 2022). During that same time frame, only 9 per cent (821 instances) of police reported hate crimes labelled as a non-violent violation were located at a religious institution (Statistics Canada, 2022).

While not discounting the impact each individual instance has on the communities targeted, to suggest any type of security infrastructure funding program will help keep various communities safer is a bit disingenuous, as the data suggests religious institutions are not a major site for where hate-motivated offenses occur. According to the data, most hate crimes (both violent and non-violent) occur in open areas, such as "parking lots, streets, roads, and highways, construction sites, and other open areas" (Statistics Canada, 2022). Interestingly, in July, and shortly after the discovery of 215 bodies at the site of a Kamloops residential school in May 2021, numerous Catholic churches were targeted through arson and other forms of vandalism. Calgary, for example, had 11 Catholic churches vandalized with orange and red paint, some using the tagline "OUR LIVES MATTER" on Canada Day (Herring, 2021), potentially connecting the vandalism to the rightful anger and resentment from members of Canada's Indigenous communities. Mere weeks later, the UCP government announced the establishment of the security infrastructure program to help "beef up" security at various religious institutions across the province. While the data is not available publicly, it would be interesting to assess how many Catholic churches received the grant in relation to other religious institutions such as synagogues, mosques, and temples. Perhaps these punitive approaches to countering race-based hate crime should not come as a surprise, particularly from a government who draws some parallel ideological motivation from the Harper government. Studies have found that when crime becomes an issue of public discourse — which hate crime certainly has in Alberta — the mood turns punitive, and the attention turns to [aggressive] policies (Baumgartner, Daniely, Hunang, Johnson, Love, May, Mcgloin, Swagert, Vattikonda, and Washington, 2021). More problematic, however, is when the government ignores the counterargument, which is oftentimes more evidence-based, the dramatic solution is often put forward as the remedy to solve the issue (Baumgartner et. al, 2021).

Alberta-based racial discrimination. Alberta has a racism problem, and the approaches highlighted above will not get the province to a space that can productively counter the issue. As a 2021 report titled *On the Rebound? Perceptions of Discrimination in Alberta Society* found, the impact of ongoing Indigenous activism and the Black Lives Matter movement on public perceptions of racial discrimination in Alberta remained strong in early 2021 (Wagner, 2021). This research, which draws upon a three-year time frame, found that half (50.1 per cent) of Albertans feel the Indigenous community experiences a great deal of racial discrimination in the province.

This study also found the degree to which respondents believed racism exists in Alberta depended on their political affiliation. NDP supporters, for example, "continued to be far more likely than supporters of other partisan groups and non-partisans to believe Indigenous people face discrimination. In 2021, two-thirds (69 per cent) of NDP partisans agreed that Indigenous people face a lot or a great deal of discrimination" (Wagner, 2021). UCP supporters showed the biggest drop in perceptions of anti-Indigenous discrimination over the last year, going from 39.9 per cent in 2020 to 32 per cent in 2021 (Wagner, 2021). When it comes to anti-black racism, almost half of NDP supporters thought the issue remained strong in 2021, while only 15 per cent of UCP supporters perceived a high degree of anti-black racism in Alberta (Wagner, 2021).

This finding is interesting, as it highlights a larger ideological underpinning possibly driving the UCP's position on anti-racism initiatives. This claim can be illustrated through an initial assessment of the UCP response to the proposed Bill 204, the Anti-Racism Act. This bill, introduced by NDP MLA David Shepherd in April 2022, would have established an anti-racism office within the Government of Alberta, as well as require all departments to collect data aimed at spotting racial disparities in government programs and policies (Wakefield, 2022). Additional purposes of this bill would also include the implementation of the recommendations of the Anti-Racism Advisory Council (an initiative spearheaded by the previous NDP government) and the reporting on key outcomes and performance indicators of racial equity in Alberta through the establishment of data standards. Unfortunately, "a UCP-dominated committee put an end to the proposed NDP bill by recommending it not be debated in the legislature" (Joannou, 2022). As a result of this step, the bill did not gain any further traction, sending a significant message to Albertans that the UCP government does not take the issue of racism — systemic racism, specifically — seriously. As Shepherd professes, by not even agreeing to debate the item at the committee level,

> what they [the UCP government] are saying is that they don't feel that addressing this issue is worthy of their time and effort. They are saying that it is not worth taking the time to debate and discuss this on the floor of the legislature. They are saying that there is no value in the members of the assembly sitting down and having actual fulsome debate on this issue (Joannou, 2022).

Conclusion

This chapter began with a quote from the late Dr. Darren E. Lund, reminding us that governments' ideological underpinnings in creating and supporting policy and interventions to counter race-based hate may cause further harm than good. As Lund surmises, "a more immediate need for Alberta's diversity activists is to confront the intolerance promoted by its own government" (Lund, 2006). This statement is harsh but does highlight an ongoing theme that has hovered over the UCP party since its establishment in 2017. As outlined in this chapter, the UCP approach to counter race-based hate has been reactive, with little to no effort being paid to long-term systemic change. The examples shared in this chapter, such as the creation of a provincial hate crime co-ordination unit, the increase of funding to enhance security infrastructure at faith-based institutions, and advocating for a lethal weapon to be an approved measure to combat race-based hate, speak volumes to an ideological underpinning which minimizes and perhaps even denies the existence of racism as a real social issue impacting Albertans. For some, this is no surprise. Kenney has a long and interesting history at both the provincial and federal level in politicizing social identity markers to try to drum up support from as broad a Conservative voter base as possible, particularly during election time. Sometimes, this means engaging in "dog whistle" politics — coded racial appeals that carefully manipulate hostility toward non-whites (Haney-Lopez, 2014). In the 2015 federal election, for example, Kenney heavily promoted the ill-advised and racist "reporting barbaric cultural practices tip line" as an attempt to sway white voters who were afraid of "Canadian values" being taken away by non-white immigrants. The premise of this tip line was to provide all Canadians a venue to report "barbaric cultural practices," namely forced marriages of young girls (Powers, 2015), to the RCMP.

The other "barbaric practice" this tip line would also address is the wearing of a niqab, a face veil worn by Muslim women as a sign of faith observation. While this ban was introduced by Kenney during his time as immigration minister as a mechanism to support equality during citizenship oath-taking, the issue dovetailed into an election strategy, where other conservative candidates would lump this practice as a "barbaric cultural practice" that the 2015 federal Conservative government would commit to eradicating if elected to power. When the Federal Court of Appeal struck down this ban, Kenny, who introduced this bill, voiced his displeasure, citing that this ruling "goes against the democratic will of Canadians, and to long standing Canadian values of openness and equality of women and men" (CBC News, 2015).

Kenny's flirtation with "dog whistle politics" at the federal level is what made him appealing to some provincial voters (and candidates) during his time as UCP leader, especially during the 2019 Alberta provincial election. Some writers have even suggested that Kenny's Build That Pipeline campaign slogan — a nod to getting support from the oil and gas industry — was akin to Donald Trump's mantra of "build that wall" (CBC News, 2015). This sentiment, combined with strong Conservative values, and a desire to Make Alberta Great Again, attracted other provincial candidates, who also flirted with polarizing ideas in the hopes of securing votes from what Conservative radio host and commenter Charles Adler called Alberta's "knuckle-draggers" (Wright, 2019), a derogatory reference to under-informed Conservative voters who may be receptive to the dog whistle strategy mentioned above. Failed UCP candidates such as Caylan Ford, who publicly questioned the demographic replacement of white people in their home lands, or Eva Kiryakos, who claimed Muslims used forced breeding as a form of Christian genocide (Wright, 2019), are a few of a handful of examples of candidates who were attracted to Kenney's UCP party because they felt these views would acquiesce with a specific conservative voter base. To some degree, they were right. During the campaign period for the 2019 provincial election, members of the anti-immigrant and anti-Muslim group Soldiers of Odin started to circle around certain UCP candidates, as they felt this party best aligned with their sentiments regarding the aforementioned identities." As one UCP candidate explained, after being question on why this hate group was at his "meet and greet" night, "people have a constitutional right to voice their opinions and I'm not going to deny them that" (McMillan, 2018).

While dog-whistle style strategies can help a party get elected, its sporadic use while a party is in power can also have strategic advantages to try and garner further ideological support. One example of this can be found in Kenney's response to the 2020 Liberal Throne Speech, where he criticized the prime minister of being "too woke." Then Kenney "slammed the throne speech, suggesting the emphasis on systemic racism and women came at the cost of attention to the oil and gas industry and transfer payments" (Woods, 2020). This sentiment is revealing, as it highlights the lack of importance the UCP government has placed on addressing systemic racism. As University of Calgary professor Melanee Thomas explained of Kenney's comments, "it's either gross ignorance on his part, or he's actually saying that he doesn't want to take systemic racism seriously" (Woods, 2020, para 20). Another example of the use of dog whistle politics can be found in recent discussion and debates regarding the revamp of the Alberta K-12 school curriculum. In a bid to hold on to his leadership seat, Kenney delivered a speech claiming he was defending the province's school children from the Alberta NDP's secret agenda to teach "woke ideology" in Alberta schools (Magusiak, 2022). This is a tactic straight out of American-style Trumpism where,

> fear mongering over "woke ideology" and "critical race theory" has become a favorite tactic of American Republican figures like former President Donald Trump, Florida Governor Ron DeSantis and Fox News' Tucker Carlson, who have used it to justify attempts to shut down free speech on racial justice and gender equality (Magusiak, 2022, para 3).

As Kenny highlights in his speech, "instead of divisive woke-left ideology like critical race theory, cancel culture, and age inappropriate sex education, we are putting kids and the authority of parents back in charge of our education system" (Magusiak, 2022, para 6).

The UCP's ideological influences are of great importance when considering how to approach long-term and meaningful actions in addressing the systemic racism that contributes to overt acts of race-based hate. Quite simply, if a government does not believe systemic racism and discrimination are real issues, the only path presented will be filled with reactive approaches that do little to address the embedded nature of race-based hate in our province. Ironically, in 2021, the provincial anti-racism advisory council put forward a report with recommendations on how the province can better address race-based hate, with many of their recommendations reflecting actions that would contribute to addressing systemic layers of racism. To date, the UCP government has made no commitment to take any action on any of the recommendations presented. The current version of the UCP government does not seem interested in pursuing anti-racism work in any meaningful way, and if that is the case, then perhaps we should all be prepared for a time when the provincial government will provide pepper spray for all as the only solution to address race-based hate.

References

Anand, Sanjeev. (1998). Expressions of racial hatred and racism in Canada: An historical perspective. *The Canadian Bar Review* 77, 181-197.

Alberta. (n.d.). *Alberta Security Infrastructure Program Grant*. Retrieved March 4, 2022, from https://www.alberta.ca/alberta-security-infrastructure-program-grant.aspx

Alberta Human Rights Commission. (2014). *Human Rights Education and Multiculturalism Fund*. Retrieved May 6, 2022, from http://142.229.235.58/education_fund_grants.asp

Alberta Human Rights Commission. (2017). *Alberta Human Rights Information Service*. Retrieved May 6, 2022, from https://albertahumanrights.ab.ca/publications/AHRIS/2017/Pages/January_18.aspx

Alberta Labour History Institute. (n.d.). *A shocking racist event in Alberta's history: the "Aryan Fest."* Retrieved March 19, 2022, from https://albertalabourhistory.org/a-shocking-racist-event-in-albertas-history-the-aryan-fest/

Antoneshyn, A. (2022, May 11). *Alberta Security Infrastructure Program grants downsized, program made retroactive*. CTV News Edmonton. https://edmonton.ctvnews.ca/government-cuts-size-of-grants-available-to-religious-cultural-groups-for-security-upgrades-but-makes-program-retroactive-1.5898609

Baergen, W. P., & Central Alberta Historical Society. (2000). *The Ku Klux Klan in Central Alberta*. Central Alberta Historical Society.

Baumgartner, F. R., Daniely, T., Hunang, K., Johnson, S., Love, A., May, L., Mcgloin, P., Swagert, A., Vattikonda, N., & Washington, K. (2021). Throwing away the key: The unintended consequences of "tough-on-crime" laws. *Perspectives on Politics* 19(4), 1233-1246. https://www.cambridge.org/core/journals/perspectives-on-politics/article/abs/throwing-away-the-key-the-unintended-consequences-of-toughoncrime-laws/6E1206127F65C921DC9BDB0DC1C1D79F

Boyko, J. (2006). Keegstra case. *The Canadian Encyclopedia*. Retrieved March 19, 2022, from https://www.thecanadianencyclopedia.ca/en/article/keegstra-case

CBC News (2015, Sept. 15). *Niqabs OK at citizenship ceremonies, as government loses appeal*. CBC News. https://www.cbc.ca/news/politics/niqab-ruling-federal-court-government-challenge-citizenship-ceremonies-1.3229206

City of Edmonton (2022). *Brief history of racism in Alberta*. Retrieved March 20, 2022, from https://pub-edmonton.escribemeetings.com/filestream.ashx?DocumentId=129668

Cohen Committee Report. (1966). Special Committee on Hate Propaganda. Ottawa, Ontario: Queen's Printer.

Commission, A. M. (1994). *Multiculturalism...the next step*. Government of Alberta.

Cormack, E., Fabre, C., & Burgher, S. (2015). *The Impact of the Harper Government's "Tough on Crime" Strategy: Hearing from Frontline Workers*. Canadian Centre for Policy Alternatives Manitoba. https://policyalternatives.ca/sites/default/files/uploads/publications/Manitoba%20Office/2015/09/Tough%20on%20Crime%20WEB.pdf

Finkel, A. (n.d.). Alberta Labour History Institute. *Systemic racism in Alberta's history*. Retrieved March 14, 2022, from https://albertalabourhistory.org/systemic-racism-in-albertas-history/

Gibson, C. (2021, June 11). *Alberta to launch security grant to protect religious, cultural organizations from hate crimes*. Global News. https://globalnews.ca/news/7941308/alberta-premier-justice-minister-hate-motivated-crime/

Government of Canada. (2005). *A Canada for all: Canada's action plan against racism*. https://publications.gc.ca/site/eng/9.687396/publication.html

Haney-López, I. (2014). *Dog Whistle Politics: How Coded Racial Appeals Have Reinvented Racism and Wrecked the Middle Class*. New York: Oxford University Press.

Herring, J. (2021, July 1). Calgary police investigate after at least 11 Catholic churches vandalized with orange and red paint. *Calgary Herald.* https://calgaryherald.com/news/local-news/calgary-police-investigate-after-several-catholic-churches-vandalized-with-red-paint

Janhevich, D. (1997). *The criminalization of hate: A social constructionist analysis.* Unpublished MA Thesis. Accessed from: https://www.ruor.uottawa.ca/handle/10393/4162

Joannou, A. (2022, April 22). UCP-led committee recommends dropping proposed NDP bill calling for race-based data. *Edmonton Journal.* https://edmontonjournal.com/news/politics/anti-racism-bill

Johnson, L. (2021, Aug. 4). Federal ministers reject Alberta justice minister's call to legalize pepper spray. *Edmonton Journal.* https://edmontonjournal.com/news/politics/federal-ministers-reject-alberta-justice-ministers-call-to-legalize-pepper-spray

Leavitt, K. (2021, July 22). Why critics are slamming Alberta's 'ill-advised' proposal to fight hate-motivated attacks with pepper spray. *Toronto Star.* https://www.thestar.com/news/canada/2021/07/22/alberta-proposes-new-tactic-in-battle-against-hate-motivated-attacks-pepper-spray.html

Leman, M. (1999). *Canadian multiculturalism* (93-6E). Retrieved May 6, 2022, from https://publications.gc.ca/Collection-R/LoPBdP/CIR/936-e.htm

Lund, D. E. (2006). Social justice activism in the heartland of hate: Countering extremism in Alberta. *The Alberta Journal of Educational Research* 52(2), 181-194.

Magusiak, S. (2022, April 13). "Utter nonsense": Teachers say Jason Kenney's right-wing culture war rhetoric has no place in Alberta schools. *PressProgress.* https://pressprogress.ca/utter-nonsense-teachers-say-jason-kenneys-right-wing-culture-war-rhetoric-has-no-place-in-alberta-schools/

McMaster, G. (2019, Feb. 1). *Citizen historian determined to expose Edmonton's racist past to reconcile and move forward.* University of Alberta. Retrieved March 19, 2022, from https://www.ualberta.ca/arts/faculty-news/2019/february/citizen-historian-determined-to-expose-edmontons-racist-past-to-reconcile-and-move-forward.html

McMillan, A. (2018, Oct. 11). *UCP nomination candidate says he knew Soldiers of Odin were coming to party's pub night.* CBC News. https://www.cbc.ca/news/canada/edmonton/ucp-west-henday-soldiers-of-odin-photos-1.4858201

Moreau, G., & Wang, J. H. (2022, March 17). *Police-reported hate crime in Canada, 2020.* Statistique Canada. Retrieved April 4, 2022, from https://www150.statcan.gc.ca/n1/pub/85-002-x/2022001/article/00005-eng.htm

NCCM. (n.d.). *Send a letter to the leaders of Alberta on Islamophobic attacks.* National Council of Canadian Muslims. Retrieved May 24, 2022, from https://www.nccm.ca/alberta/

Powers, L. (2015, Oct. 2). *Conservatives pledge funds, tip line to combat 'barbaric cultural practices'.* CBC News. https://www.cbc.ca/news/politics/canada-election-2015-barbaric-cultural-practices-law-1.3254118

Ross, J. (1994). Hate crime in Canada: Growing pains with new legislation. In Mark Hamm (Ed.), *Hate Crime: International Perspectives on Causes and Control.* Cincinnati, Ohio: ACJS/Anderson Monograph Series.

Southern Poverty Law Center (n.d.). Retrieved March 14, 2022, from https://www.splcenter.org/fighting-hate/extremist-files/ideology/ku-klux-klan

Statistics Canada. (2022). *Table 8 locations of hate crimes, by most serious violation type and detailed motivation, Canada, 2011 to 2020.* Retrieved May 29, 2022, from https://www150.statcan.gc.ca/n1/pub/85-002-x/2022001/article/00005/tbl/tbl08-eng.htm

Statistics Canada. (n.d.). Canadian Centre for Justice and Community Safety Statistics, Incident-based Uniform Crime Reporting Survey.

Stewart, C. (2007). *Combating Hate and Bias Crime and Incidents in Alberta.* TANDIS. https://tandis.odihr.pl/bitstream/20.500.12389/20255/1/04682.pdf

Taniguchi, K. (2021, July 21). Justice Minister Kaycee Madu wants pepper spray for all. *Edmonton Journal.* https://edmontonjournal.com/news/local-news/justice-minister-kaycee-madu-wants-pepper-spray-for-all

Wagner, A. (2021, July 21). *On the rebound? Perceptions of discrimination in Alberta society.* Common Ground. Retrieved May 29, 2022, from https://www.commongroundpolitics.ca/discrimination2021

Wakefield, J. (2021a, July 19). National Council of Canadian Muslims releases report on Islamophobia with focus on Edmonton incidents. *Edmonton Journal.* https://edmontonjournal.com/news/local-news/national-council-of-canadian-muslims-releases-report-on-islamophobia-with-focus-on-edmonton-incidents

Wakefield, J. (2021b, June 11). Alberta justice minister announces plan to create hate crimes unit. *Edmonton Journal.* https://edmontonjournal.com/news/local-news/alberta-justice-minister-announces-plan-to-create-hate-crimes-unit

Wakefield, J. (2022, April 30). Supporters of race-based data bill rally after UCP nixes debate in legislature. *Edmonton Journal.* https://edmontonjournal.com/news/local-news/supporters-of-race-based-data-bill-rally-after-ucp-nixes-debate-in-legislature

Woods, M. (2020, Sept. 24). Kenney calls intersectionality in Throne speech "kooky academic theory." *HuffPost.* https://www.huffpost.com/archive/ca/entry/jason-kenney-intersectionality-throne-speech_ca_5f6d1f22c5b64deddeeb2130

Wright, S. J. (2019, April 16). Alberta election 2019: victimhood, greed and racism. *NOW Magazine*. https://nowtoronto.com/news/alberta-election-2019-jason-kenney-racism-islamophobia-homophobia

Yousif, N. (2019, Nov. 12). Alberta cancels decades-old grant for anti-racism initiatives. *Toronto Star*. https://www.thestar.com/edmonton/2019/11/12/alberta-cancels-decades-old-grant-for-anti-racism-initiatives.html

NOTES

1 Subparagraph 718.2(a)(i) of the Canadian Criminal Code, requires courts to consider, as an aggravating factor when determining the sentence for any crime, if the crime was motivated by hatred, bias, or prejudice, based on numerous criteria.

2 Examples of projects funded by the grant can be found here: http://142.229.235.58/fund/recipients.asp

CHAPTER 21

THE UNITED CONSERVATIVE GOVERNMENT, RIGHT-WING POPULISM AND WOMEN

Lise Gotell[1]

"Crises and the mechanisms to manage them often entrench the power of particular economic and gender orders and constrain the possibilities and space for contestation and critique."

— Penny Griffin (2015)

WHO CAN FORGET the photograph of United Conservative Party (UCP) MLAs romping in the fountain of the Alberta Legislature moments after passing legislation that removed protections for LGBTQ2S+ students? In Bill 8, the UCP government made good on its electoral promise to repeal the previous New Democratic Party (NDP) government's ban on parental notification for students joining Gay Straight Alliances, effectively guaranteeing that children and young people would be outed to potentially unsupportive family members. The government's celebratory romp after the end of its first legislative session was drenched in symbolism. It stood as a coded signal to the social conservative base that helped sweep the Jason Kenney-led UCP to victory in 2019. Together with the post-election elimination of a freestanding department, Status of Women Alberta, this romp in the fountain signaled the centrality of so-called anti-gender ideology (anti-feminist and anti-LGBTQ2S) (Kovats, 2018) to the UCP's particular brand of right-wing populism.

Much ink has been spilled in diagnosing the form of right-wing populism that has taken root on the Canadian political scene over the past three decades. As many scholars argue, the rise of right-wing populism needs to be seen in relation to the crisis of neoliberalism, a crisis that only deepened in the decade between the 2008 recession and the COVID pandemic (Brodie, 2018). Central to this political phenomenon now sweeping the globe is the political division of society into two imagined homogenous and opposing groups, the "pure people" and the "corrupt elite." As Cas Mudde (2004) has argued, populism is a "thin-centred ideology" that can easily be combined with other ideologies. In most of its expressions, and perhaps especially in its contemporary

European forms, right-wing populism's anti-elite conception of the "pure people" is fundamentally exclusionary, resting on an appeal to racist or nativist nationalist rhetoric that seeks to transform private economic anxieties into public animosity. But, for numerous reasons that include enduring popular support for multiculturalism, positive views on the value of immigration, the impact of the single-member plurality electoral system, and an aversion to coalition governments, such overtly nativist versions of populism have yet to take firm hold in Canada. Instead, as Brian Budd (2021) has argued, Canadian versions of populism are "maple-glazed," combining an ardent commitment to neoliberalism with an appeal to the hard-working, tax-paying "real people," undermined by a "corrupt elite" of special interests.

I argue in this chapter that too little attention has been paid to the role and significance of gender in the analysis of Canadian right-wing populism. By examining the Kenney government's record on gender, described by some commentators as the UCP's "war on women" (Payne, 2020), I will demonstrate how anti-feminist claims were strategically deployed in an effort to align feminism with the Justin Trudeau federal government, the symbolic representative of a central-Canadian elite, portrayed as threatening Alberta's economy and society. Consistent with this construction of feminism as antithetical to the interests of the "real people," the UCP government pursued degendered and individualizing policy strategies, including an emphasis on regressive law and order politics, in the face of pandemic that laid bare the fundamental intersectional inequalities facing women in Alberta. I argue that gender has been central to the Kenney government's populist playbook, and this, in turn, has resulted in intensified gender, race, and class disadvantage.

The UCP Neoliberal Populist Agenda and Gender: The Pre-Pandemic Phase

Jared Wesley (2022) argues that the UCP has cultivated a "cowboy myth," an image of the "real Albertan" as "rural, rugged, individualist, white, blue-collar, and masculine." While this image has deep roots in prairie populist soil, grounding the politics of Western regionalism for nearly a century, we cannot miss the manner in which this self-sufficient myth overlaps with the good citizen of the neoliberal order. Neoliberalism as a form of governance is fundamentally gendered; its strategies of privatization and depoliticization, its weakening of social provision and social citizenship, have all had disproportionate impacts on women, particularly those marginalized by race and class (Brodie, 2008). This governance project, in which market values are introduced into every sphere of human life, has relied upon the production of a new ideal citizen subject — the *homo oeconomicus* (Brown, 2016). In this normative construction of the ideal citizen, every "man" is made into an entrepreneur of himself, responsible for managing risk without state interference, as collective conditions become risks belonging to the individual. We are encouraged to live up to this norm of the responsible citizen who does not make demands for excessive state intervention or state spending. A prominent recent example is UCP leadership candidate Danielle Smith's claim that cancer patients, before stage 4, can manage their own diseases through health and wellness measures and avoid overburdening the (underfunded) health-care system. By contrast, those who continue to make political claims on the basis of structural disadvantage have been increasingly stigmatized, cast outside the boundaries of legitimate public discourse (Brodie, 2008).

Shortly after the 2019 election, it became increasingly clear the new Kenney UCP government was repeating the playbook put in place by Stephen Harper Conservatives after coming to power federally with a minority government in 2006 (Friesen, 2021). As Janine Brodie (2008) describes, the federal Conservatives pursued a deliberate strategy aimed at disassembling the remnants of social liberalism that informed equality-based claims-making on the state, with feminism as one of the principal targets. Feminist organizations, once considered legitimate actors in the policy process, were increasingly delegitimized as "special interest groups," standing outside and in opposition to the interests of ordinary, taxpaying Canadians whose interests were aligned with the market and a small state. Kenney, as parliamentary secretary to Harper, no doubt sat around the decision-making tables to defund feminist advocacy, to remove equality from the mandate of Status of Women, and, through budget cuts, to erode the gender-based policy machinery in the federal state. Brodie (2008, p. 167) refers to this as the three Ds: "the delegitimization of feminism, the dismantling of gender-based policy units, and the disappearance of the gender subject of public policy." Harper, in uniting the Canadian Alliance with the Conservatives, is widely credited for the successful neoliberalization of right-wing politics, a strategy that relied upon selective uses of populism (Budd, 2021). The core ideology of neoliberal populist governments is rooted in economic liberalism; they advance anti-egalitarian measures, aim to reduce state intervention, and defend the "ordinary people" against an allegedly "corrupt elite" (see Harrison, Chapter 5). Attacks on feminist advocacy and expertise played a crucial role in the cultural project of Harper-era policies, disarming a vocal political constituency tied to social liberalism and to demands for social citizenship (Brodie, 2008).

One of the Kenney government's first acts was to begin dismantling the gender expertise that had only recently been established within the Alberta public service. In 2015, Premier Notley's NDP had created a freestanding department, Status of Women Alberta, and appointed a minister responsible for the status of women, the first in almost 25 years. This dedicated policy machinery was intended to guide the government's strategy in three key policy areas: Alberta's large gender gap in wages, high levels of gender-based violence, and women's under-representation in government and in the corporate sector (Thomas, 2019). To move forward in these priority areas, the ministry led consultations with civil society organizations and critically, trained more than 1,500 provincial civil servants in gender-based analysis plus (GBA+), a "gender-mainstreaming" policy tool that begins from the assumption that policy impacts are shaped by different social locations (Thomas, 2019). As an approach to public policy, GBA+ is based on a concept of substantive, rather than formal equality, in which it is assumed that equal treatment does not necessarily result in equal outcomes. GBA+ seeks to make visible intersecting inequalities and to consider how public policies may exacerbate or ameliorate social disadvantage. Status of Women Alberta also engaged in community capacity-building through grants and outreach, seeking to build a network of groups to provide it with political currency and policy advice (Thomas, 2019). As Melanee Thomas (2019, p. 261) argues, "[the ministry's] presence demonstrated strong commitment to gender equality on the part of the Notley government, which was prepared to use the institutional apparatus of the provincial government to ensure pro-equality outcomes."

From the moment the Kenney UCP government was elected, this autonomous gender policy machinery and the networks it had cultivated were in the crosshairs. In forming his first cabinet, Kenney combined the freestanding department with Culture and Multiculturalism, depriving Status of Women of a dedicated voice around the cabinet table. The new department would now be responsible for the Francophonie, heritage, creative and cultural industries, and multiculturalism. In the March 2020 budget, the government singled out the expanded ministry for a damaging and disproportionate reduction of 33 per cent, to be phased in over four years (McIntosh and Hussey, 2020). This cut resulted in a shrunken policy unit within the status of women portfolio.

Notably, there were also deliberate moves to shut down GBA+ within the policy process. Minister Leela Aheer is reported to have told staff that GBA+ "made people in the government defensive" (Paradis, 2021). In a 2020 interview with *The Sprawl*, Janis Irwin, NDP Status of Women Critic, stated she had tried to "get more information on where [gender-based analysis plus] is at to no avail...so I can only interpret from [the lack of response] that this government has fully abandoned any attempt at intersectionality" (quoted in Paradis, 2021). In its 2019-2020 *Annual Report* (Alberta, 2020a), the Ministry of Culture, Multiculturalism and Status of Women indicated Minister Aheer had directed staffers to "redesign" GBA+ as "Inclusive Diversity and Equality Analysis" (p. 23), though there is no information here nor in subsequent reports about what this new approach actually entails. Ministry staffers reported the word "intersectionality" began to be removed from research reports (Paradis, 2021).

These bureaucratic moves were combined with a heightened political rhetoric that aimed to delegitimize feminism and any residual claims about gender equality as a focus of government policy. Premier Kenney aggressively criticized the Trudeau federal government's use of the feminist concept of intersectionality in its 2020 Throne Speech, characterizing this as a "kooky academic theor[y]," arguing the speech had laid out a what was a "fantasy plan for a mythical country that only exists, apparently, in the minds of Ottawa liberals and like-minded Laurentian elites" (Woods, 2020). Here the anti-feminist threads of UCP populism were in bold display. Feminism is rhetorically aligned with the Trudeau Liberals, who are together constructed as a threat to Alberta's economy and way of life. Doubling down on this hyperbolic attack after critics accused Kenney of dismissing systemic racism, Matt Wolf, the premier's chief of staff, tweeted a video by right-wing American political commentator Ben Shapiro. In it, Shapiro claims that intersectionality is a dangerous "victim" narrative that privileges voices based on how many victim groups they are a part of.

One wonders why feminist expertise became such a political target for the UCP, why GBA+ was purged from the policy process, or why even the Government of Alberta's equity statement for appointments was replaced with a new statement emphasizing formal liberal equality, erasing the need for proactive measures to increase diversity ("It is recognized in Alberta as a fundamental principle and as a matter of public policy that all persons are equal in regard to race..."). Some have speculated this has to do with the premier's personal antifeminism, recalling his campaign as student senator at the University of San Francisco to have the women's law forum shut down for its support of reproductive rights, or his role in founding the parliamentary pro-life caucus (see Chapter 1). Over the first years of the Kenney government, political analysts convincingly opined about the UCP's "war on women;" and there were scores of aggressive political attacks on women professors, doctors, and activists, often led by an army of UCP secretaries on social media platforms like Twitter (Payne, 2020).

These misogynist attacks even spilled onto the floor of the legislature. In one prominent instance, Kenney dismissed criticisms made by University of Calgary political scientist Melanee Thomas, denouncing her on the legislature floor as an "NDP candidate," in a move clearly threatening to academic freedom (Jeffrey and Yousif, 2019). Over the first year of the UCP government, it became commonplace to attack women experts of all kinds as "NDP" or "socialist." I experienced this auto-ethnographically, when the press secretary to the minister of advanced education tweeted my own university to complain about my critiques of the government. Much like the relentless gender politics in the aftermath of the Harper government's election in 2006 (Brodie, 2008), the shift away from the gender equality commitments of

Alberta's NDP relied on the targeted dismissal and silencing of oppositional voices.

To fully understand these attacks on feminism and feminist expertise, it is important to interrogate the political purposes they served. The downsizing and reorganization of Alberta's gender bureaucracy was never explicitly attributed to the UCP government's emphasis on austerity. The erosion of gender policy capacity instead served an instrumental goal for a government deeply committed to neoliberal governance. Such moves make gender disadvantage and intersectional inequalities magically disappear. One strategy used to disarm feminist claims and to get past demands that the government address persistent problems like gender-based violence or the gender wage gap is to render such claims irrelevant by making the problems themselves disappear (Brodie, 2008).

The UCP's efforts to undermine feminism and the government's own gender expertise must be situated in relation to its formulation of populism. While unconcealed racism and nativism are constrained within the electoral calculus of "maple-glazed populism" in Alberta (Friesen, 2021; Budd, 2021), antifeminism and gender politics can be strategically marshalled in the construction of the "real Alberta" under threat from a "corrupt elite." The masculinist construction of the "real Albertan," what Wesley (2022) refers to as the "cowboy myth," is juxtaposed against the so-called "Laurentian elite," represented as a Central Canadian cabal that both exploits Alberta's resource economy and, via its globally-influenced environmentalism and left-wing politics, threatens economic prosperity and the Albertan way of life. This populist antagonism, or what some have called "extractive populism" (Friesen, 2021), is reinforced through gendered discourses. The Trudeau government, often represented as if married to the Notley NDP, comes to symbolize the "woke elite." The Trudeau Liberals are feminized, in part through representations of Trudeau himself as a lightweight drama teacher, and notably by turning the federal government's own self-labelling as feminist against it. Anti-feminism bolsters UCP populism, and in this way, feminist expertise comes to be viewed as an imposition of the "Laurentian elite." In this discourse of drawing boundaries, right-wing populism contributes to the self-affirmation of masculinity, even reinforcing Kenney's own image as a populist strong-man leader, by offering points of reference for the re-establishment of traditional gender constellations and thus, for the abolishment of gender equality policies.

COVID 19 and the UCP Agenda: "#Brocovery" in Response to a "She-cession"

In 2020, as the new reality of the pandemic was just taking hold, policy staffers in the Department of Culture, Multiculturalism, and the Status of Women began to collect emerging studies pointing to the gendered and intersectional impacts of COVID-19, developing what was, in essence, a GBA+ policy analysis for the government (Paradis) 2021. The United Nations (2020) and the Organization of Economic Cooperation and Development (OECD) (2020) issued early warnings about how the pandemic was deepening pre-existing gender inequalities on a global scale, which were, in turn, intensifying the impacts of the pandemic. According to Danielle Paradis' (2021) report in *The Sprawl*, the ministry's research, adding up to more than 200 pages of documents, was shared with the government's pandemic response team. A briefing note, describing how "these events [disasters] impact men, women, and gender diverse people differently," was circulated to multiple ministries. The UCP government was therefore made well aware of emerging evidence of COVID-19's disproportionate gendered effects, yet it was highly resistant to this message about systemic inequality. According to Paradis' (2021) review of documents obtained through a freedom of information request, for example, the word "intersectionality" was purged from later versions of the ministry's briefing note.

Pre-pandemic, Albertan women, and in particular racialized, Indigenous and immigrant women, already faced steep economic barriers. Calgary and Edmonton were the second and third worst places for women to live in a study that ranked Canadian cities on measures of gender inequality (including economic security, leadership, health, personal security, and education) (McInturff, 2014). Alberta had the second highest gender pay gap of all Canadian provinces (Pelletier, Patterson, and Moyser, 2019). This pay gap was even larger for Indigenous and racialized women (Scott, 2021). A Parkland Institute study found the difference in total market incomes between men and women was an astonishing 50 per cent (Lahey, 2016). In part, this is due women's relatively high rates of part-time work, with lack of access to subsidized childcare cited as the main reason for not working full time (Lahey, 2016). The highly gender-segregated nature of Alberta's resource economy also contributes to wage differentials, with women clustered in jobs in the service and care sectors of the economy, while higher paying resource sector jobs are dominated by men (Lahey, 2016). High male wages and restricted access to childcare have also meant Alberta has had the most stay at home mothers of any Canadian province (Statistics Canada, 2018).

That the UCP would fail to appropriately respond to deepening gender and intersectional inequalities in the wake of the pandemic seemed written in the cards. Early on its mandate, the Kenney government announced it would kill one of the NDP government's most significant attempts to redress gendered economic inequality, a pilot $25/day childcare project creating 7,300 subsidized spaces, which the NDP had promised to extend province-wide in the 2019 campaign (Smith, 2019; see Cake, Chapter 16). In a tweet defending the decision to eliminate the pilot as a deficit reduction measure, then-Minister of Children's Services Rebecca Shulz responded to

Notley: "The fiscal hole your government put us in makes province-wide $25/day childcare completely impossible." Furthermore, Shulz argued "[the program] didn't help parents who need it most," capping fees for all parents, rather than being delivered as a subsidy to low-income earners.

Underlying this swift cancellation was opposition to public provision and to universal forms of funding. The UCP had campaigned against such a system, despite clear economic evidence that "Quebec's universal access to low fee childcare" system had resulted in an increase in women's labour force participation, a related increase in economic productivity, and growth in taxation revenue that exceeded the cost of the program (Fortin, Godbout, and St-Cerny, 2012). The UCP resisted framing childcare as a gender equality measure, consistent with its efforts to erase the systemic problem of women's disadvantage. As became clearer in its later opposition to the Trudeau government's childcare deal, the Kenny government clearly endorsed a marketized, "flexible and adaptive" childcare system with minimal state interference, constructing parents (mothers) as consumers of a service (Alberta, 2020b). Many critics speculated the UCP was being driven by a "retrotopian" (Roth, 2021) social conservative vision of gender, idealizing the heterosexual nuclear family. This is implicit in the UCP government's repeated mantra emphasizing "parental choice." Neoconservative discourses that justify the family as the appropriate site of care work legitimize the privatization of social reproduction, in which the family and women's unpaid work serve to absorb the social impacts of neoliberal austerity (Brodie, 2008).

The government's efforts to erase gender equality as a policy problem meant it was ill-equipped to deal with the significant gendered and intersectional impacts of the pandemic. In fact, many of the government's early pandemic actions would make the devastation of COVID-19 so much worse for many groups of Albertan women. In March 2020, during the first wave of school closures, the government issued a shocking Saturday press release laying off 26,000 educational workers — educational assistants, substitute teachers, school support staff — the vast majority of whom were women. This was the largest mass layoff in Alberta's history, and one that made the COVID-imposed home-schooling struggles of parents so much more gruelling. The well-being of these workers, and of parents and children, was sacrificed on the neoliberal altar of austerity, at a time when governments around the world were embracing the necessity of stimulus spending.

The government also astonishingly went to war with Alberta health-care workers in the middle of a pandemic. Women, who make up the majority of the health-care workforce, have led the health response to COVID, exposing themselves — and their loved ones — to a higher risk of infection. As other chapters in this book discuss in more detail (see Church, Chapter 13), the government ripped up an agreement with the Alberta Medical Association, also threatening to lay off 750 nurses and roll back their wages, while castigating the United Nurses of Alberta for refusing to delay contract negotiations until "after" the pandemic (Climenhaga, 2020). Furthermore, then-Minister of Health Tyler Shandro proceeded with an aggressive plan to lay off between 9,700 and 11,000 health-care workers employed as laboratory, linen, cleaning and food services. This outsourcing plan targeted low-wage women workers, many of whom are racialized. As for Alberta teachers, the government sought to roll back their wages and forced them to return to in-classroom teaching with reduced cleaning staff in schools (a legacy of the March 2020 cuts), and without limiting class sizes or implementing other COVID mitigation measures, such as improved classroom ventilation (Stirling, Chapter 12). In effect, the female-dominated care sector of Alberta's economy both stood at the front lines of the pandemic and also bore the brunt of these cruel UCP austerity moves. As Wendy Brown (2016, p. 10) has argued, in a context of neoliberal politics, citizens are expected to sacrifice themselves: "[T]he responsibilized citizen tolerates insecurity, deprivation and extreme exposure to maintain the productivity, growth, fiscal stability...." Essential women workers in Alberta became the ultimate "sacrificial citizens" during COVID-19.

The effects of the pandemic recession on women's unemployment during the first months of COVID-19 were so pronounced, economists referred to it as a "she-cession." Nationally, women's rate of employment dropped by more than 10 per cent in two short months (Scott, 2021), while in Alberta the drop was five per cent (Baker, Koebel, and Tedds, 2021). Compared with men, Alberta women experienced disproportionate economic losses because of the highly gendered character of the labour market and their concentration in sectors and occupations that are vulnerable to the pandemic. The industries most affected by the pandemic closures and restrictions (leisure, hospitality, and education) were those in which women, and especially racialized and poor women, predominate, often in low-wage jobs. As a result, the pandemic hit those least able to withstand the economic uncertainty (Baker et al., 2021). Moreover, waves of school and daycare closures caused many Alberta women to take time away from paid work, or else to struggle under the triple burden of remote work, home-schooling, and domestic labour, resulting in extreme stress, as well as reductions in working hours. The health of Alberta's economy and society during the pandemic was very much predicated on unpaid domestic labour; parents, and especially mothers, were expected to be flexible as caregivers, responding to waves of the virus and to the UCP's often chaotic efforts to manage "*livelihoods* and lives" (with an emphasis on the former). As Banu Ozkazanc-Pan and Alison Pullen (2021, p. 4) argued, "mothers oscillate in-between being beneficial yet undervalued, essential to the functioning of societies, yet omitted from economic discourses on recovery from COVID-19."

The androcentric character of Alberta's COVID recovery plans completely failed to acknowledge these profound gendered and intersectional effects. In fact, there seemed to be an active resistance to a gender-aware analysis of the social and economic costs of the pandemic. In June 2020, in the depths of the she-cession, and before women's employment began to rebound later that year, Kenney announced the government's plans for a recovery based on a massive $10-billion investment in infrastructure spending (Alberta, 2020c). The masculine "cowboy myth" grounding the UCP's populist construction of the "real Albertan" is discursively imprinted in this lengthy statement that lauds the "personal responsibility...hard-wired into this province" (p.4) and that hangs its hat on infrastructure-led job creation. Women's service sector job losses get a brief mention, and the gesture of a $5,000 relaunch grant for businesses impacted by COVID closures, while the bulk of funding is to be directed at male-dominated job sectors, particularly in the energy and construction sectors. Perhaps this should not have come as a surprise in the context of Alberta's particular brand of right-wing populism. As Kyle Friesen (2021) writes, "the UCP establishes 'the people' as a cross section of working Albertans and the extractive oil sector through "symbolic nationalization," whereby the interests of for-profit oil extraction become one and the same as Albertans: the prosperity of the former equating to the prosperity of the latter." The masculinity of this idealized real Albertan became apparent in what I have described (in my Twitter feed) as the UCP's "#brocovery agenda, on sharp display in repeated social media "jobs and economy" announcements featuring male workers in hard hats or cowboy hats, with Kenney and other cabinet ministers smiling proudly in similar attire.

Choosing to focus on job sectors in which women are so underrepresented guarantees the #brocovery agenda will leave women behind. Energy is the least representative sector of the economy when it comes to gender (Hughes, 2022). In fact, mining, quarrying, and oil and gas extraction industries were among the largest drivers behind the national wage gap (Pelletier et al., 2019). Moreover, an emphasis on infrastructure directs recovery funds toward workers in the skilled trades where women are particularly underrepresented. Only 10 per cent of Alberta apprentices are women, and using 15 per cent as a metric, only 39 or 48 skilled trades have a critical mass (Cameron, Morin, and Tedds, 2020). The UCP "Skills for Jobs" agenda (see Spooner, Chapter 14) has emphasized the importance of exposing women to trades, including by providing generous funding for the organization Women Building Futures, and also by creating a modest new scholarship of $2,500 for women pursuing programs in science, technology, engineering and mathematics (Alberta Innovates, 2020). In a rare statement on women's equality, perhaps in response to gendered critiques of the recovery agenda, then-ministers Aheer and Doug Schweitzer (Jobs, Economy and Innovation) emphasized that "the recovery plan was created to help *all* Albertans recover from our current situation and return to prosperity"

(Alberta Innovates, 2020). However, the answer they provided, programs to attract women to male-dominated job sectors, is highly unlikely to secure true gender equity. Women in hard hats have become more common in the UCP ads promoting the recovery agenda. In UCP discourse, the answer to gendered economic inequality becomes a self-improvement project. The systemic nature of this disadvantage — biases, discrimination, gender segregation, women's disproportionate responsibility for domestic labour — gets erased. The thrust of the #brocovery agenda for women becomes clear. Women are expected to become, in Wendy Brown's (2016, p. 8) words, "isolated bits of self-investing human capital, a process that both makes them more governable and integrates them into the project of economic growth…."

Policies that might have addressed pandemic-amplified gender inequalities were often resisted by the UCP government, particularly when initiatives were driven by the Trudeau government. Throughout COVID-19, the Kenney government downplayed the source of the federal support funds delivered by the province. The Trudeau government's efforts to ensure the well-being of the population throughout this period of profound crisis were often stigmatized and criticized, as driving up the national debt and as intruding in areas of provincial jurisdiction. This performance was made necessary by the political logic of UCP populism, one that focuses on the Trudeau government as the symbolic representative of a corrupt "global woke elite" threatening the interests of real Albertans (Wesley, 2022). The pandemic thus put the UCP in a difficult position. To adhere to its populist we/them rhetoric, the government had to continue to reinforce the message that the "corrupt Trudeau elite" is the main source of the province's economic problems, even as it accepted the funding that was so desperately needed by Albertans.

The foot-dragging on how to distribute federal funding for essential private-sector workers, an initiative that would benefit low-wage women workers, was a case in point. It took the government until just before the funding expired to announce how it would structure the 75/25 cost-shared Critical Worker Benefit, with $347 million provided by the federal government (Johnson, 2021). The government chose to disburse this support as a $1,200 lump sum payment, when what women workers in sectors like long-term care or retail most need is pay equity and a re-valuation of their work in relation to male-dominated job sectors, policy moves that would have been staunchly resisted by the business sector. The requirement that a claimant must have worked at least 300 hours between October 2020 and January 2021 excluded many part-time workers, who are, of course, highly likely to be women. Women who had left paid employment to cope with childcare and remote schooling were also excluded from this benefit. The purging of GBA+ from the policy process, however, meant these gendered effects were simply ignored and obfuscated by the government.

Perhaps most egregious, from the perspective of women, were the UCP government's machinations around the Trudeau Liberal's national childcare program, the first new shared-cost social program in decades (see Cake, Chapter 16). While bilateral deals were concluded with eight other provinces prior to the 2021 federal election, the UCP's foot-dragging nearly risked $3.8 billion in federal funding which would ultimately reduce fees to $10 per day by 2026, create more than 26,000 new spaces, and improve wages for childcare workers (Bennett, 2021). Kenney argued that Alberta was being unfairly treated by the Trudeau government in the negotiations. He focused on how Quebec had secured unconditional funding, ignoring how its childcare plan had served as a model for the national plan, and how it in fact already exceeded the federal requirements (Bennett, 2021). Childcare became a site for contesting the federal government's imagined antipathy to Alberta's interests and for making a stand for provincial autonomy. In its ideological adherence to the market, the government sought to structure a program that could be delivered through subsidies, rather than through public provision, and that would allow maximum scope for for-profit operators. In its hardline posturing, the UCP government appeared almost indifferent to the critical childcare needs of Albertans with young children and to the significant benefits for gendered economic equality that accessible and affordable childcare would create. Rahki Pancholi, NDP Children's Services Critic, suggested that pushback from citizens finally drove the Alberta government to sign on to a bilateral childcare agreement (Bartko and Mertz, 2021).

In sum, the pandemic intensified gender and intersectional inequalities in Alberta, but the ideological purging of GBA+ from the policy process made these effects disappear. The UCP's mismanagement of COVID-19, together with its exclusionary #brocovery agenda, have created additional barriers for women, and, in particular, for marginalized women, that were offset by the federal government's pandemic supports and by the national childcare program.

The UCP's Penal Populism, Gender-based Violence and Gendered Colonial Violence

Once again pulling from the Harper Conservative populist playbook, the UCP has made extensive uses of penal populism, drawing the social and systemic problems of gender-based violence and gendered colonial violence deeply into a carceral embrace. As critical criminologists have argued, neoliberal governance has meant the increased deployment of law-and-order politics, as carceral approaches increasingly supplant previous regimes that were organized around the provision of material welfare (see, for example, Wacquant, 2009). As a right-wing populist strategy, punitive politics have an obvious popular appeal. "Tough on crime" rhetoric can bolster a populist government's image as strong and masculine against opponents, in this case, the "Trudeau-Notley"[2] alliance, who are represented as weak and feminine by comparison. Indeed, as Elizabeth Bernstein (2012, p. 238) has argued, neoliberalism can be "... described as a remasculinization of the state in which its soft "social bosom" is transformed into a hard "penal fist"...." The UCP's focus on problems such as rural crime and its proposal to develop an Alberta police agency must be seen as penal populist moves. The government's consistent efforts to reframe gender-based violence and gendered colonial violence as issues of crime control have the effect of de-gendering and individualizing what are complex systemic issues.

While embracing gender-inclusive frameworks capable of encompassing violence against trans and non-binary people, and that can acknowledge men's experiences of victimization, feminists have long argued we need to treat these problems as gendered expressions of domination, rooted in power disparities between men and women, dominant norms of heterosexuality and in hegemonic forms of masculinity (Gavey, 2019, pp. 227-259; see also Chaudhry, Chapter 20). Women experience twice the incidence of "violent" victimization as men, driven by rates of sexual assault that are five times those of men (Statistics Canada, 2021). Department of Justice data show women also face much higher rates of intimate partner violence and are much more likely to experience chronic and severe forms of violence (Canada, 2020). Alberta has among the highest rates of sexual violence and intimate partner violence in the country (McInturff, 2015), perhaps rooted in the failures of decades of conservative governments who have neglected to take these social problems seriously. Nationally, a campaign led by Women's Shelters Canada, and engaging Alberta organizations like the Alberta Council of Women's Shelters (ACWS), has laid out strategic road map for a National Action Plan — a co-ordinated multi-jurisdictional social investment in gender-based violence as a fundamental problem of gender inequality (Maki and Dale, 2021). As Jan Reimer, Executive Director ACWS insists (Author's Personal Interview, 2022),

> In our province, just like everywhere else in the country, domestic violence is at the hub of many other issues. There is policing, and there is the court. And those are different departments, provincial and federal. And then [there are the] housing issues women have. And how much economic support women have. And child custody, and the child welfare concerns, mental health, PTSD, and the effect of exposure to violence on children. So, it spans almost any provincial department.

In a structural context of ongoing colonization and racism, Indigenous women experience violent victimization and sexual violence at many times that of non-Indigenous women; almost 6/10 have experienced physical assault and almost half have experienced sexual assault (Heidinger, 2022, p. 5). Indigenous feminist scholars have emphasized how colonial relationships are gendered and sexualized, and how gender-based violence functions as a tool of settler colonialism, not merely as a means of patriarchal control (see, for example, Belanger and Newhouse, Chapter 18; Kuokannen, 2008). Experiencing targeted forms of gendered colonial violence, Indigenous women and girls are far more likely than other Canadian women to encounter violence, to be "disappeared," or to be killed, often in circumstances that involve sexual assault (Heidinger, 2022). Alberta is an epicentre in the ongoing national tragedy of missing and murdered women and girls. The National Inquiry into Missing and Murdered Indigenous Women and Girls (2019, pp. 584-594) found there is a correlation between the work camps that characterize resource extraction industries like oil and gas, and extremely high rates of violence against Indigenous women. In its analysis of root causes, the Inquiry condemned the institutional complicity of the health care, child welfare, and policing and justice systems, and called for sweeping structural reforms in its 231 "Calls to Action," which were directed at the federal, provincial, and territorial governments.

The UCP have largely ignored the comprehensive policy strategies laid out in the National Action Plan and by the National Inquiry. Instead, gender-based violence and gendered colonial violence have been squeezed into a narrow crime control framework consistent with UCP penal populism. This approach, seeking to contain gender-based violence within discourses of abstract risk and individuated criminal responsibility, is exemplified by Claire's Law, often held up by the government as the centerpiece of its response to what it refers to as "family violence." Passed with much fanfare in 2019, but only proclaimed in 2021, Clare's Law, or the Disclosure to Prevent Domestic Violence Law, aims to prevent the perpetration of violence between intimate partners through the sharing of information about prior histories of violence. Modelled on a law originally passed in England and Wales, the premise of Clare's Law is that "individuals should have access to information to help them make informed choices and reduce the risk of harm" (preamble, Alberta, 2019). The law allows those who fear intimate partner violence to find out about their partner's previous history of violence, and enables police to proactively inform an applicant of this history.

The discursive framing of Clare's Law is important. As with other aspects of the UCP government's agenda, this initiative is framed in entirely gender-neutral terms, contributing to the erasure of "gender" in gender-based violence. In announcing Clare's Law, for example, Kenney deliberately avoided using the word "women," emphasizing that the law was meant to "keep vulnerable Albertans safer" (Russell, 2019). Likewise, data released in 2022 on applications and outcomes fails to specify the gender of the applicants or of their partners (French, 2022). Combined with this gender-neutral framing is the way this policy individualizes a systemic social problem. Clare's Law, much like the UCP's "women should just take up a trade" response to gendered economic inequality, is heavily reliant on the neoliberal technique of responsibilization. It makes women, as potential victims, responsible for preventing intimate partner violence by appropriately evaluating their own risk and by seeking information on their partners (Koshan and Wiegers, 2019). In this way, it shifts responsibility from perpetrators to women as individual risk managers, and obscures social responsibility for preventing gender-based violence. And this, in turn, opens the door for victim-blaming: women who fail to seek information and suffer abuse can be subjected to blame (why didn't you ask?), and those who fail to leave abusers when they obtain information under the law can be held responsible (why didn't you just leave?).

As critics have argued (Koshan and Wiegers, 2019; Robinson, 2020), Clare's Law's almost singular focus on individual risk management ignores the conditions that may prevent women from making applications, particularly those from racialized and Indigenous communities who are more likely to experience intimate partner violence and who have good reasons to distrust the police. An individualized focus on women as risk managers also ignores the obstacles women face when they do seek to leave an abusive partner, as well as how leaving is quite often a trigger for more violence. The UCP has failed to invest in necessary social supports for women seeking to escape violence, most importantly, through appropriately funding women's shelters, second-stage shelters and affordable housing. As Reimer has emphasized, for example, the UCP government has still not announced its plans for new federal funding under the Rapid Housing Initiative, which requires at least 25 per cent of funds to be directed at women-focused housing initiatives (2022). No doubt because of such barriers, very few people have made applications under Clare's Law — only 440 as of June 2022. There have also been significant problems with the implementation of the law, with a backlog that saw applicants waiting up to three months for risk assessments, rendering this mechanism of protection almost useless (French, 2022).

The incorporation of gender-based violence and gendered colonial violence into the UCP's law and order agenda has been highly performative, focused on piecemeal gestures and "announcements" bolstering penal populism. For example, the UCP passed legislation requiring that provincially appointed judges receive training on sexual assault law (Alberta, 2022a). Yet the UCP government has appointed very few new judges. This requirement is therefore unlikely to have any immediate effects on the frequency with which Alberta judges seem to reinforce rape myths or misunderstand the law of consent. Yet this performative move firmly situates the problem of sexual violence as an issue only of crime control and creates the illusion of decisive action.

Gender-based violence has also been managed through a "government by panel" approach that has characterized the Kenney government's entire policy agenda. The UCP's response to the crisis of violence against Indigenous women and girls is a case in point. After a lengthy process led by Indigenous experts, with hearings held in Alberta, the National Inquiry into Missing and Murdered Indigenous Women and Girls (MMIWG) issued a transformative set of recommendations aimed at addressing the gendered effects of intergenerational trauma and marginalization, as manifested in poverty, inadequate housing and health care, and barriers to education and cultural supports. The National Inquiry's "calls to action," issued to federal, provincial, and territorial governments, demanded an action plan from the Alberta government. Instead, the UCP established a Joint Working Group (of Indigenous leaders and MLAs) on MMIWG, whose mandate was to provide direction for the government's proposed actions. The Working Group took almost two years to report, recently issuing a "roadmap" with 113 Pathways to Justice (Alberta, 2022). These pathways reiterate many of the recommendations made by the National Inquiry. To date, the only concrete government actions in response were the creation of Premier's Council with Indigenous women representatives, intended guide implementation of the roadmap, as well as the proclamation of Sisters in Spirit Day. These initiatives are obviously more performative than substantive, as Indigenous women critics like Michelle Robinson have astutely noted, "It's great to have ceremony and pomp, [but] [w]hat we need is action and a budgeting line item in order to be successful." Moreover, Robinson raises the real danger that appointment to a Premier's Council could end up co-opting the very people who might try to hold the government accountable for its actions (Lachacz, 2022). Meanwhile, critical programs that were helping to address this crisis have fallen victim to UCP cuts. In 2019, SNUG, a highly effective program providing support for Indigenous girls in Edmonton involved in sex work, saw its funding cut, perhaps because its harm reduction approach conflicted with the government's social conservatism (Short, 2020).

Alberta's women's shelters and sexual assault centres on the frontlines of the struggle against gender-based violence have faced incredible resource constraints and growing demands for support, particularly in the face of what has been called the "shadow pandemic" of violence against women — escalating violence during COVID-19, intensified by isolation and by restrictions affecting core services. Survivors contacting sexual assault centres for support currently face delays of up to a year. Shelters, as Reimer (2022) has emphasized, "are still turning down thousands and thousands of women." The UCP has resisted calls to meaningfully invest in these important frontline systems. The government even raided the "Victims of Crime Fund" to backfill cuts to its justice budget. The Victims of Crime Fund, built from victim surcharges imposed on those convicted of crimes, is intended to assist victims and to fund support services, such as shelters and sexual assault centres. Mired in bureaucracy, however, the fund had accumulated a significant surplus. In 2020, the Minister of Justice announced that the purpose of the fund would be expanded to include public safety initiatives (Bellefontaine, 2022). Until 2022, when the government indicated the fund would return to its original purposes, the UCP diverted it to finance core elements of its law-and-order agenda, including additional police to combat rural crime, and hiring more crown prosecutors (Bellefontaine, 2022). The UCP's penal populist agenda quite literally swallowed up much needed interventions in response to the pervasive social problems of gender-based violence and gendered colonial violence.

Conclusion

In June 2022, after Kenney was defeated in the UCP leadership review and before a new leader was chosen, the Speaker of the Legislature quietly announced the winners of the "Her Vision Inspires" essay contest, posted for display on the Legislative Assembly webpage. Open to Alberta women between 17 and 25, the contest required the submission of a short essay explaining what submitters would do if elected as a Member of the Legislative Assembly of Alberta. Her Vision Inspires was a partnership between the Alberta Legislature and the Commonwealth Women Parliamentarians Canadian Region, an organization working to improve women's political representation. Given this liberal feminist mandate, one might have expected that the winners, selected by UCP Members of the Legislative Assembly, would advance a vision that centres gender equality. Instead, the third-place winner laid out, in stark terms, the anti-gender ideology that many believe has driven UCP policy on women. The essay begins by defining women's role as reproductive vessels, then arguing that "promot[ing] that women break into careers that men traditionally dominate is not only misguided, but it is harmful" (quoted in Climenhaga, 2022). Like European versions of right-wing populism, the essay veers into the explicit racism and nativism that grounds so-called "white replacement theory"; "it is sadly popular nowadays to think that the world would be better off without humans, or that Albertan children are unnecessary as we can import foreigners to replace ourselves, this is a sickly mentality that amounts to a drive for cultural suicide" (as quoted in Climenhaga, 2022). In making the case that women should be rewarded for fulfilling their destiny as mothers of the nation, the essay evoked direct comparisons with Nazi Germany (Climenhaga, 2022). The selection of this essay as a "winner" in a contest sponsored by the Legislative Assembly of Alberta, the apparent "brainchild" of then-Associate Status of Women Minister Jackie Armstrong-Homeniuk, was so shocking in its racism and misogyny that it made national and even international news.

Armstrong-Homeniuk initially sought to explain away the selection of this winning essay as "reflect[ive] of a broad range of opinions" (as quoted in Climenhaga, 2022). Its selection, however, must be seen as a kind of dog-whistle to the UCP's social conservative base, laying out, in raw terms, the anti-gender threads of UCP populism. Anti-feminism has been central to the UCP's agenda. The Kenney government engaged in a Harperesque project of what Brodie (2008, p. 167) has called the three "Ds": delegitimizing feminism, the dismantling of gender-based policy units, and the disappearance of the gender subject of public policy through purging GBA+ from the policy process. This anti-gender ideology has resulted in heightened gendered and intersectional inequalities, amplified in the context of the pandemic's she-cession. The #brocovery agenda, as I have argued, will only deepen these inequalities, while subsuming of gender-based violence and gendered colonial violence into a penal populist agenda accomplishes a magical disappearing act that erases the systemic nature of complex social problems.

The recent election of libertarian Danielle Smith as leader, whose victory is indebted to the party's social conservative base, makes the UCP's populist rhetoric even more central to its agenda. Smith has signalled that anti-Trudeau and pro-Alberta autonomy politics will drive her policies in the months leading up to the 2023 election. The UCP's three D anti-gender agenda discussed here appears institutionally cemented by Smith's decision to completely eliminate the status of women portfolio, thereby absolving the government of the need to develop a policy mandate that addresses gender equality. The UCP agenda, and its androcentric approach to COVID-19, has left Alberta women poorer and less equal, and has erased gendered systemic problems, such as intersectional economic inequalities and gender-based violence and gender colonial violence. The stark gender gap in voting intentions revealed by recent polls suggests that Alberta women will not endorse an agenda that so undermines their status (Dryden and Marcusoff, 2022). This provides us with some measure of optimism that political change may be on the horizon.

References

Alberta. (2019). *Bill 17: Disclosure to Protect Against Domestic Violence (Clare's Law).* First Session, 30th Legislature, 68 Elizabeth II. https://docs.assembly.ab.ca/LADDAR_files/docs/bills/bill/legislature_30/session_1/20190521_bill-017.pdf

Alberta. (2020a). *Annual report: Culture, Multiculturalism and Status of Women 2019-2020.* chrome-extension://efaidnbmnnnibpcajpcglclefindmkaj/https://open.alberta.ca/dataset/4a9716c2-e826-4bdd-bcdd-8aefd8e9fc12/resource/2b1ec7aa-ec4a-40ef-8ef7-66f209c1052a/download/cmsw-annual-report-2019-2020.pdf

Alberta. (2020b, October). *Childcare Consultation Report*. Ministry of Children's Services. https://open.alberta.ca/dataset/ba16226e-719c-4a57-b448-126d59057903/resource/22644725-b176-4cc9-ae59-9c50ff8c6f87/download/cs-child-care-consultation-report-2020.pdf

Alberta. (2020c). *Alberta's Recovery Plan*. https://mcusercontent.com/f4402c1734c0069635018cbf1/files/063592f2-3fdb-4560-a6d6-066eecd1a0d1/alberta_recovery_plan.pdf?utm_source=Prairie+Sky+Strategy&utm_campaign=764b98e765-EMAIL_CAMPAIGN_2020_02_27_10_16_COPY_01&utm_medium=email&utm_term=0_5e3355e3df-764b98e765-329672821

Alberta. (2022a). *Enhancing Sexual Assault Law Education*. https://www.alberta.ca/enhancing-sexual-assault-law-education.aspx

Alberta. (2022b). *113 Pathways to Justice: Recommendations of the Joint Working Group on Missing and Murdered Indigenous Women and Girls*. https://open.alberta.ca/dataset/9fb69695-3796-4fe3-99b5-4b18916ca4cb/resource/cd21d777-66e9-4ec6-9295-b5c40d2ef6ec/download/ir-113-pathways-to-justice-recommendations-ajwg-on-mmiwg-2021.pdf

Alberta Innovates. (2020, Oct. 14). Statement from ministers Schweitzer and Aheer. https://albertainnovates.ca/impact/newsroom/statement-from-ministers-schweitzer-and-aheer/

Author's Personal Interview. (2022). Interview with Jan Reimer, August 31.

Baker, J., Koebel, K., & Tedds, L. M. (2021). *Gender disparities in the labour market? Examining the COVID-19 pandemic in Alberta*. https://papers.ssrn.com/sol3/papers.cfm?abstract_id=3862950

Bartko, K., & Mertz, E. (2021, Nov. 14). *Trudeau, Kenney promise $10-a-day child care across Alberta within next 5 years*. Global News. https://globalnews.ca/news/8373579/alberta-child-care-announcement/

Bellefontaine, M. (2022, July 19). *Alberta victim services fund will now be used solely to support victims of crime*. CBC News. https://www.cbc.ca/news/canada/edmonton/alberta-victim-services-fund-will-now-be-used-solely-to-support-victims-of-crime-1.6525492

Bennett, D. (2021, Nov. 15). Alberta premier snipes at Trudeau as province signs on to $10-day child-care deal. *The Star*. https://www.thestar.com/politics/2021/11/15/alberta-to-announce-child-care-deal-with-federal-government.html

Bernstein, E. (2012). Carceral politics as gender justice? The "traffic in women" and neoliberal circuits of crime, sex, and rights. *Theory and Society* 41(3), 233-259.

Brodie, J. (2008). We are all equal now: Contemporary gender politics in Canada. *Feminist Theory* 9(2), 145-164.

Brodie, J. (2018). Inequalities and social justice in crisis times. In J. Brodie (eds.), *Contemporary Inequalities and Social Justice in Canada*, 3-25. Toronto: University of Toronto Press.

Brown, W. (2016). Sacrificial citizenship: Neoliberalism, human capital, and austerity politics. *Constellations* 23(1), 3-14.

Budd, B. (2021). Maple-glazed populism: Political opportunity structures and right-wing populist ideology in Canada. *Journal of Canadian Studies* 55(1), 152-176.

Cameron, A., Morin, V. & Tedds, L. (2020). The gendered implications of an infrastructure-focused recovery: Issues and policy thoughts. *Economic Policy Trends*. Calgary: School of Public Policy, University of Calgary. https://papers.ssrn.com/sol3/papers.cfm?abstract_id=3781960

Canada, Department of Justice. (2020). *State of the Justice System 2020: Focus on Women*. https://www.justice.gc.ca/eng/cj-jp/state-etat/2021rpt-rap2021/pdf/SOCJS_2020_en.pdf

Climenhaga, D. (2020, Oct. 13). In the middle of its war on doctors, Alberta's UCP picks a fight with nurses too. *rabble.ca*. https://rabble.ca/politics/canadian-politics/middle-its-war-doctors-albertas-ucp-picks-fight-nurses-too/

Climenhaga, D. (2022, Aug. 9). Her vision inspires … outrage! Fury greets UCP Government prize for essay that calls women best suited as vessels for babies. *Alberta Politics*. https://albertapolitics.ca/2022/08/her-vision-inspires-outrage-fury-greets-ucp-government-prize-for-essay-that-calls-women-best-suited-as-vessels-for-babies/

Dryden, J., & Marcusoff, J. (2022, Nov. 3). Danielle Smith's rough first impression puts Alberta NDP in likely majority territory: new poll. *CBC News*. https://www.cbc.ca/news/canada/calgary/danielle-smith-rachel-notley-ucp-ndp-alberta-janet-brown-1.6638402

Fortin, P., Godbout, L., & St-Cerny, S. (2012). Impact of Quebec's universal low fee childcare program on female labour force participation, domestic income, and government budgets. *Working Paper, University of Sherbrooke, Research Chair in Taxation and Public Finance* at the University of Sherbrooke, 607-615. https://www.researchconnections.org/childcare/resources/24672

French, J. (2022, June 16). *Alberta government says it has tackled backlog of Clare's Law applications*. CBC News. https://www.cbc.ca/news/canada/edmonton/alberta-government-says-it-has-tackled-backlog-of-clare-s-law-applications-1.6491756

Friesen, K. R. (2021). Reimagining populism to reveal Canada's right-wing populist zeitgeist. *Inquiries Journal* 13(01).

Gavey, N. (2019). *Just Sex: The Cultural Scaffolding of Rape*. Oxfordshire, U.K.: Routledge.

Griffin, P. (2015). Crisis, austerity and gendered governance: A feminist perspective. *Feminist Review* 109(1), 49-72.

Heidinger, L. (2022). *Violent Victimization and Perceptions of Safety: Experiences of First Nations, Métis and Inuit Women in Canada*. Canadian Centre for Justice Statistics, Statistics Canada, Catalogue no. 85-002-X. https://www150.statcan.gc.ca/n1/en/pub/85-002-x/2022001/article/00004-eng.pdf?st=pvlNj_wq

Hughes, L. *Gender Inequity Problems are Flying Under the Radar in Alberta's Energy Sector*. Pembina Institute, Jan 17. https://www.pembina.org/op-ed/gender-inequity-problems-are-flying-under-radar-albertas-energy-sector

Jeffrey, A., & Yousif, N. (2019, Nov. 29). Why experts say Kenney's critique of a Calgary professor is a strike against academic freedom. *The Star*. https://www.thestar.com/calgary/2019/11/26/jason-kenney-criticism-melanee-thomas-academic-freedom.html

Johnson, L. (2021, Feb. 11), Alberta frontline workers to receive one-time $1,200 payment from government. *Edmonton Journal*. https://edmontonjournal.com/news/politics/alberta-workers-covid-kenney-copping

Koshan J. & Weigers, W. (2019, Oct. 19). *Clare's Law: Unintended consequences for domestic violence victims?* ablawg.co. https://ablawg.ca/2019/10/18/clares-law-unintended-consequences-for-domestic-violence-victims/

Kováts, E. (2018). Questioning consensuses: Right-wing populism, anti-populism, and the threat of "gender ideology." *Sociological Research Online* 23(2), 528-538.

Kuokkanen, R. (2008) Globalization as racialized, sexualized violence. *International Feminist Journal of Politics* 10(2), 216-233.

Lachacz, A. (2022, June 26). *Alberta report shares 113 recommendations to stop violence, improve safety for Indigenous women and girls*. CTV News. https://edmonton.ctvnews.ca/alberta-report-shares-113-recommendations-to-stop-violence-improve-safety-for-indigenous-women-and-girls-1.5931691

Lahey, K. (2016). *Equal worth: Designing Effective Pay Equity Laws for Alberta*. Edmonton, AB: Parkland Institute. chrome-extension://efaidnbmnnnibpcajpcglclefindmkaj/https://d3n8a8pro7vhmx.cloudfront.net/parklandinstitute/pages/341/attachments/original/1457119686/equalworth.pdf?1457119686

Maki, K., & Dale, A. (2021). *A Report to Guide the Implementation of a National Action Plan on Violence Against Women and Gender-Based Violence*. Women's Shelters Canada. https://nationalactionplan.ca/wp-content/uploads/2021/06/NAP-Final-Report.pdf

McIntosh, A., & Hussey, I. (2020, Feb. 28). *What you need to know about Alberta Budget 2020*. Parkland Blog. https://www.parklandinstitute.ca/what_you_need_to_know_about_alberta_budget_2020.

McInturff, K. (2015). *The best and worst places to be a woman: The gender gap in Canada's 25 biggest cities*. Canadian Centre for Policy Alternatives. chrome-extension://efaidnbmnnnibpcajpcglclefindmkaj/https://policyalternatives.ca/sites/default/files/uploads/publications/National%20Office/2015/07/Best_and_Worst_Places_to_Be_a_Woman2015.pdf

Mudde, C. (2004). The populist zeitgeist. *Government and Opposition* 39(4), 541-563.

National Inquiry into Missing and Murdered Indigenous Women and Girls. (2019). *Reclaiming power and place: Final report*. https://www.mmiwg-ffada.ca/wp-content/uploads/2019/06/Final_Report_Vol_1a-1.pdf.

Organization of Economic Cooperation and Development. (2020). *Women at the Core of the Fight Against COVID-19 Crisis*. https://www.oecd.org/coronavirus/policy-responses/women-at-the-core-of-the-fight-against-covid-19-crisis-553a8269/

Ozkazanc-Pan, B., & Pullen, A. (2021). Reimagining value: a feminist commentary in the midst of the COVID-19 pandemic. *Gender, Work, and Organization* 28, 1-7.

Paradis, D. (2021, July 27). The UCP know the inequitable impacts of COVID-19 all along: Women have it harder and need more support. *The Sprawl*. https://www.sprawlcalgary.com/ucp-knew-the-inequitable-impacts-of-covid-19-all-along

Payne, S. (2020, Oct. 6). The UCP's war on women (and why it might cost them the next election). *Progress Report*. https://www.theprogressreport.ca/the_ucp_war_on_women

Pelletier, R., Patterson, M., & and Moyser, M. (2019). *The Gender Wage Gap in Canada*. Statistics Canada, Catalogue number 77-004-M-2019004. https://www150.statcan.gc.ca/n1/en/pub/75-004-m/75-004-m2019004-eng.pdf?st=1dBmg3MU

Robinson, J. (2020). Clare's legacy travels to Alberta: Undesired consequences and repercussions of bill 17. *INvoke, 6*.

Roth, J. (2021, Oct. 10). *Retrotopian desires and gender in right-wing populism*. CIPSBlog. https://www.cips-cepi.ca/2021/10/10/retrotopian-desires-and-gender-in-right-wing-populism/

Russell, J. (2019, Oct. 19). Alberta introduces "Clare's Law" bill in attempt to curb domestic violence. *CBC News*. https://www.cbc.ca/news/canada/edmonton/alberta-clare-s-law-domestic-violence-jason-kenney-1.5323439

Scott, K. (2021). *A Bumpy Ride: Tracking Women's Economic Recovery Amid the Pandemic*. Ottawa, ON.: Canadian Centre for Policy Alternatives. https://policyalternatives.ca/publications/reports/bumpy-ride

Short, D. (2020, March 2). "SNUG" outreach helping sex trade workers since 2005 folding due to funding loss. *Edmonton Journal*. https://edmontonjournal.com/news/local-news/snug-program-among-metis-services-cut-after-funding-expired

Smith, M. (2019, March 25). Alberta NDP proposes provincewide cap for child care at $25 per day. *The Star*. https://www.thestar.com/calgary/2019/03/25/alberta-ndp-proposes-provincewide-cap-for-child-care-at-25-per-day.html

Statistics Canada. (2018). *Changing Profile of Stay-at-Home Parents*. https://www150.statcan.gc.ca/n1/pub/11-630-x/11-630-x2016007-eng.htm

Statistics Canada. (2021, Aug. 25). Criminal victimization in Canada. *The Daily*. https://www150.statcan.gc.ca/n1/en/daily-quotidien/210825/dq210825a-eng.pdf?st=dOFswmKx

Thomas, M. (2019). Governing as if women mattered. In S. Bashevkin (Ed.), *Doing Politics Differently?: Women Premiers in Canada's Provinces and Territories*, 250-271. University of British Colombia Press.

United Nations. (2020). *Policy brief: The impact of COVID-19 on women*. chrome-extension://efaidnbmnnnibpcajpcglclefindmkaj/https://www.unwomen.org/sites/default/files/Headquarters/Attachments/Sections/Library/Publications/2020/Policy-brief-The-impact-of-COVID-19-on-women-en.pdf.

Wacquant, L. (2009). *Punishing the Poor: The Neoliberal Government of Social Insecurity*. Durham, North Carolina: Duke University Press.

Wesley, J. (2022, Oct.19). Danielle Smith's populist playbook: make the dominant feel marginalized. *CBC News*. https://www.cbc.ca/news/canada/calgary/opinion-danielle-smith-populism-playbook-1.6617059

Woods, M. (2020, Sept. 24). Jason Kenney on Throne Speech: Intersectionality a "kooky academic theory." *HuffPost*. https://www.huffpost.com/archive/ca/entry/jason-kenney-intersectionality-throne-speech_ca_5f6d1f22c5b64deddeeb2130

NOTES

1 I wish to thank my research assistant Olesya Kochkina for her work on this project.

2 Of late, and especially in the context of deal struck between the federal NDP leader Jagmeet Singh and the Trudeau Liberal minority government (to keep the government in power in exchange for commitments to advance policies like pharmacare and dental care), the "corrupt" central Canadian elite is increasingly defined as the "Trudeau-Singh-Notley alliance."

CHAPTER 22

CONCLUSION

Ricardo Acuña and Trevor Harrison

"The fault, dear Brutus, is not in our stars, But in ourselves."

— William Shakespeare (1599), from *Julius Caesar*, Act 1, Scene 2

THIS BOOK HAS EXAMINED the years of United Conservative Party (UCP) government in Alberta, beginning with the 2019 election and ending in the fall of 2022 when the party membership elected Danielle Smith as the party's new leader and premier, replacing Jason Kenney. We have argued that Alberta's problems are largely self-inflicted; that the endless seeking for external enemies has been a ploy designed by an entrenched political elite to augment and sustain their power. As we write, Alberta's political future is uncertain, about which we make no predictions.

We conclude the book, however, with some broad observations on the nature of current politics in Alberta, Canada, and perhaps elsewhere; leftovers, if you will, in our efforts to understand the particular case of Alberta's UCP.

The Moving Spectrum

Recent years have brought repeated assertions by media and some academics that politics in Alberta has become polarized between "extremes" on the left and right. We suggest, however, this is not an entirely accurate portrayal of the political dynamic in Alberta over the last decade.

On the left of the political spectrum, for example, there is no longer any major political party embracing nationalization of the oil and gas industry, capturing 90 to 100 per cent of excess oil and gas profits, or significantly increasing personal and corporate taxes. Yes, the New Democratic Party (NDP) frequently speaks up against racism, misogyny, and homophobia, and acknowledges that climate change is real, but these are hardly radical or extreme left positions. The reality is that, over the last decade, but especially since 2015, the Alberta NDP has moved to the centre of the political spectrum in search of the big tent of Alberta voters.

The NDP's ability to occupy the centre was facilitated by the virtual collapse of the Alberta and Liberal parties and, most especially, by the neutering of the last vestiges of the Progressive Conservative (PC) party in 2017 when it was "merged" (i.e., consumed) by the Wildrose Party and became the UCP. A major consequence of these shifts along the spectrum is that many centrist and even some centre-right Albertans feel politically abandoned. The NDP has been more than happy to move into that space in search of their votes.

Unlike the former PCs, who dominated Alberta politics for 44 years, the UCP appears to have no interest in moderating its policy proposals to appeal to the broad middle of the political spectrum. Instead, since its birth, the party has moved unapologetically to the right, seemingly happy to abandon the historic base that had embraced the conservative policies of Lougheed, Hancock, Prentice, and even Redford in Alberta in favour of that more radical brand of neoliberal prairie populism espoused by Preston Manning, Stephen Harper, and Pierre Poilievre.

One result of this political shift is the gap between the province's two main parties is less today than it was in 2015, and perhaps even less than it was a decade ago. As a result of Alberta's entire political spectrum moving to the right, the most glaring political gap in the province has been to the left of the NDP. This also speaks to the success of Alberta's far right in continuing to move the Overton window in their direction over the last two decades (see Acuña, Chapter 3).

Repeated surveys show that perceptions of a polarized political landscape reflect not the values and beliefs of Albertans, but rather the rhetoric espoused by the political parties themselves. The UCP, in particular, directs its appeal to the centre-right not by moderating its policies or discourse, but by over-the-top labels of the provincial NDP and the federal Liberals as out of touch, woke, anti-freedom, job-killing, Stalinist radicals. The hope is that such rhetoric will trigger the anti-Ottawa anti-Liberal NDP-as-socialist frames established over the years by prairie conservatives and their allies at the Fraser Institute, the Canadian Taxpayers Federation, and the Frontier Centre. By raising the spectre that the NDP will kill the economy and destroy the oil and gas industry, the UCP hopes to scare the moderate right into voting for them, or at the very least, not voting at all. The strategy of painting the NDP as extremists has the added benefit of drawing attention away from those racist, misogynist, and homophobic elements that sequester within the UCP base.

The communications team at the Alberta NDP clearly sees the UCP approach as potentially successful and has responded by visibly and loudly moving to the right in hopes of inoculating themselves against the various accusations thrown at them by the UCP and the far right. By consistently moving right along the political spectrum, the NDP hopes to appear safer and more electable to that politically homeless centre-right.

But the NDP is not blameless in crafting perceptions that Alberta politics are polarized. As the party has moved right along the political spectrum, it has left much of the traditional political left and centre-left feeling abandoned and homeless. Just as the UCP has done with the traditional right, the NDP is not trying to win the votes of its traditional base with leftist policies, but rather through fear — highlighting the privatizing, anti-poor, anti-worker nature of the UCP and drawing attention to that party's growing alt-right element.

These rhetorical strategies by the political parties have been exacerbated by the changing nature of leadership races and party conventions. The move to one-member-one-vote leader selection has turned what used to be a democratic representative process by engaged members and activists into what amounts to a competition for who can sell the most memberships. Increasingly, this means leadership candidates have no interest in appealing to the people who have historically done the policy, recruitment, and electoral work within the party, or are the custodians of institutional memory. Instead, they play to well-organized (and financed) groups on the fringes who may lack allegiance or past involvement with the party, its values, or its policies. In tailoring their pitch to the particular views, interests, animosities, and fears of minority party members — sometimes termed the "selectorate" — such candidates are likely unrepresentative not only of party supporters, but the electorate as a whole, and thus restrict the party's broader appeal.[1] In the case of the UCP, for example, evidence suggests party members are far to the right of either party supporters or most Alberta voters (see Brown and DeCillia, Chapter 2; also Markusoff, J., 2022).

By the same token, the parties have responded to social media, live tweeting, and streaming at conventions by turning them into tightly produced and stage-managed circuses. Where once policy conventions were a venue for lively and broad-ranging debate on the policies that would form the core of future party platforms, they are now largely just a venue for jingoistic cheerleading and congratulatory back-patting. The real policy work and platform development these days is largely done by the leader's inner circle of strategists and communications specialists who are far more interested in responding to the strategic imperatives identified above than to the folks who have historically made up their membership.

The combination of the entire political spectrum moving right with the amped up polarizing rhetoric of the political parties has ultimately resulted in a large group of voters in both the centre-right and the traditional left of the political spectrum wandering lost, looking for the least offensive electoral alternative — orphaned in the words of Brown and DeCillia (Chapter 2). Will these voters, perceiving a real threat from the far right, vote for the NDP in 2023? Or will they stay home and not vote at all, leaving the election in the hands of the 15 to 18 per cent of voters that each of the major parties counts as its committed base?

The Battle for Jurisdiction

The Smith government's introduction of The Alberta Sovereignty within a United Canada Act is political in the sense of attempting to shore up the UCP's support for an upcoming election. At the same time, however, its premise harkens back to jurisdictional arguments at the heart of Canadian federalism and how disputes that naturally arise should be resolved.

The Constitution Act of 1867 (a.k.a. the British North America Act) lays out clearly the jurisdiction powers of the federal government and the provinces. The federal government was given powers over such things as national defense, postal services, currency and banking, navigation and shipping, fisheries, criminal law, the regulation of trade and commerce, and taxation. Provincial governments were given powers over two areas of minimal importance at the time, but hugely important later on, health and education, as well as generally local matters, including municipal governments.

The Constitution Act of 1982 updated the previous act and added the Charter of Rights and Freedoms, while also introducing the controversial Notwithstanding Clause, by which provincial governments might in certain circumstances absent themselves from federal laws. The changes made in 1982 reflected changes in Canadian society, but also the belief that the constitution must be a living document. The contrary position, widely held among conservatives in the United States but only a minority in Canada, is that of originalism; that the first constitution — in Canada's case, 1867 — should be read as inviolate. This appears to be the belief of the Smith government.

The jurisdictional confines of 1867 held together until the Great Depression, at which time provincial governments proved unable to deal with the human destruction. In part also realizing the Depression's misery contributed to social and political unrest, and eventually war, every western country began devising the post-war welfare state — a host of programs that included, among other things, unemployment insurance, public assistance, and health. Inevitably, the federal government was involved, not only because of its fiscal capacity but because a modern society required integrated social and economic policies. This also meant, in turn, that the federal government became involved in areas of strict provincial jurisdiction.

The shared nature of these programs remains a point of discussion and frequent conflict. The many First Ministers conferences, begun in the mid-1950s, provided a forum for conflict resolution, but have dropped from sight in recent years. Many provinces, and the federal government as well, have chosen instead to meet one-on-one. The federal government, for its part, has come to view the conferences as primarily a venue for provincial grand-standing.

While self-serving, this argument is not entirely wrong. Every provincial government has long known that running against Ottawa brings political support back home. This calculation has a particularly long pedigree in Alberta, going back to Social Credit in the 1930s. But past conservative premiers such as Lougheed, Klein, and Kenney did not openly proclaim themselves to be Alberta Firsters, as has the current premier.

This change in stance dates back to the Reform Party in the late 1980s and subsequent partisan frustrations among conservatives in the west every time a Liberal government is elected in Ottawa; hence, the famous Firewall Letter of 1991, of which the current Sovereignty Act is an offspring. Germane to this time is the rise of right-wing populism whose chief fuel is unrelenting anger.

Much of the current anger is directed at the federal Liberals and, no doubt, that government has at times contributed to the anger. But it must also be understood as highly partisan and politically self-serving, aimed not at some lofty defense of people's rights, but rather, as Harrison notes (Chapter 5), increasing the power of the provincial state. Referring to both Saskatchewan's Saskatchewan First Act and Alberta's Alberta Sovereignty Within a United Canada Act, a recent *Globe and Mail* editorial (2022a) states, "the Prairie premiers … conflated what are ordinary provincial-federal policy debates and disagreements with accusations of Ottawa's undue constitutional overreach."

Canada and the world face an onrushing set of pressures, economic, social, and environmental, of which the recent COVID-19 epidemic was only one. But addressing these assorted crises requires coordinated action, much in the manner of how Canada dealt with the Great Depression and the Second World War. Creating provincial fiefdoms is not the way to deal with these crises.

As in the 1930s, many provincial governments today seem increasingly unwilling or unable to deal with the accumulating crises, their only apparent solution being to demand more power with which they can protect their own narrow interests. Consider the recent pandemic. While the federal government was not without its failings in dealing with the COIVD-19 pandemic, the performance of several provincial governments — notably, the conservative ones of Alberta, Manitoba, and Ontario — was particularly less than stellar. In the pandemic's aftermath, and as the country's population ages, Canada's national health system is going to require not just reorganization — everybody's favourite mantra — but a massive financial reinvestment in staff training and recruitment. While many of Canada's richest provinces have fiscal capacity to do this, they are not doing so. Instead, as a *Globe and Mail* (2022b) editorial recently intoned, provincial governments are engaging in what it terms a "populist grift": Saskatchewan's Moe government has given $500 to everyone over 18; Quebec's Premier François Legault the same; and Ontario's Ford government $120 in rebate cheques; while the Alberta governments of Kenney and Smith forgo the provincial gasoline tax. Such "gifts" not only reduce the government treasury in the name of cheap populist tricks and prevent substantive public investments in public programs, such as health, but are also potentially inflationary. Fiscal recklessness aside, however, one should worry even more about the casualness with which some provincial governments now speak of using the Notwithstanding Clause.

Let us engage in a thought-experiment. Realizing the inability or unwillingness of many provinces to play a constructive role in building a coherent and functioning country, were this 1867, would the jurisdictional powers be divided in the same way as then?

Let's think about it another way. The prominent American sociologist Daniel Bell (1996, p. 362) has written, "the national state has become too small for the big problems of life, and too big for the small problems." At what level should different powers to act be housed? Federal powers speak to a need to co-ordinate actions for the larger collective. Municipal governments deal with the immediate and local. What functions do provinces in their intermediate position now actually perform?

Consider the following. Canada's population is just over 38 million. In 2021, nearly three in four Canadians (73.7 per cent) lived in one of Canada's Census Metropolian Areas (StatsCan, 2022). If all urban centres are included, the percentage of Canadians living in urban environments increases to over 81 per cent.

As Ben Henderson argues in Chapter 17, the majority of people today — in Alberta, above all — reside in urban areas. The things that matter to them on a day-to-day basis are municipal. But municipalities also increasingly have to deal with social, educational, and health issues which fall constitutionally within provincial jurisdiction.

It gets worse. Cities and towns have few sources of funding, mainly property taxes. On the one hand, municipal councils are the focus of complaints by local citizens as taxes go up; the City of Lethbridge has recently approved a four-year operating budget that will result in an annual residential property tax increase of 5.1 per cent. On the other hand, they also face complaints when sewers back up, potholes emerge, parks are unkempt, and crime and homelessness proliferate.

The problems faced by many municipalities everywhere are the result of provincial governments starving them to make themselves look good by lowering provincial income and corporate taxes. But, again, these problems can be traced to Canada's constitution, which gives municipalities no fiscal room and, indeed, no genuine status. They are solely the creation of provinces and are thus open to fiscal and political coercion — as we are likely to witness by the Smith government under its Alberta Sovereignty within a United Canada Act.

This is, again, a thought experiment. A change in the status of municipalities is unlikely to occur. The provinces would never willingly give up their power. Yet, the principle of subsidiarity — a favourite idea of neoliberal economists — demands that, unless there is a valid reason otherwise, state functions should be exercised by the lowest level of government (Courchene, 1997). Maybe it's time for such a re-thinking.

Conservatism's Crisis; Conservatism's Alberta Crisis

The arrival of Jason Kenney in 2016 was meant to heal a rift between warring factions within Alberta's conservative parties. The resultant pact — the ironically-named United Conservative Party — has apparently broken down. Before attempting to answer the question "Why?" we first have to ask, "What is conservatism?"

The question is not easily answered. For one thing, conservatism is time and country specific. Grant (1965) famously differentiated Canada's British-infused Tory conservatism from conservatism in the United States which he viewed as not conservative at all, but merely a variation on liberalism. Micklethwait and Wooldridge (2005) make a similar distinction.

Citing Edmund Burke (p. 13), they define classical (British) conservatism as exhibiting six principles: "A deep suspicion of the power of the state; a preference for liberty over equality; patriotism; a belief in established institutions and hierarchies; skepticism about the idea of progress; and elitism." By contrast, they argue, modern American conservatism exaggerates the first three of these and contradicts the last three, adding (pp. 13-14):

> The American Right takes a resolutely liberal approach to Burke's last three principles: hierarchy, pessimism, and elitism. The heroes of modern American conservatism are not paternalistic squires but rugged individualists who don't know their place; entrepreneurs who build mighty businesses out of nothing, settlers who move out West and, of course, the cowboy.

Traditionalists of twenty-five years ago would have found modern conservatism unrecognizable (see Tanenhaus, 2010). Consider the following contrasts:

- Traditional conservatives valued institutions and believed in a positive, if limited, role for the state; modern conservatives often seem like anarchists, skeptical of institutions and quite willing to tear them down.
- Traditional conservatives valued knowledge and expertise; modern conservatives view experts as self-interested elites, and reject the possibility of rational decision-making, a view that brings modern conservatives into close alignment with post-structuralists.
- Traditional conservatives believed that, while change was possible and sometimes even desirable, changes should be brought about cautiously;

modern conservatives are believers in Joseph Schumpeter's dictum of "creative destruction."

The broad conservative tent today comprises a confusing but growing hodge-podge of players — traditionalists, economic liberals, social conservatives, populists, QAnon conspiracists, and libertarians (see Harrison, Chapter 5) — whose primary glue is that of hating the Liberals, or anything vaguely defined as liberal. Conservatives spend an inordinate amount of time, energy, and money ramping up the anger machine, and otherwise waiting for the Liberal party federally to self-destruct, which it dutifully does at some point.

Political parties today are businesses. They package and sell slogans — loosely termed "ideas" — to unsuspecting consumers, some of whom proudly wear the company brand. Brand loyalty is part of their supporter's identity.

When it comes to branding, no jurisdiction surpasses Alberta. "Conservatism" owns a majority of Albertans, and has done so since the late 1950s, the Diefenbaker years. Loyalty to the brand is passed down from generation to generation. Right-wing "identity politics" dominates Alberta politics. Conservative parties have carefully nurtured adherence to their particular brand, as described by political scientist Lisa Young (2022, p. 30).

> What has developed over the past three decades is an overt effort to cultivate a regional identity grounded in a shared political ideology. What is distinctive about Alberta, in this view, is its conservatism. And when Canada rejects conservatism, it rejects Alberta. And this conservatism can't be disentangled from the oil industry.

The term "political ideology" is perhaps too strong, for conservatism in Alberta today is not based on a set of coherent ideas, but on an emotional response to real and imagined threats. This was not always the case. The Lougheed conservatives had ideas — lots of them — built around what was termed at the time "province building." The Klein conservatives also had an idea, albeit a small one, adapted from the time's prevailing neoliberal ethos of free markets, low taxes, and small government. Since the 2008 recession, however, Alberta's conservatives have increasingly substituted anger, intransigence, sloganeering, and nostalgia for thoughtful ideas.

At the institutional level, this change has made Alberta's conservatives intellectually lazy. They belong to a comfortable club of insiders who eschew real challenge. In many ridings outside the larger cities, conservative candidates win nominations unopposed, knowing that success at this stage secures a life-long seat in the legislature. The acclaimed candidate in many cases seems to bring little to the table. But the problem is even worse at the leadership level.

Political scientist Max Weber wrote there are three ways to power: class, status, and party. Class, or financial means, remains a major pathway to power, while status, unless you are a pop star or athlete, has declined. But party as a means to power today has increased in importance, especially among populist-inspired conservative parties. Recent years have seen a parade of individuals who, beginning in their late teens and early twenties, and devoid of any solid achievement outside of politics, work their way to prominence as conservative party apparatchiks; to name a few: Jason Kenney, Andrew Scheer, and Pierre Poilievre. The style of politics is faux-populist, grounded in manufactured outrage. It is a style with a long pedigree in Western Canada. It has come to dominate conservatism in the west and at the federal level as well.

Conservative parties federally can reliably count in every election upon a swath of electoral seats in Western Canada, especially in Alberta. This gives the province enormous influence over federal conservative politics, including its leadership: Poilievre's victory in the 2022 Conservative leadership race is very much attributable to Alberta delegate support. There is nothing *per se* wrong with this. Quebec and Ontario have long had similar influence. But Alberta's current populist-libertarian style, combined with hostility to health care, and seeming obliviousness to developing a workable climate policy, translates poorly to the national level. Conservatives at the national level are loath to criticize their provincial counterparts, but the Alberta wing's policies and profile work against forging a saleable national alternative on the right. Stated succinctly, the strength of conservatism in Alberta is conservatism's weakness at the national level. The initial responses by many prominent conservatives to Smith's Alberta Sovereignty within a United Canada Act reveals starkly this dilemma.

Anger and Angst

Alberta's PC Party took office in 1971 and remained in power for 44 years, facing a threat of defeat on only two occasions, 1993 and 2012. They gained and remained in power the first 20 years by working to reflect the changing political consensus in the province. Much like Social Credit before them, the PCs were able to link their party label with the province's emerging identity.

When the underlying consensus eroded in the early 1990s, the party in 1993 found shelter in Klein's firm embrace and his unapologetic implementation of neoliberal policies. The suite of pro-privatization, anti-tax, low-royalty, anti-government, pro-corporate measures proved immediately successful, but were ultimately responsible for the seemingly unending crises of the past few years. The gutting of public health care, education, and social services; the shrinking of government; a reluctance to introduce sustainable and predictable revenue streams; and a hands-off approach to an economy that made Alberta increasingly dependent upon a single resource sector: all of these things meant the government — and Alberta — had no way out when the global economy crashed in 2008, a crash that, itself, was caused by those very same neoliberal policies, particularly deregulation of the banking sector and corporate wealth concentration, south of the border.

Neither the economy nor the PCs fully recovered after 2008. The jobs never fully came back, and there was no practical reinvestment in public services and social supports. The fallout from the crash gave rise to the Wildrose Party and Smith, who ably directed much of people's anger at the PCs themselves, and not the neoliberal regime they had put in place. The weaponization of public anger against the PCs did not immediately gain Wildrose power, though it got the party close in 2012. Instead, it helped facilitate the election of Notley and the NDP in 2015. But when global energy prices collapsed again, resulting in further job losses and more despair, it was easy for the public's smoldering discontent to be rekindled, directed this time against a government portrayed by the corporate and media elite as socialist. In 2019, Kenney and the UCP rode anger and fear to victory, but were themselves trampled when COVID-19 struck in 2020. Not only did the pandemic undo Kenney, it also exposed the damage neoliberal policies had done over the years to the health care, education, and social support systems — systems upon which people rely in times of crisis. Hospitals and front-line health-care staff simply could not meet the demand, understaffed schools could not provide students with the infrastructure and support they needed, and government services lacked the ability to properly respond to growing unemployment, homelessness, and mental health issues.

The anger and angst unleashed during the pandemic has not abated. It ended Kenney's tenure as premier, but in its wake has emboldened racism, homophobia, misogyny, and alt-right movements, comfortable with bold and ugly displays of hatred. At the same time, however, there seems little willingness on the part of the province's major parties to name neoliberal economics and its policies as the culprit behind the various crises under which Albertans are living — the NDP because they do not want to appear radical, and the UCP because, ideologically, the only problem they see with neoliberalism is that it has not been radical enough in dismantling the social state.

It is in this context that Albertans will head to the polls in 2023; and it is also in this context that the electoral result is unlikely to yield anything resembling political peace or stability in the province. Anger and angst are in the saddle and show no sign of dismounting.

References

Bell, D. (1996). *The Cultural Contradictions of Capitalism*. 20th anniversary edition. New York: Basic Books.

Courchene, T. (1997). *The Nation State in a Global/Information Era: Policy Challenges*. Kingston: John Deutsch Institute for the Study of Economic Policy, Queen's University.

Globe & Mail, Editorial Board. (2022a, Oct. 22). Scott Moe tries to stand up to Ottawa but fails a basic math test.

Globe & Mail, Editorial Board. (2022b, Aug. 27). Can't get that operation you need? Don't worry, the premier just sent you a $500 cheque.

Grant, G. (2002). *Lament for a Nation: The Defeat of Canadian Nationalism*. 40th anniversary edition. Montreal and Kingston: McGill-Queen's University Press.

Helm, T., & Savage, M. (2022, Sept. 9, pp. 10-12). Mission impossible. *The Guardian Weekly*.

Markusoff, J. (2022, Aug. 25). *Why choosing Alberta's next premier largely lies in the hands of folks in Rimbey, Strathmore and Three Hills*. CBC News.

Micklethwait, J., & Wooldridge, A. (2004). *Right Ration: Conservative Power in America*. New York: Penguin.

Statistics Canada. (2022, Feb. 9). *Canada's Large Urban Centres Continue to Grow and Spread*. https://www150.statcan.gc.ca/n1/daily-quotidien/220209/dq220209b-eng.htm

Tanenhaus, S. (2010). *The Death of Conservatism: A Movement and its Consequences*. New York: Random House.

Weber, M. (1958). Class, status, party. In H. Gerth and C. W. Mills, *From Max Weber: Essays in Sociology*,180-191. New York: Oxford University Press.

Young, L. (2022, December). Danielle Smith's populist wave. *Alberta Views*, 28-31.

NOTES

1 We note the ironic similarity of Danielle Smith being chosen UCP leader (and premier) by party members at nearly the same time as Conservative party members in the United Kingdom chose Liz Truss as party leader and prime minister, a choice that proved ultimately disastrous (see Helm and Savage, 2022).